Statistical Methods for Geographers

W. A. V. Clark
University of California, Los Angeles

P. L. Hosking
University of Auckland, New Zealand

JOHN WILEY & SONS
New York • Chichester • Brisbane • Toronto • Singapore

For Irene and Carol

Library of Congress Cataloging in Publication Data:

Clark, W. A. V. (William A. V.)
Statistical methods for geographers.

Includes index.
1. Geography—Statistical methods. I. Hosking, P. L.
II. Title.

G70.3.C55 1985 910′.21 85-20309
ISBN 0-471-81807-0

Printed in the United States of America

10 9 8 7 6 5 4 3 2 1

Preface

Few geographers in the early 1950s would have envisaged the changes that were to take place in their discipline over the following twenty years—changes that were probably greater than in any period in the history of geography. Although these changes reflected, in part, new thinking about the philosophy of the subject, primarily they involved a methodological revitalization. In the move to a more "scientific" approach, geographers came to rely heavily on the methods of mathematics and statistics—to describe and test the concepts that previously they were satisfied to formulate subjectively. Although many branches of mathematics contributed ideas to this change, none did so more than the field of statistics, which in its widest sense can be considered as an analysis of information about real-world phenomena as an aid in their description, interpretation and prediction. This change in methodology was not without its problems. Examples of ill-conceived analyses, overexuberance in the use of some methods, and gross errors can be found in abundance in the geographic literature.

The early part of this twenty-year period was marked by a major debate on the implications of the use of mathematical methods, evoking emotional pleas supporting one side or the other. But by the 1970s the arguments had ceased, to be replaced by other more fundamental arguments relating to the philosophy of the discipline; and mathematics—especially statistical methods—had become an accepted and essential tool in geographic research. With a twenty-year background in applying techniques developed by mathematicians (and statisticians in related disciplines), geographers at last could turn to expanding their methodologic armory by developing techniques specifically formulated to handle those problems arising from the very nature of their discipline—the distribution of phenomena over space. The period since the early 1970s has been marked by major developments in what we could simply call *spatial analysis,* and unlike the changes that occurred in the previous twenty years, much of the original development has been initiated through the efforts of geographers themselves.

One of the side-effects of this methodologic change has been an increasing emphasis on a 'modeling approach' in analyzing geographic problems. Many of these so-called models are simply reformulations of ideas presented in less rigorous or less systematic form several years earlier. As such, they may involve only a change in terminology or in presentation. However, the emphasis on models has forced investigators in geography to emphasize the formulation and conceptualization of the problem in which they are interested. Geographers have been forced to search for relevant factors and discard the irrelevant. Tests and verification of hypotheses have become an essential element of geographic investigation. The interest in verification has resulted in both an increased use of numerical methods and a demand for more meticulously constructed research designs. The importance of this preliminary step of preparing a sound research design before data collection begins cannot be overemphasized.

This book provides insights into research design as well as basic methods for testing hypotheses and analyzing functional relationships, very much like other statistical texts. However, there are some important differences. First, despite several introductory books by geographers, there is a need for a comprehensive book that focuses on geographic examples and spatial problems. When using a book written for, say, sociologists, many students have difficulty transferring the statistical concepts to the types of geographic problems they are meeting in their own work or that are being analyzed in their geography courses. In this book, there is a strong emphasis on the presentation of specific methods related to spatial data and perhaps more importantly, an emphasis on the problems of applying standard statistical methods to data distributed over space.

Second, this book emphasizes the use of statistical routines which are now readily available on most computers. The major emphasis is on SPSS and SAS, although BMD is introduced for specific problems. The book includes examples of raw data, statistical set-ups and the output of actual runs. For the advanced techniques in the text, this is a particularly useful way of grasping the nuances of their application to geographic problems.

The decision to use SPSS and SAS (occasionally BMD) statistical packages also requires a comment. With the proliferation of personal computers, one approach would be to gear the text to the uses of these machines with interactive statistical packages. We believe, however, that we are presently in a transition phase and that the large packages such as SPSS and SAS will be "down-loaded" to the newer and faster personal computers in the near future (SPSS is available for the new IBM PC-AT). Thus, a background in SPSS and SAS will provide a basis for their use on smaller machines and in an interactive mode.

The book is designed to be used at the upper division undergraduate level and first year graduate level in North American, British, Australian and New Zealand universities. The first nine chapters could form the basis for a one quarter or one term course introducing geographers to the whole field of statistical methods, to be followed by a further course on multivariate methods (Chapters 10–13); or the book could be used as a one-quarter "review and extension" course for those who have been introduced to statistical methods in a statistics course in mathematics or statistics. These courses would provide the basis for graduate courses on probability-based models and spatial statistics. The book emphasizes verbal interpretation of equations. Mathematical derivations and developments are carefully explained. No mathematical background is assumed, and mathematical concepts are introduced as they are needed.

The first three chapters introduce terms, and ways of displaying and describing distributions. Chapters 4 and 5 focus on probability and sampling methods and are followed by three chapters on statistical testing. Chapter 9 introduces simple linear regression. Chapters 10 to 13 are concerned with extensions of the simple linear model to multiple regression and its assumptions, stepwise logit regression, canonical and discriminant analysis. The focus throughout the more advanced topics is on one or more dependent variables as they relate to sets of independent variables and for this reason, we have not included the data exploratory techniques of component and factor analysis.

We would like to acknowledge the help of many colleagues and students, both present and past, who have contributed directly and indirectly to this book. The material presented here has been worked out over a number of years in presentations to classes both in the United States and New Zealand. We would like to thank specifically Michael Hollis for running the programs and analyzing the examples used in the book, and Maggi Sokolik for typing and proofreading the many drafts that this book has gone through. We are also grateful to the Literary Executor of the late Sir Ronald A. Fisher, F.R.S., to Dr. Frank Yates, F.R.S., and to Longman Group Ltd. for permission to reprint Table V from their book, *Statistical Tables for Biological, Agricultural and Medical Research* (6th Edition, 1974).

W. A. V. Clark
P. L. Hosking

Contents

CHAPTER 1
An Introduction to Statistical Methods

In a book about the use of statistical methods in geography, it is appropriate to begin with a few brief definitions. The word *statistics* can take on a variety of meanings. In everyday usage, it simply refers to a set of data; but in the physical and social sciences, *statistics* also refers to a body of knowledge—a part of the field of mathematics or, more particularly, applied mathematics. In this sense, statistics or statistical methods has been defined as being concerned with the analysis of information about real-world phenomena as an aid in their description, interpretation, and prediction. More specifically, statistics involves the analysis of *distributions* of *scores* for *variables* derived from the *measurement* of *elements* from a *population* or *sample*. All these italicized terms require formal definitions—and to a considerable extent, that is what this introductory chapter is about—but for most of them, we probably already have some intuitive feeling about what is implied in their use.

The material that makes up the subject of statistics can be subdivided in a number of ways. First, we note a very broad separation into two fields: *descriptive statistics,* concerned with describing phenomena (leading ultimately to methods of interpretation and prediction); and *inferential* or *inductive statistics,* a set of methods to enable conclusions to be made about populations from a subset of that data—a sample. Another way of looking at statistics is in terms of the complexity of the problem—whether we are interested in just the description of a single variable (*univariate statistical methods*) or whether we are interested in the interaction (relationship) of two variables (*bivariate statistical methods*) or more than two variables (*multivariate statistical methods*).

1.1 POPULATIONS AND SAMPLES

Before attempting to outline even the most elementary statistical methods, we must examine carefully some basic concepts required to set up a piece of research involving statistical analyses. We emphasize again and again throughout this book the need to employ a careful, systematically derived

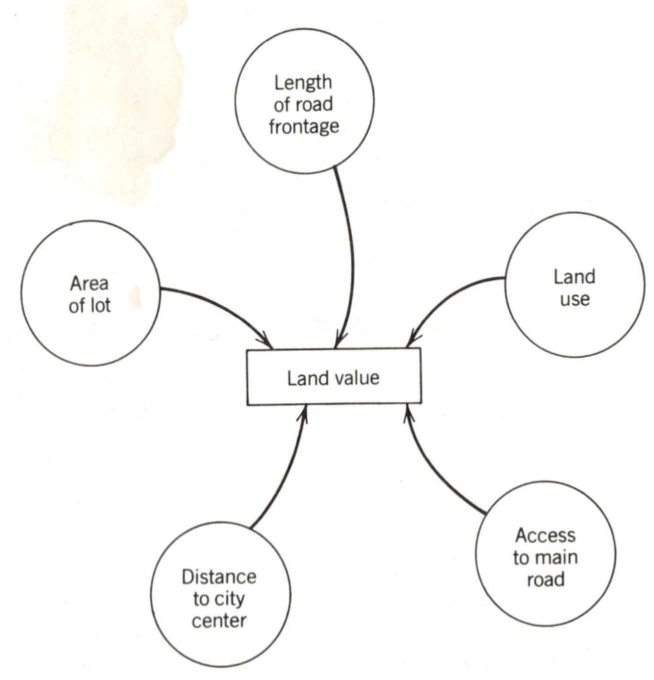

Figure 1.1 Conceptual model of land value relationships.

research design. To assist in this process, we use *operational definitions*—carefully constructed descriptions or sets of rules outlining explicitly how each step or process in an analysis is to be performed. Throughout this introductory chapter, we use a single example—an example that although simple, illustrates one type of research problem faced by geographers. In a small city of, say, 15,000 people, we are interested in how land value varies throughout the central area of the city. We want to describe and analyze this variation and to attempt to explain the pattern of variation. We refer to this example as the *Urban Land Value Survey.* As a first step in the analysis, a simple conceptual model of possible relationships between land value and other potential influencing factors is prepared (Figure 1.1). We postulate that land value is a function of land use, distance to city center, and so on, where the changing values for one variable—land value in this case (called the *dependent variable*)—is dependent on (is a function of) the changing values for the other variables (called the *independent variables*). Statistical methods can be used to describe and test various parts of simple models of this sort.

Elements and Populations

Applying statistical methods to describe real-world phenomena is based on the process of *measurement*—the conversion of some characteristic of a phenomenon into symbols (numbers). As we will see, this measurement process can be interpreted in a much wider sense in statistics than in its normal usage. The measurement process is directed at an individual entity—the smallest sampling unit about which measures are to be obtained and to which we will give the general name *element.* Alternative names for the element found in many texts are *case, observation,* or *observational unit.* An element may be a readily observed entity: for example, a person, a plant, a town, or a rainfall station; but in some situations the identity of the element that is being analyzed may not be as apparent. In geographic work, it is common to find that the element under consideration consists of a collection of a number of *subelements,* each of which has to be subjected to some form of actual measurement. Some examples might be the study of sediment size over a beach where the elements are individual sediment samples consisting of individual grains (subelements) or a study of the population characteristics of a city based on census information where the element would be the census subdivision (census tract) consisting of a specified number of individual people (subelements). In these situations, measuring the characteristics of the element would be derived from measuring the individual subelements. The distinction between these two types of elements (those of indivisible entities and those with subelements), although not always readily apparent, is of critical importance in preparing the research design. We defer a detailed discussion of the grouping of subelements and its implications until much later in the text.

Statistics, then, is concerned with the analysis of elements. The nature of the analysis is determined by the characteristics of the elements to be measured (the *variables*—examined in Section 1.2) and by the "area of interest" of the analysis. This area of interest may be defined as the *universe* or *population*—the totality of all individual elements being examined.

Characteristics of Populations

A population consists of a set of *elements*. We need an operational definition of the element under study. For example, an operational definition for an element in the Urban Land Value Survey might be:

> **the recorded *lot* as specified on the map of the City of Woodford (Valuation map 82/01 Sheet 7) dated May 24 1982.**

The population itself requires an operational definition, in conjunction with and directly related to the defined element. For example:

> **all recorded lots within the area demarcated as Central Business District Core by the Woodford City Planning Department and recorded on Planning Map C/4 dated January 1 1982.**

A population is also defined by its size—the number of included elements. The population size, usually designated by uppercase N, may be infinite, finite and known, or finite and unknown. In the Urban Land Value Survey it is finite and, using the operational definitions of an element and the population outlined above, $N = 86$.

The above examples using the Urban Land Value Survey are simple, but they illustrate the sort of problem frequently faced by geographers. We have not yet mentioned the type of information that is to be measured from the population of elements, but this obviously would have been determined before decisions were made as to the operational definitions outlined. The full implications of these operational definitions will become more apparent when the whole survey design is discussed in Section 1.4. Some examples of populations and their elements are illustrated in Figure 1.2.

Sampling from Populations

In situations where the size of the population is infinite, obviously not all elements can be measured, and some means of selecting a set of elements—a *sample*—from the population must be devised. Even with finite populations, it may not be possible to examine all the elements of a population for reasons of time, cost, or inaccessibility. In fact, it may not even be desirable to examine a complete population, because statistical methods are so powerful that valid conclusions can be "inferred" about a population simply from information obtained from a sample.

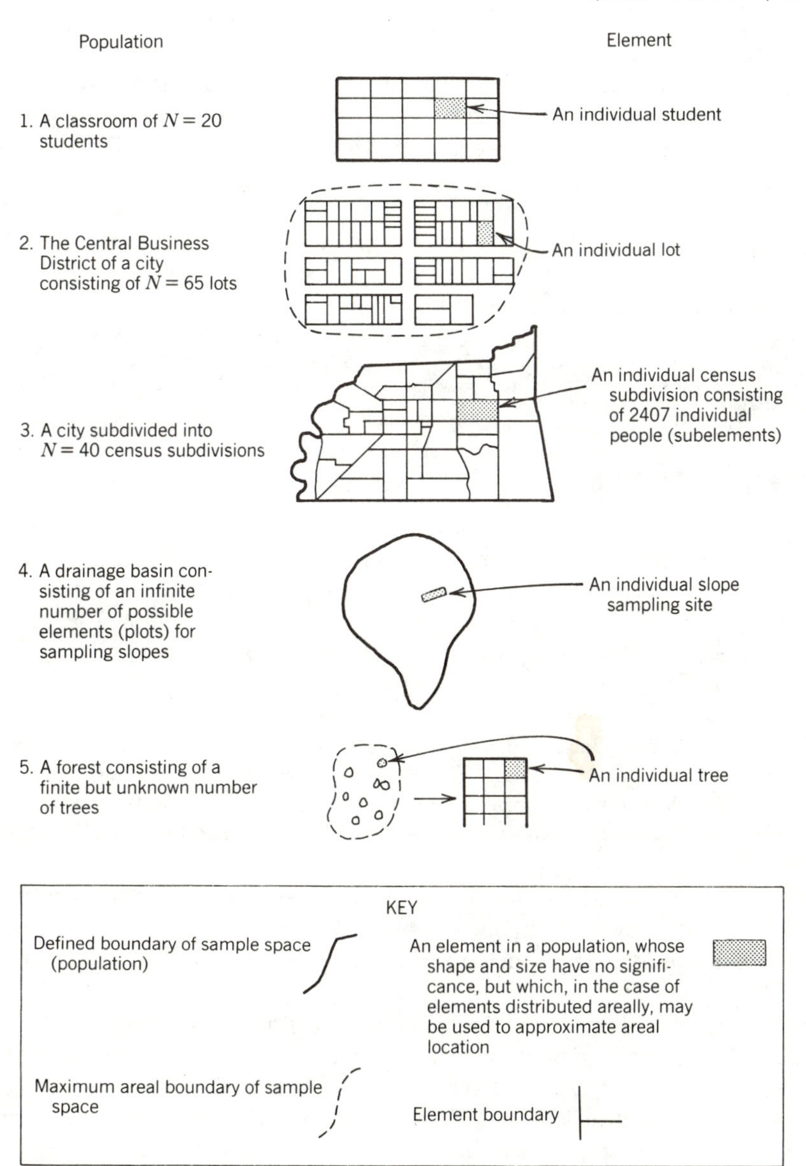

Figure 1.2 Diagrammatic representation of some populations and elements.

If we define a sample as a subset of elements from the population, then the interpretation of the information obtained from the sample is dependent on the method of selection of the subset of elements. Purposeful sampling may be defined as the procedure whereby elements are selected from a population because they have certain characteristics in which we are interested. For example, in a study of urban labor force characteristics, 5 cities from a population of 50 cities are selected because they are in a region of interest, because they are the largest, because they are growing the fastest, etc. The extent to which we can regard this sample as being "representative" of the population depends entirely on our definition of the population. As a general statement, however, we can say that the method just outlined would produce a sample from which information obtained could refer only to the sampled elements. More usually, our aim would be to select a sample from which information could be used to make conclusions about the entire population from which the sample was selected. Such an inference requires that the sample was based on probability sampling methods (to be discussed in Chapter 4), or, more specifically, a simple random sampling method, where each element of the population has an equal chance of being selected into the sample.

Characteristics of Samples

Samples consist of elements drawn from a population. Thus, we need operational definitions of an element and of the population itself.

The sample, or more accurately, the sampling method, requires an operational definition. For example, in the Urban Land Value Survey, if we are interested in obtaining a representative sample of 20 elements, an operational definition might be:

> **the 86 lots in the Central Business District Core were numbered 1 to 86 on the plan, and two-digit numbers were drawn from a random numbers table without replacement; the first 20 valid (between 01 and 86) two-digit numbers drawn were used to locate the 20 sampled lots.**

We note that a random numbers table is a special table of numbers having the property that every item has the same probability of occurrence; "without replacement" means that once an item has been chosen it cannot be chosen again.

A sample has a specified size, usually designated by lowercase n—the number of selected elements. If the population size is known or approximately known, then the relative sample size, usually expressed as a percentage, should also be specified. For the Urban Land Value Survey, a sample of size $n = 20$ would give a relative sample size of 20/86 or a 23% sample.

1.2 VARIABLES AND THEIR MEASUREMENT

A variable may be defined as any characteristic (or attribute) of an element of a population that can be measured in some form. Although an element of a population will obviously have innumerable characteristics, we are usually interested in only one or a few of these. These relevant characteristics—the variables—are the ones that would have been used to assist us in our definition of the population (Section 1.1).

Characteristics of Variables

Each variable should be carefully constructed and the measurement process fully described in an operational definition. For example, from the Urban Land Value Survey, a variable describing the area of the lot might have the following as its operational definition:

> **from the City of Woodford Valuation records held in the City Assessor's Land Tax Office, the lot area (in square meters) was extracted. The City Assessor gave the date of measurement as May 1982.**

In statistical work, we frequently need to refer to variables, and some form of abbreviated description is often required. It is possible to distinguish four levels of description, which in increasing order of abbreviation are as follows:

A full *operational definition,* as used in the foregoing example, which should provide a complete description of the variable and its measurement.

A *descriptive title,* which might be used in the text of an analysis, for example: "Area of Lot."

A short *one-word acronym,* which gives an indication of what is being measured and would be used in computer analyses where it would appear in summary tables and results. For example, the Statistical Package for the Social Sciences (SPSS) computer programs (Section 1.4) provide a maximum of eight alphabetic or numeric characters for variable description at this level, for example: AREA.

A *single symbol* representing a variable is an advantage in computational work. The symbols used include the whole range of alphabetic characters (including Greek letters), but the most common are X and Y. In a particular calculation, each symbol used would be defined using the descriptive title. For example, X where X = Area of Lot.

Using the operational definition of the variable, each element is assigned a *score*. This is the process of *measurement*. For each variable in a population or sample, there will be a set of scores constituting the input data for an analysis, one score for each element. Such a set of scores can be termed a *distribution of scores*. Where some symbol, such as X, can be used to describe the entire set of scores for the variable, subscripts can be used to refer to a particular score for that variable. Thus, X_4 would refer to the score for the variable X obtained from the fourth element. Alphabetic subscripts (especially lowercase i, j, and k) are frequently used to refer to any, or every, score for a variable. Thus, X_i refers to any score for the variable X, and the X_i would mean the whole distribution of scores for the variable X. Distributions can take on a number of forms, but there are two forms in particular.

Discrete distributions consist of scores that can take on only fixed numeric values with no intermediate values possible and usually consist of a set of *integers* (1,2,3. . .) and *continuous distributions,* where the scores theoretically can assume an infinite number of values between any two fixed points (the so-called real numbers). In practice, the values for the scores are determined by the precision of measurement.

Measurement of Observed Variables

In discussing variables and scores for elements, the term *measurement* has been used in a much wider sense than in normal practice. It is intended to include any process whereby a score (in whatever form) is obtained from an element with respect to a variable. An understanding of the various types of measurement process is essential in statistical analysis in that it is the measurement process that determines the type of distribution of scores and dictates which statistical methods are applicable in their analysis. The terminology used for differentiating types, scales, or levels of measurement is inconsistent and often confusing. We will attempt to separate four major processes of measurement for observed (or raw) variables, and to this we will add a fifth process of derivation, where scores for a new variable are derived from one or more observed variables. We will examine each of the four basic processes (classification, ordination, enumeration, and metrication) in detail.

Classification

In this process, the operational definition for the variable outlines a classification system. Each element is examined with reference to the operational definition and is assigned to a class; the score for the element is the name of the class. The classes must be mutually exclusive. We can distinguish two types of variables based on the method of classification.

The simplest level of measurement—where the scores (classes) have only the property of being equal or not equal one to another—are designated as *nominally classified.* Normal arithmetic operations cannot be meaningfully applied. The score for an element is simply the class name, although for convenience, they may be numbered 1,2,3. . . , remembering of course, that these symbols have no inherent value. An example from the Urban Land Value Survey was the variable Land Use that was constructed as a nominally classified variable with nine classes, coded 1,2,3, . . . , 9 (see Section 1.4).

A special form of nominally classified measurement occurs when there are only two classes. These variables are referred to as *binary* or *dichotomized* variables and are usually based on a present/absent-, true/false-, male/female-type classification. An example from the Urban Land Value Survey is the variable Access to Main Road, which was classified and coded as:

1 = the lot has direct frontage on the main road

0 = the lot does not have direct frontage on the main road

Binary variables are of considerable importance in that, in a number of situations, they may be treated in a similar fashion to variables based on higher levels of measurement and thus can be subjected to a wider range of statistical methods.

When the scores (classes) not only have the properties related to equalities but also may be ordered so that one class is greater than or less than another class, although the magnitude of the difference is not specified, they are designated *ordinally classified.* The classes may be coded 1,2,3. . . , as before. Elements would be classified into one of a small number of ordered classes either objectively or, more usually, subjectively. An example might be for a variable measuring *suburban status* in a socioeconomic survey of a city, where the classes are based on a subjective impression gained from observation and may be coded:

1 = low class
2 = middle class
3 = high class

The scores 1, 2, and 3 have little inherent meaning and the coding could just as easily have been 3, 2, 1 for the same order of classes.

Ordination or Ranking

Although similar to ordinal classification, ordination is concerned with the ranking of all elements with respect to each other. The individual elements are ranked from "smallest" to "largest" or vice versa. Scores would range from 1 to N, the size of the population (or n the size of the sample). Tied ranks (elements that have the same value or rank) can be handled in a number of ways, but usually they are given the same score, and the following rank (or ranks) is omitted. Commonly, the ranking process is subjectively based—especially where human respondents are asked to rank a number of items in some order. However, in some instances, ordination is employed on a set of data that has been metrically measured but in which we do not have much confidence (or interest) in the absolute value of the scores—and thus a rank-score is used in place of the original score.

Enumeration

Enumeration, or counting, is one of the most widely used forms of measurement employed by geographers. By definition, each element must contain some form of subelement structure (Section 1.1), which is used in the counting process. Each subelement would either have or not have a particular characteristic, and the number of subelements that does have the particular characteristic is the score for that element. Variables derived directly from census information would employ this type of measurement. For example, a census subdivision area (an element of a population of census subdivision areas) contains a specified number of people (subelements), some number of which (the score for the element) have the characteristic of being aged 0 to

4 years (the variable). Scores for enumerated variables are limited to the set of integers, usually only the positive integers, and we note that the principles of constant intervals and constant ratios allow the application of the full range of arithmetic capabilities. In fact, enumerated variables differ from those measured by metrication (to be described shortly) only in that their distribution of scores is discrete rather than continuous. This, however, is an important distinction and one that limits the range of techniques that are directly applicable to enumerated variables. Nevertheless, when the enumerated scores are large (say, with an absolute range from smallest to largest in excess of 100), then it would be reasonable to ignore the distinction in many circumstances. It should be apparent that large enumerated scores are little different (apart from precision of measurement) from rounded metric scores. We examine this concept again later in this chapter.

Metrication

There is no generally accepted term for this, the highest level of measurement, although the descriptive terms *ratio* or *interval* measurement are often used. However, we will use *metrication* and *metric scores* to describe a measurement system that is based on an equal interval of measurement and a continuous distribution of scores. Two forms of metrication can be separated: *interval-metric variables* have an arbitrary zero point, and thus the principle of constant ratios does not apply, whereas *ratio-metric variables,* with an absolute zero point, have defined ratios of scores. Fortunately, very few variables employed by geographers (with the major exceptions of temperature and calendar date) are measured on an interval-metric scale, for the distinction is quite important. Interval-metric variables should not be manipulated or transformed by any method involving multiplication—thus, for example, the use of transformations of scores (square root, logarithmic, etc.), geometric means, and the coefficient of variation are all inappropriate. With this distinction noted, from here on we refer only to metric variables with the assumption that they are derived by ratio-metric measurement. Care should therefore be exercised in ascertaining that a variable being used is ratio metric, or if not, that the statistical method being employed is appropriate. In the Urban Land Value Survey, the variables measuring Area of Lot, Length of Road Frontage, and Distance to City Center are all examples of ratio-metric variables.

Measurement of Derived Variables

In the preceding discussion, we outlined all the major methods of measuring variables originating from observation. Yet, an examination of the many statistical analyses that have been performed by geographers would show that it is rare for these observed or raw variables to be used directly in an analysis, except perhaps in the initial stages. More commonly, these observed variables are used to produce new sets of variables that we will simply call derived variables. Frequently, the reason for doing this is to standardize the data in some way. This is especially true when dealing with data collected in areal units where the influence of varying size of units is considered to distort the pattern of the variable, and thus the variable is standardized for area.

There is considerable confusion in the terminology used to describe these types of modification—terms such as *ratios, rates, indices, proportions,* and *percentages* often seem to be used interchangeably. In fact, these modifications can all be considered as *ratios* of one form or another—the score of an observed variable (which must be either enumerated or metric) is divided by (i.e., as a ratio of) "another score" to produce the derived score. In some situations, this simple ratio may be modified by further multiplication or by addition. In Table 1.1 we describe four types of ratios by the nature of the numerator and denominator making up the ratio.

1. ***The denominator in the ratio consists of some total for the element and the numerator is a contributor to this total.*** For example, the area of arable land in a county may be divided by the total area of the county to produce a measure of the relative importance of arable land in that county. The derived score is expressed as a proportion (retaining the simple division) or, more commonly, as a percentage (by multiplying by 100).

2. ***The denominator in the ratio is the total for all elements in the population*** that have the particular characteristic being measured by the original variable. For example, the area of arable land in a county is divided by the total area of arable land in all the counties being considered as the population. Again, the score may be expressed as a proportion or a percentage. The distinction between this type of ratio and the foregoing one is very important, because the derived variables measure completely different phenomena. In the first case (1) the derived variable is measuring the relative importance of the variable (arable land) in the element (the county) compared to other possible land uses; in the second case (2) the derived variable is measuring the relative importance of the element (the county) for the variable (arable land) compared to other elements in the population.

3. ***The denominator in the ratio is some standardized base number*** to which each element's score can be compared. Such a ratio is usually called an index, and values of 1.0 or 100.0 (a common modification) indicate that the element has an original score identical to some base time period, base area, or national average. The index itself frequently employs derived variables in its construction; for example, the ratio of percentage arable area in a county to an overall national percentage arable (the population) provides a measure of the importance of arable land in a county relative to its average importance in the nation as a whole.

4. ***The denominator in the ratio is a further enumerated or metric variable,*** but it is not directly related to the variable in the numerator (compare with 1 above). Usually expressed as simple ratios, these derived variables stem

TABLE 1.1 Examples of Derived Variables

Element	Numerator	Denominator	Derived Variable
1. DENOMINATOR IS A TOTAL FOR THAT ELEMENT			
Census subdivision	Number of males	Total number of people	Percent male
Census subdivision	Number of unemployed	Total number of people	Percent unemployed
Vegetation plot	Area in grass	Total area	Percent grass
Factory	Number of unskilled workers	Total employment	Percent unskilled
County	Arable area	Total county area	Percent arable
Farm	Income from nonfarm sources	Total farm income	Percent nonfarm income
2. DENOMINATOR IS A TOTAL FOR WHOLE POPULATION			
Census subdivision	Number of unemployed	Total number of unemployed	Percent of unemployed
County	Arable area	Total arable area	Percent of arable area
City	Number employed in manufacturing	Total number employed in manufacturing	Percent of manufacturing employment
3. DENOMINATOR IS A STANDARDIZED BASE NUMBER			
Month	Measure of price	Price at base month	Price index
County	Proportion of area arable	Proportion of nation area arable	Arable index
City	Proportion of people in manufacturing	Proportion of population's total number of people in manufacturing	Manufacturing employment index
4. DENOMINATOR IS A SECOND ENUMERATED OR METRIC VARIABLE			
Census subdivision	Number of males	Number of females	Male/female ratio
Drainage basin	Length	Width	Length/width ratio
Country	Number of live births in a year	Total number of people at midyear	Crude birthrate
Property lot	Unimproved land value	Area	Land value/m^2
Census subdivision	Number of people	Area	Population density

from two aims; first, to form significant ratios, for example, a male/female ratio, and second, as a standardizing method to reduce to units of time, length, area, volume, weight, etc.

Geographers rely heavily on derived variables, especially as a means of standardizing for differing areas of elements. Derived variables, regardless of the process of measurement for the original variable or variables, have all the characteristics of variables measured on a ratio-metric scale. The only exception to this is that the variables that are expressed as proportions or percentages have a range from only 0 to 1 or 0 to 100, respectively. However, there are four potential problems in the use of derived variables. First, with ratios derived from two observed variables, the precision of the new variable is a product of the precision of the two contributing variables. Second, when two variables are divided, any relationship between the two variables is masked in the process—a result that could distort conclusions drawn from the analysis. Third, the distribution of scores for derived variables may bear little relationship to the distribution of the original observed variable and, more often than not, requires some form of transformation before statistical tests can be applied. This applies in particular to proportion (or percentage) data. Fourth, problems in analysis and interpretation may be encountered when looking at relationships among variables, when proportioned (or percentaged) variables are included that together total to unity (a so-called closed number system).

Variable Measurement and Errors

The measurement process for a variable assumes that there exists for each element a "true" value that the process is attempting to determine. In many circumstances, the recorded score will not equal this true value but will depart from it to a greater or lesser degree. This departure can be described by a simple statistical model—the measurement-error model—that states

$$Y = X + E$$

where Y is the (unknown) true value, X is the recorded value, and E is the (unknown) amount of departure or *error* of measurement. Statistical methods are concerned to a very large extent with analyzing such errors. A discussion of measurement error could also be approached through the more formal examination of the estimation of parameters (Chapter 6), but consideration of this process must be delayed until a number of other topics have been examined. It is convenient at this stage in our discussion to separate out four basic sources of error.

Gross error is a mistake performed by the person collecting or recording the measurements. It is essential to eliminate these errors. It cannot be emphasized too strongly that data derived from measurement should be checked meticulously before processing begins.

Errors of method are concerned with poorly constructed variables. This is a rather nebulous but very important concept. A particular characteristic of a population can be measured in a variety of ways; that is, there is a number of alternative forms of setting up an operational definition for a variable. The final decision may well influence the adequacy of the functional relationships that are investigated.

Systematic errors occur when the measurement process results in scores that are consistently higher or lower than the assumed true value. The degree of departure is referred to as *bias*. If detected, systematic errors can usually be traced to instrumentation problems—from a poorly calibrated, observation-data-recording instrument such as a slope recorder, thermometer, or traffic counter to an incorrectly programmed calculator or computer. Again, it is important to eliminate systematic errors as much as possible by careful checking at both the recording and processing stages.

If we repeatedly measured the same element, we would probably obtain a slightly different score each time. These fluctuations around the assumed true value are termed *random errors* in that in the absence of the other three sources of error, it can be assumed that these fluctuations are attributable to chance. The measurement-error model, with the error term E defined as a random variable, forms the basis of much statistical testing. The closeness of repeated measurements to each other—the degree of variability of the scores—we call *precision*.

Assuming there are no gross or method errors, an adequate measurement process produces scores that are precise; and if obvious systematic errors can be eliminated, precision of measurement will automatically lead to accuracy. Thus, in regard to measurement error, we are ultimately concerned with precision. We will examine the concept of precision in the measurement process in relation to both observed and derived variables, but before we do so, we need to comment briefly on methods of recording scores.

Levels of Precision and the Recording of Scores

A useful concept, and one that forms the basis of a discussion of precision, is that of the significant digits in a number. Let us examine a statement such as "the length of road frontage of a property is 20 meters." Because variables measuring length will form continuous distributions, this statement can be interpreted in a number of ways. It should be apparent that it is extremely unlikely that the road frontage is exactly 20 meters long, no more no less. Twenty meters refers to an approximate length, and represents a range of possible scores whose implied limits will be determined by the precision of the measuring process. Some possible interpretations of the above statement are illustrated in Table 1.2, reflecting the number of significant digits and a more correct specification for recording the score. The important point is that the recording of scores should be such that the last digit of the number specifies the level of precision. Thus, as another example,

TABLE 1.2 Possible Interpretation of a Score of 20 Meters

Precision Standard	Approximate Implied Limits (in meters)		Number of Significant Digits	Preferred Form of Recorded Score	
To nearest 10 meters	15	to 25	1	2	(in 10 meters)
To nearest meter	19.5	to 20.5	2	20	(in meters)
To nearest 0.1 meter	19.95	to 20.05	3	20.0	(in meters)
To nearest centimeter	19.995	to 20.005	4	20.00	(in meters)
To nearest millimeter	19.9995	to 20.0005	5	20.000	(in meters)

a score for a variable measuring the area of a property lot of 678 square meters implies that the true area of the lot is between 677.5 and 678.5 square meters.

With observed variables, the measurement process consists of recording scores from observations. How precisely should we measure these scores? First, we note that the measurement processes of classification, ordination, and enumeration produce scores that *are* precise and are not subject to errors of imprecision. An agricultural county, for example, either has or does not have exactly 2703 dairy cows. Apart from the possibility of gross or method error, this score must be absolutely accurate and precise. This does not mean, however, that scores must be retained in an analysis in their original precise form. Very large enumerated scores, for example, could be rounded off to multiples of 10, 100, etc., to retain a more meaningful range of values.

It is with metric variables (variables forming continuous distributions) that problems of recording precision arise. It is important to remember that it is the precision of measurement utilized in the initial recording phase that will be retained throughout a statistical analysis. In some instances, precision will be determined by the instruments used or the form of archival data, but in other instances, some decision on precision will be necessary. Several factors can contribute to this decision. A too imprecise measurement might jeopardize the conclusions from a study, whereas excessively precise measurement is often unwarranted. Measuring Length of Road Frontage to the nearest centimeter is unnecessary, for example. As a general guideline, it has been suggested that the range from smallest to largest recorded score should be between 30 to 300 units (measurement steps) apart. This suggests that scores would be recorded only to two or three significant digits. In the Urban Land Value Survey, the variable Length of Road Frontage, for example, ranged from a minimum of 5.7 to a maximum of 181.8 meters. This range represents 1761 unit steps of 0.1 meter, suggesting that a precision level of 1 meter with 176 units would have been adequate. However, this suggested guideline is by no means a rule to be followed in all instances. Distributions of scores that are skewed (the scores concentrate at one or the other end of the range of scores), as is the case in this example, would frequently require more than 300 units to accommodate their widely varying scores.

When dealing with derived variables, the situation is even more problematic. The precision of derived variables will be based on the precision of the original variables from which they were formed. Thus, for derived variables based on enumerated variables, the scores will also be precise and not subject to errors of imprecision. In the case of new variables derived from metric variables, however, the precision of the scores will decrease significantly. Let us take a simple example (outlined in Table 1.3) from the Urban Land Value Survey to illustrate the increase in possible error. One element in the survey had a recorded score of 66 for Unimproved Land Value (measured in $1000s) and 402 for Area of Lot (measured to the nearest square meter). A derived variable—Land Value per Square Meter—is recorded as 164 (to the nearest dollar). But an examination of Table 1.3 shows that its true score could range from less than $163 to nearly $166. Thus, we were not justified in recording this score with any greater apparent precision (e.g., $164.18).

1.3 THE NATURE OF GEOGRAPHIC DATA

With some understanding of populations, samples, variables, and levels of measurement, we can discuss the nature and sources of the data collected by geographers, the aims involved in data collection, the data collection process, and the methods of analysis. We wish to emphasize that there are certain statistical techniques that form the basis of statistical enquiries in

TABLE 1.3 Example of Imprecision of Derived Scores

			Denominator Variable Area of Lot (in meters)		
			Minimum	Recorded	Maximum
			401.5	402	402.5
Numerator variable: Unimproved Land Value (in $000)	Maximum	$66,500	165.63	165.42	165.22
	Recorded	$66,000	164.38	164.18	163.98
	Minimum	$65,500	163.14	162.94	162.73

Land Value per Square Meter is recorded to two decimal places.

geographic research, and these techniques are the methods that were developed and have been widely used by researchers in both the physical and social sciences. It is the aim of this book to outline these classical statistical methods and to apply them, sometimes with necessary extensions, to geographic examples.

The Spatial Component of Geographic Data

We have described statistical methods as being concerned with the analysis of information about populations and samples. For many sciences, it is not too difficult to define their field of information—the nature of the particular phenomena to be subjected to statistical analyses. However, if we attempt to apply this principle to geography, we encounter considerable difficulty and some explanation is necessary. In a rather crude fashion, we could state that any information to be subjected to statistical processing can be described as being three dimensional—thematic information (the phenomenon of interest) collected from elements described in space and time. Because of geography's extensive field of interest, there is no specific thematic information that would separate it from other subjects; thus, from one point of view, statistical analyses employed by physical geographers may differ very little in subject material from those of other earth scientists, and similarly, analyses employed by human geographers might resemble those of other social scientists. But, although the subject material may appear very similar, there is often one very important distinction. For many scientists, statistical methods are concerned with the analysis of thematic information that happens to be collected from elements distributed over space and time; whereas for the geographer, this spatial component is not just of indirect significance, but often reflects the primary purpose of the analysis. Although statistical methods can be considered as the analysis of distributions of scores—how (and why) one score differs in magnitude from another—from the spatial point of view, the geographer is also interested in distributions of scores over space—how (and why) one score at a point in space differs in magnitude from that at other points. This additional component of interest can be illustrated by referring to the Urban Land Value Survey discussed earlier. The aim of that survey was to examine how land value varied throughout the central area of a city; thus, it was concerned not only with how the variable Unimproved Land Value varies (how the scores differ in magnitude), but also how this variability is reflected in space (spatial patterns).

The data with which geographers are concerned in statistical analysis, therefore, may have an additional component of variation that is usually not of major concern to other scientists. Data with this additional spatial component have been variously referred to as *areal data, geographic data,* or *spatial data.* We will simply call any data collected by geographers *geographic data;* and geographic data with a significant spatial component will be referred to as *spatial data.* A useful distinction in terminology could be made among different spatial bases used for measuring spatially distributed phenomena. Thus *linear data* are measured along one-dimensional lines (e.g., stream flow along a river); *areal data* are measured on a two-dimensional surface (e.g., population distributed in a city); *spatial data* are measured in three-dimensional space (e.g., temperature distribution in the atmosphere.) Unfortunately, these distinctions in terminology are not universally accepted and in this text, following common practice, the term *spatial data* is used to describe any of the above spatial situations. The distinction between spatial and nonspatial data is one of purpose rather than character. Spatial data defines a set of information in which the location of an element defined for measurement is not just an indirect consequence of the definition of the population. In discussing statistical methods, distributions of scores over space will be described as *spatial distributions* to differentiate them from *statistical distributions* of scores.

While recognizing that geographers are always concerned in one way or another with the analysis of phenomena distributed over space, it is possible to separate four reasonably distinct approaches to the use of this spatial component of data in the statistical analysis of geographic data.

First, in many of the statistical analyses performed by geographers, the spatial component is of little or no significance. From the definition outlined above, the data are, then, nonspatial. Although the data may be collected from spatial elements, the aim of the analysis is to determine empirical relationships among phenomena regardless of their spatial base. As such, the methods of analysis are those of classical statistics that would be employed by any researcher.

Second, in other statistical analyses, the spatial distribution of scores merely forms the medium from which statistical information is obtained and has little direct relevance to the analysis. Summary information is provided for the location from which the data were obtained, and this information can be compared to other locations or the same location for different times. Because interest is not in the variability of the scores over space, the data are basically nonspatial, and the methods of analysis are those of classical statistics.

Third, the analysis of the variability of a phenomenon in an area may include a specific spatial component—where location is regarded as one of the contributing factors to this variation. Geographers frequently engage in analyses of this type, and statistical methods play an important role in the analysis of these spatial data. In applying classical statistical methods, however, considerable care must be exercised, because the spatial component of the data could lead to the violation of some of the assumptions inherent in the method. Discussion of this type of problem forms an important part of the techniques outlined in this book.

Fourth, at the other end of the continuum, the analysis of the spatial distribution of scores may be the primary purpose of the investigation—the description and analysis of the patterns of distribution of phenomena over space. In recent years, geographers have placed greater emphasis on analyses of this type and a whole new field of investigation (loosely termed *spatial analysis*) has developed. Statistical methods, especially that subset called classical statistics, constitute only one set of methods of many in the whole field of spatial analysis and the relationship between statistical methods and spatial analysis requires a brief elaboration.

Spatial Distributions, Spatial Analysis, and Statistical Methods

In the early parts of this chapter, we outlined the concepts of populations, samples, elements, and variables without reference to whether we were dealing with spatial data or not. It is important to review these concepts in relation to spatial data. While the definitions can be simply extended to the spatial situation (Table 1.4), further elaboration is necessary for some of the terms and concepts used.

Spatial elements have specific locations in space and can be described as *points, lines,* or *areas*—occupying none, one, or two dimensions respectively.

The extent of a spatial population may be defined as a *line* of measurable length or a *surface* of measurable area.

Spatial distributions of scores may be considered spatially *discrete* (defined only at certain points, lines, or areas) or spatially *continuous* (varying over the whole line or surface).

Variables with continuous spatial distributions can only be analyzed by measuring sample elements at discrete points, lines, or areas. The sampling method may be spatially regular (uniform distance or area) or spatially irregular.

The foregoing descriptions allow us to define seven types of spatial distributions, examples of which are illustrated in Figure 1.3. These distributions can be considered to form the basic spatial data for input to statistical analyses in geography. However, they also describe the basic data of the subject of spatial analysis, since the statistical analysis of spatial data is only one of many methods used in analyzing spatial phenomena. Without entering into the debate as to what specifically constitutes spatial analysis, we note the following characteristics of the subject in relation to the role of statistical methods.

First, the methods of spatial analysis are those of the geometries, but not just the well-known Euclidean geometry that forms the basis of much of statistical analysis among other things, but also non-Euclidean geometries, especially topology.

Second, the aims of spatial analysis are centered on the search for patterns in spatial phenomena, involving their description, prediction, and optimization.

Third, the description phase of the analysis of spatial patterns is concerned with (1) the geometry of the spatial elements themselves—shapes of lines, areas, etc., (2) the geometry (topologic, Euclidean) of the distribution of the

TABLE 1.4 Basic Statistical Concepts Applied to Spatial Data

Basic Statistical Concepts	Applied to Spatial Data
1. Statistical methods are concerned with the analysis of information about real-world phenomena.	The real-world phenomena are distributed over space.
2. Statistical methods are based on the process of *measurement* in some form or other. The measurement process is directed at an individual *element*—the smallest measurement (or sampling) unit.	The elements will be defined in space—forming *spatial elements.*
3. The area of interest of an analysis is defined as a *population.* A population consists of a *set of elements* making up the total sample space. The *size* of a population (the number of individual elements) may be infinite, finite and known, or finite and unknown.	A *spatial population* will have a specified areal extent and size.
4. A *sample* consists of a subset of elements drawn from the population.	Sampling methods will have to be extended to handle spatial situations.
5. A *variable* is a measured characteristic of a population. For any variable, the measurement process produces a *score* for each element.	Variables measuring spatial elements will produce *spatial scores* (spatial data).
6. A (statistical) *distribution of scores* for a variable consists of the scores for all elements in a population or sample. Most distributions of scores may be described as *discrete* or *continuous* depending on the numeric form of the scores.	Scores for spatial elements will not only have statistical distributions but also *spatial distributions.* Spatial distributions can be described as *discrete* or *continuous.*

Thus (spatial) statistical methods are concerned with the analysis of distributions of (spatial) scores for a (spatial) variable derived from the measurement of (spatial) elements from a (spatial) population or sample.

spatial elements—point pattern analysis, network analysis, etc., and (3) the geometry of the distribution of scores for the spatial elements—sometimes referred to as geostatistics.

Fourth, spatial analysis involves the use of a wide variety of methods developed in applied mathematics—statistical methods being only one. These statistical methods are used whenever the spatial analysis is concerned with

Type of Spatial Distribution	MAP	Examples of Spatial Elements
1. Distribution of discrete points on a line		Towns along a river or highway Bifurcation points along a river
2. Continuous distribution along a line	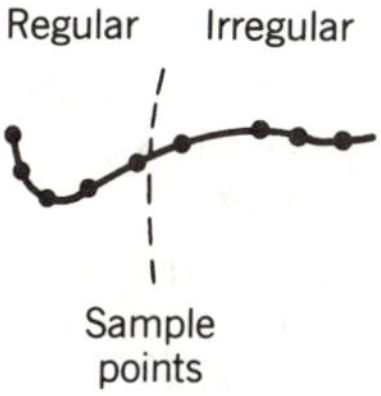 Regular Irregular Sample points	Sampled points of discharge along a river Sampled points of traffic flow along a highway
3. Distribution of discrete points on a surface	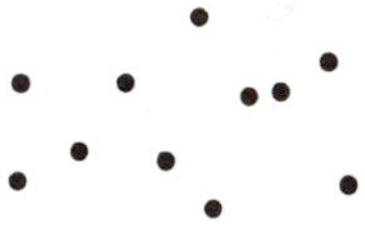	Towns in an area Volcanoes
4. Distribution of discrete lines on a surface		Highways Rivers
5. Distribution of discrete areas on a surface		Fourth-order drainage basin Areas of forest vegetation or arable farmland
6. Distribution of contiguous areas on a surface	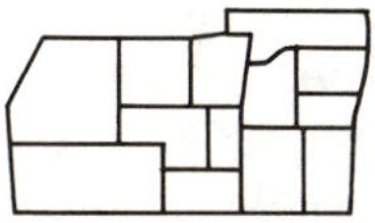	Census subdivision recording population characteristics Counties recording agricultural information
7. Continuous distribution over a surface	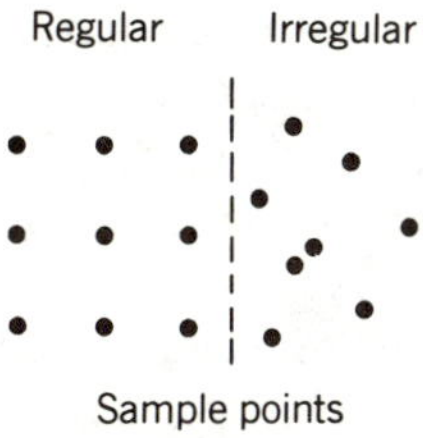 Regular Irregular Sample points	Sample points of elevation above sea level Sample points of rainfall

Figure 1.3 Spatial distributions.

measuring the variability of scores that include a random error component, and therefore statistical models can be applied.

The role of statistical methods in spatial analyses is, therefore, centered on the description phase, and particularly on the analysis of the variability of scores over space. The statistical methods that are used in this process are simply extensions of methods used in the analysis of nonspatial data. Because the aim of this book is to present the fundamental statistical methods of use to the geographer, we will outline classical statistical methods, show how these can be used when the data are collected from spatial elements, and then extend the methods to the statistical analysis of spatial distributions.

Of the seven basic types of spatial data previously outlined, statistical analysis of spatial distributions is primarily concerned with only four types.

Discrete point data are the data of the first type of spatial distribution described in Figure 1.3. The data for each point are defined by its location and its magnitude (the measured score). The points are assumed to be dimensionless in space (occupy no area), and the intervening surface between points is not part of the spatial distribution. A simple example would be the distribution of the variable Town Size.

Discrete area data are the data of the fifth and sixth type of spatial distribution outlined in Figure 1.3. The data for each area are defined by its location and its magnitude (the measured score). Usually, the areas completely occupy the surface (i.e., they are contiguous), but not necessarily so, and the phenomenon being measured is not considered to be distributed continuously over the surface. Examples are the variables measuring characteristics of the property lots in the Urban Land Value Survey.

Sample point data from a spatially continuous distribution are the data of the seventh type of spatial distribution outlined in Figure 1.3. The data for each sample point are defined by its location and its magnitude (the measured score), but because the variable is considered to be spatially continuous, our interest is on the whole surface of variability and not just the sampled points for which data happen to be available. Examples are sampled observations on rainfall and elevation above sea level.

It is possible, and in many circumstances desirable, to modify the nature of the spatial data by changing it from one form to another. This can best be described by example. Human populations consist of individual people, and each person (theoretically, at least) can be located at a particular point on a surface at a particular point in time. As such, this distribution of people consists of discrete point data. However, by enumerating the number of people within a census subdivision area (or an artificial grid square) we have a set of data that would be described as discrete area data. (Later we refer to these distributions as areal frequency distributions.) We could then reverse the process and summarize the location of each areal element by a point and attach the frequency of the areal unit as a score for that summary point—once more, discrete point data. In addition (or alternatively), we could convert the frequency measure into a density measure (people per unit area), and each summary point could then be considered as a "sample point" from a continuous distribution of the variable Population Density. The importance of transformations of this sort will become apparent in later chapters.

Sources of Data

As one would expect with such diverse subject material, geographers draw on a wide variety of sources for data collection. Without examining these in any detail, we note the following major types of data sources frequently used in geographic analyses.

Primary sources (providing raw information specifically collected for the analysis), which may be measured by either contact with direct recording—such as interviews, field observations, or instrumental measurement; or contact with remote recording—where measurement and recording are done at separate locations, such as telephone interviews, traffic counters, or automatic rainfall recorders; or noncontact methods—including specifically implemented remote sensing methods.

Secondary sources (previously collected information providing a depository from which information relevant to the analysis can be extracted), which may be either administrative data—where the data are produced as a by-product of the record-keeping functions of some government or private agency, such as birth/death records, property lot data, or rainfall records; or survey data—collected for some particular purpose, as a one-time effort or updated on a regular basis. The original survey methods are the same as those outlined for the collection of data from primary sources. Included in this category is the whole range of data, from census material to imagery obtained from general-purpose remote sensing systems.

Obviously, there are major differences in the measurement process for data obtained from these different sources, and there is an important contrast between the basic categories of primary and secondary sources. In the former, the research design for the analysis would be much simpler, but the measurement process itself would usually be more difficult and certainly more time consuming. With control over the entire data-collection process, variables could be chosen and measured in the form necessary for the proposed statistical analysis. When secondary sources are used, on the other hand, the researcher has no choice over what variables are present and how they are measured. This can create considerable problems when it comes to the analysis stage.

1.4 THE ANALYSIS OF GEOGRAPHIC DATA

The remainder of this book is concerned with the statistical analysis of data collected in situations similar to the Urban Land Value Survey example used throughout the preceding sections. The methods that are employed to assist in the description and analysis of problems of this type have been developed over many years. The mechanics of these methods are centered on the manipulation of the data through the simple arithmetic processes of addition and multiplication. In recent years, the tediousness of these manipulations has decreased as mechanical aids—especially the electronic calculator and the computer—enable a far more rapid application of the various methods. Although we will make full use of these aids in describing the techniques outlined in this book, the simple underlying principles and assumptions of any statistical process should be understood before recourse should be made to more automated processing. Thus, as a simple example, although it is possible to obtain the *mean* of a distribution of scores simply by pushing a button, the button pusher should be aware of the mathematical processes the mechanical aid is performing to obtain that mean. To do this, the researcher must have carried out the same sort of process by hand and therefore realize the implications at each step in that process. It is important to remember that a statistical analysis begins at the design stage, with the formulation of the problem and the construction of operational definitions for the population, sample, elements, and the variables. To a large degree, the success of an analysis depends just as much on its original construction as on careful measurement and analysis phases. A poor analysis is often the product of poor design.

Setting Up the Design, Coding, and Recording Scores

We can illustrate the initial steps in an analysis by referring once again to the Urban Land Value Survey. The aim of this analysis was to examine the variability of land values in the central area of the city of Woodford. Insofar as information on land value could be obtained only from the administrative records of the city assessor's office, it was decided to rely on these secondary sources of information as the basis for measuring all the variables in the analysis. The process of setting up operational definitions for the preliminary steps in the analysis, together with a listing of data obtained, follows in the next section.

Given a well-formulated set of operational definitions for the variables, the process of measurement becomes the simple task of recording the information in some convenient form. Although computer processing may not be contemplated, we will illustrate the process of coding and recording scores as if they were to be set up for computer input. In this form, they can be easily checked and corrected, and are then accessible for analysis. For the Urban Land Value Survey, the survey design led to recording for each property lot eight observed variables. To these variables was added an identification code. Thus, for an element (the property lot), there were nine pieces of information to record. Although there is a variety of ways to tabulate and record this type of information, it is usually convenient to regard the set of scores (one score for each variable) for a single element as constituting a unit of data. This unit would form a row in a data table, or a card in a

DATA FORM

DESCRIPTION WOODFORD—URBAN LAND VALUE SURVEY | DATE 8/8/83 | PAGE 1 OF 4

PUNCHED | CHECKED

LAND VALUE ($000)	AREA (m^2)	FRONTAGE (0.1m)	LANDUSE (1 TO 9)	# OF FLOORS	# OF WORKERS	AGE (1 TO 4)	ACCESS (Ø/1)	LOT CODE
66	4Ø2	33.2	3	2	6	3	Ø	14ØØ1ØØ
86	72Ø	21.6	3	2	8	3	Ø	14ØØ2ØØ
42	314	15.1	6	2	16	2	Ø	14ØØ3ØØ
97	787	39.6	7	1	Ø	1	1	14ØØ4ØØ
26	171	35.Ø	1	1	2	4	1	14ØØ5ØØ

Figure 1.4 Data coding sheet.

computer card deck, or a single-entry row on a terminal to be input to the computer. In computer terminology, they constitute a single *record* of the data file. A data coding sheet (either specifically constructed or a modification of an existing general purpose data sheet), can be used both for recording data and eventually for computer entry (Figure 1.4).

Introduction to the Use of SPSS for Statistical Analysis

There are a variety of computer packages available to perform statistical analysis. Some are in widespread use, some are more localized in application. This book provides the reader with a workable and understandable outline of the major statistical methods. To do this, it is essential to introduce computer methods as well as hand-calculated examples—and for most geographers this would mean recourse to the use of one or more of the major statistical packages available on mainframe or personal computers. Many users will have access to the latter and there is a lot to be said for introducing statistical methods on these much more personalized interactive machines. However, their variety makes them unsuitable to use for illustrating worked examples in this general introductory book. Despite rapid growth in computing speed, memory, and bulk storage capabilities, personal computers still have difficulty in handling large data analyses. Moreover, in two or three years it is likely that smaller versions of SPSS and SAS will be available for many personal computers.

Thus, we will use the two most widely used statistical packages available on mainframe computers throughout the world today—and we will mix their usage throughout the book. SPSS is used for initial processing of the Urban Land Value Survey, SAS will be introduced later. Note that this book is *not* a manual for the use of either of these two packages, and although examples of setups for running various programs will be provided, recourse will have to be made to the official manuals for each of these packages.

> SPSS—Statistical Package for the Social Sciences, version 9. Major reference: Nie et al. (1975), *SPSS Statistical Package for the Social Sciences* (Second Edition), McGraw-Hill. There is already an alternative SAS-like version of SPSS called SPSS-X.
>
> SAS—Statistical Analysis System. Major reference: SAS Institute Inc. (1982), *SAS User's Guide (1982 Edition)*, SAS Institute.

The Statistical Package for the Social Sciences (SPSS) is one of several packages in which standard statistical tests have been programmed in such a way that they can be implemented with one or two statements to access the computer, and a limited number of specifications to access the appropriate test. We will provide examples of the SPSS control statements, and students can precede these "setups" with the appropriate statements to access the computer. Some familiarity with computers and format statements is required. The actual specification of how many of these statements are used can be acquired from the SPSS manual or from a reduced version of that manual, *The SPSS Primer*.

The statements required for inputting the information about the survey, the variables, and the data are shown in Table 1.5. We have deliberately provided this example with very full documentation to illustrate some of the concepts discussed in this chapter. Normally, the setup would be briefer, although there is a lot to be said for recording all the decisions that had to be made during the design phase. An additional variable—Land Value per Square Meter—was created using the COMPUTE statement, which with the eight observed variables, plus the lot identification code and the three sequencing and weighting variables generated automatically by SPSS, give a total of 13 variables in the SPSS computer file labeled ULVS. Using the DOCUMENT and LISTFILEINFO statements of SPSS, the list of variables and their documentation are provided in Tables 1.6 and 1.7. Finally, so the data can be easily checked, the LIST CASES statement is used to provide a listing of the observed and derived variables (Table 1.8).

The Mechanics of Statistical Analysis

Throughout this book, we attempt to outline the actual processes involved in a statistical analysis step by step. In the early sections of the book, the processes are outlined in detail, but as the techniques become more complex, the data manipulations more involved, and the reliance on mechanical aids

TABLE 1.5 SPSS: Computer Setup for Entering the Urban Land Value Survey Data

```
RUN NAME         ULVS(1) - BASIC DATA SET UP
FILE NAME        ULVS  URBAN LAND VALUE SURVEY FOR THE CITY OF WOODFORD
VARIABLE LIST    VALUE AREA FRONTAGE LANDUSE FLOORS WORKERS AGE ACCESS LOTCODE
INPUT MEDIUM     CARD
N OF CASES       86
INPUT FORMAT     FIXED (2F9.0,F9.1,5F9.0,F8.0)
VALUE LABELS     LANDUSE (1) SHOPS (2) OFFICES (3) MULTI-COMMERCIAL
                 (4) WAREHOUSES (5) HOTELS (6) ADMIN-COMMUNITY
                 (7) RECREATIONAL (8) INDUSTRIAL (9) RESIDENTIAL/
                 FLOORS (1) ONE-FLOOR (2) TWO-FLOORS (3) THREE-FLOORS
                 (4) FOUR-FLOORS/
                 AGE (1) NEW (2) MODERN (3) POST-WAR (4) OLD/
                 ACCESS (0) NO ACCESS MAIN ROAD (1) ACCESS MAIN ROAD/
PRINT FORMATS    FRONTAGE(1)
COMPUTE          UNITVALU=RND(VALUE/AREA*1000)
VAR LABELS       VALUE      UNIMPROVED LAND VALUE/
                 AREA       AREA OF LOT/
                 FRONTAGE   LENGTH OF ROAD FRONTAGE/
                 LANDUSE    DOMINANT LAND USE OF LOT/
                 FLOORS     MAXIMUM HEIGHT OF BUILDINGS/
                 WORKERS    EMPLOYMENT NUMBERS ON LOT/
                 AGE        AGE OF DOMINANT BUILDING/
                 ACCESS     ACCESS TO MAIN RD GLADSTONE RD/
                 LOTCODE    VALUATION CODE FOR LOT/
                 UNITVALU   LAND VALUE PER SQUARE METER
DOCUMENT         SOURCE OF INFORMATION:     (As listed in Table 1.7)

LIST FILEINFO    COMPLETE DOCUMENTS
LIST CASES       CASES=86/VARIABLES=VALUE TO UNITVALU
FREQUENCIES      INTEGER=ACCESS(0,1)
READ INPUT DATA

                 (86 data records as in Table 1.8)
```

TABLE 1.6 SPSS: Variable List for the Urban Land Value Survey

```
DOCUMENTATION FOR SPSS FILE 'ULVS      URBAN LAND VALUE SURVEY

LIST OF THE    1 SUBFILES COMPRISING THE FILE

ULVS      N=    86

DOCUMENTATION FOR THE    13 VARIABLES IN THE FILE 'ULVS

REL  VARIABLE  VARIABLE LABEL                                    MISSING PRT
POS      NAME                                                     VALUES FMT

  1  SEQNUM                                                       NONE    0
  2  SUBFILE                                                      NONE    A
  3  CASWGT                                                       NONE    4
  4  VALUE     UNIMPROVED LAND VALUE                              NONE    0
  5  AREA      AREA OF LOT                                        NONE    0
  6  FRONTAGE  LENGTH OF ROAD FRONTAGE                            NONE    1
  7  LANDUSE   DOMINANT LAND USE OF LOT                           NONE    0
                            1. SHOPS
                            2. OFFICES
                            3. MULTI-COMMERCIAL
                            4. WAREHOUSES
                            5. HOTELS
                            6. ADMIN-COMMUNITY
                            7. RECREATIONAL
                            8. INDUSTRIAL
                            9. RESIDENTIAL
  8  FLOORS    MAXIMUM HEIGHT OF BUILDINGS                        NONE    0
                            1. ONE-FLOOR
                            2. TWO-FLOORS
                            3. THREE-FLOORS
                            4. FOUR-FLOORS
  9  WORKERS   EMPLOYMENT NUMBERS ON LOT                          NONE    0
 10  AGE       AGE OF DOMINANT BUILDING                           NONE    0
                            1. NEW
                            2. MODERN
                            3. POST-WAR
                            4. OLD
 11  ACCESS    ACCESS TO MAIN RD GLADSTONE RD                     NONE    0
                            0. NO ACCESS MAIN ROAD
                            1. ACCESS MAIN ROAD
 12  LOTCODE   VALUATION CODE FOR LOT                             NONE    0
 13  UNITVALU  LAND VALUE PER SQUARE METER                        NONE    0
```

more essential, only an outline of the processes can be attempted. In designing this book for a reader with little mathematical background, the mathematics of the processes will be introduced when they are required. Reflecting this decrease in detail of the techniques involved will be a corresponding change in the detail of the examples provided. In the earlier chapters, we will make considerable use of a few selected examples, which we will specify fully. The Urban Land Value Survey, for instance, provides us with a set of data to be used for all examples in Chapter 2. In later chapters, our number of examples will increase but the level of detail for each example will decline. Our aim is simply to provide a working knowledge of the various statistical techniques to be discussed using examples that are as realistic as possible.

TABLE 1.7 SPSS: Documentation for the Urban Land Value Survey

```
ULVS(1) - BASIC DATA SET UP                                                   08/13/85
FILE   ULVS    (CREATION DATE = 08/13/85)   URBAN LAND VALUE SURVEY FOR THE CITY OF WOODFORD
DUMP OF DOCUMENTARY INFORMATION..
08/13/85       SOURCE OF INFORMATION:

               THE CITY OF WOODFORD LAND TAX OFFICE PROVIDED THE FOLLOWIN
               INFORMATION:

                1. DESCRIPTION OF EACH PROPERTY LOT IN THE CITY RECORDING
                   INFORMATION ON LAND VALUE AND LOT CHARACTERISTICS BASED ON A
                   SURVEY DATED MAY 1982. DATA FOR EACH LOT WAS FILED ON A
                   SEPARATE SHEET AND THE SHEETS WERE FILED IN NUMERIC ORDER
                   BASED ON PROPERTY LOT NUMBER

                2. A MAP OF THE CITY OF WOODFORD SHOWING PROPERTY LOTS AS OF MAY
                   24 1982 (VALUATION MAP 82/01 SHEET 7) WHICH PROVIDED THE
                   INDEX FOR (1) ABOVE

                3. A MAP OF THE CITY OF WOODFORD PREPARED BY THE CITY PLANNING
                   DEPARTMENT (PLANNING MAP C/4) DATED JANUARY 1 1982
                   DEMARCATING THE AREA DESIGNATED AS CBD FOR PLANNING
                   PURPOSES

               DEFINITION OF POPULATION AND ELEMENTS:

                1. THE ELEMENT IN THE SURVEY WAS THE RECORDED LOT AS SPECIFIED
                   ON THE MAP OF THE CITY OF WOODFORD (VALUATION MAP 82/01 SHEET
                   7) DATED MAY 24 1982

                2. THE POPULATION CONSISTED OF ALL RECORDED LOTS WITHIN THE
                   AREA DEMARCATED AS CDB CORE BY THE WOODFORD CITY PLANNING
                   DEPARTMENT AND RECORDED ON PLANNING MAP C/4 DATED JANUARY
                   1 1982. THE SIZE OF THE POPULATION WAS N = 86

               DEFINITION OF RECORDED VARIABLES:

                1. VALUE       UNIMPROVED LAND VALUE
                   RECORDED FROM SOURCE (1) ABOVE AND CODED IN UNITS OF $000
                   RETAINING THE ORIGINAL LEVEL OF SUBJECTIVE APPROXIMATION
                   LEVEL OF MEASUREMENT: RATIO METRIC
                   PRECISION: SUBJECTIVE +/- $500

                2. AREA        AREA OF LOT
                   RECORDED FROM SOURCE (2) ABOVE AND MEASURED IN SQUARE METERS
                   ALTHOUGH IT WAS NOTED THAT THE ORIGINAL MEASUREMENT WAS IN
                   SQUARE FEET WHICH  HAD BEEN CONVERTED TO METERS AT THE TIME
                   OF THE VALUATION SURVEY
                   LEVEL OF MEASUREMENT: RATIO METRIC
                   PRECISION: ASSUMED +/- 0.5 SQUARE METERS

                3. FRONTAGE    LENGTH OF ROAD FRONTAGE
                   RECORDED FROM SOURCE (1) ABOVE WHERE ROAD FRONTAGE WAS
                   LISTED IN METERS TO TWO DECIMAL PLACES. HOWEVER THE ORIGINAL
                   SURVEY HAD BEEN IN FEET AND INCHES THUS THE DATA WAS
                   ROUNDED AND RECORDED AT ONLY ONE DECIMAL PLACE
                   LEVEL OF MEASUREMENT: RATIO METRIC
                   PRECISION: ASSUMED +/- 0.05 METERS

                4. LANDUSE     DOMINANT LAND USE OF LOT
                   FROM SOURCE (1) ABOVE THE ORIGINAL DESCRIPTIONS OF DOMINANT
                   LAND USE WERE CLASSIFIED AND CODED NUMERICALLY INTO 9
                   CLASSES
                   CLASS 1 COMMERCIAL - SHOPS, SHOWROOMS. ETC
                   CLASS 2 COMMERCIAL - OFFICES, BANKS. ETC
                   CLASS 3 COMMERCIAL - MIXED SHOPS/OFFICES
                   CLASS 4 COMMERCIAL - WAREHOUSES, BULK STORES, ETC
                   CLASS 5 COMMERCIAL - HOTELS, HOSTELS, ETC
                   CLASS 6 ADMINISTRATION/COMMUNITY SERVICES
                   CLASS 7 SOCIAL AND RECREATIONAL
                   CLASS 8 INDUSTRIAL
                   CLASS 9 RESIDENTIAL (EXCLUDING HOTELS ETC)
                   LEVEL OF MEASUREMENT: NOMINALLY CLASSIFIED

                5. FLOORS      MAXIMUM HEIGHT OF BUILDINGS
                   RECORDED FROM SOURCE (1) ABOVE AS NUMBER OF FLOORS OF
                   DOMINANT BUILDING ON LOT
                   LEVEL OF MEASUREMENT: ENUMERATED

                6. WORKERS     EMPLOYMENT NUMBERS ON LOT
                   RECORDED FROM SOURCE (1) ABOVE WHICH INDICATED THAT THE DATA
                   WAS OBTAINED BY INTERVIEW DURING THE SURVEY
                   LEVEL OF MEASUREMENT: ENUMERATED

                7. AGE         AGE OF DOMINANT BUILDING
                   FROM SOURCE (1) ABOVE THE DATES OF INITIAL BUILDING AND/OR
                   RECONSTRUCTION WERE EXAMINED AND THE DOMINANT BUILDING ON
                   THE LOT WAS CLASSIFIED AS PREDOMINANTLY
                   CLASS 1 NEW - 1980 OR LATER
                   CLASS 2 MODERN - 1965 TO 1979
                   CLASS 3 POST-WAR - 1945 TO 1964
                   CLASS 4 OLD - PRE 1945
                   LEVEL OF MEASUREMENT: ORDINALLY CLASSIFIED

                8. ACCESS      ACCESS TO MAIN ROAD (GLADSTONE RD)
                   FROM OBSERVATIONS OF THE MAP (SOURCE (2) ) EACH LOT WAS
                   CLASSIFIED AND CODED AS
                   CLASS 0 NO FRONTAGE ON MAIN ROAD
                   CLASS 1 FRONTAGE ON MAIN ROAD
                   LEVEL OF MEASUREMENT: BINARY

                9. LOTCODE     VALUATION CODE FOR LOT
                   RECORDED AS LISTED FROM SOURCE (1) ABOVE FOR IDENTIFICATION

               DEFINITION OF DERIVED VARIABLE:

                1. UNITVALU    LAND VALUE PER SQUARE METER
                   UNIMPROVED LAND VALUE WAS RATIOD AGAINST AREA OF LOT AND
                   ROUNDED TO THE NEAREST $
                   LEVEL OF MEASUREMENT: DERIVED (RATIO METRIC)
                   PRECISION: +/- $3.00
```

TABLE 1.8 Urban Land Value Survey for City of Woodford

FILE ULVS CREATION DATE = 08/13/85 URBAN LAND VALUE SURVEY FOR THE CITY OF WOODFORD

CASE-N	VALUE	AREA	FRONTAGE	LANDUSE	FLOORS	WORKERS	AGE	ACCESS	LOTCODE	UNITVALU
1	66.	402.	33.2	3.	2.	6.	3.	0.	1400100.	164.
2	86.	720.	21.6	3.	2.	8.	3.	0.	1400200.	119.
3	42.	314.	15.1	6.	2.	16.	2.	0.	1400300.	134.
4	97.	787.	39.6	7.	1.	0.	1.	1.	1400400.	123.
5	26.	171.	35.0	1.	1.	2.	4.	1.	1400500.	152.
6	84.	598.	15.4	1.	1.	3.	3.	1.	1400602.	140.
7	171.	1489.	39.9	3.	3.	17.	2.	1.	1400601.	115.
8	110.	976.	26.3	3.	2.	6.	2.	1.	1400700.	113.
9	92.	817.	22.1	1.	1.	2.	4.	1.	1400800.	113.
10	49.	466.	13.2	1.	1.	2.	4.	1.	1400901.	105.
11	40.	466.	13.2	1.	1.	2.	4.	1.	1400902.	86.
12	45.	708.	18.6	9.	1.	0.	4.	1.	1401000.	64.
13	28.	221.	8.2	2.	2.	6.	3.	1.	1401001.	127.
14	141.	1154.	33.5	3.	2.	16.	2.	1.	1401100.	122.
15	47.	313.	9.2	6.	2.	10.	3.	1.	1401101.	150.
16	66.	581.	16.5	5.	2.	10.	4.	1.	1401304.	114.
17	28.	312.	20.3	6.	4.	20.	1.	0.	1401305.	90.
18	115.	1849.	47.9	1.	1.	11.	2.	1.	1401303.	62.
19	190.	1198.	50.0	1.	1.	10.	1.	1.	1401400.	159.
20	125.	562.	18.5	7.	1.	5.	1.	1.	1401500.	222.
21	30.	315.	10.9	6.	1.	7.	2.	1.	1401600.	95.
22	40.	314.	15.7	6.	1.	5.	2.	1.	1401700.	127.
23	54.	611.	52.2	6.	1.	2.	3.	0.	1401800.	88.
24	105.	1208.	70.9	6.	1.	2.	2.	1.	1401900.	87.
25	49.	276.	20.7	2.	2.	8.	3.	0.	1800800.	178.
26	37.	245.	8.8	2.	2.	9.	3.	0.	1800700.	151.
27	35.	305.	10.5	4.	1.	8.	4.	0.	1800600.	115.
28	34.	315.	10.4	2.	1.	2.	4.	0.	1800500.	108.
29	21.	331.	9.9	2.	1.	4.	4.	0.	1800400.	63.
30	17.	245.	7.2	1.	1.	3.	4.	0.	1800300.	69.
31	55.	1093.	29.7	3.	2.	11.	4.	0.	1800200.	50.
32	18.	398.	9.7	6.	2.	19.	4.	0.	1800102.	45.
33	16.	503.	11.9	5.	2.	9.	4.	0.	1800101.	32.
34	47.	1212.	48.8	6.	3.	31.	4.	0.	2701600.	39.
35	230.	4858.	98.5	3.	1.	46.	1.	1.	1900000.	47.
36	46.	238.	8.3	1.	1.	2.	4.	1.	2000100.	193.
37	47.	401.	7.4	1.	1.	5.	4.	1.	2001400.	117.
38	53.	197.	5.7	1.	1.	5.	4.	1.	2000200.	269.
39	31.	181.	6.5	2.	1.	3.	4.	1.	2000300.	171.
40	98.	496.	11.0	1.	2.	7.	3.	1.	2000400.	198.
41	45.	221.	7.5	2.	2.	8.	3.	1.	2000500.	204.
42	55.	201.	6.6	1.	2.	12.	3.	1.	2000600.	274.
43	48.	244.	38.4	2.	3.	10.	3.	1.	2000700.	197.
44	24.	199.	9.3	5.	3.	12.	4.	0.	2000800.	121.
45	53.	454.	12.3	1.	1.	3.	3.	0.	2000900.	117.
46	45.	514.	10.6	1.	1.	6.	4.	0.	2001000.	88.
47	77.	1216.	27.7	1.	1.	7.	3.	0.	2001100.	63.
48	36.	561.	14.1	1.	1.	8.	4.	0.	2001201.	64.
49	34.	429.	10.8	9.	1.	0.	4.	0.	2001202.	79.
50	99.	994.	61.7	9.	1.	0.	4.	0.	2001300.	100.
51	75.	291.	40.6	3.	2.	10.	3.	1.	2100100.	258.
52	79.	405.	12.7	3.	3.	25.	3.	1.	2100200.	195.
53	43.	276.	8.7	3.	2.	12.	3.	1.	2100300.	156.
54	27.	309.	10.3	3.	1.	6.	3.	0.	2101700.	87.
55	36.	309.	10.3	6.	1.	10.	4.	0.	2101600.	117.
56	35.	309.	10.3	1.	1.	5.	4.	0.	2101500.	113.
57	65.	554.	21.1	1.	1.	8.	4.	0.	2101300.	117.
58	226.	829.	57.8	6.	4.	40.	1.	1.	2100509.	273.
59	29.	273.	10.3	1.	1.	2.	4.	0.	2100700.	106.
60	28.	273.	10.3	1.	1.	1.	4.	0.	2100800.	103.
61	27.	273.	10.3	5.	1.	7.	3.	0.	2100900.	99.
62	26.	273.	10.3	7.	1.	0.	1.	0.	2101000.	95.
63	26.	347.	9.5	7.	2.	0.	1.	0.	2101100.	75.
64	315.	3597.	181.8	8.	1.	84.	4.	0.	2101200.	88.
65	73.	314.	41.7	3.	2.	12.	1.	1.	2200100.	232.
66	11.	127.	13.3	1.	1.	1.	4.	0.	2201400.	87.
67	96.	511.	16.8	3.	1.	10.	1.	1.	2200200.	188.
68	100.	1238.	38.2	6.	1.	34.	2.	0.	2201300.	81.
69	29.	187.	6.4	1.	1.	4.	4.	1.	2200300.	155.
70	30.	293.	9.7	1.	1.	6.	3.	1.	2200400.	102.
71	72.	332.	42.2	6.	1.	7.	3.	1.	2200500.	217.
72	27.	508.	19.9	7.	1.	0.	4.	0.	2200600.	53.
73	20.	282.	10.5	9.	1.	4.	4.	0.	2200700.	71.
74	170.	2246.	83.8	1.	1.	18.	2.	0.	2200900.	76.
75	54.	1648.	115.3	4.	1.	16.	4.	0.	2201100.	33.
76	160.	1512.	134.9	3.	2.	35.	2.	1.	2300100.	106.
77	39.	554.	20.2	2.	1.	12.	2.	0.	2301100.	70.
78	38.	554.	49.8	5.	1.	9.	2.	0.	2300300.	69.
79	37.	554.	20.2	4.	1.	2.	3.	0.	2301000.	67.
80	36.	554.	20.2	4.	1.	3.	3.	0.	2300400.	65.
81	19.	554.	20.2	7.	1.	0.	2.	0.	2300900.	34.
82	18.	277.	10.2	4.	2.	10.	4.	0.	2300501.	65.
83	17.	277.	10.2	4.	1.	2.	3.	0.	2300502.	61.
84	74.	1246.	70.8	7.	1.	3.	4.	0.	2300800.	59.
85	34.	583.	20.5	7.	1.	0.	4.	0.	2300600.	58.
86	34.	647.	52.2	8.	1..	24.	4.	0.	2300700.	53.

1.5 THE COMPASS OF QUANTITATIVE GEOGRAPHY

This book is focused on classic inferential statistics. We have done this, as we noted earlier, because we believe that students need a fundamental grounding in inferential statistics before they move on to the many other applications of statistical methods in geography. However, it is not our intention to slight the important body of work in geography—a body of work that is developing rapidly—which is focused on issues of interaction modeling, the modeling of dynamic systems, extensions of classical statistical methods to analyze spatial processes, and time series analysis. These approaches make up another course or courses; we have always believed that such extensions of statistical approaches need a basis in classical statistical methods. However, for the interested student, we provide a few paragraphs and some references to these exciting developments in geographical analysis.

Research on spatial flows that began with gravity models has now developed into an extensive body of work on interaction modeling and models of interaction in space. The recent interest in spatial interaction modeling was stimulated by Wilson's (1970) *Entropy in Urban and Regional Modelling*. Since that time, issues of interaction and location allocation modeling have been the subject of extensive research. Some of that literature is surveyed in Haggett, Cliff, and Frey (1977). There are also extensions of the interaction studies to problems of aggregation, the size of units of analysis, and the impact of network geometry on levels of interaction.

A second important research focus has been the development of what can be generally called spatial models. These are attempts to provide models that directly address the nature of spatial processes. Much of this research has focused on issues of autocorrelation and testing for autocorrelation, but it has been increasingly recognized that geographers need to develop specific models that are applicable to spatial processes (Cliff and Ord, 1981). A good general survey of the discussions of these kinds of issues is contained in Haggett, Cliff, and Frey (1977) where there is a specific focus on the issues of modeling spatial processes, especially those related to the diffusion of epidemics. There has also been extensive research on regional dynamics (see Hepple in Wrigley, 1979), and the application of time series models to phenomena of geographic interest. The most complete of the works is Bennett's *Spatial Time Series: Analysis-Forecasting-Control* (1979). Some sense of the developing applications and wide range of statistics in geography is contained in Wrigley's (1979) *Statistical Applications in the Spatial Sciences*.

We want to emphasize here that statistical methods in geography operate at two levels: (1) the application of statistical and inferential techniques to specific geographic problems, and (2) the attempt to design spatial models to understand location-allocation problems, changes of phenomena over time, the impact of spatial units on processes, and in particular, the important issues created by attempts to provide models that reflect spatial autocorrelation. In this introduction, we wish only to establish that we are clearly aware that this is an important field of statistical work in geography and that we see it as a second tier after the establishment of a basic core understanding of inferential statistics. There is exciting work going on in the development of the modeling of spatial processes. This research is being undertaken by people who are well trained in classic statistical methods and in mathematics. For the beginning and even more advanced student, an initial statistical methods course is a good place to begin.

References and Readings

1. General Introduction to Methodology, Models, and Statistics.

Ackoff, R. L., S. K. Gupta, and J. S. Minas (1962) *Scientific Method: Optimizing Applied Research Decisions,* Wiley: New York.

Harvey, D. W. (1969) *Explanation in Geography,* Arnold: London.

Krumbein, W. C. and F. A. Graybill (1965) *An Introduction to Statistical Models in Geology,* McGraw-Hill: New York.

2. The Theory of Measurement

Hodge, G. (1963) "The use and mis-use of measurement scales in city planning," *Journal of the American Institute of Planners* 29:112–121.

Krantz, D. H., R. D. Luce, P. Suppes, and A. Tversky (1971) *Foundations of Measurement* Volume 1, Academic Press: New York.

Siegel, S. (1956) *Nonparametric Statistics for the Behavioural Sciences,* McGraw-Hill: New York.

Stevens, S. S. (1946) "On the theory of scales of measurement," *Science* 103:677–680.

3. Spatial Data and Analysis

Abler, R., J. S. Adams, and P. Gould (1971) *Spatial Organization,* Prentice-Hall: Englewood Cliffs, N.J.

Nystuen, J. D. (1963, reprinted 1968) "Identification of some fundamental spatial concepts," in B. J. L. Berry and D. F. Marble (eds.), *Spatial Analysis: A Reader in Statistical Geography,* Prentice-Hall: Englewood Cliffs: N.J.

Papageorgiou, G. T. (1969) "Description of a basis necessary to the analysis of spatial systems," *Geographical Analysis* 1:213–215.

4. Computer Statistical Packages

Dixon, W. J. (ed.) (1983) *BMDP Statistical Software,* University of California Press: Los Angeles.

Hull, C. H. and N. H. Nie (1981) *SPSS Update* (Version 7–9), McGraw-Hill: New York.

Klecka, W., N. H. Nie, and C. H. Hall (1975) *SPSS Primer,* McGraw-Hill: New York.

Nie, N. H., C. H. Hull, J. Jenkins, K. Steinbrenner, and D. Brent (1975) *SPSS Statistical Package in the Social Sciences*—Second Edition, McGraw-Hill: New York.

SAS Institute (1982) *SAS User's Guide, Basics,* SAS Institute Inc.: New York.

SAS Institute (1982) *SAS User's Guide, Statistics,* SAS Institute Inc.: New York.

5. Computers and Geographic Research

Gould, P. R. (1970) "Computers and spatial analysis: Extensions of geographic research," *Geoforum* 1:53–69.

Haggett, P. (1969) "On geographical research in a computer environment," *Geographical Journal* 135:497–507.

MacDougall, E. B. (1976) *Computer Programming for Spatial Problems,* Edward Arnold: London.

Mather, P. M. (1976) *Computers in Geography. A Practical Approach,* Basil Blackwell, Oxford: UK.

Monmonier, M. S. (1982) *Computer-Assisted Cartography,* Prentice-Hall: Englewood Cliffs, N.J.

Tarrant, J. R. (ed.) (1976) *Computers in Geography,* Geo Abstracts, Norwich: UK.

6. The Compass of Quantitative Geography

Bennett, R. J. (1979) *Spatial Time Series: Analysis-Forecasting-Control,* Pion: London.

Cliff, A. and J. K. Ord (1981) *Spatial Processes, Models and Applications,* Pion: London.

Haggett, P., A. Cliff, and A. Frey (1977) *Locational Methods,* Volumes 1 and 2, Halstead Press, Edward Arnold: London.

Wilson, A. (1970) *Entropy in Urban and Regional Modelling,* Pion: London.

Wrigley, N. (ed.) (1979) *Statistical Applications in the Spatial Sciences,* Pion: London.

CHAPTER 2

The Display of Distributions

Displaying the characteristics of distributions of scores for a single variable may be the final stage in a simple statistical analysis, or it may constitute a preliminary step before moving to more sophisticated analyses of the data. In any event, data display methods fill an important role in statistical analysis. As part of the final stage of an analysis, data display methods provide the major means of presenting the results on which conclusions may be based; as a preliminary step, they present information about the data that will form the basis of much of the later analysis. Yet despite this importance, and despite the simplicity of the methods, an equal number of inaccuracies and misconceptions are generated from data display methods as from any other form of statistical analysis. Insofar as data display methods provide (or should provide) a means of presenting distributions of scores in a manner that can be readily perceived by anyone, it is imperative that this visual presentation does, in fact, accurately present the characteristics of the distribution.

The methods for displaying distributions are of three distinct types. First is listing the distribution of scores in a table in some organized manner (*tabulation*), second is portraying the distribution visually on a graph (*graphic display*), and third is providing a three-dimensional portrayal of the spatial distribution of scores (*cartographic display* or *mapping*). This last group of methods is very different from the other two in that it is not only concerned with showing how the scores vary but also where the higher and lower scores are found (providing the third dimension to the map). In geography, the three methods of portrayal are frequently used together because they complement rather than duplicate each other. Major methods for the tabulation, graphic display, and mapping of distributions will be outlined in Sections 2.1 to 2.3 of this chapter. The applicability of the various techniques is very much dependent on the level of measurement, and the relationships of the tabulation and graphic display techniques to the different levels of measurement are outlined in Table 2.1 (for mapping methods, this presentation is delayed until Section 2.3). No mention is made in this chapter of variables measured by the ranking (ordination) level of measurement. For the most part, ranked data cannot be displayed by the techniques outlined.

2.1 TABULATION

Tabulation is concerned with displaying the distribution of scores usually by means of summary tables. The summary tables may take on a variety of forms and be prepared for a number of different reasons, but central to their production is the need to show in a single table the characteristics of the

TABLE 2.1 Techniques for Tabulation and Graphic Display

	Level of Measurement			
	Binary	Nominally Classified	Ordinally Classified	Enumerated and Metric
TABULATION				
Ranked array	—	—	X	X
Frequency distribution	X*	X*	X*	X*
Relative frequency distribution	X	X*	X*	X*
Cumulative frequency distribution	—	—	X	X*
GRAPHIC DISPLAY				
Bar graph	X	X*	X*	—
Histogram	—	—	—	X*
Frequency polygon and curve	—	—	—	X*

X = technique used.
* = more important method.

distribution of scores. These characteristics include the range of the scores, their variability, and the dominance of scores in certain parts of the distribution. For a variable such as Land Value per Square Meter in the Urban Land Value Survey, the simplest way of doing this is to rank (order) the scores from lowest to highest (or vice versa) and present them as a ranked array (Table 2.2). Some characteristics of the distribution are immediately apparent, but even with a relatively small total of only 86 scores, very little pictorial description is provided. What is needed is a method that summarizes this distribution even further, and although some of the detail will be lost, a general picture of the distribution can be presented. A variety of methods are available but they are all centered on the production of *frequency distributions* (sometimes referred to as *tabulated arrays*) that are simply tables listing the *frequency* or number of occurrences of a particular score or group of scores.

However, before we examine the construction of frequency distributions for data similar to that displayed in Table 2.2, we will look at simpler examples for variables based on the classification levels of measurement.

Frequency Distributions for Classified Variables

The measurement process of classification is usually based on only a small number of classes. The score for each element is simply the class within the classification to which it belongs. In the case of the nominally classified variable Dominant Land Use of Lot in the Urban Land Value Survey, there

TABLE 2.2 Ranked Array for Land Value per Square Meter

32	65	90	117	159
33	67	95	117	164
34	69	95	117	171
39	69	99	117	178
45	70	100	119	188
47	71	102	121	193
50	75	103	122	195
53	76	105	123	197
53	79	106	127	198
58	81	106	127	204
59	86	108	134	217
61	87	113	140	222
62	87	113	150	232
63	87	113	151	258
63	88	114	152	269
64	88	115	155	273
64	88	115	156	274
65				

TABLE 2.3 Frequency Distribution for Dominant Land Use of Lot

Class	Count	Frequency
1. Shops	卌 卌 卌 卌 卌	25
2. Offices	卌 IIII	9
3. Multicommercial	卌 卌 IIII	14
4. Warehouses	卌 I	6
5. Hotels	卌	5
6. Administration-community	卌 卌 III	13
7. Recreational	卌 III	8
8. Industrial	II	2
9. Residential	IIII	4
TOTAL		86

Source: Urban Land Value Survey.

were nine classes (different types of land use), and each property lot was allocated to one of these classes. A frequency distribution is constructed by counting up the number of individual scores that fall into each of the classes, and presenting them in a form similar to Table 2.3.

In a few instances, especially when dealing with much larger populations or samples, the number of classes may be quite considerable, and a simple frequency distribution may not summarize the distribution sufficiently. This means that the classification has too many classes for a simple visual portrayal, and some sort of grouping of classes is necessary. For nominally classified variables, this would mean grouping classes together that have some common attribute—for example, the commercial land use classes for the variable Land Use (Table 2.3) could easily be grouped together. Adjacent classes for an ordinally classified variable could be grouped in a similar manner.

Frequency Distributions for Enumerated and Metric Variables

Enumerated and metric variables do not have classes already established and classes have to be constructed in some "meaningful" manner. While it is possible to plot the actual scores for the variable Land Value per Square Meter (Figure 2.1, Column 1), a frequency distribution based on the actual scores does not satisfy the main aim of tabulation—to portray a distribution so that its characteristics can be readily perceived.

There are two basic problems: how many classes to differentiate, and what boundaries to use to separate classes. Obviously, the two are closely related and if we decide on one, then determining the other is considerably simplified. No hard-and-fast rules can be provided. However, in regard to the number of classes, we can note the following points.

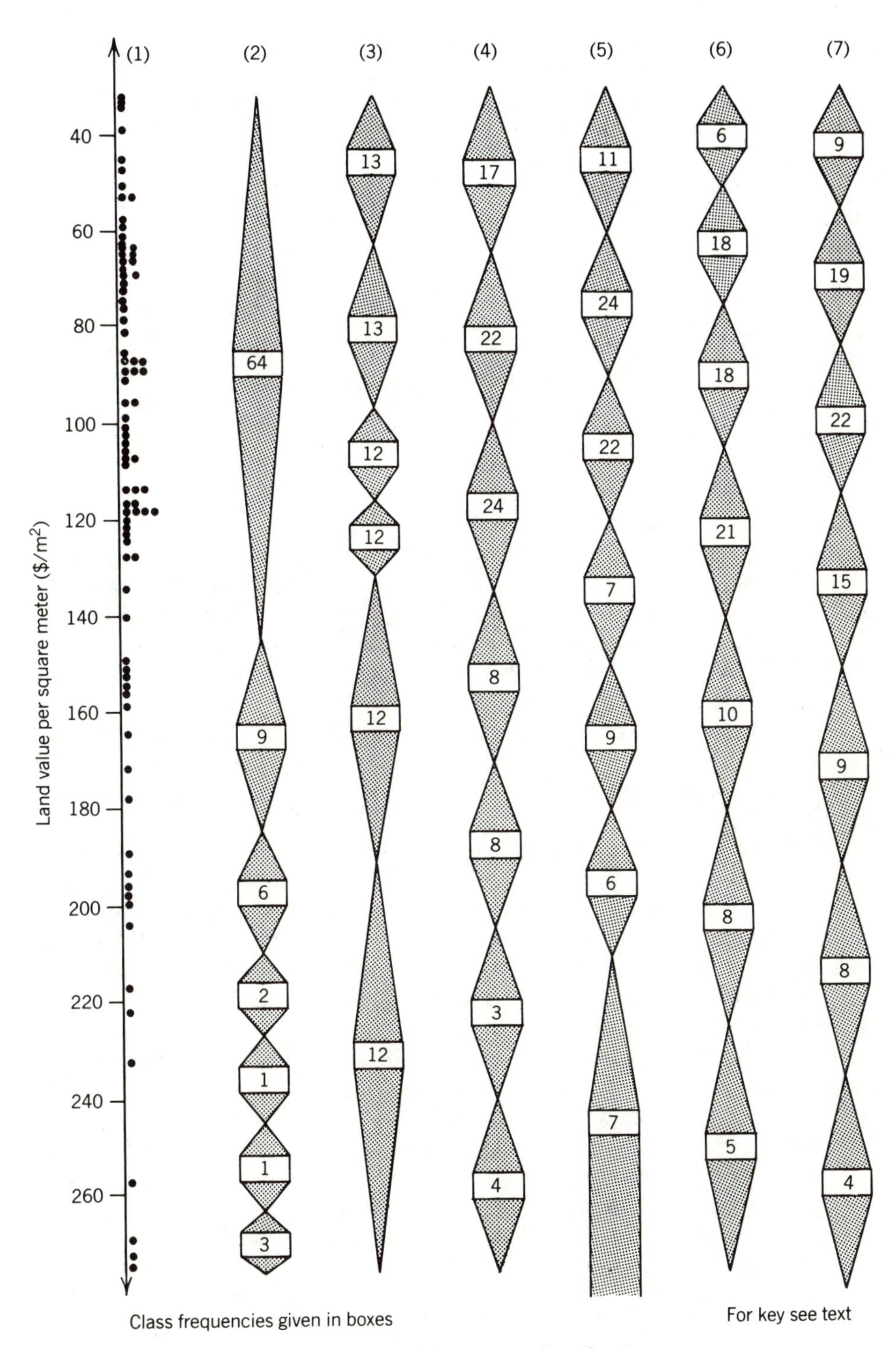

Figure 2.1 Illustration of a variety of frequency distributions.

First, it is generally recognized that the number of classes necessary to portray a distribution adequately is related to the population or sample size—the larger the number of scores, the larger the number of classes. Some researchers suggest no fewer than 5, no greater than 20. An often-quoted (although not so frequently used) suggestion determines the approximate number of classes (k) based on the number of scores (N) using the following relationship:

$$k = 1 + 3.3\ (\log_{10} N) \tag{2.1}$$

where $\log_{10} N$ is the logarithm to the base 10 of the total number of cases. For the Urban Land Value Survey with 86 scores, this would suggest

$$\begin{aligned} k &= 1 + 3.3\ (\log_{10} 86) \\ &= 1 + 3.3\ (1.9345) \\ &= 7.381 \end{aligned}$$

which would be rounded to 7 classes.

Second, in planning for later statistical testing, some of the requirements of these tests would have to be considered when setting up the frequency distribution. Such requirements are centered on having a "reasonable" number of classes (commonly not fewer than 5) and a "reasonable" frequency in each class (again not fewer than 5 or sometimes 3). It is also important to avoid "gaps," that is, empty classes in the distribution.

Third, if the classes are going to be used for mapping purposes—to produce what we will call a choropleth map—then the human eye can readily discern only about six or seven different black to white shadings (more when color is used).

In regard to setting up boundaries to divide up the distribution of scores, there are a wide variety of methods and combinations of methods. Instead of examining these in detail, we will simply outline some of the major methods and illustrate these with reference to the same set of data—the distribution of scores for Land Value per Square Meter (Figure 2.1). We will assume for our presentation that the decision has been made to divide the distribution into seven classes (using Equation 2.1).

In a few instances, certain types of variables have established boundaries, making the process of tabulation almost automatic. Age data are usually classified into five-year (or multiples of five years) intervals; income data into $1000 (or multiples of $1000) intervals. Such a convention is not present for the variable we are examining.

Two methods are based on the distribution of scores without referring to their magnitude, producing *irregular* classes (i.e., the difference in magnitude of scores within a class—the class interval—is not mathematically related to the intervals of the classes on either side). Column 2 in Figure 2.1 is based on "natural" breaks in the distribution, and the six largest "gaps" (to give seven classes) were used as the basis for division. Another division, often used for mapping distributions, is based on approximately equal frequencies per class (a frequency of 12 or 13 per class in our example—Column 3).

The simplest (and most common) system is with a *uniform* (or *constant*) class interval. An example with a constant interval of 35 and starting the first class from a minimum value of 30 is illustrated in Column 4 of Figure 2.1. Examining the frequency distribution shows that there is a dominance of scores at the lower end of the distribution with nearly one-half of the scores (39 of 86) falling into the first two of the seven classes. Such a distribution is described as skewed to the right—"stretched out" on the right-hand side of the distribution. A frequent modification to the method just outlined to handle skewed distributions of this sort is to use a smaller class interval but leave the highest class open ended. An example starting at a value of 30, with a class interval of 30, and an open-ended highest class, is shown in Column 5 of Figure 2.1.

An alternative method of creating classes systematically increments the size of the interval from class to class. Column 6 in Figure 2.1 illustrates an *arithmetic* incremental classification starting with a value of 30 and a class interval of 20 and then incrementing each successive interval by +5. A similar incremental structure could be produced but with a multiplication factor rather than addition. Such *geometric* incremental systems can allow for a greater degree of skewness. Column 7 uses a class interval of 25, and an incremental factor of × 1.25.

Even though there is a wide variety of classification approaches (and there are many others not discussed here), it is suggested that constant class intervals be used wherever possible, with the number of classes determined by Equation 2.1. This method usually provides a reasonable portrayal of the distribution and at the same time, provides a simple structure for later analysis. Thus, following this suggestion, the frequency distribution illustrated in Column 4 of Figure 2.1 was retained as best reflecting the distribution of the variable Land Value per Square Meter (Table 2.4).

Properties of Frequency Distribution Classes

Frequency distributions form an important initial step in any statistical analysis and some of their properties need to be examined in greater detail. Inherent in the previous discussion has been the concept that classes are mutually exclusive. That is, an individual score can fall into one—and only one—of the classes. Thus, the *boundaries* between the classes would be defined as lying between the highest possible score of one class, called its *upper limit* (U) and the lowest possible score of the next highest class, called its *lower limit* (L).

These class limits are defined with reference to the precision of the original measurement. The stated class limits may not necessarily be the true class limits. For example, taking the variable Land Value per Square Meter, all

TABLE 2.4 Frequency Distribution for Land Value per Square Meter (in dollars per square meter)

Class	Frequency
30–64	17
65–99	22
100–134	24
135–169	8
170–204	8
205–239	3
240–274	4
TOTAL	86

Source: Urban Land Value Survey.

the 86 scores were recorded in dollars, but remember that these were rounded off from a more detailed calculation used in their derivation. If we had not been involved in the measurement process for this variable (if, for example, we had obtained the data from some source exactly as we had recorded it in Table 2.2), we would not be aware of the method used in the rounding-off process. However, most data in this form would have been rounded off to the nearest dollar (as we did) or truncated to whole dollar values. These two methods would produce different true class limits (Table 2.5)—a small difference, but it does influence later calculations.

As already stated, the *frequency* (f) of a class refers to the number of individual scores falling within that class, and may be written, for example, as

$$f_{100-134} = 24$$

The *size* (or *range*) of a class interval, or simply the *interval* (I), is found by subtracting the lower limit from the upper limit of that class, which using the two possibilities for the definition of true upper and lower limits would be, for example, either

$$I_{100-134} = 134.999 - 100.0 = 34.999$$

TABLE 2.5 Upper and Lower Limits for the Class \$100 to \$134

	Possible True Limits	
Stated Limits	(in whole dollars)	(to nearest dollar)
Upper (U) 134	134.999. . .	134.499. . .
Lower (L) 100	100.000	99.500

or

$$I_{100-134} = 134.499 - 99.5 = 34.999$$

For practical purposes, this interval would be regarded as being 35.0.

The *midpoint* (M) of a class is situated halfway between the lower and upper limits

$$M_j = \frac{U_j + L_j}{2} \text{ for any class } j \tag{2.2}$$

where U_j and L_j represent the true upper and lower limits of the jth class. Thus, for example, either

$$M_{100-134} = \frac{134.999 + 100.0}{2} = 117.4995 = \$117.5$$

or

$$M_{100-134} = \frac{134.499 + 99.5}{2} = 116.9995 = \$117.0$$

Note carefully, therefore, that the value of the midpoint depends on the true class limits, although in our example, we know that the second case is correct. This distinction is important in analyses in which the midpoint is used as a summary measure for all individual scores falling within that class.

Classes may be either *open* or *closed*. A closed class has both a fixed upper and lower limit, whereas an open class, which could occur at either end of the distribution (rarely at both), would have just one fixed limit, either the upper or lower. The example in Column 5 of Figure 2.1 used an open class at the upper end of the distribution and the definition of this class would be ≥ 210 (greater than or equal to \$210). Open classes are handled in the same way as closed classes, except that their interval size and their midpoint are not defined. This distinction is frequently ignored and the interval size and midpoint are calculated assuming that the open end has an upper (or lower) limit in a corresponding position to the other classes (an upper limit of approximately \$240 in the example cited). That is, it is treated as if it were a closed class. The underestimation that possibly results is often of minor importance and could be ignored. However, in some circumstances, it is necessary to take the range of scores within the open class (if these are known) into account and, for example, the interval size could be defined as the difference between the lower limit (or upper limit) of that class and the highest (or lowest) score recorded. In the foregoing example, this would be $274 - 210 = \$64$. The midpoint would be halfway between the two scores (\$242). An alternative method is to use the arithmetic mean (see Section 3.2 of the next chapter) of all the scores within that class as an estimate of the midpoint.

Relative and Cumulative Frequency Distributions

A frequency distribution provides a reasonable summary, in tabular form, of the distribution of scores, but further modifications to this distribution will be of considerable use in later analyses. These modifications have two forms. One is to convert the absolute frequency for each class into a *relative frequency* (*Rf*) by calculating the frequency as a percentage (or proportion) of the total number of scores (N)

$$Rf_j = \frac{f_j}{N}(100) \qquad \text{or} \qquad Rf_j = \frac{f_j}{N} \text{ for each class } j \qquad (2.3)$$

Apart from rounding errors, relative frequencies will total 100 if the percentage form is used, or 1.0 if the relative frequencies are simply expressed as proportions. A relative frequency distribution for the Land Value per Square Meter data is given in Table 2.6. Frequencies in this form can then be compared with other frequency distributions regardless of the total population or sample size.

A second modification is to *cumulate* the frequencies, *Cf*, (or the relative frequencies, *CRf*), starting from one end of the distribution or the other. This modification, of course, only applies to enumerated or metric variables. Cumulating from the lower end of the distribution produces a *less-than cumulative frequency distribution* (if the lower limit of each class is used), or a *less-than-or-equal-to cumulative frequency distribution* (if the upper limits are used). For the Land Value per Square Meter, an example of the first of these limits is outlined in Table 2.7*a*. The choice of which method to use depends on which class boundaries best represent the distribution and really reflect whether we are using the upper or lower limits of a class to describe the assumed boundary between classes. In our example, the lower limits reflect the boundaries, and thus the "less-than" type is preferred.

TABLE 2.6 Relative Frequency Distribution for Land Value per Square Meter

Class ($/m²)	*f*	*Rf* (unity)	*Rf* (percentage)
30–64	17	.20	20
65–99	22	.26	26
100–134	24	.28	28
135–169	8	.09	9
170–204	8	.09	9
205–239	3	.03	3
240–274	4	.05	5
TOTAL	86	1.00	100

Source: Urban Land Value Survey.

TABLE 2.7 Cumulative Frequency Distributions for Land Value per Square Meter

(a) Less Than			(b) Greater Than or Equal to		
Class ($/m²)	*Cf*	*CRf*	Class ($/m²)	*Cf*	*CRf*
<65	17	.20	≥30	86	1.00
<100	39	.45	≥65	69	.80
<135	63	.73	≥100	47	.55
<170	71	.83	≥135	23	.27
<205	79	.92	≥170	15	.17
<240	82	.95	≥205	7	.08
<275	86	1.00	≥240	4	.05

Source: Urban Land Value Survey.

From the other end of the distribution, a *greater-than* or *greater-than-or-equal-to cumulative frequency* (or relative frequency) *distribution* is produced (Table 2.7*b*). As with relative frequencies, these cumulative frequency distributions and cumulative relative frequency distributions not only add to the explanation but will be of considerable use in later analyses. If an open class occurs at the upper end of the distribution, that class will not be defined in a cumulative (less-than) frequency distribution, and the same applies to an open class at the lower end of a distribution in a cumulative (greater-than) frequency distribution.

2.2 GRAPHIC DISPLAY

While the tabulated frequency distribution is adequate for many purposes, its translation into a graph will provide a better visual impression of the distribution. Bar charts, histograms, and frequency curves are the commonly used graphic methods. The basis of all major methods of graphic display is the rectangular coordinate system.

The Rectangular Coordinate System

A graph is simply a diagrammatic display of sets of data, in our case restricted to two dimensions and thus to two sets of data. These two sets of data can either be scores for individual elements on two variables (a bivariate situation—examined in Chapters 8 and 9) or the magnitude of classes from a distribution and the frequencies of those classes. The most commonly encountered graph form is the rectangular (or Cartesian) coordinate system. This graph scales the two sets of scores along two axes at right angles to each other (*orthogonal* axes). The two axes are labeled the x axis (horizontal) and the y axis (vertical). In the complete rectangular coordinate system, the axes intersect at a point (the origin) where both sets of data have zero magnitude. This divides the graph into four *quadrants*, based on values for the data scaled along the x axis and the y axis. The complete graph system is shown in Figure 2.2. A few points should be noted to assist in the construction of graphs of this type.

First, most geographic variables (but not all) have a range of positive values only, thus only the first quadrant of the graph would need to be constructed.

Second, the data are plotted in the appropriate quadrant in a position determined by their x value (or *abscissa*) plotted parallel to the y axis, and their y value (or *ordinate*) plotted parallel to the x axis. An example is point A (Figure 2.2) with an x value of 3 and a y value of 4. Points are usually represented by either "." or "x."

Third, the x value and y value locating an individual point are referred to as the point's coordinates and can be referenced in the form (3,4), with the x value listed first.

Fourth, a graph would normally be constructed with origin (intersection of the x and y axes) at (0,0), but if the range of scores for either axis or both is far removed from zero, then the origin may be represented as in Figure 2.3. Using simple methods such as these are not essential (the origin of the drawn graph could be located at any value), but they do avoid a lot of confusion over interpretation of graphed results.

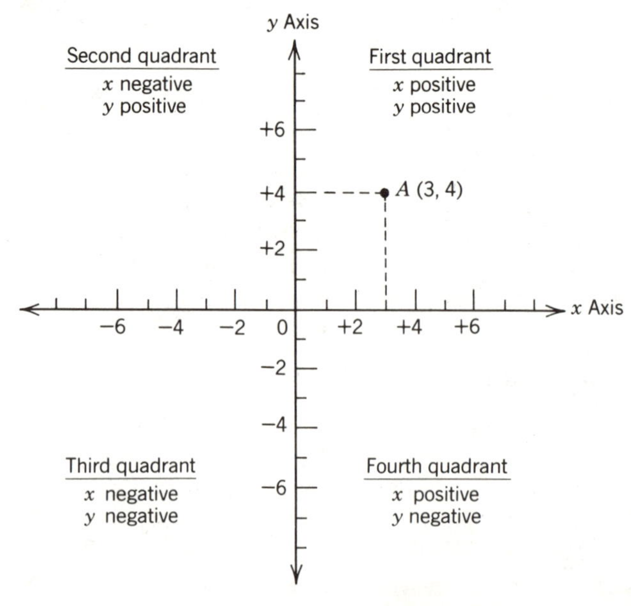

Figure 2.2 The rectangular coordinate system.

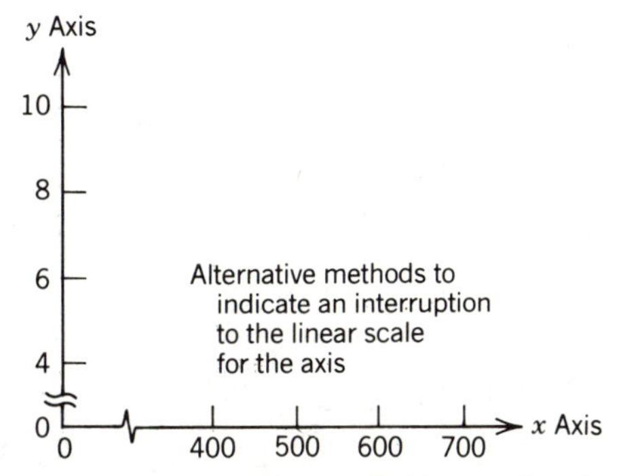

Figure 2.3 Simple methods to maintain (0,0) origin.

Fifth, when the two sets of scores can be structured in such a way that it can be said that the values in one set of scores are dependent in some way on values in the other set (a *functional relationship*), it is customary to plot the dependent scores along the y axis and the independent scores along the x axis.

Sixth, the scales of the x and y axes would be selected with reference to the range and distribution of scores for each of the sets of data, allowing enough room for the largest scores to be plotted. Usually the scores are scaled in an arithmetically regular manner (an arithmetic scale), although in certain instances other scales (such as logarithmic) are used. The latter scales will be introduced when they are needed, and for the moment we will restrict the scaling to the arithmetic metric. The scales of the two axes need not be—and are only rarely—the same.

Bar Charts and Histograms

The rectangular coordinate system is the underlying principle for the portrayal of frequency distribution data. We have two sets of data: one a set of classes, the other a set of frequencies. The frequencies, in a sense, are dependent on the classes in that for a given class there is associated with it a specific frequency. What we are defining here is a simple form of a functional relationship. For enumerated and metric variables, this functional relationship can be described by a statistical frequency model, where y, the frequencies, are a function of x, the arithmetically scaled variable. Because frequencies are dependent scores, they are scaled along the y axis.

For variables measured by classification, the classes are plotted along the x axis, but as the scores for these variables are not based on an arithmetic scale, no mathematically meaningful scaling of the x axis is possible. The class names are simply used as labels along the axis, although in the case of ordinally classified variables, these presumably would be listed in an ordered sequence. The frequency for each class can be plotted using the scaled

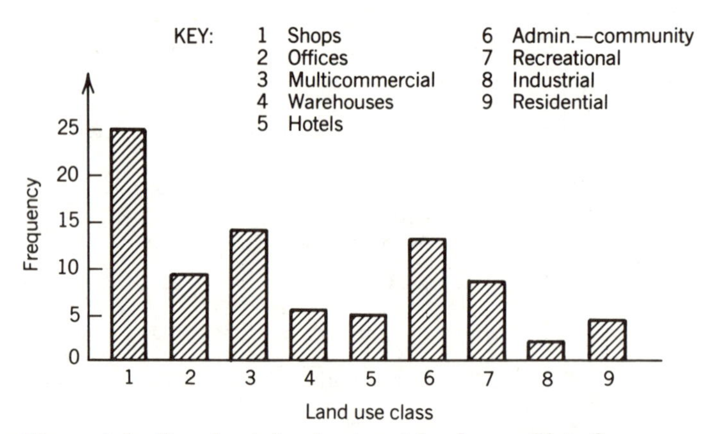

Figure 2.4 Bar chart for dominant land use of lot. *Source:* Urban Land Value Survey.

y axis, and a bar or a line constructed with height representing the frequency. Such a graph is called a *bar chart* or a *bar graph*. An example for the variable Dominant Land Use of Lot in the Urban Land Value Survey is given in Figure 2.4. When the *x* axis is not scaled, it is usual to draw attention to this by leaving spaces between the bars to reduce the impression of a sequence of classes. Bar graphs can, of course, be drawn for both absolute and relative frequency distributions, and the appearance of the graph would be identical with the only difference being the scaling of the *y* axis, either in absolute frequencies or in relative frequencies.

The graphic portrayal of enumerated and metric data follows the same

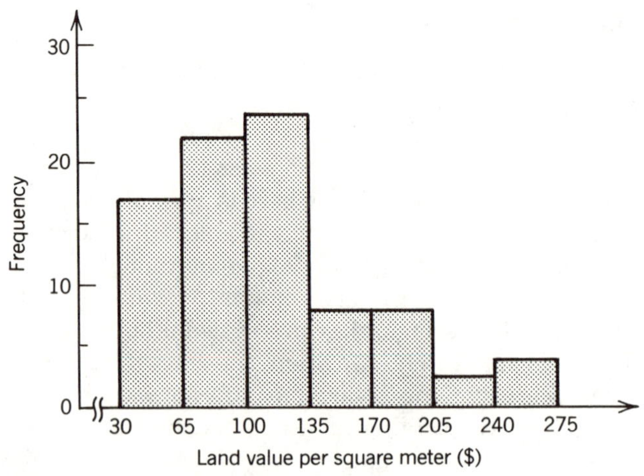

Figure 2.5 Histogram for land value per square meter. *Source:* Urban Land Value Survey.

sort of procedure except that, in this instance, the x axis will actually be scaled. The resulting graph, where the bars will be placed adjacent to each other, is referred to as a *histogram*. The bars can be located and labeled either by the midpoint of the class or by using the upper or lower limits of the classes as boundaries. Histograms can be based on absolute frequencies, relative frequencies, or cumulative frequencies (Figures 2.5 and 2.6). The resulting histograms provide a visual impression of the distribution of scores; however, they do tend to place an emphasis (visually) on the boundaries between classes where sharp changes of frequency occur. Although this is quite correct, remember that the appearance of the histogram can be drastically varied by changing the position or size of the class intervals. The selection of classes is of considerable importance. Where class intervals are unequal, then either the x axis should be scaled in a nonarithmetic manner, or the width of the bars in the histogram should vary with the size of the class interval, so that their area gives an accurate depiction of their relative frequency.

Frequency Polygons and Frequency Curves

One important further modification in graphic portrayal can also be attempted. This is to try to summarize the frequency distribution of scores by a single line and therefore decrease the visual emphasis on boundaries. This is possible only for enumerated and metric variables. As a first approximation, a *frequency polygon* can be constructed for histograms drawn from absolute frequencies by connecting each midpoint of the classes to the adjacent class midpoints (Figure 2.7). Note that we have continued the curve to the midpoints of the intervals on either side of the extreme values. For

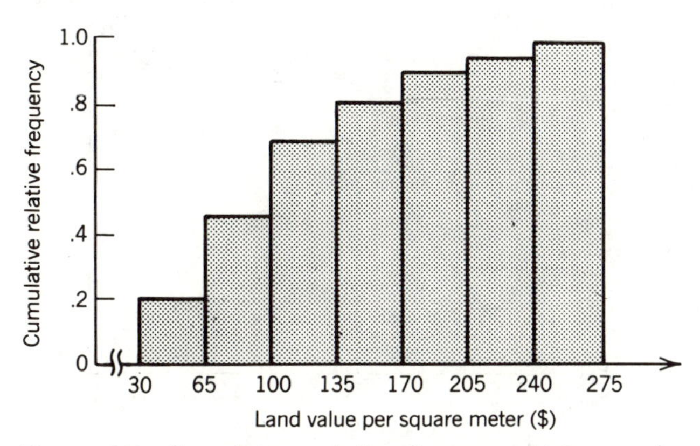

Figure 2.6 Cumulative relative frequency histogram for land value per square meter. *Source:* Urban Land Value Survey.

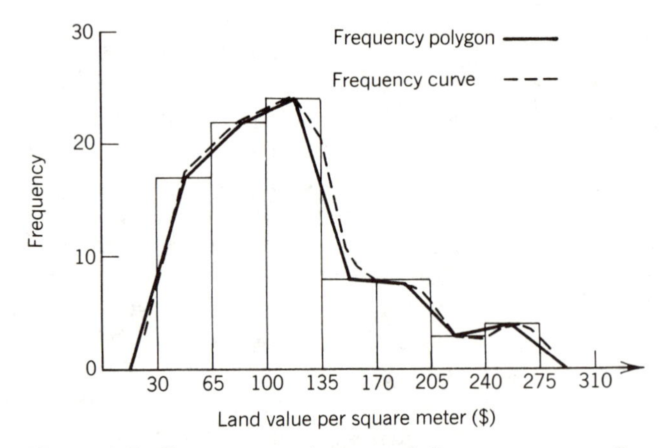

Figure 2.7 Frequency polygon and frequency curve for land value per square meter. *Source:* Urban Land Value Survey.

cumulative frequency histograms, the construction is the same except that rather than use the midpoints, the boundaries that separate the classes should be used to construct the polygon (Figure 2.8). Note that this appears to produce an underestimation of frequencies by "cutting off" part of the tops of the bars. This is not the case. Cumulative histograms do not give an accurate portrayal of the complete distribution of scores, but simply illustrate the cumulative frequencies at the boundaries between classes. Frequency

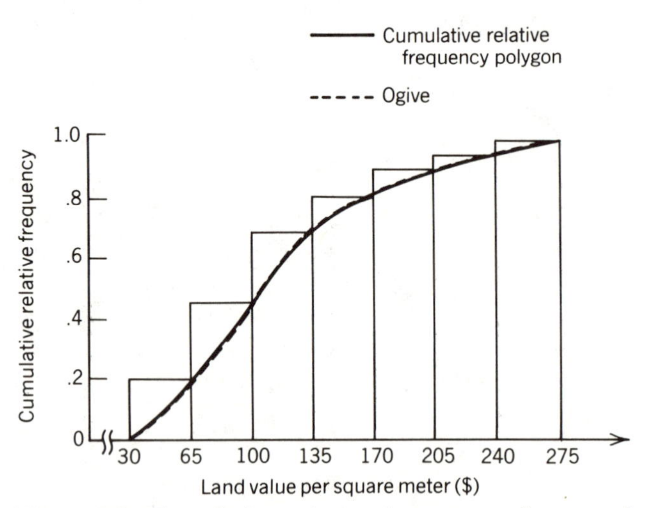

Figure 2.8 Cumulative relative frequency polygon and ogive for land value per square meter. *Source:* Urban Land Value Survey.

polygons (and frequency curves) attempt to describe the complete distribution, providing an estimate of cumulative frequency at any point on the continuous linear scale of the x axis. For a less-than cumulative frequency polygon, this would mean using the right-hand top corner of the bars, and for a greater-than cumulative frequency polygon, using the left-hand top corner of the bars.

As a further attempt to summarize a continuous distribution of scores, the frequency polygon can be smoothed out to form a *frequency curve*. This is usually done by eye. Frequency curves have been plotted on the frequency polygons for the Land Value per Square Meter data in Figure 2.7 and 2.8. The frequency curve for the cumulative distribution is often referred to as an *ogive*.

We have now described a set of methods that has reduced an array of scores into a single smooth line that by itself provides a simple visual summary of the distribution. Later, when we have combined this graphic portrayal with a mathematical description of the smooth curve, we have an extremely powerful descriptive model.

2.3 MAPPING SPATIAL DISTRIBUTIONS

Much of the work of the geographer is concerned with the analysis of the spatial distribution of scores for variables. As a basic tool, geographers use maps to provide a means of summarizing these distributions, to assist in the determination of patterns of distribution, and to facilitate comparison of one distribution with another. There are a wide variety of types of maps that portray spatial distributions, and only those that are of direct relevance to our discussion of statistical methods will be outlined here. Several elementary cartography texts provide elaboration on the methods.

The cartographic portrayal of distributions is accomplished by representing a score or a group of scores by a symbol. It is convenient to describe cartographic symbols as being point, line, or areal symbols, although the distinction is somewhat arbitrary because, for example, areal symbols must be bounded by lines that themselves can be considered as cartographic line symbols. The types of cartographic symbols to be discussed and their application to the three different types of spatial data being examined (Section 1.3) are outlined in Table 2.8. Apart from the isarithmic symbol that can be considered as both a line and an areal symbol, line symbols will not be examined. Note that the applicability of these various techniques also depends on the type of measurement process used.

We briefly examine the use of each of these cartographic symbols, illustrating the method wherever possible by reference to the data obtained in the Urban Land Value Survey. The survey used as its basic spatial elements, 86 property lots, and, as such, can be described as consisting of areally discrete spatial data. For illustrative purposes, however, it is feasible to

TABLE 2.8 Major Techniques for Cartographic Display

	Types of Spatial Data Examined								
	Discrete Points			Discrete Areas			Sampled Points		
	Classified	Enumerated	Metric	Classified	Enumerated	Metric	Classified	Enumerated	Metric
POINT SYMBOLS									
Varying	X*	X	X	—	—	—	X	X	X
Graduated	—	X*	X	—	—	—	X	X	X
Dot	—	—	—	—	X*	—	—	—	—
AREAL SYMBOLS									
Choropleth	—	—	—	X*	X*	X*	—	—	—
Dasymetric	—	—	—	—	X	X	—	—	—
Isarithmic	—	—	—	—	—	—	—	X*	X*
Proximal	—	—	—	—	—	—	X*	X	X

X = technique used.

* = more important method.

transform this set of data into a set of discrete point data by using a summary point to represent the location of each property lot. In some of the examples that follow, we will use the data in this form. The problem of where to locate summary points of this type within an areal unit is one that is frequently encountered in geographic research. Insofar as we are attempting to summarize some of the characteristics of the whole property lot by a discrete point, using an approximate center of the lot will suffice. In other situations, the location of the summary point may be guided by additional factors. We also note that for the variable Land Value per Square Meter, these summary points can be considered as sample points from a continuous distribution of land values over the central city area, and thus, methods for analyzing spatially continuous data can be applied to some degree. The fact that the sample points were derived from discrete areas that were noncontiguous (they were separated by the street pattern) detracts from this assumption but does not invalidate it.

The Mapping of Distributions Using Point Symbols

Three methods will be discussed: varying point symbols, graduated symbols, and dot distributions. Mapping with *varying point symbols* simply involves allocating an individual symbol to each class of a distribution, then plotting these symbols at the points on the map where each score falls. Treating the Urban Land Value Survey data as discrete point data, an example for the variable Dominant Land Use of Lot is shown in Figure 2.9. Obviously, the method is mainly used to portray data derived by classification, especially when there is only a small number of classes and a large number of elements; for higher levels of measurement other methods are more applicable.

For enumerated and metric data, a *graduated point symbol* provides a better means of portrayal, where the magnitude of the score is represented by different sizes of the same symbol. The usual method is to employ graduated circles where the area of the circle is in some scaled proportion to the magnitude of the score. Geometrical shapes other than circles have been used, and properties other than area have been used in the scaling process. Construction would usually be based on isolating the maximum size circle that can be adequately portrayed on the selected map scale, taking into account the number of discrete points and the possibility of overlap, any planned photographic reduction of the map size, and the minimum size of circle that can realistically be presented. Once the radius of the circle (r_{max}) to represent the largest score (X_{max}) has been determined, the radii (r_i) for all other scores (X_i) can be found by

$$r_i = \frac{X_i}{X_{max}} (r_{max}) \tag{2.4}$$

An example for the Employment Numbers on Lot is given in Figure 2.10.

Dot distribution maps are widely used to portray enumerated variables

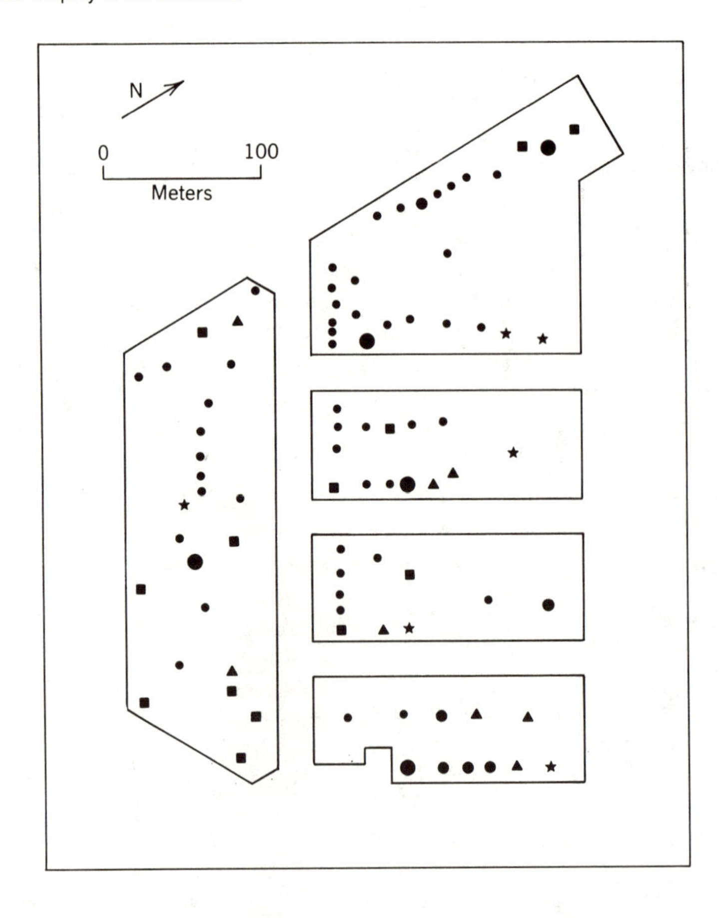

KEY:
- • Commercial—shops and offices (1, 2, 3)
- ● Commercial—warehouses (4)
- ⬤ Commercial—hotels (5)
- ■ Administration—community (6)
- ▲ Recreational (7)
- ★ Other (industrial (8) and residential (9))

Figure 2.9 Woodford—dominant land use of lot. *Source:* Urban Land Value Survey.

because they provide a good visual picture of a spatial distribution at the same time as retaining accuracy of information. Because their basic construction involves the use of points, we discuss them as a point symbol, but in reality they provide a method to portray discrete area data, and the *density* of dots within an areal unit may be considered as an areal symbol summarizing that area. A uniform-sized dot is allocated a certain value (the *dot value*) and the magnitude of the score for an element is represented by a

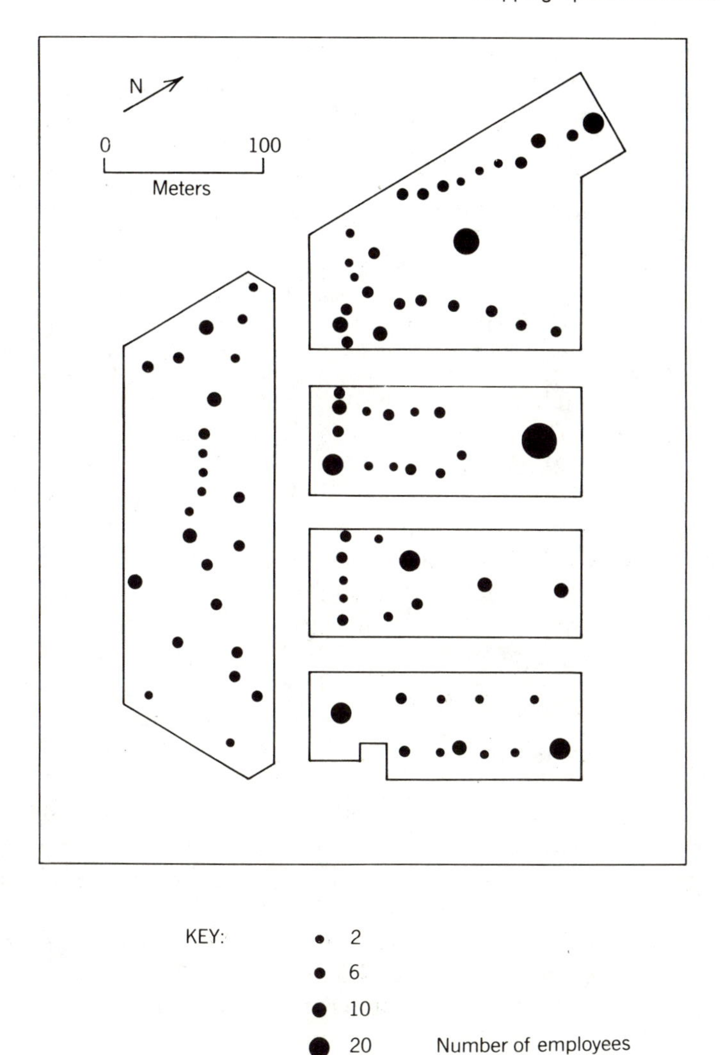

Figure 2.10 Woodford—employment numbers (on lot). *Source:* Urban Land Value Survey.

specified number of dots—the rounded-off multiple of the dot value. Three problems are involved in the construction of dot distribution maps: what to use as the dot value, what size (diameter) to use for the dot, and where to place the dots within the areal unit. A correct cartographic balance between dot value and size of the dot is mainly a matter of experience but guidelines are available from most cartographic texts. Placing the specified number of dots within the areal unit is more problematical and generally, unless a nonuniform distribution is suggested by the nature of the data and unless there is substantial evidence to assist in placing the dots in a nonuniform manner, a uniform pattern of distribution is employed. Figure 2.11 provides an example of a dot distribution map for the Employment Numbers on Lot, using a dot value of two employees, a dot size (when constructed) of 1 mm, and a uniform distribution within areal units.

Mapping Distributions Using Areal Symbols

Geographers make wide use of the techniques of portraying spatial distributions by areal symbols, especially if we include under areal symbols the line symbol *isarithm,* separating areas of differing magnitude. Two forms of isarithm are often differentiated: the *isometric line* for portraying distributions of scores actually observable at a point (e.g., annual rainfall), and the *isopleth* summarizing distributions of derived scores that cannot exist at a point (e.g., population density). Although their interpretation is very different, they are identical in terms of construction, and we will simply refer to them as *isarithms.* An isarithm is, therefore, a continuous line joining points of equal magnitude and between which the phenomenon being mapped is assumed to change uniformly in magnitude from one isarithm to the next. Isarithms can map distributions by the line symbol itself, or by focusing attention on the areas between isarithms. It is convenient to discuss them from the latter point of view.

The techniques discussed in this section are spatially equivalent to fitting histograms and frequency curves to enumerated and metric frequency distributions. In the spatial domain, each areal element can be considered a class of an areal frequency distribution. The "class" is located in the distribution by its position on the surface, or more simply, by its summary point (the spatial equivalent of the midpoint). Each class has a magnitude given by its frequency (if we are dealing with enumerated data) or by its measured score. In the Urban Land Value Survey example, we are dealing with a map based on 86 discrete areal units, the property lots, and for most of the variables we have been examining, the data describing these areal units are assumed to represent a particular characteristic of that property lot. For each property lot, we could draw a column whose height is proportional to the magnitude of the score. Such a three-dimensional map is called a *stepped statistical surface* and is the spatial equivalent of the histogram. This would be a tedious and difficult task and although computer programs are available for producing these maps, the scores are more commonly represented by different levels of shading to produce the simpler *choropleth map.*

However, when we examine the variable Land Value per Square Meter, we are assuming that this variable measures not only the characteristics of each property lot, but also information about basic patterns of variation over the whole central area of the city. From this point of view, the variable Land

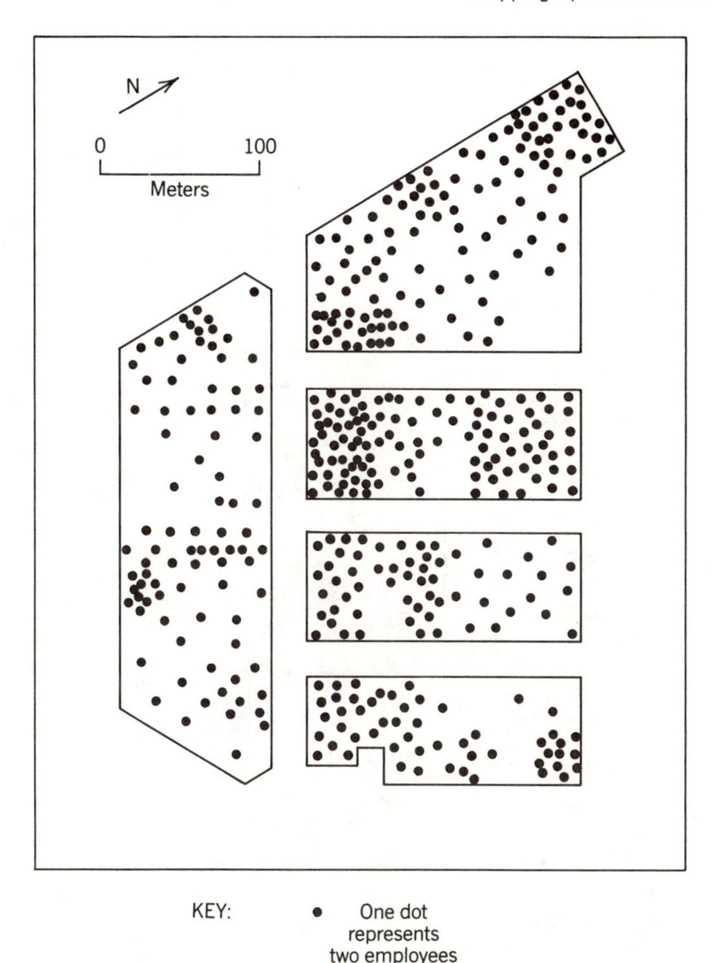

Figure 2.11 Woodford—employment numbers (on lot). *Source:* Urban Land Value Survey.

Value per Square Meter can be considered to be spatially continuous, and mapping it as discrete areas may mask the pattern of gradual change over space. We can "redistribute" areal data more realistically across areal unit boundaries in the same manner that we would treat sample points from spatially continuous data such as annual rainfall or elevation above sea level. The smoothing of a stepped statistical surface is accomplished in basically the same way as frequency curves are derived from histograms. These *smoothed statistical surfaces* provide an excellent visual presentation of the distribution, but their construction usually requires computer routines. In most cases, it is simpler to use *isarithms* (contours) to summarize the elevations of this surface. Thus, the isarithmic map is the spatial equivalent of the frequency curve.

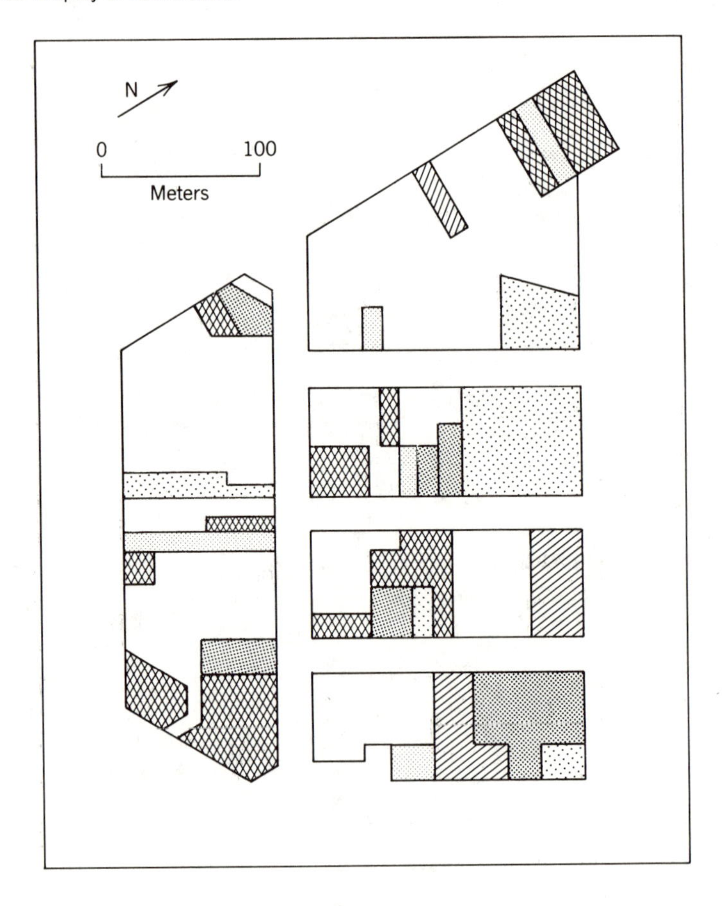

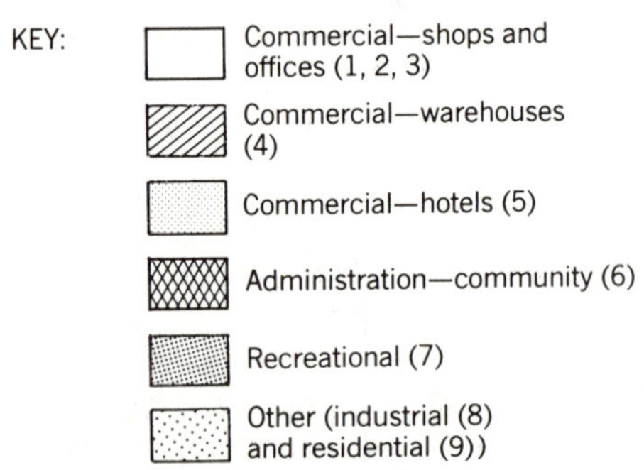

Figure 2.12 Woodford—dominant land use of lot. *Source:* Urban Land Value Survey.

Choropleth mapping is simply the mapping of a distribution of classes, the classes being determined in the same manner as for frequency distributions. The score for each areal unit determines its class, and the areal unit is then shaded by some symbol (or color) representing that class. Three examples from the Urban Land Value Survey may be compared with alter-

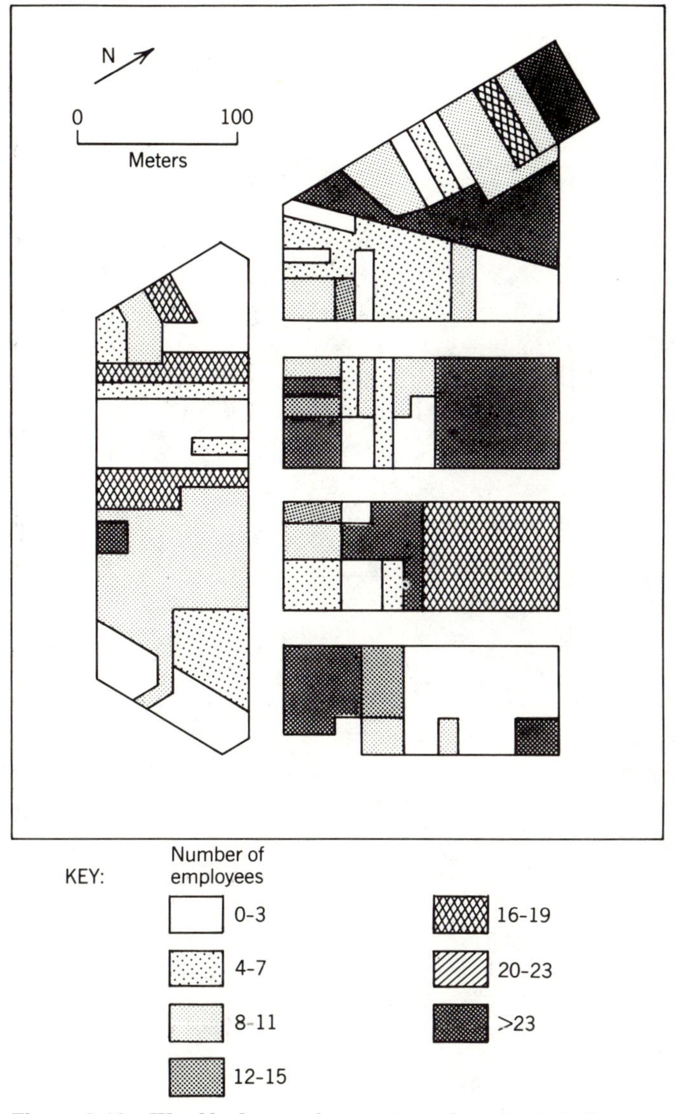

Figure 2.13 Woodford—employment numbers (on lot). *Source:* Urban Land Value Survey.

native forms of protrayal: Dominant Land Use of Lot (compare Figure 2.12 with Figure 2.9), Employment Numbers on Lot (compare Figure 2.13 with Figures 2.10 and 2.11), and Land Value per Square Meter (compare Figure 2.14 with Figure 2.15). Note that the varying symbols and choropleth maps (Figures 2.10 and 2.13) portray the same set of data as the dot distribution map (Figure 2.11) but express it very differently. The last map portrays the actual spatial distribution of employees, whereas the former two show the

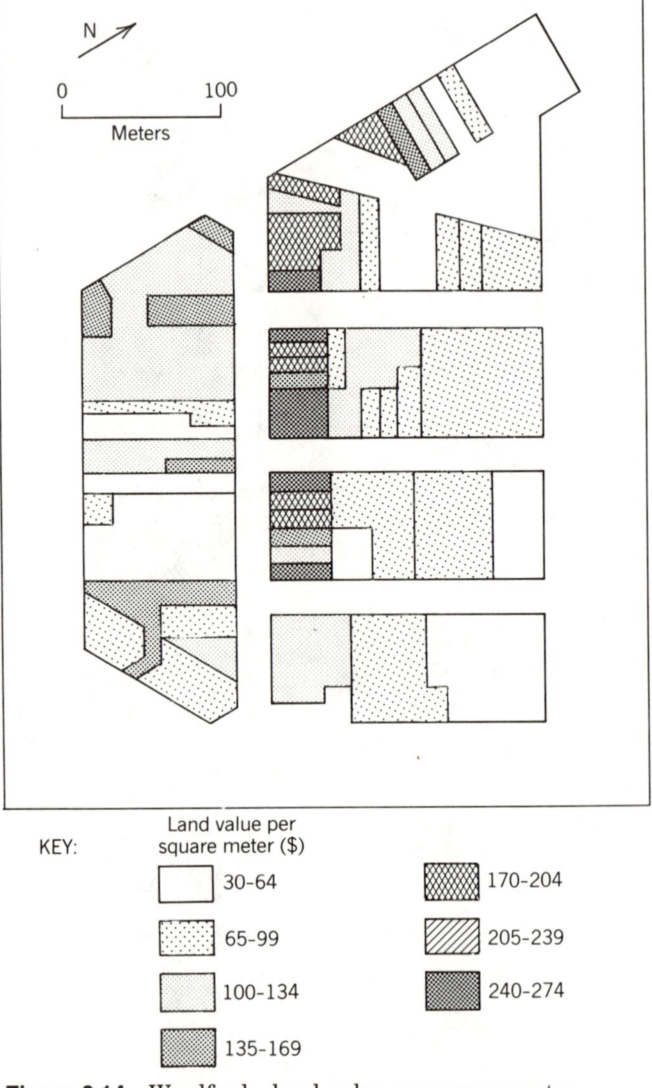

Figure 2.14 Woodford—land value per square meter.

distribution on a per lot basis and therefore the size of the lot is emphasized in the visual appearance. With a preponderance of data collected by areal units, geographers rely heavily on the choropleth map as a means of cartographic portrayal—it retains the form of the data base (the areal units) and thus portrays the information accurately; the only subjective operation in its preparation is the choice of classes for enumerated and metric data. Its one major "fault" is that it is dependent on the character of the areal units themselves. When the information being mapped cannot be considered

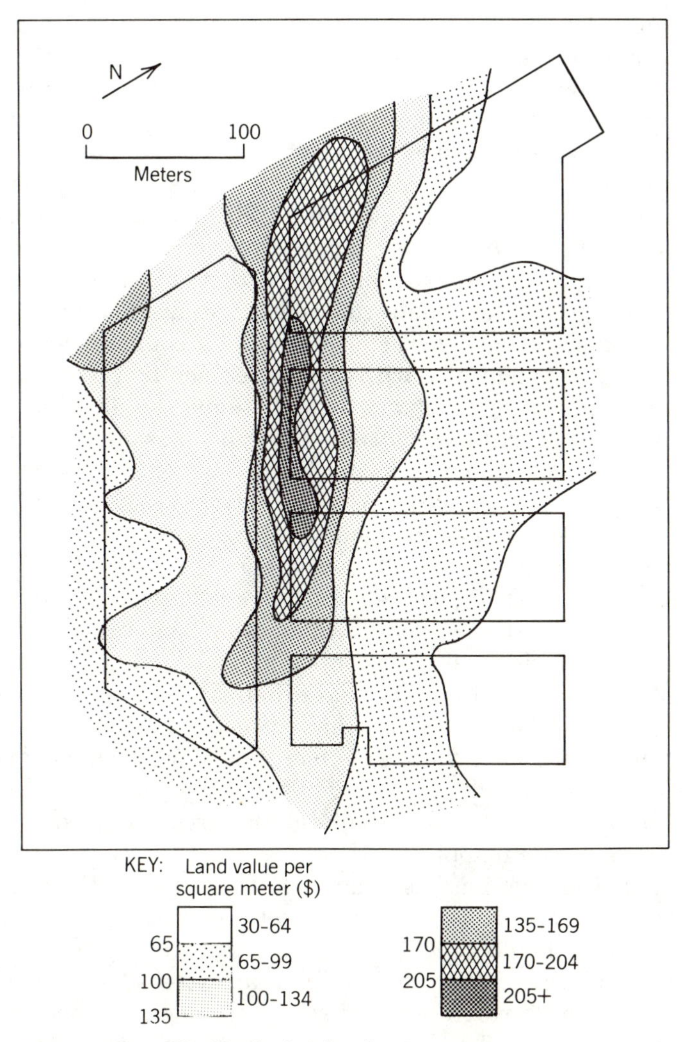

Figure 2.15 Woodford—land value per square meter.

to apply to an areal unit as a whole (there is known variability within the areal unit), and where the choropleth technique is assumed to focus too great attention on abrupt boundaries, then other forms of portrayal have to be considered (including dasymetric mapping).

For reasons outlined in the preceding paragraphs, areally discrete data are sometimes treated as if they were measuring a spatially continuous phenomenon using a summary point for each areal unit as if it were a sample point. This allows the application of the major method to portray spatially continuous data—*isarithmic mapping*. Isarithmic maps form a basic method of cartographic portrayal of distributions and within the confines of the techniques used in their construction, they provide an accurate visual portrayal of a continuous distribution. Through their "data smoothing" process, isarithmic maps remove possible "random" fluctuations in scores; they provide the best visual portrayal of areal patterns if such a pattern exists; and they are capable of mathematical description. Unfortunately, their method of construction is quite complex, tedious, and, to some degree, subjective, although many of these problems are partially overcome by use of computer "contouring" techniques. Apart from the location of summary points (if we are dealing with continuous data converted from discrete areal data), the method of construction involves determining what boundaries to use for the isarithms, locating potential isarithms between points (the major problem, for which a variety of methods are available) and the drawing of the isarithms themselves. Isarithmic maps can be simply presented as line symbols (the isarithms) or isarithmic areas may be defined as shading between the isarithms or between major isarithms. In the latter case, the same limitations on number of classes that is common to all frequency distributions would occur. The only example to be illustrated here is that for Land Value per Square Meter (Figure 2.15), where a considerable degree of data smoothing has been used to illustrate average changes of land value over the city, emphasizing the dominance of the central area.

Isarithmic mapping is limited to data that can be assumed to vary continuously over a surface. It is also limited to variables derived by metric measurement or, in some cases, by enumeration. For classified variables, no isarithms could be defined, and the equivalent method to portray a continuous distribution is to extrapolate information from sampled points to adjacent areas (proximal mapping).

The preceding discussion has done little more than outline some of the methods available for the cartographic display of spatial distributions. The interested reader should consult a text on cartography to obtain further information on these and other topics. This is an extremely important subject to geographers, and cartography and statistical methods constitute two of the most important tools for conducting geographic research. In recent years, the complementarity of these two apparently different approaches has been emphasized.

2.4 COMPUTER METHODS

With only 86 elements (observations), data analysis for the Urban Land Value Survey can be easily handled using manual methods. Frequency distributions, bar charts, histograms, and even maps can be constructed with little effort for the few variables involved. However, when the number of variables increases, or more importantly, when the number of elements

TABLE 2.9 SPSS Setup for Frequency Distributions for Urban Land Value Survey

```
RUN NAME          ULVS(2) - FREQUENCY DISTRIBUTIONS
FILE NAME         ULVS  URBAN LAND VALUE SURVEY FOR THE CITY OF WOODFORD
VARIABLE LIST     VALUE AREA FRONTAGE LANDUSE FLOORS WORKERS AGE ACCESS LOTCODE
INPUT MEDIUM      CARD
N OF CASES        86
INPUT FORMAT      FIXED (2F9.0,F9.1,5F9.0,F8.0)
VALUE LABELS      LANDUSE (1) SHOPS (2) OFFICES (3) MULTI-COMMERCIAL
                  (4) WAREHOUSES (5) HOTELS (6) ADMIN-COMMUNITY
                  (7) RECREATIONAL (8) INDUSTRIAL (9) RESIDENTIAL
COMPUTE           UNITVALU=RND(VALUE/AREA*1000)
RECODE            UNITVALU (30 THRU 64=1)(65 THRU 99=2)(100 THRU 134=3)
                  (135 THRU 169=4)(170 THRU 204=5)(205 THRU 239=6)
                  (240 THRU 274=7)
VALUE LABELS      UNITVALUE (1) 30 - 64(2) 65 - 99(3)100 - 134(4)135 - 169
                  (5)170 - 204(6)205 - 239(7)240 - 274
VAR LABELS        LANDUSE  DOMINANT LANDUSE OF LOT/
                  UNITVALU  LAND VALUE PER SQUARE METER
FREQUENCIES       INTEGER=LANDUSE(1,9) UNITVALU(1,7)
OPTIONS           8
```

increases substantially, the task becomes tedious and prone to error. It is in these situations that processing by computer is useful, if not essential.

The process is illustrated with reference to two variables: Dominant Land Use of Lot and Land Value per Square Meter using the SPSS file we created earlier. The construction of frequency distributions is handled using the FREQUENCIES procedure and, in the case of enumerated and metric variables, use of the RECODE data modification operation to create classes. The setup is displayed in Table 2.9 and the frequency distributions in Tables 2.10 and 2.11 resulted. It should be noted that with RECODEd variables that have been created by COMPUTE statements (such as Land Value per Square Meter) although here we were careful to round the result to whole dollars, the stated upper limit of one class should be set at the lower limit

TABLE 2.10 SPSS Frequency Distribution for Dominant Land Use of Lot

LANDUSE DOMINANT LANDUSE OF LOT

CATEGORY LABEL	CODE	ABSOLUTE FREQUENCY	RELATIVE FREQUENCY (PERCENT)	ADJUSTED FREQUENCY (PERCENT)	CUMULATIVE ADJ FREQ (PERCENT)
SHOPS	1	25	29.1	29.1	29.1
OFFICES	2	9	10.5	10.5	39.5
MULTI-COMMERCIAL	3	14	16.3	16.3	55.8
WAREHOUSES	4	6	7.0	7.0	62.8
HOTELS	5	5	5.8	5.8	68.6
ADMIN-COMMUNITY	6	13	15.1	15.1	83.7
RECREATIONAL	7	8	9.3	9.3	93.0
INDUSTRIAL	8	2	2.3	2.3	95.3
RESIDENTIAL	9	4	4.7	4.7	100.0
	TOTAL	86	100.0	100.0	

TABLE 2.11 SPSS Frequency Distribution for Land Value per Square Meter

UNITVALU LAND VALUE PER SQUARE METER

CATEGORY LABEL	CODE	ABSOLUTE FREQUENCY	RELATIVE FREQUENCY (PERCENT)	ADJUSTED FREQUENCY (PERCENT)	CUMULATIVE ADJ FREQ (PERCENT)
30 - 64	1	17	19.8	19.8	19.8
65 - 99	2	22	25.6	25.6	45.3
100 - 134	3	24	27.9	27.9	73.3
135 - 169	4	8	9.3	9.3	82.6
170 - 204	5	8	9.3	9.3	91.9
205 - 239	6	3	3.5	3.5	95.3
240 - 274	7	4	4.7	4.7	100.0
	TOTAL	86	100.0	100.0	

of the next higher class to avoid accidentally omitting scores falling between the stated limits. When these two limits are the same, the SPSS program only interprets the lower limit as being absolutely correct. Crude graphic displays of the frequency distributions (using the lineprinter) can be created as optional printout. SPSS does not differentiate between bar charts and

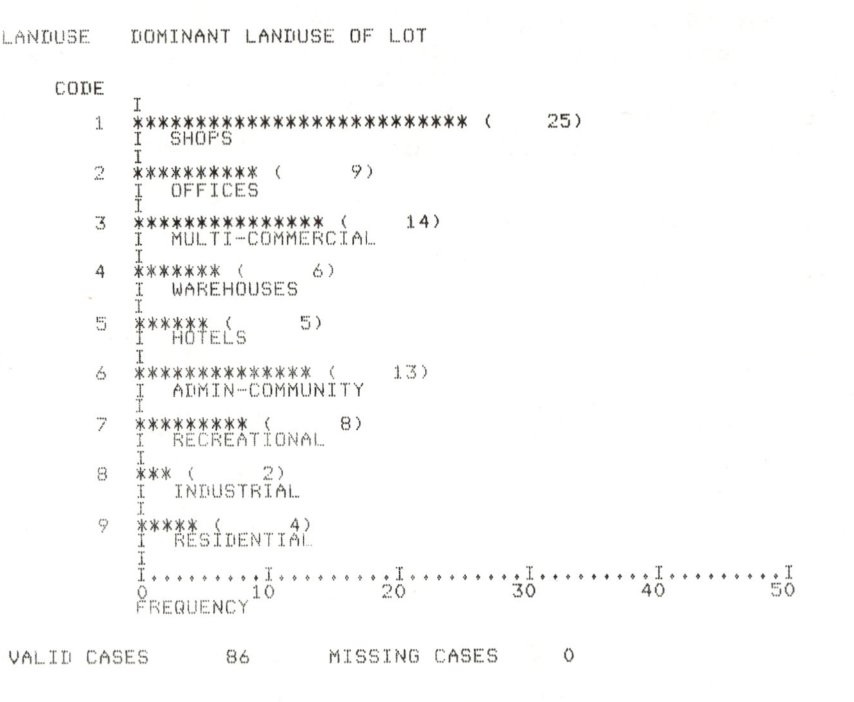

Figure 2.16 SPSS Histogram for dominant landuse of lot.

histograms but simply displays the frequencies class by class in sequence down the page (Figure 2.16).

SPSS is a product of the mid-1960s, and the version used here, although current, has changed little (apart from an expansion of the range of techniques) since that time. It is based on lineprinter output—cheap and very efficient, but also very crude when it comes to displaying data. The strength of the package, and the reason it is retained today, is its ease of use and its efficiency in handling large sets of data, especially the complex data sets derived from social science research. Extensions to SPSS including higher-quality graphics are available but are not widely implemented. SAS is almost the direct opposite. Its strength derives from the wide variety of processing packages that can be linked to a SAS data file. No examples will be given here, and it is sufficient to note that similar frequency distribution methods are available in addition to a wide range of graph plotting techniques producing high-quality screen and plotter output.

There are a large number of specific mapping packages available, including the now-dated but widely used SYMAP lineprinter mapping system and its various derivatives. SAS itself is rapidly developing mapping procedures within SAS/GRAPH, which can be easily linked to SAS data sets developed for statistical analysis. Discussion of these applications is largely outside the scope of this text.

References and Readings

1. Data Display Methods

Dickinson, G. C. (1963) *Statistical Mapping and the Presentation of Statistics,* Arnold: London.

Krumbein, W. C. and F. A. Graybill (1965) *An Introduction to Statistical Models in Geology,* McGraw-Hill: New York.

Tufte, E. R. (1983) *The Visual Display of Quantitative Information,* Graphics Press: Cheshire, Conn.

2. Cartographic Methods

Monmonier, M. S. (1982) *Computer-Assisted Cartography,* Prentice-Hall: Englewood Cliffs, N.J.

Robinson, A., R. Sale, and J. Morrison (1979) *Elements of Cartography,* Fourth Edition, Wiley: New York.

CHAPTER 3
Statistical Summaries of Distributions

In Chapter 2, we outlined a variety of methods for portraying a distribution of scores for a single variable. These were centered on the use of the frequency distribution as a summary method, and, in particular, on the frequency distributions that result when the variable can be considered as numeric (enumerated or metric measurement). In addition, the relationship between frequencies and the magnitude of the x scores can be described by frequency models. It is important to reiterate the reasons for using frequency distributions to portray a set of data. Frequency distributions provide an easily visualized summary of the distribution of scores and prepare the data for further processing. In constructing frequency distributions, we automatically decrease the emphasis on the characteristics of individual scores and instead present a "smoothed" description of the distribution. Starting with the frequency distribution, we outlined additional methods to display the distribution, each increasing the amount of smoothing and decreasing the amount of original detail. These culminate in the frequency curve and the ogive, where the distribution was represented by a smooth curve. In this way, we achieved our first aim by presenting the distribution in an easily visualized form. This chapter continues the process of smoothing by providing methods that further summarize this distribution through the use of key descriptors.

It may seem strange that we are placing such heavy emphasis on methods that decrease the amount of detail obtained from the data, but we do so for a very good reason. The aim of statistical methods is to assist in the description of real-world phenomena. Thus, our interest is not so much in the characteristics of any individual element but in the general character presented by a whole set of elements. To achieve this, we must summarize the data and mask the detail. This does not mean that we should ignore the character of individual elements but that these elements contribute only a small part to the general picture of the whole data distribution. This concept is especially important when we consider a data distribution for a variable obtained by *sampling*. In this situation, individual elements appear in the distribution only by chance, and thus their precise and absolute score is of little direct interest. Some form of smoothing out of these scores is essential. Frequency distributions, and more particularly frequency curves, provide the basis for doing so.

This chapter emphasizes the description of summary methods that utilize the complete distribution of scores. This form of data is called *ungrouped data* to differentiate it from *grouped data* that are data in class frequency form. Most data collected by geographers are in ungrouped form and when ungrouped data are available, it is usual to calculate statistical summary measures based on this complete enumeration. However, equivalent methods for grouped data will also be presented for three important reasons. First, following the discussion in the previous paragraph, when we wish to mask the detailed characteristics of the scores, such as might occur with sampled data, then we can replace the collected ungrouped data by a frequency distribution of grouped data, and use this in our calculations. In doing so, we ignore the individual scores and assume that the scores are evenly distributed within a class and may be summarized by the midpoint of that class. A second reason, which may not become apparent until later in this book, is that certain statistical manipulations, especially the fitting of continuous distribution functions (Chapter 6), require the data in grouped form. The final reason is that some data are only available in grouped form. Variables obtained from census reports, published in grouped form to facilitate presentation and to preserve anonymity, are the most common example. Table

3.1 outlines the summary measures to be discussed in the first sections of this chapter, and indicates their appropriateness at the different levels of measurement.

3.1 CURVE DESCRIPTION

Although our immediate concern is with the statistical summary of a distribution of scores, it is convenient to begin the discussion by enumerating methods used to describe frequency curves and ogives. Frequency curves provide a simple method for data summary and description. They provide a ready portrayal of the distribution of scores, and distributions can be compared one to another through their curves. This is particularly so when the frequencies have been converted to relative frequencies and plotted either as relative frequency curves or especially as ogives (cumulative relative frequency curves). The basis for much of the curve description and for a number of the summary measures derives from characteristics of cumulative relative frequencies.

Percentiles and Cumulative Relative Frequency Distributions

In Chapter 2 we outlined methods for constructing tables and graphs based on cumulative relative frequencies. For our example, Land Value per Square Meter from the Urban Land Value Survey (Table 3.2), we could then make statements about the distribution of scores relative to the boundaries between the classes. Thus 20% of the scores were less than $64.5 (the true boundary between classes 1 and 2); 83% were less than $169.5 (the true boundary between classes 4 and 5); and so on. More importantly, when these cumulative relative frequencies were plotted on a graph and the curve smoothed to produce an ogive, not only could we make statements about the boundaries but we could also estimate other cumulative relative frequencies from between class boundaries.

These selected points from a cumulative relative frequency distribution (table or graph) are called *percentiles*. We provide the following simple definition: P_v = the vth percentile = the value of X below which v% of the scores lie. By implication, there is $(100 - v)$% of the scores with values greater than P_v. Values of v would usually be integers between 1 and 100. From the above example,

$$P_{20} = \$64.5$$

meaning 20% of the scores in the distribution have values less than $64.5, and 80% have values greater than $64.5. For a ranked distribution, we can define the vth percentile as

$$P_v = \left[\frac{v}{100}(N + 1)\right]\text{th score} \tag{3.1}$$

TABLE 3.1 **Univariate Summary Measures**

Summary Measure	Level of Measurement					
					Enumerated and Ratio Metric	
	Nominally Classified	Ordinally Classified	Binary	Interval Metric	Grouped	Ungrouped
MEASURES OF CENTRAL TENDENCY						
Mode and modal class	X*	X*	X	X	X	X
Median		X*	X	X	X*	X
Arithmetic mean			X	X*	X*	X*
Geometric mean					X	X
Harmonic mean					X	X
MEASURES OF VARIATION						
Absolute range		X		X		X
Interquartile range		X		X	X	X
Quartile deviation		X		X	X	X
Mean deviation (around mean)			X	X	X	X
Mean deviation (around median)				X	X	X
Standard deviation/variance			X*	X*	X*	X*
Coefficient of variation					X*	X*
MEASURES OF SKEWNESS						
Coefficient of skewness					X	X
Quartile skewness					X	X
Standardized third moment (*Sm*)					X*	X*
MEASURE OF KURTOSIS						
Standardized fourth moment (*Km*)					X*	X*
MEASURES OF RELATIVE POSITION						
Rank				X		X*
Standard scores						X*

X = technique used.
* = more important measure.

TABLE 3.2 Frequency Distributions for Land Value per Square Meter

Class	Frequency	Relative Frequency	Class	Cumulative Frequency	Relative Cumulative Frequency
30–64	17	.20	<65	17	.20
65–99	22	.26	<100	39	.45
100–134	24	.28	<135	63	.73
135–169	8	.09	<170	71	.83
170–204	8	.09	<205	79	.92
205–239	3	.03	<240	82	.95
240–274	4	.05	<275	86	1.00
TOTAL	86	1.00	—	—	—

Source: From Tables 2.4, 2.6, and 2.7.

where

$$v = \text{the required percentile}$$

$$N = \text{the total number of scores}$$

If P_v is not an even score, then the fraction part of the result indicates the proportional distance from the score indicated to the next highest score. For example, for Land Value per Square Meter, the 15th percentile is

$$P_{15} = \left[\frac{15}{100}(86 + 1)\right]\text{th score}$$

$$= [13.05]\text{th score}$$

That is, the 15th percentile is located 5/100ths of the distance between the 13th and 14th score. From the ranked array (Table 2.2), the 15th percentile lies 5/100ths of the way between 62 (the 13th score) and 63 (the 14th score).

$$P_{15} = 62 + \frac{5}{100}(63 - 62)$$

$$= \$62.0 \text{ (to 1 decimal place)}$$

For data in grouped form, an equivalent method for estimating the position of a percentile can be found by locating the class containing the required percentile, then estimating how far from one end of the class the percentile lies.

The cumulative relative frequency curve (the ogive) was presented in Chapter 2 as the last step in the graphic display of a frequency distribution. Note that the scale of the y axis on this graph is a scale of percentiles. This gives us a basis for curve description and a method for creating summary measures. As we will see, percentiles are used widely in constructing certain

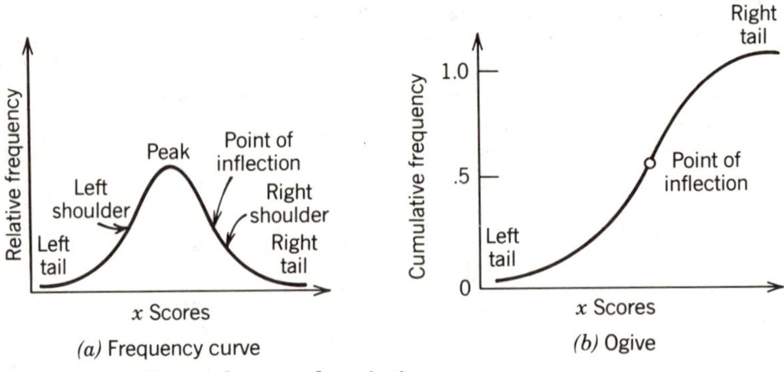

Figure 3.1 General curve description.

summary measures—those originally based on the visual extraction of information from the ogive.

Curve Shapes and Characteristics

The frequency curve (or the ogive) summarizes the distribution of scores for a variable. It is a simple matter to provide a description of that curve by comparing its shape to shapes of curves that are often encountered in statistical analysis. A curve can be described by the position and shape of certain curve descriptors (Figure 3.1), or simply by stating its generalized curve shape. Many frequency curves are characterized by a peak (maximum frequency) in one part of the distribution, usually near the center. Much of statistical analysis has been concerned with these curves, called *unimodal curves*. However, other curve shapes occur. Figure 3.2 illustrates some of the most common shapes, in both frequency and cumulative frequency (ogive) form. A number of the terms used in the description of these curves are defined later in this chapter.

Although a general description of the curve provides a simple means of summarizing the curve shape, there might be considerable variability between curves roughly described by the same shape. Thus, general shape description can be considered only as an approximate summary of a frequency distribution. It is necessary to supplement such a description with a statistical description of the curve, using one or more statistical summary measures. The first type of statistical summary measure, *measure of central tendency,* is especially for unimodal curves. This measure defines the peak of the curve by selecting a score near the center, or by finding some sort of "average" or "typical" value. Although these measures of central tendency locate the center of the distribution and summarize it in a single score, they give no idea of how close the scores in the distribution are to this centrality measure. Frequency distributions with similar central tendency measures may have widely differing sets of values, and additional summary measures are required. *Measures of variability* (or *dispersion*) describe the spread of

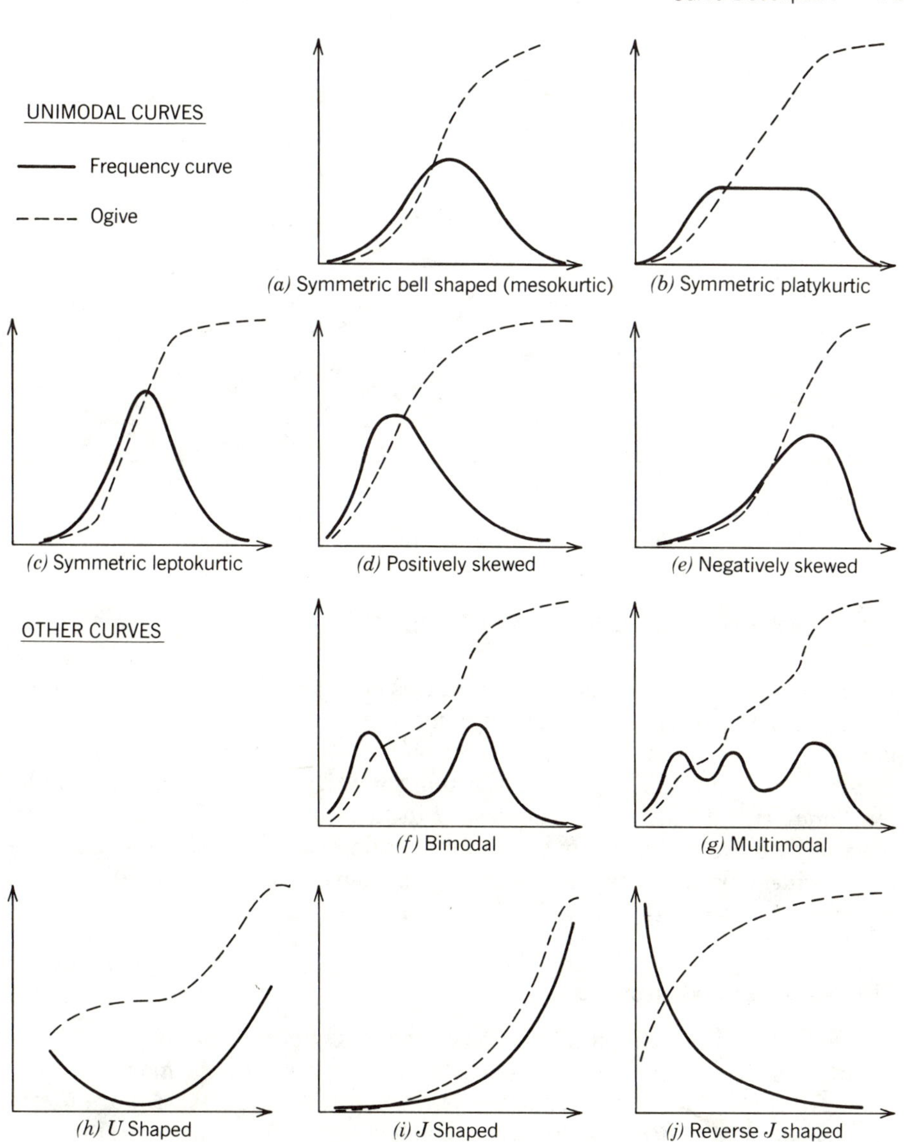

Figure 3.2 Some important curve shapes.

the distribution of scores around the central measure, *measures of skewness* provide an indication of how symmetrical the curve is, and *measures of kurtosis* look at the degree of peakedness or flatness of the curve. In some situations a different type of summary measure is required to describe the relative position of one score in a distribution to another score or to some summary measure. These *measures of relative position* will prove useful constructs in later analyses. The different summary measures are illustrated in Figure 3.3.

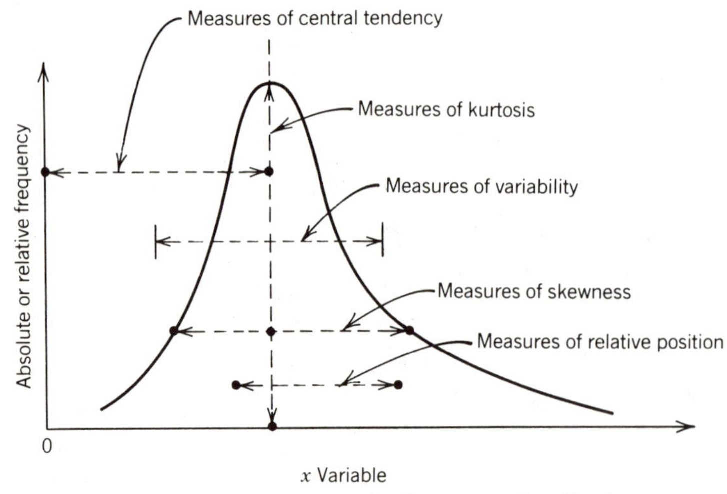

Figure 3.3 Summary measures for frequency distributions.

3.2 MEASURES OF CENTRAL TENDENCY

The most important summarizing measures are those that locate the "center" of the distribution. Not only do these measures provide a single score that is more representative (a better summary) of all the scores than any other single value, but they assist in the mathematical characterization of the frequency curve that will be necessary in later analyses. For the moment, however, we are interested in their summarizing capabilities. Although there are a wide variety of central tendency measures, we limit our discussion primarily to the *mode,* the *median,* and the *arithmetic mean.*

The Mode and Modal Class

The *mode* (*Mo*) is defined as the most frequently occurring value in a distribution of scores. However, it is possible that there could be more than one mode. In the Urban Land Value Survey, the Land Value per Square Meter has a mode of $117 because this score has the highest frequency (4). However, the scores of $87, $88, and $113 all have frequencies of only one less than four (see Table 2.2). Thus, for a distribution in which there is a wide range in values and only a few scores with the same value, the mode as just defined is a poor measure of central tendency.

One alternative is to define the *modal class* of a frequency distribution—the class with the highest frequency. The classes must have an equal interval size for the modal class to be valid. The concept of the modal class can be applied to all levels of measurement, although for classified data (where the mode would be defined as the score for the modal class), we are measuring the most characteristic or frequently occurring score. Although probability concepts (Chapter 4) are not explicitly considered here, the reader should note throughout this chapter the implied use of some probability ideas. Thus, to say that the modal class contains the greatest number of individual elements implies that if any one property lot from the 86 in our survey is selected, the chances are (it is most probable) that it came from the modal class. That is, there is a higher probability that it came from that class than from any other single class. In this sense, the measures of central tendency may be considered as attempts to define the most probable or the *expected* value of the distribution (see Section 4.4).

For enumerated and metric data, having determined the modal class, we need to locate the mode within that class. The standard method uses the frequencies of the classes on either side of the modal class to estimate the probable concentrations of scores within the modal class.

$$Mo = L_{Mo} + \left[\frac{f_{Mo} - f_{Mo-1}}{(f_{Mo} - f_{Mo-1}) + (f_{Mo} - f_{Mo+1})}\right] I \qquad (3.2)$$

where

Mo = the mode

L_{Mo} = the true lower limit of the modal class

f_{Mo} = the frequency of the modal class

f_{Mo-1} = the frequency of the class one lower than the modal class

f_{Mo+1} = the frequency of the class one higher than the modal class

I = the interval size

For Land Value per Square Meter, the modal class (Table 3.2) is 100 to 134 and the mode may be found by

$$Mo = 99.5 + \left[\frac{24 - 22}{(24 - 22) + (24 - 8)}\right] 35$$

$$= 99.5 + \left(\frac{2}{18}\right) 35$$

$$= \$103.4$$

As an alternative, the *midpoint* of the modal class could be used, which in our example would give

$$Mo = \frac{99.5 + 134.5}{2}$$

$$= \$117.0$$

The mode is the only measure of central tendency defined for nominally classified variables, and except in special circumstances, it is rarely used at other levels of measurement. Nevertheless, it does have some important properties and uses, as will be seen later. However, we note here that because most frequency distributions are characterized by a single peak (a unimodal distribution), the mode provides an important method of locating this peak. Where more than one modal class occurs (bimodal and multimodal distributions), an individual mode cannot be defined.

The Median

If a ranked array has been prepared, the simplest centrality measure would be the middle score of the array. The *median* (Md) is defined as the value with an equal number of ranked scores above and below. That is,

$$f_{>Md} = f_{<Md}$$

This indicates the 50th percentile. Thus, from Equation 3.1,

$$Md = \left[\frac{50}{100}(N + 1)\right]\text{th score} = \left[\frac{N + 1}{2}\right]\text{th score} \tag{3.3}$$

For Land Value per Square Meter, with 86 scores, the median would be the 43.5th score, or halfway between the 43rd and the 44th scores (see Table 2.2).

$$Md = \frac{105 + 106}{2}$$

$$= \$105.5$$

Although the median is not the major central tendency measure for enumerated or metric data, it does have considerable importance in statistical description as a result of two statistical properties. First, its value is unaffected by extreme values at one end of the distribution, and thus it is often used to measure central tendency of distributions that are skewed by a few extreme values. A second property, one that is not so apparent, concerns the deviations of scores about the median ($X_i - Md$). The sum of the absolute values (signs ignored) of the deviations about the median is less than the sum of the deviations around any other score. More will be said of this later.

When data are available in grouped form only, two methods are available to estimate the median. If a cumulative relative frequency curve (an ogive) has been constructed, then the median (as with any percentile) can be estimated directly from the graph. The median corresponds to a relative frequency of 0.5. For the Land Value per Square Meter data, this would give an estimate of the median at approximately \$108 (from Figure 2.8). This value is close to the calculated median of \$105.5. Estimation in this manner requires a carefully drawn ogive.

An estimate of the median can also be made by calculation from a frequency distribution or a cumulative frequency distribution. First, we locate the *median class,* the class containing the median. Then

$$Md = L_{Md} + \frac{[(N + 1)/2 - Cf_{<Md}]}{f_{Md}} \cdot I_{Md} \qquad (3.4)$$

where

L_{Md} = the true lower limit of the median class

N = the total number of scores

$Cf_{<Md}$ = the cumulative frequency up to but not including the median class

I_{Md} = the interval (size) of the median class

f_{Md} = the frequency of the median class

To illustrate its use, we apply it to the grouped data for Land Value per Square Meter (Table 3.2).

$$Md = 99.5 + \frac{[(86 + 1)/2 - 39]}{24} \cdot 35$$

$$= 99.5 + \frac{4.5}{24} \cdot 35$$

$$= \$106.1$$

The results are again quite close to the median from the original scores.

The Arithmetic Mean

Although the median is based on all scores, it is defined only with reference to the relative position of the scores. The magnitude of the scores does not affect its location. The *arithmetic mean,* in contrast, is based on the magnitude of all scores. To compute the arithmetic mean, we sum all the scores and divide by the number of observations.

The arithmetic mean is the most frequently used measure of central tendency. A variety of symbols can be used to represent the arithmetic mean, but we will use the most frequent *bar-x* ($\overline{X}$). Although in later discussion we simply call this measure the *mean,* we retain the adjective "arithmetic" to distinguish it from a variety of other forms of mean. However, the context normally makes its use clear. To reiterate, we define the arithmetic mean as the sum of all the individual scores divided by the total number of scores (N)

$$\overline{X} = \frac{\sum_{i=1}^{N} X_i}{N} \qquad (3.5)$$

For Land Value per Square Meter, we sum the scores for all the property lots and divide by the size of the population (N), to get

$$\overline{X} = \frac{9907}{86}$$

$$= \$115.2$$

Note that we have used a shorthand mathematical notation for the process of summing a large set of numbers. The Greek symbol Σ (uppercase *sigma*) is used to represent a summation process. This symbol is used extensively in this chapter and throughout the book. It is used in place of a series of numbers to be added, or a long verbal description of how a set of numbers is to be added. By definition,

$$\sum_{i=1}^{N} X_i = X_1 + X_2 + \ldots + X_N$$

where X (for example) is a variable whose scores are numbered in a systematic manner, and each score is represented by using a subscript, i, for example. All the sequential scores for the variable X are added up, starting with the first score, or *lower limit* of summation, and continuing to the Nth score, the *upper limit* of the summation. For example,

$$\sum_{i=1}^{4} = X_1 + X_2 + X_3 + X_4$$

The lower and upper limits are often omitted in situations where the range of scores over which the summation is to take place is clear from the context. Manipulating summation terms in a mathematical expression can be accomplished through three rules.

First, the summation of two or more added terms equals the summation of the separate parts (the property of associativity).

$$\sum_{i=1}^{N} (X_i + Y_i) = \sum_{i=1}^{N} X_i + \sum_{i=1}^{N} Y_i \tag{3.6}$$

Second, the summation of a constant (c) times a variable, equals the constant times the summation of the variable (the property of distributivity).

$$\sum_{i=1}^{N} cX_i = c \sum_{i=1}^{N} X_i \tag{3.7}$$

Third, the summation of a constant (c) over the number of elements (N) equals the constant times the number of elements

$$\sum_{i=1}^{N} c_i = N \cdot c \tag{3.8}$$

where

$$\text{all } c_i = c$$

The arithmetic mean has two important properties. First, with the deviation about the mean defined as the difference between the mean and an individual score, the sum of these deviations about the mean is zero.

$$\sum_{i=1}^{N} (X_i - \overline{X}) = 0 \qquad (3.9)$$

In some texts, the symbol x_i is used to represent $(X_i - \overline{X})$.

Second, the sum of the squared deviations about the mean is a minimum. In other words, the sum of the squared deviations about the mean is less than the sum of the squared deviations about any other number.

$$\sum_{i=1}^{N} (X_i - \overline{X})^2 < \sum_{i=1}^{N} (X_i - A)^2 \qquad (3.10)$$

where

$$A = \text{any number other than the mean}$$

For data in class frequency form, we usually assume that the scores are distributed evenly within the class and they then can be summarized by the midpoint. The alternative is to use a formula that corrects for the frequencies of the classes on either side. We will retain the simpler form. In the summing process, therefore, we obtain the estimate of the *class* total by multiplying the midpoint (M_j) by the class frequency (f_j).

$$\overline{X} = \frac{\sum_{j=1}^{k} M_j f_j}{N} \qquad (3.11)$$

where

$$k = \text{the number of classes}$$

$$M_j = \text{the midpoint of class } j$$

$$f_j = \text{the frequency of class } j$$

$$N = \text{the total number of scores}$$

As an example, for Land Value per Square Meter, Table 3.3 outlines the calculations to give

$$\overline{X} = \frac{9817}{86}$$

$$= \$114.2$$

TABLE 3.3 Work Table for Calculation of Mean and Variance for Land Value per Square Meter

Class (j)	Frequency (f_j)	Midpoint (M_j)	(2) × (3) (f_jM_j)	(4) × (3) $(f_j)(M_j)^2$
30–64	17	47.0	799	37,553
65–99	22	82.0	1804	147,928
100–134	24	117.0	2808	328,536
135–169	8	152.0	1216	184,832
170–204	8	187.0	1496	279,752
205–239	3	222.0	666	147,852
240–274	4	257.0	1028	264,196
TOTAL	86	—	9817	1,390,649

Although the arithmetic mean is usually regarded as *the* mean, there is a variety of other types of means, all of which are based on averaging all scores. Two of the most common are defined here. Their use as measures of central tendency is a rather indirect one, and they are employed only infrequently as simple summary measures of distributions. When we wish to transform variables by changing their units of measurement (to be examined later) we may wish to obtain means of the transformed values. The arithmetic mean of the transformed scores will have a different value to the transformation of the arithmetic mean of the raw scores. Each type of transformation (and there are many) will have its own special type of mean.

When a *logarithmic* transformation is used, the *geometric mean* (Mg) is defined as the antilog of the arithmetic mean of the transformed scores

$$Mg = \text{antilog}\left[\frac{\sum_{i=1}^{N} \log X_i}{N}\right] \tag{3.12}$$

Because the addition of logarithms is equivalent to the multiplication of their antilogs, in untransformed form, this is

$$Mg = \sqrt[N]{X_1 \cdot X_2 \cdot X_3 \dots \cdot X_N} \tag{3.13}$$

The *harmonic mean* (Mh) is defined as the reciprocal of the arithmetic mean of scores that have been transformed into their reciprocals.

$$Mh = \left(\frac{\sum_{i=1}^{N} X_i^{-1}}{N}\right)^{-1} = \frac{N}{\sum_{i=1}^{N} \frac{1}{X_i}} \tag{3.14}$$

In addition to its use as a summary measure for data that have been transformed by reciprocals, the harmonic mean also is useful as a measure of central tendency in spatial distributions.

Selection of a Central Tendency Measure

For enumerated and metric data, three major alternative measures of the central tendency of a distribution were defined. In terms of frequency of use, the arithmetic mean far exceeds the median, which is more common than the mode. However, it is worthwhile to examine some characteristics of these three alternative measures that may assist in selecting the most appropriate measure in given situations. It will be seen that the arithmetic mean has a considerable number of advantages over both the median and the mode.

First, with reference to a smoothed unimodal frequency curve, regardless of its shape, the mode defines the value at which the curve peaks; the median divides the area under the curve into two equal parts; and the arithmetic mean divides the curve into two equally balanced parts through the center of gravity.

Second, for a truly symmetric unimodal distribution, the arithmetic mean, the median, and the mode would be identical. However, if the distribution were skewed, although there is no change in the mode, the median will be displaced in the direction of the skew, and the arithmetic mean will be changed even more because it is influenced by the actual values of the scores (Figure 3.4). This situation would be true both for major degrees of skew and even for just a few isolated extreme values. Thus, in skewed distributions, the arithmetic mean will be highly influenced by the degree of skew and will cease to describe a "central" or even "typical" value. The median would be affected to a much lesser extent, and the mode not at all.

Third, the arithmetic mean can be easily calculated from both grouped and ungrouped data (the order of the scores is irrelevant); the calculation

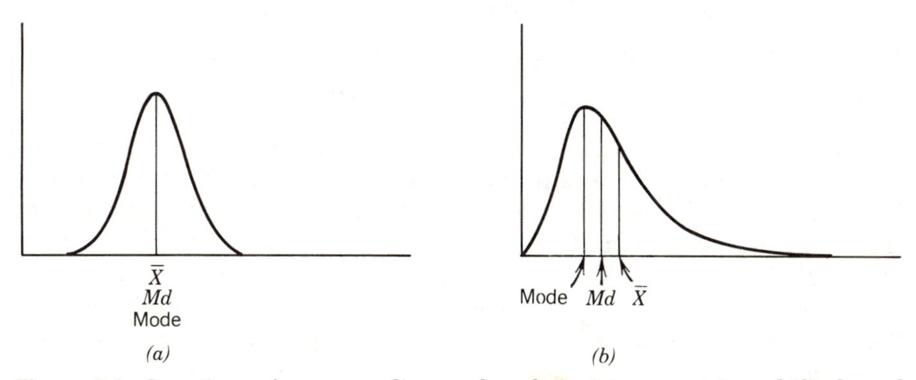

Figure 3.4 Locations of mean, median, and mode in (*a*) symmetric and (*b*) skewed unimodal distributions.

of the median can be performed on both grouped and ungrouped data, but the data have to be ranked; the mode is really only adequately defined for grouped data or for ungrouped data with a low range of values but a high frequency.

Fourth, unequal class intervals in grouped data do not hinder the calculation of the arithmetic mean or the median but severely limit the calculation of the mode.

Fifth, the presence of an open-ended class at either or both ends of a frequency distribution do not usually affect the calculation of the median or the mode but severely limit or even invalidate the calculation of the arithmetic mean.

3.3 VARIABILITY, SKEWNESS, AND KURTOSIS

A measure of central tendency, which locates the center of the distribution, should be complemented by a measure of the spread of the scores. This spread can be considered in terms of how variable the scores are in magnitude (measures of *variability*), whether the spread is the same on either side of the central measure (measures of *skewness*), and whether the scores are uniformly concentrated around the central measure (measures of *kurtosis*).

Variability can be measured in two ways: by the range or spread of values of the scores, or by the deviation or average departure from around a centrality measure.

Variability Measurement by Range

The simplest measure of variability is to state the *absolute range* (R) of the distribution—the difference between the largest (Xmax) and the smallest (Xmin) scores

$$R = X\text{max} - X\text{min} \tag{3.15}$$

For the Land Value per Square Meter, this gives

$$R = 274 - 32 = \$242$$

The absolute range (usually just referred to as the range) is strongly influenced by extreme values at either end of the distribution. For this reason it is only infrequently used as a measure of variability. Instead, various range measurements have been devised that avoid the extreme values and use other selected *percentiles* of the distribution. Any percentile position could be chosen but one of the most frequently used involves the *quartiles* (the 25th and 75th percentiles). The *interquartile range* (Qr) uses the 25th and 75th percentiles.

$$Qr = P_{75} - P_{25} \tag{3.16}$$

For the Land Value per Square Meter, this gives us

$$P_{75} \text{ (the third quartile)} = 150.2$$

$$P_{25} \text{ (the first quartile)} = 69.0$$

therefore

$$Qr = \$81.2$$

Variability Measurement by Deviation

More frequently used measures involve measurement of variability around some measure of central tendency. The *quartile deviation* (Qd) summarizes the average deviation on either side of the median and is simply the interquartile range divided by 2. Using the quartiles just defined, we get a quartile deviation of \$40.6 about the median of \$105.5.

Other measures of variability are based on averaging the deviations of all scores about some measure of central tendency. The mean deviation is the average of the absolute values of the deviations of each score from a selected central tendency measure. This measure is usually defined with reference to deviations about the arithmetic mean, where the use of absolute values is necessary to overcome the first property of the mean, which stated that the sum of the deviations about the mean totaled zero. The *mean deviation around the arithmetic mean* ($V_{\bar{X}}$) is defined as

$$V_{\bar{X}} = \frac{\sum_{i=1}^{N} |X_i - \bar{X}|}{N} \tag{3.17}$$

The *mean deviation around the median* (V_{Md}) is defined in a similar manner:

$$V_{Md} = \frac{\sum_{i=1}^{N} |X_i - Md|}{N} \tag{3.18}$$

The one important property of this last measure is that the mean deviation about the median is less than the mean deviation around any other score including the arithmetic mean.

The most common measure of variability is the *variance*, or the average squared deviation, around the arithmetic mean.

$$s^2 = \frac{\sum_{i=1}^{N} (X_i - \bar{X})^2}{N} \tag{3.19}$$

The *standard deviation*, which has practical advantages in many circumstances, removes the effect of the squaring of the deviations and returns the measure to its original units by taking the positive square root of the variance.

$$s = \sqrt{s^2} = \sqrt{\frac{\sum_{i=1}^{N} (X_i - \overline{X})^2}{N}} \tag{3.20}$$

The importance of the variance (and the standard deviation) as measures of variability will be seen later in this section. In most circumstances, their use with enumerated and metric data is preferred. For the Land Value per Square Meter variable, we obtain values of 3354 for the variance and \$57.9 for the standard deviation. When there is a large number of scores, calculation by hand or by calculator can be very time consuming. Equation 3.20 can be manipulated into a number of alternative forms using the rules of summation.

$$s^2 = \frac{\sum_{i=1}^{N} X_i^2 - \dfrac{\left(\sum_{i=1}^{N} X_i\right)^2}{N}}{N} \tag{3.21}$$

and

$$s^2 = \frac{\sum_{i=1}^{N} X_i^2}{N} - \overline{X}^2 \tag{3.22}$$

where

$$\sum_{i=1}^{N} X_i^2 = \text{the sum of the squared } X_i \text{ values}$$

For data in class frequency form, Equation 3.19 can be modified using the midpoint and frequency in the summary process

$$s^2 = \frac{\sum_{j=1}^{k} f_j (M_j - \overline{X})^2}{N}$$

where

k = the number of classes

f_j = the frequency of class j

M_j = the midpoint of class j

Or, in a more manageable form,

$$s^2 = \frac{\sum_{j=1}^{k} f_j (M_j)^2 - \frac{\left(\sum_{j=1}^{k} f_j M_j\right)^2}{N}}{N}$$

where

$$f_j (M_j)^2 = (f_j)(M_j)(M_j) \text{ or } (f_j M_j)M_j$$

Using the Land Value per Square Meter variable as an example, the steps in the calculation are outlined in Table 3.3, giving

$$s^2 = \frac{1{,}390{,}649 - (9817)^2/86}{86}$$

$$= 3139.8$$

and, therefore,

$$s = \$56.0$$

As with the calculation of the arithmetic mean, short-cut methods are available.

Variability and Sum of Squares

Before continuing the discussion of summary measures, let us reexamine the variance measure just defined. In doing so, we introduce two new measures of variability that will prove very useful later. One way to look at variability of the scores for a variable X is to consider *total variability* as being the sum of the squared values of the variable X—that is, the squared deviations about zero. Thus,

$$\text{Total variability} = \sum_{i=1}^{N} X_i^2$$

$$= \text{the } (\textit{uncorrected}) \textit{ sum of squares of } X \qquad (3.23)$$

Then considering deviations about the arithmetic mean, rather than about zero, we define the *variation* of X as

$$\sum_{i=1}^{N} (X_i - \overline{X})^2 = \text{the } \textit{corrected sum of squares of } X \qquad (3.24)$$

By expanding the summation operation, as in Equation 3.21, we get

$$\text{Variation} = \sum_{i=1}^{N} X_i^2 - \frac{\left(\sum_{i=1}^{N} X_i\right)^2}{N}$$

$$= \textit{(uncorrected) sum of squares of } X \; - \textit{ correction for the mean} \qquad (3.25)$$

Thus, we can consider the variation of a variable as the total sum of squares of the X, minus a correction for the mean. The variance, of course, is the variation divided by N or the "average variation."

The Coefficient of Variation

The arithmetic mean and the variance (or standard deviation) provide the basic measures of central tendency and variability. Their values, however, are dependent on the units with which they were measured, and, in addition, the magnitude of the variance and the standard deviation are dependent on the magnitude of the mean. Thus, the degree of variability is not readily apparent, and comparison from one distribution to another is difficult. The *coefficient of variation* (CV) provides a *relative* measure of variability by expressing the standard deviation as a ratio of the mean, or more usually, as a percentage of the mean.

$$CV = \frac{s}{\overline{X}} \quad \text{or} \quad \frac{s}{\overline{X}} \cdot 100 \qquad (3.26)$$

Note that the coefficient of variation is defined only when the scale of measurement is enumerated or ratio metric. Because of the arbitrary zero point for interval-metric data, a ratio of the standard deviation and the mean has little validity. An example of the coefficients of variation for three variables for the Urban Land Value Survey is given in Table 3.4. The three dimensionless coefficients allow comparison of the relative degrees of variability. For two of the variables, the standard deviation is numerically greater than the mean (a coefficient in excess of 100%). Because negative numbers cannot occur with these variables, this indicates that there is a very high degree of skew to the right.

Measures of Skewness and Kurtosis

The measurement of the degree of skewness and kurtosis in a distribution of scores is only infrequently encountered in geographic analysis. This could be due to the belief that most distributions are adequately described by the arithmetic mean and standard deviation. However, many distributions display some degree of departure from symmetry and the omission of estimates of skewness and kurtosis may seem surprising.

TABLE 3.4 Coefficients of Variation for Three Variables

Variable	Units	Arithmetic Mean	Standard Deviation	Coefficient of Variation
Land value per square meter	Dollars	$115.2	$57.9	50.3%
Length of road frontage	Meters	27.7 m	29.6 m	106.9%
Employment numbers on lot	Enumerated	9.6 people	12.1 people	126.0%

Source: Urban Land Value Survey.

Skewness measures the degree of asymmetry in a distribution around a measure of central tendency. A wide variety of different measures have been proposed. One of the most common, derived from the work of Karl Pearson, takes into account the relative position of the mean and the median in a skewed distribution. Pearson's *coefficient of skewness* (Sk) is

$$Sk = 3\,\frac{(\overline{X} - Md)}{s} \tag{3.27}$$

The Sk values provide a relative measure of skewness, with a positive sign indicating a skew to the right (a *positive skew*) and a negative sign a skew to the left (a *negative skew*), with Sk having a score of zero when the distribution is symmetrical.

A number of skewness measures are used based on percentile ranges. One of these is,

$$\text{Quartile skewness} = Sq = \frac{P_{25} + P_{75} - 2Md}{2} \tag{3.28}$$

An alternative form, the most common measure when computer methods are used, is based on the *moments* of a distribution (to be discussed in Section 3.6).[1]

$$Sm = \frac{\left[\sum_{i=1}^{N} (X_i - \overline{X})^3\right] \Big/ N}{(s^2)^{3/2}} \tag{3.29}$$

For the Land Value per Square Meter, this gives us

$$Sm = \frac{192{,}689}{194{,}243}$$

$$= .992 \text{ (positively skewed)}$$

Again, interpretation is the same as for Sk.

[1]The skewness and kurtosis measures Sm and Km are the standardized third and fourth moments.

Kurtosis measures how pointed or flat the frequency curve appears, and the most common measure is *Km:*

$$Km = \frac{\sum_{i=1}^{N} (X_i - \overline{X})^4/N}{(s^2)^2} - 3 \tag{3.30}$$

Negative values of *Km* indicate a *platykurtic curve* (flattened); positive values represent *leptokurtic curves* (peaked or pointed); and a value of 0 represents a *mesokurtic curve* or, more particularly, the bell-shaped curve of the *normal distribution* (to be discussed in Chapter 6). In fact, the 3 is subtracted to give the value of 0 to normal curves. Note, however, that some computer programs *do not* subtract 3, thus values of <3 would indicate platykurtic and values >3 would indicate leptokurtic curves. For the Land Value per Square Meter variable, this gives

$$Km = \frac{5{,}202{,}246}{11{,}249{,}316}$$

$$= .462 \text{ (leptokurtic)}$$

Computer Results

Using the SPSS setup in Table 3.5 we produce the summary results for several of the variables from the Urban Land Value Survey (Table 3.6). Note that the hand-calculated values for the variance, standard deviation, skewness and kurtosis for the variable Unimproved Land Value are slightly smaller than the computer calculations. This is because the computer routines used division by n-1 rather than N. This difference, and the standard error values will be discussed in Chapters 5 and 6.

3.4 MEASURES OF RELATIVE POSITION

Through the use of the arithmetic mean, the standard deviation, and the other measures of central tendency—variability, skewness, and kurtosis—

TABLE 3.5 SPSS Setup for Summary Statistics

```
RUN NAME        ULVS(3) - SUMMARY STATISTICS
FILE NAME       ULVS  URBAN LAND VALUE SURVEY FOR THE CITY OF WOODFORD
VARIABLE LIST   VALUE AREA FRONTAGE LANDUSE FLOORS WORKERS AGE ACCESS LOTCODE
INPUT MEDIUM    CARD
N OF CASES      86
INPUT FORMAT    FIXED (2F9.0,F9.1,5F9.0,F8.0)
COMPUTE         UNITVALU=RND(VALUE/AREA*1000)
VAR LABELS      VALUE     UNIMPROVED LAND VALUE/
                AREA      AREA OF LOT/
                FRONTAGE  LENGTH OF ROAD FRONTAGE/
                WORKERS   EMPLOYMENT NUMBERS ON LOT/
                UNITVALU  LAND VALUE PER SQUARE METER
CONDESCRIPTIVE  VALUE AREA FRONTAGE WORKERS UNITVALU
STATISTICS      ALL
READ INPUT DATA
```

TABLE 3.6 Summary Statistics for Urban Land Value Survey

```
VARIABLE   VALUE       UNIMPROVED LAND VALUE

MEAN            63.395          STD ERROR        5.780          STD DEV        53.603
VARIANCE      2873.254          KURTOSIS         6.424          SKEWNESS        2.313
RANGE          304.000          MINIMUM         11.000          MAXIMUM       315.000
SUM           5452.000

VALID OBSERVATIONS -       86              MISSING OBSERVATIONS -      0
- - - - - - - - - - - - - - - - - - - - - - - - - - - - - - - - - - - - - - - - - - - - - - - - - - - - -
VARIABLE   AREA        AREA OF LOT

MEAN           653.430          STD ERROR       75.591          STD DEV       701.002
VARIANCE    491403.354          KURTOSIS        17.794          SKEWNESS        3.733
RANGE         4731.000          MINIMUM        127.000          MAXIMUM      4858.000
SUM          56195.000

VALID OBSERVATIONS -       86              MISSING OBSERVATIONS -      0
- - - - - - - - - - - - - - - - - - - - - - - - - - - - - - - - - - - - - - - - - - - - - - - - - - - - -
VARIABLE   FRONTAGE    LENGTH OF ROAD FRONTAGE

MEAN            27.703          STD ERROR        3.210          STD DEV        29.771
VARIANCE       886.339          KURTOSIS         9.830          SKEWNESS        2.811
RANGE          176.100          MINIMUM          5.700          MAXIMUM       181.800
SUM           2382.500

VALID OBSERVATIONS -       86              MISSING OBSERVATIONS -      0
- - - - - - - - - - - - - - - - - - - - - - - - - - - - - - - - - - - - - - - - - - - - - - - - - - - - -
VARIABLE   WORKERS     EMPLOYMENT NUMBERS ON LOT

MEAN             9.628          STD ERROR        1.313          STD DEV        12.179
VARIANCE       148.331          KURTOSIS        16.657          SKEWNESS        3.474
RANGE           84.000          MINIMUM          0.0            MAXIMUM        84.000
SUM            828.000

VALID OBSERVATIONS -       86              MISSING OBSERVATIONS -      0
- - - - - - - - - - - - - - - - - - - - - - - - - - - - - - - - - - - - - - - - - - - - - - - - - - - - -
 VARIABLE   UNITVALU    LAND VALUE PER SQUARE METER

 MEAN           115.198          STD ERROR        6.282          STD DEV        58.255
 VARIANCE      3393.690          KURTOSIS         0.551          SKEWNESS        1.008
 RANGE          242.000          MINIMUM         32.000          MAXIMUM       274.000
 SUM           9907.000

 VALID OBSERVATIONS -       86              MISSING OBSERVATIONS -      0
```

distributions can be summarized and compared one with another. In certain circumstances, some method of comparing the position of individual scores within a distribution or between distributions may also be required. In the Urban Land Value Survey we may want to compare the position of two property lots in the distribution of a variable, or compare the relative position of a single property lot for two different variables. Two basic methods are available to measure relative position: *ranks* within a distribution or *ratios* about selected positions in, or derived from, the distribution.

Measurement by Rank

The *rank* (R_i) of a score in a distribution (the position of the score when all the scores are arranged in numeric order) has already been used as a method of examining distributions (the ranked array), and it provides a simple means of showing the relative position of individual scores within a single distribution. It will also prove to be a very useful method for comparing the relative position of scores between two distributions.

Measurement by Ratio

For enumerated and metric variables, the relative position of an individual score can be assessed by expressing that score as a *ratio* of some other selected score or value. This selected score could be the mean, the maximum or minimum score, or some other summary measure. From the Urban Land Value Survey, for example, lot 13 has a Land Value per Square Meter of 127/115.2 or 1.1 times the mean Land Value per Square Meter, or 0.46 (127/274) of the maximum Land Value. When the sum of the scores for a variable has some inherent meaning (as in the case of the variable Employment Number where the total represents the total employment in the central area of the city), then we can structure the ratio as a proportion or a percentage. Thus, property lot 13 has 6/828 or 0.7% of the total central city employment numbers. These methods of relative position description are the same as outlined in the discussion of derived variables (Section 1.2) using ratios, proportions, percentages, and index numbers.

An important special type of ratio, the *standard score* (or *Z score*) allows the comparison of relative position regardless of the magnitude and measurement units of the scores. From this point of view, it is comparable to the coefficient of variation. For any score on a variable (X_i) its standard score (Z_i) is defined by

$$Z_i = \frac{X_i - \overline{X}}{s} \tag{3.31}$$

That is, the standard score for an element is its difference (deviation) from the arithmetic mean in terms of units of standard deviation (*standard deviation units*). Positive Z scores indicate X scores greater than the mean, negative Z scores indicate X scores less than the mean, and a Z score of zero would represent an X score equal to the mean. For example, a Z score of $+1.0$ would represent an X score equivalent to the arithmetic mean plus one standard deviation.

In deriving classes for frequency distributions (introduced in Section 2.1), one further possible method involves setting up boundaries between classes using the mean and standard deviation. Thus, if the aim of a frequency distribution is to show how elements are distributed around the mean, classes would be set up using some proportion or multiple of standard deviation units on either side of the mean. The classes could be structured so that all classes are either less than or greater than the mean (the mean forms one of the class boundaries), or a central class could be formed uniformly around the mean. In the Land Value per Square Meter example used to illustrate class construction in Chapter 2, a constant class interval of two-thirds of a standard deviation unit, with a central class about the mean, can be used to divide the distribution into seven classes. The resulting frequency distribution is illustrated in Table 3.7.

3.5 SEDIMENT SIZE ANALYSIS

The summary measures that have been discussed are the standard ones used in most analyses. There are, however, a large variety of other measures—measures created for special situations or as arguably better alternatives than some of the measures we have considered. Within individual fields of research, the use of certain sets of summary measures becomes established as part of the measurement process and are used automatically without

TABLE 3.7 Frequency Distribution using Standard Scores: Land Value per Square Meter

Approximate Class (Z_i Scores)	Approximate Class (X_i Scores)	Frequency (f_j)
-1.67 to -1.0	19–57	9
-1.0 to -0.33	58–95	29
-0.33 to $+0.33$	96–134	25
$+0.33$ to $+1.0$	135–173	9
$+1.0$ to $+1.67$	174–211	7
$+1.67$ to $+2.33$	212–250	3
$+2.33$ to $+3.0$	251–288	4
TOTAL		86

$\bar{X}$ = \$115.2
s = \$57.9.

much regard to alternatives. Thus, for example, we find the median used as a central tendency measure for age and income data; or the *midrange* $Mr = (X\text{min} + X\text{max})/2$ used in the averaging process for temperature data. The methods are still standard; the measures used are well known; the emphasis derives simply from a preference for certain measures and historical usage dictates their continuation.

There is one research field, however, where the derivation of a group of summary measures have so removed the measures from their standard counterparts that some specific attention is warranted. The analysis of the size of sediment particles is a key technique in many fields of research in physical geography, especially geomorphology and pedology. Average size of particles and the characteristics of the size distribution are fundamental descriptors of the material, leading to explanation and prediction. Deposits of sediment have two major features that make statistical summaries of size difficult. First, the number of elements (sediment particles) is extremely large, inhibiting individual measurement; and the size of the smaller particles make individual measurement almost impossible. Second, the distribution of sediment sizes is almost always highly positively skewed—that is, a lot of small particles and a small number of larger particles. Over the years, a variety of suggestions have been made as to how these difficulties might be overcome and a reasonable degree of uniformity of method is now established. To overcome the first problem, sediment size itself is not measured, instead, the sediment is split into size classes (by sieving or other methods) and the *weight* of sediment in each size class is used as a measure of the relative dominance of sediment of that size. To handle the high degree of skew, summary measures are usually based on transformations of the straight arithmetic scale of sediment size (defined as the intermediate diameter of the particle). The most common scale is that derived by Krumbein (1936), and described fully by Folk (1964), as the ϕ scale (Greek lowercase *phi*).

$$\phi = -\log_2 b \tag{3.32}$$

where b is the diameter (intermediate axis) of the particle in millimeters. By using base 2 logarithms, ϕ units change one unit for every *doubling* in size of the diameter of the particle (which therefore follows standard size classes) and by setting the transformation negatively, the ϕ values *increase* as sediment size gets *smaller*.

In a sediment size analysis, the weight of sediment in each size class is obtained by using a variety of methods. These values are then converted into cumulative relative form and plotted as percentiles on either linear graph paper or onto graph paper where the y axis is scaled following a normal probability (Section 6.3). Selected percentile values can then be read from the graph. The emphasis on graphic interpretations rather than calculation of percentiles directly from the data derives from the belief that the researcher can extract more realistic values subjectively off the graph than can be derived by calculation, especially when the sediment size occurs is nonsimple. More recent development of curve-fitting techniques has overcome this deficiency to a large degree.

A variety of measures has been suggested to describe the mean, variability (sorting), skewness, and kurtosis of the sediment sample. The ones described here derive from Folk and, using the selected percentiles, are as follows:

$$\text{Graphic mean} = \frac{P_{16} + P_{50} + P_{84}}{3} \tag{3.33}$$

$$\text{Inclusive graphic standard deviation} = \frac{P_{84} - P_{16}}{4} + \frac{P_{95} - P_{5}}{6.6} \tag{3.34}$$

$$\text{Inclusive graphic skewness} = \frac{P_{16} + P_{84} - 2P_{50}}{2(P_{84} - P_{16})} + \frac{P_{5} + P_{95} - 2P_{50}}{2(P_{95} - P_{5})} \tag{3.35}$$

$$\text{Graphic kurtosis} = \frac{P_{95} - P_{5}}{2.44(P_{75} - P_{25})} \tag{3.36}$$

Figure 3.5 shows an example of a possible summary coding sheet and the resulting summary measures.

3.6 STATISTICAL MOMENTS OF A DISTRIBUTION

Although it may seem that this section digresses from the general examination of statistical summary measures, the topic does have considerable relevance. The concept of statistical moments is the basis of many of the summary measures discussed, it is used in the estimation of summary parameters from sample data, and it has direct application in the summary of spatial data. The concept of moments derives from mechanics. Moments are measurements of force applied to a mass with reference to their tendency to produce rotation of the mass, and they can be defined as the magnitude of the force multiplied by the distance between its line of application and the axis of rotation—forces applied farther from the axis of rotation having a greater rotational effect.

In statistical moments, we are concerned with distributions of numeric scores, and we can define moments around any selected "axis of rotation"—a selected X value. However, the process can most easily be illustrated by looking at moments about an X value of zero (Figure 3.6). The axis of rotation passes through the origin, the frequencies in grouped data are the "forces" (forces are unity in ungrouped data—that is, each score has a frequency or force of 1), and the value of the X score (ungrouped data) or the midpoint of a class (grouped data) is the "distance" from the axis of rotation. In defining

Sample Description: S127 (Nanumaga #5)
Sediment Weights: Total Sample Dry Weight = 66.32 g

Sieve Size	Weight (g)	Weight (%)	Cumulative %
−2φ	0.16	0.2	0.2
−1φ	1.91	2.9	3.1
0φ	7.56	11.3	14.4
1φ	34.86	52.2	66.6
2φ	13.21	19.8	86.4
3φ	6.45	9.7	96.1
4φ	2.61	3.9	100.0
Pan	0.03	0.0	100.0
TOTAL	66.79	100.0	
SIEVE GAIN/LOSS	0.47 g gain		

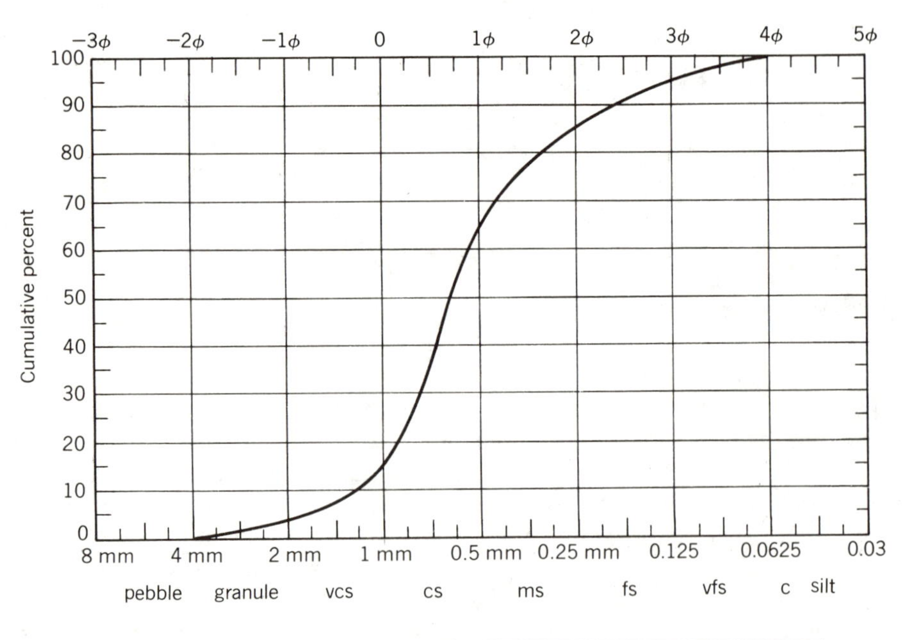

Selected Percentiles	φ	Summary Measures		Description
P05	−0.75	Mean	.88φ	Coarse sand
P16	0.05	Standard deviation	1.01φ	Poorly sorted
P25	0.25			
P50	0.70	Skewness	.25	Fine skewed
P75	1.35	Kurtosis	1.34	Leptokurtic
P84	1.90			
P95	2.85			

Figure 3.5 Sieve analysis result sheet.

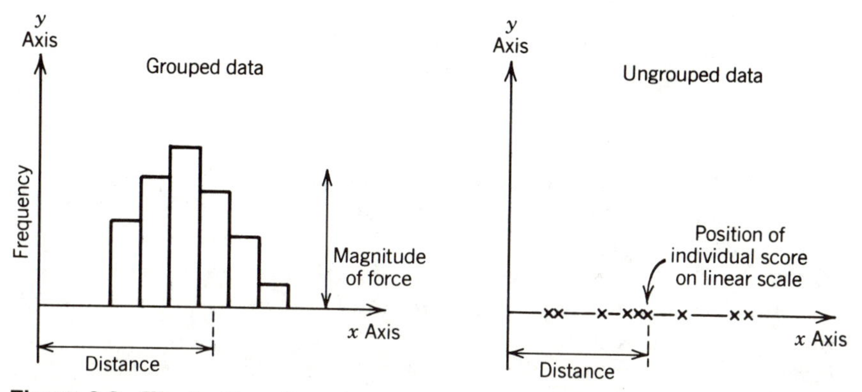

Figure 3.6 Illustration of concept of statistical moments.

statistical moments, one fundamental difference from their counterpart in mechanics is employed in that the total calculated moment is divided by the total number of elements N to produce an average measure in place of a net total.

A system of moments involves the construction of a set of moments at different powers of the distance from the selected value of X (which we call x). For grouped data, the nth statistical moment, π_n (where π = Greek lowercase *pi*), about any point x is given by

$$\pi_n = \frac{\sum_{j=1}^{k} [f_j(d_{jx})^n]}{N} \tag{3.37}$$

where

f_j = the frequency of class j

k = the number of classes

d_{jx} = the distance from the midpoint of class j (M_j) to the given point (x) = ($M_j - x$)

N = the total number of elements = $\sum_{j=1}^{k} f_j$

For ungrouped data, this is simply

$$\pi_n = \frac{\sum_{i=1}^{N} (d_{ix})^n}{N} \tag{3.38}$$

where

$$d_{ix} = \text{the distance from any score } (X_i) \text{ to the given point}$$
$$(x) = (X_i - x)$$

Different characteristics of a distribution of scores can be measured by employing moments with the "distance" value raised to different powers (n). Thus, we can have an infinite number of *positive* moments π_1, π_2, π_3, etc., where d_{jx} or d_{ix} is raised to the first, second, third, etc. power; or *inverse* moments π_{-1}, π_{-2}, π_{-3}, etc., where d_{jx} or d_{ix} is raised to the minus first, minus second, minus third, etc. power (i.e., $1/d_{ix}$, $1/d_{ix}^2$, $1/d_{ix}^3$.)

However, as the value of the exponent n increases, more weight is given to extreme scores in the distribution and, for practical purposes only, the smaller positive exponents and the first one or two negative moments are used. In statistical use it is preferred to retain the original units of measurement. Thus, moments are usually rooted by the appropriate power; for example, the preferred form for the second moment (which involves squaring the "distance") is its square root. When moments are applied to statistical distributions, we are dealing with deviations of scores from a selected score, and thus with a set of positive and negative deviations. In the powering process on these deviations, odd-numbered moments will, therefore, provide measures of skewness around the selected value (with positive and negative values). However, even-numbered moments will only have positive values and measure the degree of variability around the selected value, regardless of the direction.

Summary Measures by Moments

The concept of moments was illustrated by constructing moments around zero. These, however, have little direct application in statistics. Of more use is the set of moments constructed around the selected value of the arithmetic mean. The first of these central moments about the mean would, of course, equal zero because we know that

$$\sum_{i=1}^{N} (X_i - \overline{X}) = 0$$

But the second, third, and fourth positive moments describe the basic measures of variance, skewness, and kurtosis, respectively. Another more fundamental concept concerns the calculation of moments around each score in a distribution (or midpoints in grouped data) and then comparing those values to find the minimum value. The location of this minimum value provides a measure of central tendency, and the minimum value itself is a measure of dispersion about that central value. Minimum values can be located for any order of moment, but two in particular have wide applications in the summary of statistical and spatial distributions. The first moment

about any point in a distribution is simply the arithmetic mean of the deviations of each score from that point. For ungrouped data:
For any point x,

$$\pi_1 = \frac{\sum_{i=1}^{N} (X_i - x)}{N} \tag{3.39}$$

This property is called the *average deviation* around the point x. It has already been shown that the point where the minimum average deviation occurs is the arithmetic mean of the scores themselves. However, if we look at deviations in absolute terms (and the importance of doing so will become apparent when we examine spatial distributions), we can redefine the first moment as being:

For any point x,

$$\pi_1 = \frac{\sum_{i=1}^{N} |X_i - x|}{N} \tag{3.40}$$

In this situation, the minimum value does not occur at the arithmetic mean of a distribution, but rather at the median. Thus, the *median* (*Md*) is the measure of the central tendency based on the first moment, and the *mean deviation about the median* defines the minimum deviation about any score in the distribution.

Measures based on the second moment form the basis of much of the statistical analysis of linear distributions. In its preferred form:

For any point x,

$$\sqrt{\pi_2} = \sqrt{\frac{\sum_{i=1}^{N} (X_i - x)^2}{N}} \tag{3.41}$$

The measure so defined for any point x is the *root-mean-square deviation* (often abbreviated to RMS). The root-mean-square deviation provides an important measure of the variability of all points (scores) around any particular point. Its minimum occurs when the point selected is the *arithmetic mean* ($\bar{X}$) of the distribution, and the minimum root-mean-square deviation is known as the *standard deviation* (s). This was stated as the second property of the mean (Equation 3.10). The mimimum π_2 is, of course, the variance (s^2). The measure of central tendency (the arithmetic mean), and its measure of dispersion (the standard deviation) based on the second moment, are the key summary measures for linear distributions because they provide the simplest measurement of variability of a distribution.

The Method of Moments

The preceding discussion has shown that several of the summary measures used for describing linear distributions are defined with reference to the system of statistical moments. The reasons for introducing this concept at this stage are to show how some of these measures originated and to justify why some summary measures are preferred over others. The usefulness of the concept will become apparent in the next section where summary measures based on moments have an even greater application in the analysis of spatial distributions. In later chapters, when we attempt to fit mathematical functions to sampled distributions, we will also find that the method of moments forms one of the major methods for estimating the parameters of these functions.

3.7 MEASURES FOR SPATIAL DISTRIBUTION

The techniques that have been outlined in this chapter have been concerned with summarizing the character of data distributions, centered on methods describing the frequency curve. For spatially distributed variables, in addition to providing summary measures of the statistical distribution of scores, we may also be interested in providing descriptors of their spatial distribution. The spatial equivalent of the frequency curve is the smoothed statistical surface, usually represented by an isarithmic map. For this surface, we wish to provide measures of central tendency and variability. As with statistical distributions, these summary measures have greatest application when the spatial distribution is characterized by a single peak, a situation that does not occur as often as it does with statistical distributions.

The three types of spatial data being examined (discrete point data, discrete area data, and sample point data from a spatially continuous distribution) can all be described as consisting of a set of spatially located points (for discrete area data these are the summary points within the areas), each with a certain magnitude (the score for the spatial element). Using this set of points, there are two basic methods for deriving spatial summary measures (Figure 3.7).

One approach is to locate each point in the spatial distribution by coordinates defined by an x axis and a y axis. These axes could measure longitude and latitude, or, more usually, they are arbitrary axes established around the population area being examined. Each point in the spatial distribution is, therefore, three dimensional with two location measures (x and y coordinates) and a measure of magnitude (the score for the variable). An alternative approach is to replace the two dimensions of location (x and y axes) by a single dimension of *relative position,* measuring the *distance* between points in some preferred length unit (e.g., meters or kilometers). The apparent disadvantage of this approach (the loss of absolute location and there-

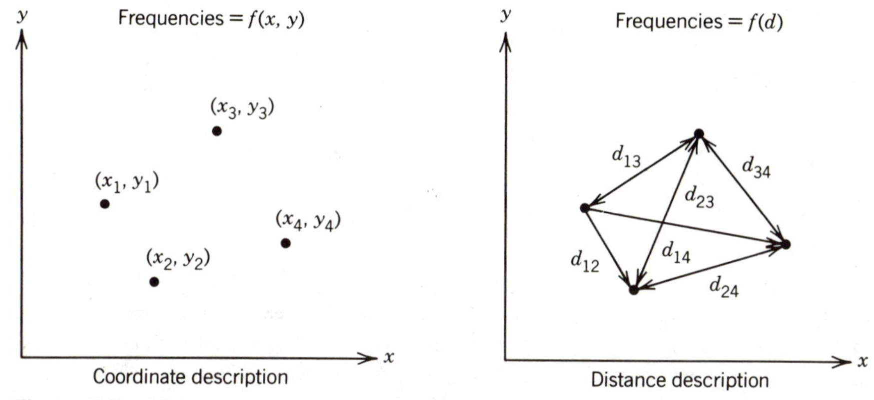

Figure 3.7 Alternative methods for point location.

fore *directional relationship*) is outweighed by the advantages of divorcing the measurement process from an arbitrary grid system and allowing the use of a two-dimensional approach rather than a three-dimensional one. However, in some situations, the two alternative approaches can be related to each other, because the squared distance between two points is equal to the sum of the squared distance differences along the x and y axes, based on the Pythagorean theorem (Figure 3.8). This will have important practical implications, although it should be noted that this relationship would only be true when we are dealing with squared distances. In the methods outlined here, we will assume that we are dealing with "small" areas and therefore we can ignore the fact that the earth's surface is curved, which would invalidate the Pythagorean theorem; Neft (1966) discusses alternative distance measurement principles.

The two-dimensional approach allows application of the methods already

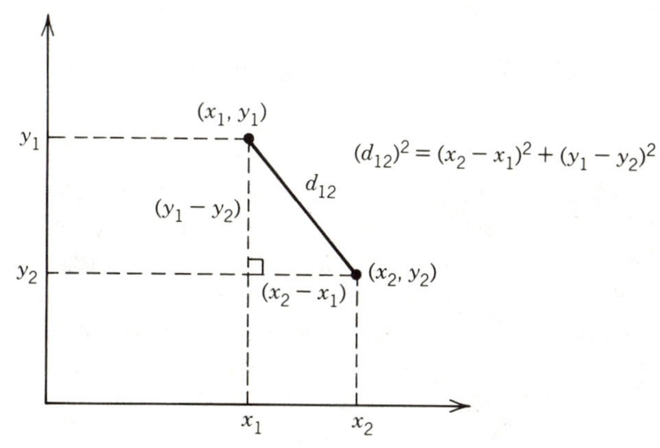

Figure 3.8 Distance related to coordinate differences.

outlined for statistical distributions to spatial distributions. This extension can best be illustrated by referring to the system of moments. Discrete area data may be considered as the areal equivalent of the frequency distribution—the areal element is the "class" of the distribution, the "summary point" of the areal element is the midpoint of the class, and the magnitude of the score for the areal unit is the frequency of that class. Thus, the methods describing statistical moments for grouped data, outlined in the preceding section, can be applied directly to this areal frequency data. In the same way as equal class intervals are desirable in the analysis of linear distributions, equal area (and shape) of the areal "classes" are desirable, but unfortunately only infrequently possible. This limits, but does not invalidate, the correspondence. Discrete point data and sampled point data can be handled in a roughly similar manner, with distances between points providing the location variable and the magnitude of the score being the equivalent of the "frequency" of the score at the midpoint of the class.

Not all the summary measures for spatial distributions are based on the system of moments, but the emphasis in this section will be on those that are, because these have proved to be the most useful. Their derivation is based on calculating moments about every point and then using the principle of minimum value to locate points of central tendency and define measures of dispersion about that point. The pattern of variability can be summarized by the single measure of dispersion and can be described by mapping concentric circles around the point of central tendency based on this measure of variability or alternatively, by mapping the actual distribution of the moments themselves using isarithms. The central tendency measures may not locate exactly at a point in the distribution but will probably occur somewhere between points. This results in a very tedious process of derivation and it is amenable to computer processing. The analysis of the variability of spatial distributions (geostatistics) is an important topic and only an outline is provided here. The interested reader is referred to Neft (1966) for a more detailed discussion.

Measures of Central Tendency and Variability

The *arithmetic mean center* (*AMn*) is the spatial equivalent of the arithmetic mean and defines the point of minimum squared distance deviation (the *root-mean-square distance deviation*). The arithmetic mean center defined by the second positive moment has been widely used as a method of describing central tendency in spatial distributions, but we note that its position is greatly affected by extreme locations and that, as a measure to describe the peak of a statistical surface, it is only applicable when the surface is unimodal and evenly distributed around that point. The same properties, of course, are characteristic of the arithmetic mean in statistical distributions. Because the arithmetic mean center is strongly influenced by values near the margins of the distribution it is, however, extremely useful in portraying the movement of the center of a spatial distribution over time. The measure of dispersion around the arithmetic mean center is the standard distance deviation (SD) commonly abbreviated to *standard distance* (as introduced by Bachi (1968)). Mapping dispersion around the arithmetic mean center can be accomplished by using the standard distance deviation to construct concentric circles of increasing variability; or alternatively, the root-mean-square distance deviation at each point can be used to produce isarithms of variability—these also would map as concentric circles around the arithmetic mean.

Because we are dealing with the second moment based on squared distance deviation, these measures can also be defined using bivariate theory, and this provides a practical method of calculation, which, for small examples, allows computation without resort to computer analysis. By constructing arbitrary x and y axes around the areal distribution, and locating each point by its x and y coordinates, the arithmetic mean center is defined at the point of the mean of the x and y coordinates, weighted by the frequency or magnitude.

$$\text{Position of } AMn = (\overline{X}_x, \overline{X}_y) \tag{3.42}$$

and the standard distance is

$$SD = \sqrt{\frac{\sum_{i=1}^{N} d_{ic}^2}{N}} \tag{3.43}$$

where

$$d_{ic}^2 = (x_i - \overline{x})^2 + (y_i - \overline{y})^2$$

or alternatively,

$$SD = \sqrt{s_x^2 + s_y^2} \tag{3.44}$$

where s_x^2 and s_y^2 are the variances of the x and y coordinates, respectively.

The mean center of a spatial distribution is illustrated by calculations from a map of major metropolitan areas in California. By imposing an arbitrary grid over the map of California, we can derive x and y coordinate values for each metropolitan area (Figure 3.9 and Table 3.8). The mean of the x coordinates is:

$$\frac{\Sigma x_i}{N} = \frac{78}{8} = 9.8$$

and the mean of the y coordinates is

$$\frac{\Sigma y_i}{N} = \frac{99}{8} = 12.4$$

Finding this position on the overlaid graph locates the (spatial) mean center.

TABLE 3.8 Coordinate Locations for California Metropolitan Areas

	x Coordinate	y Coordinate	Population (Millions)	Weighted x	Weighted y
Sacramento	7	27	1.0	7.0	27.0
San Francisco/ Oakland	3	24	3.3	9.9	79.2
Fresno	9	18	0.5	4.5	9.0
Oxnard/Ventura	9	9	0.5	4.5	4.5
Los Angeles	11	7	7.5	82.5	52.5
Riverside/ San Bernardino	14	7	1.6	22.4	11.2
Anaheim	12	5	1.9	22.8	9.5
San Diego	13	2	1.9	24.7	3.8
TOTAL	78	99	18.2	178.3	196.7

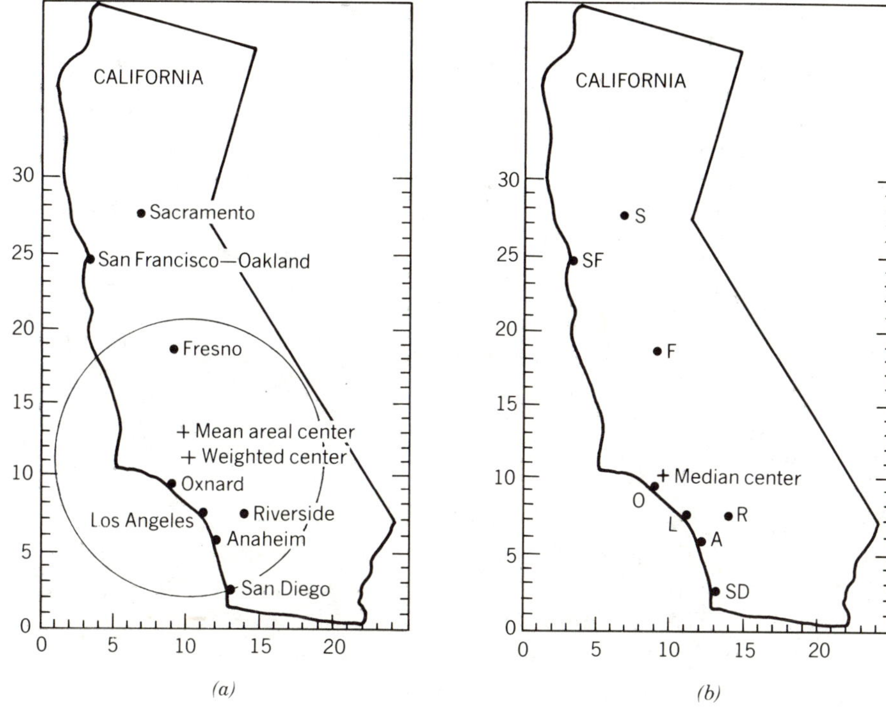

Figure 3.9 Spatial means and medians.

The position of the axes has no influence on the location of the mean center of the distribution. Although the mean center can be calculated for any distribution, it is particularly useful as a descriptor of the centroid of a population distribution. This could be done by locating the coordinates of every person in California and computing the center of that distribution, or taking many very small areas of the state and assuming that the population of these small areas was located at the center of small areas and computing the mean center of all the cells weighted by the population in each small cell. As a simplified example, imagine that the population of each of the metropolitan areas was located at the coordinates of the metropolitan areas just examined. Clearly, this is a fairly large generalization, but, in fact, it only affects the detail of the location of the mean population center.

For the unweighted mean center, the standard distance can be calculated from Table 3.9. Thus,

$$SD = \sqrt{\frac{701.4}{8}}$$

$$= 9.36$$

TABLE 3.9 Calculation of Standard Distance

	$(x_i - \bar{x})^2$	$(y_i - \bar{y})^2$	d_{ic}^2
Sacramento	7.84	213.16	221.0
San Francisco/ Oakland	46.24	134.56	180.8
Fresno	0.64	31.36	32.0
Oxnard/Ventura	0.64	11.56	12.2
Los Angeles	1.44	29.16	30.6
Riverside/ San Bernardino	17.64	29.16	46.8
Anaheim	4.84	54.76	59.6
San Diego	10.24	108.16	118.4
TOTAL			701.4

The standard distance can be illustrated in Figure 3.9.

Greater detail in the locations means greater precision of the final mean center. In this instance, weight the coordinate locations by the population from Table 3.8. The data are in millions to simplify calculations. Thus,

$$\bar{x}_w = \frac{\Sigma x_i}{N} = \frac{178.3}{18.2} = 9.8$$

$$\bar{y}_w = \frac{\Sigma y_i}{N} = \frac{196.7}{18.2} = 10.8$$

Note that we divide by the weighted N rather than the simple N. The new weighted mean is pulled (naturally) towards the largest population centers in Southern California (Figure 3.9). Because the eight metropolitan centers contain approximately 75% of the population of California, the weighted mean is a good estimate of the center of the distribution of population in that state.

It is often necessary to know the center of a geographic area—a state or country, for example. A straightforward method utilizes the approach just described to determine the center of the metropolitan areas. Overlay a grid of any accuracy over the shape, and compute x and y values for sufficient *equally spaced* points on the boundary to accurately represent the shape and compute $\bar{x}$ and $\bar{y}$ (Figure 3.10). This is the point at which the area would balance if it were cut out of firm cardboard or plastic and balanced over a pencil (the center of gravity). Clearly, the calculation of spatial means is best undertaken with a computer routine.

One of the more interesting spatial measures is the median center. In the descriptive statistics presented thus far, we have used the median as the point that divides the distribution in half. In spatial data, the median center is the point of minimum aggregate travel, a finding of considerable interest

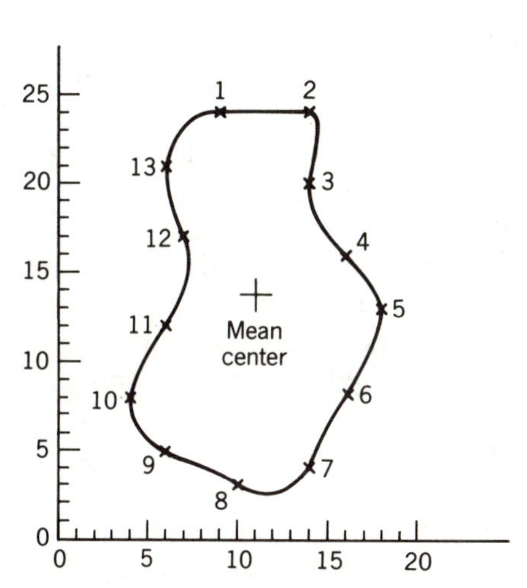

Figure 3.10 Calculation of the mean center.

	x	y
1	9	24
2	14	24
3	14	20
4	16	16
5	18	13
6	16	8
7	14	4
8	10	3
9	6	5
10	4	8
11	6	12
12	7	17
13	6	21
	140	175

$\bar{x} = 10.8$

$\bar{y} = 13.5$

to human geographers. It is the median and not the mean that minimizes the distance to all points in the distribution.

The *median center* and the *mean distance deviation* (the equivalent of the median and the mean deviation around the median) describe the minimum value of the *average distance* (or average travel distance) between points. The median center is the point of *minimum average travel* for all points in the distribution and has obvious applications in location theory. While the mean distance deviation can be used to map concentric circles around the median center, a map of average distances provides a better portrayal of the distribution as it need not necessarily map as concentric circles.

Although the median is calculated for a statistical distribution relatively easily, the same ease of calculation is not true for the spatial median. It is *not* the point located by the median values of the x and y coordinates (a median center so defined would be dependent on the arbitrary position of the axes). Instead, resort must be made to computer programs that attempt to provide approximate solutions for locating the median center in what is usually referred to as the generalized Weberian problem.

A simplified manual procedure for a chosen spatial scale grids the area within the extreme locations of interest (with the chosen spatial scale) and computes

$$\sqrt{\sum_{i=1}^{N} [(x_i - x_0)^2 + (y_i - y_0)]^2}$$

and chooses the $x_0\ y_0$ with the smallest value. This is the point of minimum aggregate travel *at that scale.* A grid of greater detail would yield a more accurate median center.

Figure 3.9*b* has a grid at 19-mile intervals and is gridded for the area within the extremes defined by San Francisco, Sacramento, Riverside, and San Diego. The median will be at one of the grid intersections. Table 3.10 gives the calculations of several choices. From the calculations, we can identify the coordinate location 10,10 as the median center. It is the point of minimum aggregate travel to the eight metropolitan centers at that scale. A quick check will indicate that the mean center produces a value of 68.4, which is larger than the median center at 10,10. Finally, it is worth reiterating that although the position of the axes does not influence the mean center, it does affect the location of the median center within the spatial grid chosen for the analysis.

All of the measures based on the three moments that have been described are constructed in the units for measuring distance; they are *absolute measures* and would obviously be strongly influenced by the size of the area being studied. *Relative measures* of dispersion can be found for any of the three variability measures by ratioing them to some base measurement in the same units. When human populations are examined, the most common form of ratioing is to divide the measure by a linear form of the population area (or the effectively settled area)—the radius of a circle with the same area (r_a).

$$r_a = \sqrt{\frac{A}{\pi}} \tag{3.45}$$

where

A = the population (or the effectively settled area)
π = approximately 3.142

For variables other than human population (such as economic measures), the ratio could be achieved by dividing the dispersion measure by the same dispersion measure calculated for the human population distribution itself.

Other Summary Measures

There are a wide variety of other summary measures based on the system of moments. Equivalent measures for skewness and kurtosis, for example, can be derived from their statistical counterparts, although as all distances in spatial distributions will be positive, odd-numbered moments will not measure the same properties as for statistical distributions. Neft (1966) provides an outline for such measures. Also, in the discussion of the summary measures for statistical distributions, a number of measures were introduced that were not based on the system of moments and for each of these a spatial

TABLE 3.10 Calculations for the Median Center

Grid Location (x_0,y_0) to Each Metropolitan Area														
9,	**11**		**10,**	**10**		**10,**	**11**		**10,**	**12**		**10,**	**15**	
(1)	(2)	(3)	(1)	(2)	(3)	(1)	(2)	(3)	(1)	(2)	(3)	(1)	(2)	(3)
4	256	16.1	9	289	17.3	9	256	16.3	9	225	15.3	9	144	12.4
36	169	14.3	49	196	15.7	49	169	14.8	49	196	15.7	49	81	11.4
0	49	7.0	1	64	8.1	1	49	7.1	1	36	6.1	1	9	3.2
0	4	2.0	1	1	1.4	1	4	2.2	1	9	3.2	1	36	6.1
4	16	4.5	1	9	3.2	1	16	4.1	1	25	5.1	1	64	8.1
25	16	6.4	16	9	5.0	16	16	5.7	16	25	6.4	16	64	8.9
9	36	6.7	4	25	5.4	4	36	6.3	4	49	7.3	4	100	10.2
16	81	9.8	9	64	8.5	9	81	9.5	9	100	10.4	9	169	13.3
		Σ = 66.9			Σ = 64.5			Σ = 65.9			Σ = 69.4			Σ = 73.6

For each grid location (x_0,y_0), (1) = $(x_i - x_0)^2$, (2) = $(y_i - y_0)^2$, and (3) = $\sqrt{(x_i - x_0)^2 + (y_i - y_0)^2}$.

counterpart can be derived. The only one that is worth mentioning is the mode. The *modal center* is an important measure of central tendency, but it is only applicable (as before) when the spatial distribution is characterized by a single peak. In this situation, the modal center is the point with the highest frequency or magnitude.

References and Readings

1. Statistical Summary Methods

Chorley, R. J. (1966) "Application of statistical methods to geomorphology," in G. H. Dury (ed.) *Essays in Geomorphology,* Heineman: London.

Conrad, V. and L. W. Pollak (1962) *Methods in Climatology,* Harvard University Press: Cambridge, Mass.

Folk, R. C. (1964) *Petrology of Sedimentary Rocks,* Hemphill: Austin, Tex.

Krumbein, W. C. (1936) "Application of logarithmic moments to size-frequency distribution of sediments," *Journal of Sedimentary Petrology* 6:35–47.

2. Spatial Distributions

Bachi, R. (1963, reprinted 1968) "Statistical analysis of geographical series," in B. J. L. Berry and D. F. Marble (eds.), *Spatial Analysis: A Reader in Statistical Geography,* Prentice-Hall: Englewood Cliffs: N.J.

Caprio, R. J. (1970) "Centrography and geostatistics," *Professional Geographer* 22: 15–19.

Deskins, D. R. (1976) "Race residence and workplace in Detroit, 1880–1965," *Economic Geography* 52:79–94.

Neft, D. S. (1966) *Statistical Analysis for Areal Distributions,* Regional Science Research Institute: Philadelphia, Penn.

CHAPTER 4
Probability and Probability Functions

Probability is concerned with the determination or estimation of the likelihood of a particular result (or outcome) from a set of possible results. These results may be derived, for example, from an experiment, from the sampling of a population or from drawing a card out of a deck of cards. Simple laws of probability can be used to draw conclusions (make inferences) about the character of a population as a whole when a sample is drawn from the population such that each element of the population has an equal chance of selection (a random sample). Because probability is the basis for inferential analyses, a review of the major concepts of probability theory is essential background for inferential statistical methods. But probability concepts are not just the basis for inferential statistical methods—they have other useful applications as well. Frequency distributions that are fitted to distributions from an experiment or a sample can be compared with frequency distributions constructed according to a given probability function and the resulting comparison can lead to the prediction of future outcomes or patterns.

In this chapter, we examine some of the main concepts of probability theory. Much of the current research work in geography requires a far greater knowledge of probability methods than is provided in this simple introduction, and the interested reader will have to delve much further into the literature on probability. However, by examining the definitions of probability (Section 4.1) and some of the simpler mathematical properties (Section 4.2), we provide a basis for the methods used in the remainder of this book, as well as introduce one of the most important and rapidly expanding fields in geographic research. In Sections 4.1 and 4.2, the examples used to illustrate concepts of probability are concerned with a sample space that is finite and completely specified. In an extension to a more general level, we introduce the ideas of random variables and probability functions (Section 4.3). The extension will greatly simplify problems where the sample space is infinite or the probabilities are not completely specified. We discuss only the major probability functions in this text, but we also mention some of the other important functions that have been used in geography (Section 4.4). The methods of deriving summary parameters can be extended to all probability functions.

4.1 PROBABILITY DEFINITIONS

The definition of probability can be approached from three different directions: by determining the "true" probability of an outcome (*theoretical probability*), by obtaining an estimate of the true probability through the relative frequency derived from empirical results (*empirical probability*), or by estimating the true probability using intuitive knowledge of the distribution of outcomes (*subjective probability*). The use of subjective probabilities (which fall within the domain of Bayesian statistical inference) is a recent innovation to geography but its discussion is outside the scope of this book. Most of the statistical ideas with which we are concerned are based on theoretical probabilities and their estimation using empirical probabilities. The ideas we will be exploring are within the framework of classical probability theory. In the discussion of probability, we will use simple ideas from *symbolic logic* and *set theory* to illustrate the definitions and concepts.

Sample Space and Events

Our concern for the moment is with the probabilities associated with a single variable, and we will use as an example the variable Farm Type taken from a hypothetical survey of farming characteristics. To keep the discussion as simple as possible, we will assume that the population consists of only 10 farms, 4 of which are dairy farms (for convenience we will number these farms 1 to 4), and the remaining 6 are nondairy farms (numbered 5 to 10). It is thus an example for which we have complete information—we know

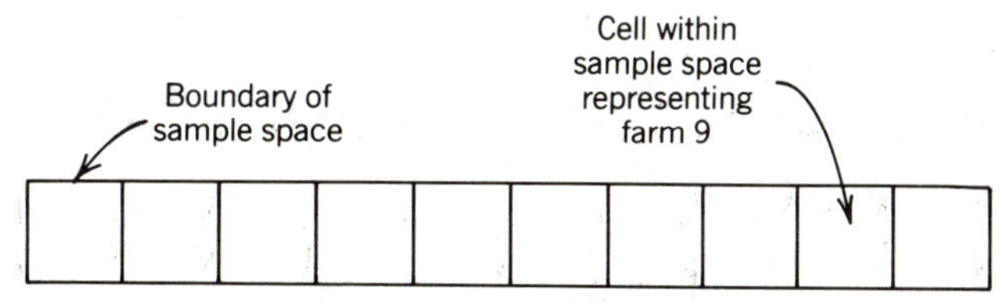

Figure 4.1 Venn diagram for sample space.

how many, and which of the 10 farms fall into each of the two classes. It is also an example that will provide an illustration of the simplest type of probability function—the *binomial* (two classes). From this population of 10 farms, we will draw samples and establish the probabilities associated with the sample selections. One of the basic initial assumptions will be that each farm has an equal chance of being selected; that is, our sampling method uses simple *random selection.*

Our first major definition concerns the population of elements that are available for selection. We define the *sample space* S as being the *set* of elements that is being considered and its *magnitude* or size $|S|$ as being the total number of elements in that set. In our example, the sample space is a population of farms and has a magnitude of 10. We can show this diagrammatically by using a *Venn diagram.* A Venn diagram can be of any shape or size, although later we will construct the diagram in a particular manner to represent the characteristics of certain distributions. A diagram made up of 10 columns, one row high, is suitable (Figure 4.1), with each *cell* of the diagram representing one element (a single farm in this instance).

We define an event E as being any *subset* of the sample space comprised of elements that share a specific property; and as a corollary definition, the *complement* of an event (written $\tilde{E}$) is the subset of the sample space that does not have that specific property. The magnitudes of these subsets would be represented by $|E|$ and $|\tilde{E}|$, respectively. For example, we consider an event as a dairy farm within the sample space of our example, and thus the complement of that event is a nondairy farm. This can be shown diagrammatically (Figure 4.2) because we know that four of the farms in the example are dairy farms, and thus

$$|E| = 4$$

$$|\tilde{E}| = 6$$

and we note that $|E| + |\tilde{E}| = 10 = S$.

Theoretical Probability

We have complete information about the variable Farm Type, and using the foregoing definitions, we are in a position to define theoretical probability.

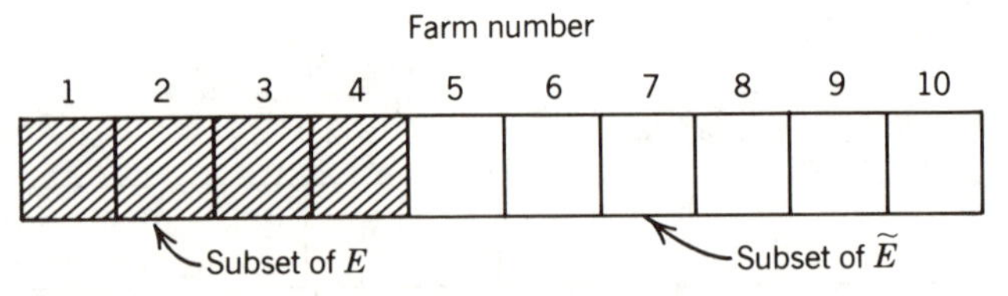

Figure 4.2 Venn diagram for event E.

In a situation where selection is random (a so-called game of chance) the probability of selecting an element belonging to an event E, written $P(E)$, is equal to the ratio of the magnitude of that event to the magnitude of the total sample space.

$$P(E) = \frac{|E|}{|S|} = p \tag{4.1}$$

In the example we have been examining, the probability of selecting a dairy farm (E) from the population of farms is 4/10, or as it would usually be expressed, .4. That is, if we select one farm randomly from the 10 farms, the chances are 4 out of 10 that it will be a dairy farm. Note that in the calculation of probabilities it is standard practice to use uppercase P to denote the probability that is being determined, and lowercase p to denote the resulting value.

$$\begin{aligned} P(E) &= p \\ P(\text{dairy farm}) &= .4 = p \end{aligned}$$

An examination of the Venn diagram (Figure 4.2) would have given the same result. The probability of the event "dairy farm" could be found by counting up the number of cells (elements) belonging to that event and comparing this number to the total number of cells (elements). Or, looking at it another way, as the cells are of equal area, we could derive the probability by comparing the *area* labeled as belonging to the event E to the total area of the sample space. The Venn diagram illustrates a further point—that the probability of selecting a dairy farm could have been found by determining the probability of selection of an individual farm, then adding up these probabilities for all the farms that are dairy farms, or more simply, multiplying this individual probability by the number of farms that were classified as dairy farms.

$$\begin{aligned} P(\text{individual farm}) &= .1 \\ P(\text{dairy farm}) &= .1 + .1 + .1 + .1 \end{aligned}$$

or

$$\begin{aligned} P(\text{dairy farm}) &= 4(.1) \\ &= .4 \end{aligned}$$

This is an important property that will be examined in Section 4.2.

Empirical Probability

Thus far, we have been able to calculate theoretical probabilities because we know the characteristics of all the farms being considered for selection. But, in many situations, we may not know the makeup of the population we are examining and can only estimate the probabilities from a sample or a series of samples. This introduces the idea of empirical probability estimates. Take the same example, but this time assume that we neither know how many farms there are in the sample space nor how many of the farms are dairy farms. Instead, select one farm at a time from the sample space (a random sample of size 1), record its farm type, and then use this result to estimate the probability. As we will see, the number of times we repeat this selection process (i.e., increase the size of our sample), the more precise the estimate of the theoretical probability will be.

We will define the process of selection of an element from the sample space as a trial, and if the outcome of the trial is the selection of an element that falls into the subset of the event being considered, we call the outcome a *success;* if it falls into the subset of the sample space which forms the complement of the event, we call the result a *failure.*

In a series of trials, the empirical probability of an event is defined as the relative frequency of success—the absolute frequency of success (f) divided by the number of times the experiment has been carried out (n).

$$P(E) = p = \frac{f}{n} \tag{4.2}$$

A random sample of one farm was taken from the population of farms (size unspecified), and the farm selected was a nondairy farm. Our empirical estimate of the theoretical probability of the event "dairy farm" is thus $0/1 = 0$. From our knowledge of the farm structure of the example, we know that this is a poor estimate. Repeating the experiment (i.e., increasing the size of our sample) will improve our confidence in the accuracy of the estimate. Each time we repeat the trial we make sure that all the farms have an equal chance of being selected—any farm that was selected in a previous experiment must have the same chance as the other farms of being selected in the current experiment.

Our second trial resulted in the selection of a dairy farm and, combining the results of the two trials, our estimate of the probability is now $p = 1/2 = .5$. Repeating the selection process, say 100 times, we see (Table 4.1 and Figure 4.3) that the estimated (or empirical) probability is slowly approaching what is the theoretical probability. That is, the "limit" of the empirical probability—as n approaches infinity—is the theoretical probability, assuming, of course, simple random sampling with replacement. The concept of determining probabilities by relative frequencies will prove to be very useful, but, for the moment, we continue our discussion of the properties

TABLE 4.1 Effects of Increasing Sample Size (Summarized in Groups of Five)

Trial Number	Success (Dairy Farm)	Failure (Nondairy Farm)	Cumulative success (f)	Total trials (n)	Estimated probability (f/n)
1– 5	3	2	3	5	.600
6– 10	2	3	5	10	.500
11– 15	0	5	5	15	.333
16– 20	2	3	7	20	.350
21– 25	0	5	7	25	.280
26– 30	1	4	8	30	.267
31– 35	3	2	11	35	.314
36– 40	2	3	13	40	.325
41– 45	3	2	16	45	.356
46– 50	4	1	20	50	.400
51– 55	3	2	23	55	.418
56– 60	1	4	24	60	.400
61– 65	2	3	26	65	.400
66– 70	1	4	27	70	.386
71– 75	4	1	31	75	.413
76– 80	2	3	33	80	.412
81– 85	1	4	34	85	.400
86– 90	2	3	36	90	.400
91– 95	1	4	37	95	.389
96–100	2	3	39	100	.390

of probabilities using theoretical probabilities only, because these are much easier to visualize. The extension to empirical probabilities is considered later in the chapter.

4.2 MATHEMATICAL PROPERTIES OF PROBABILITIES

A number of mathematical properties associated with probability concepts forms the basis of inferential statistical methods. The first two properties are obvious but very important, nevertheless. For any event E,

$$0 \leq P(E) \leq 1.0 \tag{4.3}$$

That is, the probability of an event E must lie between zero (the occurrence of the event is impossible) and 1.0 (the occurrence of the event is certain). As stated before in set theory terms, the sum of the probabilities of an event and its complement is 1.0.

$$P(E) + P(\tilde{E}) = 1.0 \tag{4.4}$$

$$P(\text{dairy farm}) + P(\text{nondairy farm}) = .4 + .6 = 1.0$$

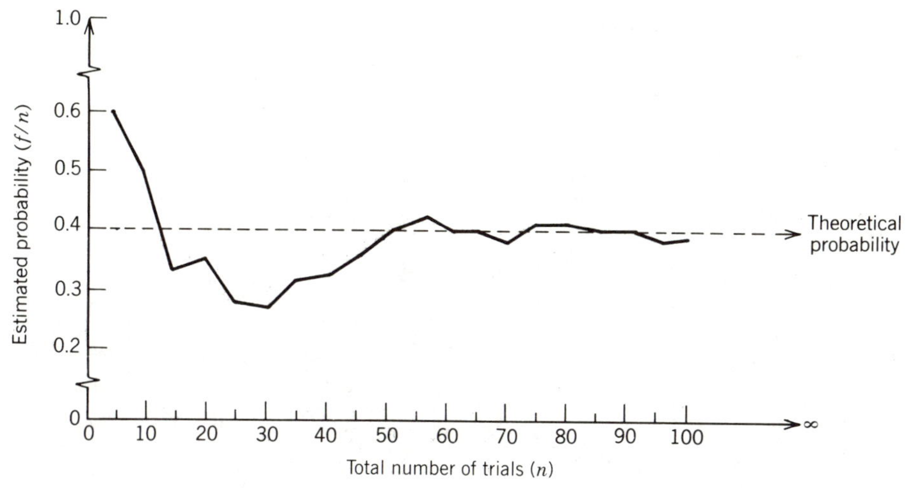

Figure 4.3 Plot of relative frequencies for the 100 experiments.

Intersections, the Multiplication Rule, and Independent Events

In the example used to define empirical probability, we were careful to make sure that at each selection (each trial), all farms had an equal chance of being selected. We illustrate the importance of equal selection with our example where the elements comprising the sample space are completely known (10 farms, 4 of which are dairy farms). In this way, we can concern ourselves with theoretical probabilities rather than estimates of the probabilities. As before, we consider an event as being a dairy farm, except that now we consider the probability of selecting a dairy farm during each of a series of successive selections (sometimes called a joint event). In the simplest case, dealing with just two trials, we determine the probabilities associated with the selection of a dairy farm in both experiments. It is convenient, then, to imagine that we are considering two separate events, and are interested in the probabilities of the event "dairy farm" in the first experiment, which we denote by E_1, and the probability of the event "dairy farm" in the second experiment (once the first has been carried out), which we denote by E_2. The example we are setting up defines two events as basically being the same—that of selecting a dairy farm. The example could be structured differently, but because our concern for the moment is with the characteristics of a single variable, we retain this simple form of the relationship of two events. Two possibilities are apparent: one where the farm selected in the first trial is retained as a member of the sample space for the second trial (*selection with replacement*) and one where a farm once selected is removed from the sample space and is not considered for further selection (*selection without replacement*).

Consider these two different possibilities of combining the two events (with replacement and without replacement) by using Venn diagrams, struc-

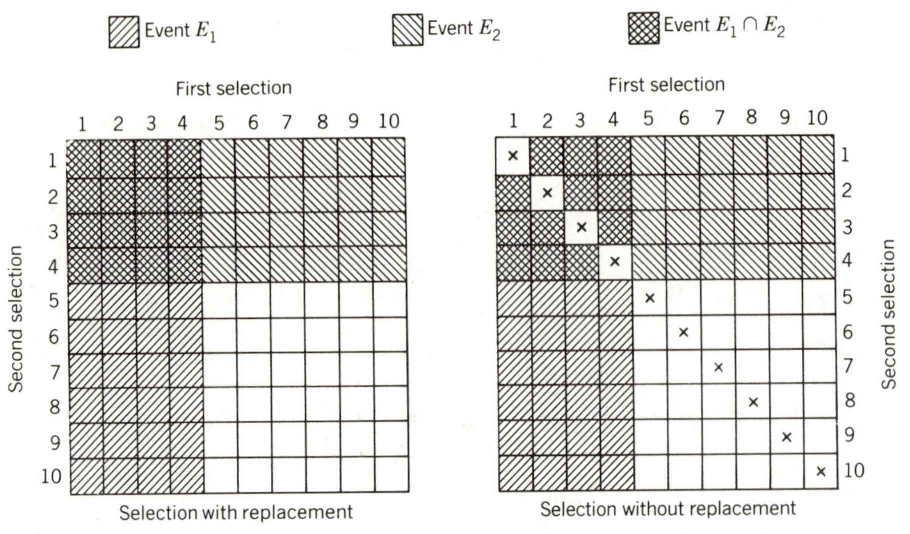

Figure 4.4 Venn diagram for the two experiments.

turing them in the form of a matrix (Figure 4.4), with columns representing the choice of farms for the first trial and rows representing the choice of farms for the second trial. The sample space for the two experiments combined is made up of 100 cells, but, in the case of selection without replacement, the diagonals of the matrix do not belong to the sample space, because if that particular farm was drawn in the first experiment, it would not be available for the second selection; that is, it is impossible to select the same farm twice. In this situation, the sample space for the two experiments combined is made up of only 90 cells (100 cells minus the 10 diagonals).

We define the *intersection* (or conjunction) of two events as the subset of the sample space whose cells belong to both the first event and the second event. In the example, where the two experiments are both concerned with the selection of dairy farms, the intersection is defined as the subset of the sample space where the event "dairy farm" occurs for both the first experiment and the second experiment. We use the symbol $\cap$ to represent an intersection, and from the Venn diagram it can be seen that the probabilities associated with the intersection are as follows:

Selection with replacement:

$$P(E_1 \cap E_2) = 16/100 = .16$$

Selection without replacement:

$$P(E_1 \cap E_2) = 12/90 = .13$$

Thus, the fact that we did not return the farm sampled in the first experiment back to the sample space to be considered again for possible selection results in a different probability of selecting a dairy farm from both attempts. It is also apparent that in the situation where selection was without replacement, the outcome of the first experiment will influence the outcome of the second experiment. This influence needs to be examined in more detail, and to do this we introduce the idea of conditional probability. The *conditional probability* of two events, where one event (E_2) is considered subsequent to the other (E_1), written $P(E_2|E_1)$, is defined as the probability of event E_2 given that event E_1 has occurred.

Examining the situation just outlined, if event E_1 has occurred (i.e., a dairy farm was selected in the first experiment), then either farm number 1, 2, 3, or 4 must have been selected. The sample space for the second selection must then be the column containing the 10 or 9 possible selections for the second experiment. Assuming that farm 3 was selected in the first experiment (any one of the farms numbered 1 to 4 would do), our calculations for the conditional probability of event E_2 can then be ascertained from an abbreviated Venn diagram (Figure 4.5).

Selection with replacement:

$$P(E_2|E_1) = 4/10 = .4$$

Selection without replacement:

$$P(E_2|E_1) = 3/9 = .33$$

Thus, as we would have expected, when the selection process involves replacement, the probability remains the same for both selections, but this is not true for selection without replacement.

We now combine the ideas of the intersection of events and conditional probability in the form of the *multiplication rule*.

$$P(E_1 \cap E_2) = P(E_1) \cdot P(E_2|E_1) \tag{4.5}$$

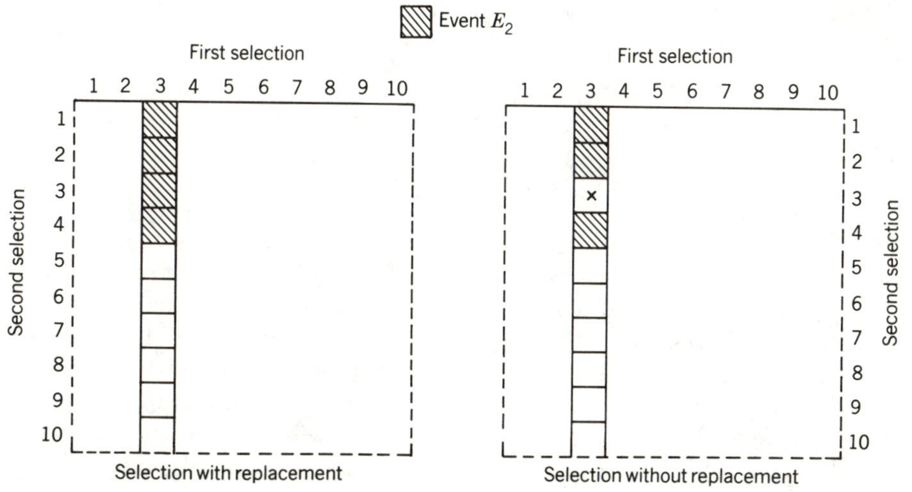

Figure 4.5 Venn diagram for second selection.

Or, in an alternative form,

$$P(E_2|E_1) = \frac{P(E_1 \cap (E_2)}{P(E_1)} \tag{4.6}$$

Illustrating this relationship using the example,

Selection with replacement

$$P(E_1 \cap E_2) = (.4)(.4) = .16$$

$$P(E_2|E_1) = .16/.4 = .4$$

Selection without replacement,

$$P(E_1 \cap E_2) = (.4)(.33) = .13$$

$$P(E_2|E_1) = .13/.4 = .33$$

The multiplication rule provides a method for deriving the probability of an intersection of two events by the multiplication of the probability of the first event by the conditional probability of the second event. It can be extended to include more than two events and, for example, for three events it becomes

$$P(E_1 \cap E_2 \cap E_3) = P(E_1) \cdot P(E_2|E_1) \cdot P(E_3|E_2 \cap E_1) \tag{4.7}$$

Using the same example (adding a third event E_3 = picking a third farm).

Selection with replacement:

$$P(E_1 \cap E_2 \cap E_3) = (.4)(.4)(.4) = .064$$

Selection without replacement:

$$P(E_1 \cap E_2 \cap E_3) = (.4)(.33)(.25) = .033$$

In the example where selection proceeded with replacement, the multiplication rule simplifies to

$$P(E_1 \cap E_2) = P(E_1) \cdot P(E_2) \tag{4.8}$$

because

$$P(E_2|E_1) = P(E_2) \tag{4.9}$$

Events that have this property (Equation 4.9) are said to be *independent*, because the selection of the first event does not influence the probability of the selection of the second. This is an extremely important property, because probabilities associated with independent events can be handled in a far simpler manner than when conditional probabilities have to be calculated. In fact, almost all inferential statistical methods are based on this property of independence of events or of independence of selection as we have illustrated it here, and thus sampling methods must be structured to incorporate this assumption.

Unions, the Addition Rule, and Mutually Exclusive Events

Having defined the probabilities associated with the intersection of one event and another, we now define another property relating two (or more) events—that of the *union* (or disjunction) of events. The union of two events is defined as the subset of the sample space whose cells belong either to the first event or to the second. The symbol $\cup$ will be used to represent these unions or additions.

Taking the same example and using the Venn diagrams (Figure 4.4) we see that the union of the two events E_1 and E_2 consists of the cells making up the first event, plus the cells making up the second event, minus the intersection of the two events (otherwise these cells would be counted twice). This relationship is defined as the *addition rule* of probabilities.

$$P(E_1 \cup E_2) = P(E_1) + P(E_2) - P(E_1 \cap E_2) \tag{4.10}$$

Applying this rule to our example,

Selection with replacement:

$$P(E_1 \cup E_2) = .4 + .4 - .16 = .64$$

That is, when the two events are independent (by selection with replacement) there are 64 chances out of 100 that one or both of the farms selected will be dairy farms.

Selection without replacement

$$P(E_1 \cup E_2) = .4 + .4 - .13 = .67$$

These can be checked by referring to the Venn diagrams (Figure 4.4), where in selection with replacement, 64 of the 100 cells ($p = .64$) belong to one or the other of the events, and in selection without replacement, 60 of the 90 cells ($p = .67$) belong to the union of the two events. The addition rule can be expanded to include any number of events and for three events it becomes:

$$\begin{aligned} P(E_1 \cup E_2 \cup E_3) = {} & P(E_1) + P(E_2) + P(E_3) - P(E_1 \cap E_2) \\ & - P(E_1 \cap E_3) - P(E_2 \cap E_3) + P(E_1 \cap E_2 \cap E_3) \end{aligned} \tag{4.11}$$

Applying this rule to our expanded example with three selections,

Selection with replacement:

$$P(E_1 \cup E_2 \cup E_3) = .4 + .4 + .4 - .16 - .16 - .16 + .064 = .784$$

Selection without replacement:

$$P(E_1 \cup E_2 \cup E_3) = .4 + .4 + .4 - .13 - .13 - .13 + .033 = .843$$

The addition rule states that in assessing the probabilities of any of a number of events occurring, or of an event occurring over a number of experiments, the individual probabilities of each event plus the probabilities of intersections have to be considered. However, events are said to be *mutually exclusive* if their intersection contains no cells, and thus the addition rule would simplify to

$$P(E_1 \cup E_2) = P(E_1) + P(E_2) \tag{4.12}$$

This is an important property, because it means that when events are mutually exclusive, we can calculate the probability of the union of these events simply by adding together their individual probabilities. In fact, we have already used this property in the calculation of the probability of selecting a dairy farm from a set of 10 farms. We know that the probability of the selection of any individual farm is $p = .1$; thus, the probability of selecting a dairy farm was

$$P(\text{farm 1} \cup \text{farm 2} \cup \text{farm 3} \cup \text{farm 4}) = .1 + .1 + .1 + .1 = .4$$

With events defined as individual farms in this instance, the fact that they are mutually exclusive events allows the use of the simplified addition rule (Equation 4.12). Again, inferential statistical methods usually assume that the events being considered are mutually exclusive and thus probabilities can be added.

The distinction between the concepts of independent and mutually exclusive events may seem confusing at first and perhaps a further illustration is warranted. These concepts are used in sampling methods and in the interpretation of samples in two different ways. Independent events refer to the situation where, if a particular event is selected in the first experiment, it will not modify (condition) the probabilities associated with a second event, and selection of the first and second events will not modify a third, and so on. This could mean that the selection of an individual element with a certain characteristic (belonging to the subset of that event) will not modify the probabilities of the selection in the next experiment or another individual element belonging to a second event. It could also mean that in the situation where there is an ongoing series of experiments, (selecting one individual at a time from a sample space), the probabilities of any individual being selected do not change from one experiment to another due to the influence of previous selections. As we have seen, this independence of events will occur only when the sampling process uses random selection with replacement.

On the other hand, mutually exclusive events refer to the probabilities of selection of events *taken at the same time*. The event (farm 1) and the event (farm 2) are mutually exclusive in that there is no possibility of intersection between these two events in any one experiment. Thus, the probability of selecting farm 1 *or* farm 2 is simply the addition of their two probabilities. As we have implied, mutually exclusive events rely on the careful demarcation of events so that no overlap occurs. Sampling in geographic problems rarely is concerned with events that are not mutually exclusive, but the assumption of independent events is sometimes violated.

4.3 RANDOM VARIABLES AND PROBABILITY FUNCTIONS

In Section 2.2, "Graphic Display," on the graphical presentation of frequency distributions, we noted that the relationship between the frequency of a class and the magnitude of the class itself (its midpoint) formed a type of mathematical function. *A mathematical function* can be defined as a measure of relationship (or correspondence or association) between one variable and another variable or variables. For any score on one variable, there is a corresponding score on the other variable. With univariate frequency distributions, the mathematical function is an expression of the relationship between scores on a variable (which we call x) and their associated frequencies. It is common practice to use lowercase symbols (such as x) when we are referring to general mathematical functions with the variables not specified exactly and uppercase symbols (such as X) when we are referring to specific variables. It is also convenient to specify any particular score for an X variable by a lowercase subscript. Thus, we derive expressions like "when ($X = X_i$)" meaning "when the variable X takes on the value X_i."

A general functional expression is written

$$y = f(x)$$

meaning, the variable y is a function of the variable x. Or, for particular scores on the variables X and Y,

$$Y_i = f(X_i)$$

the value of the ith score on Y is a function of the value of the ith score on the variable X. For a frequency distribution, we can relate the frequencies and the classes by the expression

$$Y_j = f(X_j) \tag{4.13}$$

where

Y_j = the frequency of the class j (previously abbreviated f_j)

X_j = the magnitude of the variable X for the class j, that is, the midpoint of the class (previously abbreviated M_j)

The idea of mathematical functions forms the basis for linking observed frequency distributions to a set of important functions developed from probability theory.

Random Variables

We define a *random variable* (often referred to as a *variate*), which we designate by X, as a variable that takes on numeric values ($x_1, x_2, x_3, \ldots, x_n$) within the range of all real numbers. If we examine this definition carefully, we see that we have simply redefined the term "population variable," which we have been using to describe a characteristic of a population. We have done so in a little more formal manner, and with two important additions. One is that we now refer to the variable as a *random* variable, and the other is that we have restricted its use to variables that take on *numeric* values only. These modifications require amplification.

The addition of the adjective "random" might seem to imply that the numeric values that the random variables take on are randomly distributed, but this is not quite correct. What is implied is that the values that the random variable takes on have a certain probability of occurrence—and these probabilities can be estimated from the distribution (a relative frequency) of the values taken by a set of randomly selected individual elements. In addition, it is assumed that these probabilities of occurrence of the individual numeric values may be related in some simple mathematical functional form.

The restriction to numeric values only suggests that the concept of random variables can only be applied to enumerated and metric variables where the scores for the variables are obviously numeric. However, classified variables can be considered to be random variables if we can assign realistic numeric values to the scores (illustrated shortly for binary variables), or where we can restructure the variable so that we can substitute frequency of occurrence of each of the scores in place of the scores themselves.

The example we have been using is the nominally classified variable Farm Type. In fact, it is set up as a binary variable, where each farm falls into one of two possible classes: either a dairy farm or a nondairy farm. A binary variable can be considered to be a random variable in two different ways. If we assign a numeric value of 1 when the variable takes on the score of "dairy farm," and the value of 0 when it is nondairy farm, then Farm Type is a random variable by definition, taking on the numeric values 1 and 0. The use of the binary 0/1 as valid numeric scores for nonnumeric variables has important extensions, as we will see in the discussion of multivariate techniques.

An alternative form of developing a random variable from a nominally classified or ordinally classified variable is to utilize the frequencies of a certain outcome (event) from a series of experiments. This method is capable of handling variables with more than two classes, although, for the moment, we retain for illustration the binary variable Farm Type. For example, if we carry out a random selection of a single farm from our population of 10 farms four times, then there are five possible outcomes to the series of four experiments (selections); we could select 4, 3, 2, 1, or no dairy farms. That is, we have structured a random variable, which strictly speaking should be called something like "selection of dairy farms," and which may take on the numeric values 4, 3, 2, 1, or 0.

In the same way that we divided population variables into two types, random variables can be subdivided into (1) *discrete random variables,* which may take on only a collection of fixed values, usually 0 and the set of positive integers, and (2) *continuous random variables* which may, in principle, take on any conceivable real value, although some may be bounded by an absolute maximum or minimum value. The distinction, as before, is based on whether the variable is classified or enumerated (and therefore is dealing with frequencies), or whether it is metrically measured and can therefore, theoretically at least, take on any value depending on the precision of measurement.

Probability Functions for Discrete Random Variables

For many random variables, there exist mathematical functions that describe the probabilities associated with each numeric value that the random variable can assume. These mathematical functions have been called probability laws, probability rules, or probability distributions; but, for the moment, we call them *probability functions*. Before we express these functions mathematically, it is useful to examine how these are structured in terms of the Venn diagrams we have been using. Taking the example of the binary variable Farm Type, we examine the probabilities associated with the two numeric values (0 and 1) that it can take on. Because we have complete information about the 10 farms, we know that the probability associated with the numeric value 1 (selection of the event E—dairy farm) is .4, and that associated with the numeric value 0 is .6. If we reconstruct the Venn diagram in the form of two columns, representing the numeric values of 0 and 1 (demarcated along the x axis) with the heights of the columns representing the associated probabilities (Figure 4.6), we have constructed a histogram of the distribution of probabilities.

In the same way as in our original Venn diagram (Figure 4.1), each cell of the sample space represents an individual farm and its probability, and the height of the columns represent the probabilities of each of the numeric values taken on by the random variable X (Figure 4.6*a*). Note that the widths of the columns in the Venn diagram have no significance and it would be more accurate to represent the probabilities by vertical lines (Figure 4.6*b*) rather than columns (in fact, that is the more usual method of representation). However, we illustrate them in this form because they emphasize an important property—that the total area of the histogram is made up of 10 cells of equal area, and therefore the proportion of the total area lying in each of the columns also represents the probabilities of each of the numeric values. The Venn diagram we have constructed illustrates a probability relationship we call a *probability mass function,* which is also capable of mathematical description. We use the functional symbol $f(x)$ to describe the probabilities associated with values of the random variable X, and the definition of a probability mass function becomes

$$y = f(x) = P(E) = P(X = x) \tag{4.14}$$

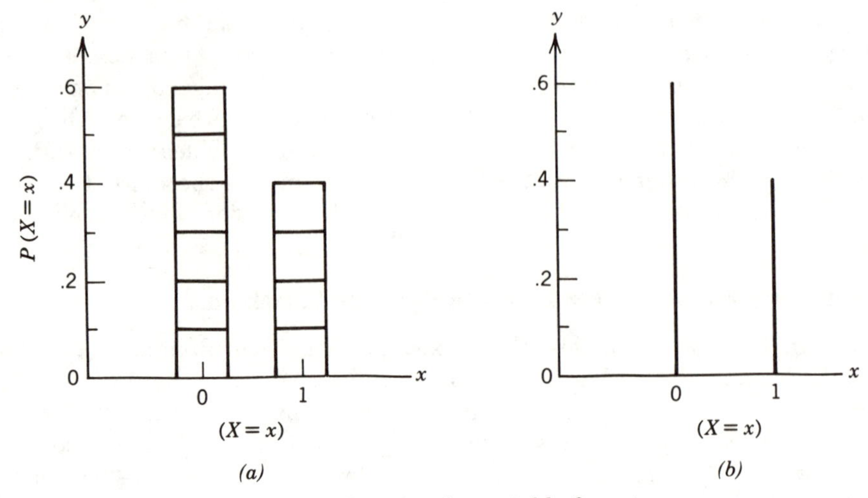

Figure 4.6 Probability mass function for variable farm type.

In our example, we need to derive a functional equation that will denote a value of

$$y = .6 \text{ when } (X = 0)$$

and

$$= .4 \text{ when } (X = 1)$$

This can be easily derived as

$$y = (.4)^x \, (.6)^{1-x} \text{ for } (x = 0,1) \tag{4.15}$$

If we use p to denote the probability of the selection of a dairy farm, and $(1 - p)$ would therefore be the probability of the selection of a nondairy farm, we can rewrite Equation 4.15 in a general form that defines a probability mass function for all binary variables of this type. Such a probability function is given the name the *point binomial probability function.*

$$y = f(x) = p^x \, (1 - p)^{1-x} \tag{4.16}$$

The point binomial function has only limited application, but its extension into the more general form of the binomial probability function is of considerable importance. This extension relates to the use of frequency of occurrence of particular outcomes from a series of experiments or selections to derive a probability function. Although we discuss this derivation using theoretical probabilities, the extension to observed frequencies (and empirical probabilities) is readily apparent. Using the same population (sample space) of 10 farms, and taking a simple example, we examine the probabilities associated with the possible outcomes of a series of four sequential trials, each consisting of the selection of one farm out of the total sample space. It is thus concerned with four independent and mutually exclusive events.

Consider first the selection process related to individual farms. In the initial selection, any of the 10 farms could be chosen; in the second selection (assuming sampling with replacement), again any of the 10 farms could be chosen; thus, the sample space of the first two selections has $(10 \times 10 =)$ 100 possible outcomes. Continuing the process for the third and fourth selections, yields $(10 \times 10 \times 10 \times 10 =)$ 10,000 possible outcomes. More generally, in n selections from a discrete and finite population of N elements, using sampling with replacement, the number of possible intersections equals N^n. We note that if we had sampled *without* replacement, the total number of possible outcomes would be somewhat smaller, given by $N!/(N - n)!$ which in our example is $10!/(10 - 4)! =$

$$\frac{10!}{6!} = 10 \times 9 \times 8 \times 7 = 5040$$

where $k!$ represents the *factorial* of the number k, and is defined as the product of all the integers from 1 up to and including k. Thus, for example, $4! = 1 \times 2 \times 3 \times 4 = 24$. By definition, $0! = 1$. The equation just given is, of course, the formula for the number of *permutations* of n objects selected at one time from a population of N objects. Sequential selection without replacement is identical to sampling of all n objects at the same time.

Obviously, the simple Venn diagrams we have been using will not be of much practical assistance in deriving probabilities, because we are dealing with large numbers of possible outcomes. Because the results of the selection process are mutually exclusive and independent (again assuming replacement), we note that the probabilities of each of these joint outcomes is identical (in fact, equal to $1/10000 = .0001$), and the probability of a particular joint outcome is simply the sum of the individual probabilities or, in other words, the probability of an intersection (.0001) multiplied by the frequency of outcomes that belong to that combination.

However, our interest is not with the probabilities related to individual farms and their selection but with the probabilities related to the selection of dairy farms versus nondairy farms. Let us look at 3 of the 10,000 possible outcomes, listing the farms in the order in which they were selected.

One possible selection:	farms 3, 9, 6, then 9
A second possible selection:	farms 9, 9, 3, then 6
A third possible selection:	farms 2, 5, 8, then 6

Examining these three results, remembering that we numbered the dairy farms 1 to 4 and the nondairy farms 5 to 10, and denoting the selection of a dairy farm by D and a nondairy farm by N, our selections are $DNNN$, $NNDN$, and $DNNN$, respectively. In fact, it is a simple matter to show that there are 16 possible distinct outcomes resulting from the selection of 4 farms from the population of dairy farms and nondairy farms. This can be illustrated by a *probability tree diagram* (Figure 4.7), and, because each experiment results in a binary outcome, the number of distinct possible joint outcomes of farm types in n selections is given by 2^n, or in our example, $2^4 = 16$.

The probability of each of these distinct joint outcomes, however, will not be the same because we know that the probability of an individual selection of a dairy farm is .4, and that of a nondairy farm is .6. Common sense tells us that it would be much more likely to have selected a *DNNN* intersection than a *DDDD*. But before we examine these probabilities, we take one further step in simplifying the process of the derivation of the probabilities associated with these selections. All three of the examples of farm selection cited above are *combinations* of one dairy farm and three nondairy farms, and as the selection process defines independent events, the order of selection will have no conditioning influence on the probability of that combination occurring, which will simply be the product of the individual probabilities of each p and $(1 - p)$. Thus, the 16 possible outcomes reduce to five combinations; or generally, the 2^n intersections reduce to $(n + 1)$ combinations. An examination of the probability tree diagram (Figure 4.7) shows that the five combinations—4 dairy farms (*DDDD*), 3 dairy farms (*DDDN*), 2 dairy farms (*DDNN*), 1 dairy farm (*DNNN*), and no dairy farms (*NNNN*)—have frequencies of possible occurrence of 1, 4, 6, 4 and 1, respectively. The probability of selection of a particular combination is thus a product of the probability of that combination and its frequency of possible occurrence.

The probability function that describes this type of situation is based on the binomial theorem. The expansion of the binomial term $(p + q)^n$, where $q = 1 - p$, and n = the number of selections or trials, provides the individual probabilities for each combination. For example, with $n = 4$, the binomial expansion is

$$(p + q)^n = (1)[p^4] + (4)[p^3q] + (6)[p^2q^2] + (4)[pq^3] + (1)[q^4]$$

where the terms in parentheses refer to the frequency of possible occurrence—the *binomial coefficients*—and the bracketed terms refer to the individual probabilities of a particular combination. In our example, with

$$p = P(\text{dairy farm}) = .4 \qquad \text{and} \qquad q = 1 - p = P(\text{nondairy farm}) = .6$$

$$1.0 = (1)(.4)^4 + (4)(.4)^3(.6) + (6)(.4)^2(.6)^2 + (4)(.4)(.6)^3 + (1)(.6)^4$$

$$= .0256 + .1536 + .3456 + .3456 + .1296$$

The values that result describe the probabilities of each combination sequentially. Thus, for example, the probability of selection of 4 dairy farms is .0256, and of 3 dairy farms (and 1 nondairy farm) is .1536. We note that this also describes the sum of the probabilities of the selection of individual farms. Of the 10,000 possible outcomes from 4 selections of the 10 different

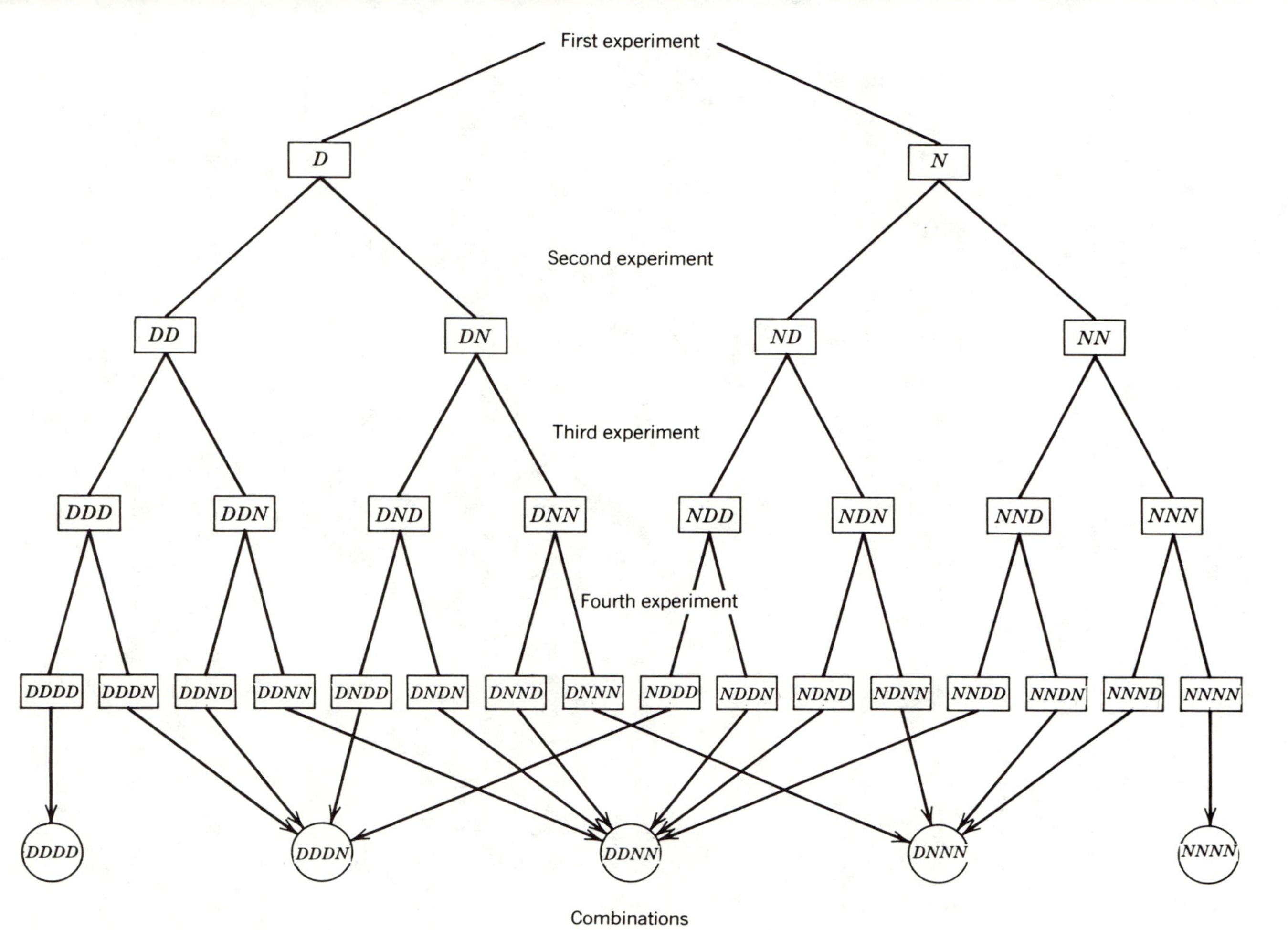

Figure 4.7 Probability tree diagram for four experiments.

farms, 256 involve only dairy farms (farms 1 to 4), 1536 involve 3 of the dairy farms and 1 of the nondairy farms, and so on.

Although the expansion of the binomial term $(p + q)^n$ provides a solution to the derivation of probabilities associated with the $(n + 1)$ possible combinations, it is not in a form that is amenable to simple calculation. For practical purposes, it can be restructured into its two separate parts of deriving the individual probabilities and of determining the frequencies of possible occurrence, and the whole process of deriving a *binomial probability function* can be set up in a series of four steps.

1. Determine the different $(n + 1)$ combinations that can occur. These will define the "classes" for the probability distribution.
2. For each combination, determine the number of possible occurrences (the binomial coefficients)—that is, the frequencies of each combination or class.
3. For each combination determine the probability of an individual outcome.
4. Calculate the total probability of a combination occurring by multiplying its individual probability by its number of occurrences.

The task has been greatly simplified. In our example, there are five different types of combinations (Step 1): combinations containing 4, 3, 2, 1, or 0 dairy farms (and, correspondingly, 0, 1, 2, 3, or 4 nondairy farms). The number of occurrences of each of these combinations (Step 2) can be found by constructing a *probability tree diagram,* by using a device known as *Pascal's triangle* (which sets up a simple relationship between the binomial coefficients at different powers of n), or by using the equation for determining numbers of possible occurrences. The latter method is the only one of practical importance.

$$\text{Number of possible occurrences of a combination} = \frac{n!}{x!(n - x)!} \tag{4.17}$$

where

n = the number of selections (4 in our example)

x = the given value of the random variable X (0, 1, 2, 3, or 4 in our example)

As an example, using Equation 4.17 to calculate the number of combinations present when $X = 3$ (3 dairy farms and 1 nondairy farm selected)

$$\text{Number of possible occurrences with } X = 3$$

$$\frac{4!}{3!(4 - 3)!} = \frac{24}{6} = 4$$

Using the fact that the events are independent (given sampling with replacement), the probability of an individual combination occurring (Step

3) can be found by multiplying the probabilities of the individual events comprising the combination. For example, for the combination of three dairy farms and one nondairy farm, this is

$$P(DDDN\text{—in any order}) = (.4)(.4)(.4)(.6)$$
$$= (.4)^3(.6)^1 = .0384$$

We note that this is equivalent to raising p (= .4) to the power of x, multiplied by $(1 - p)$ (= .6) to the power of $(n - x)$ where n equals the number of selections and x equals the given value of X. Thus, the calculation of the individual probabilities is given by

$$p^x (1 - p)^{n-x} \tag{4.18}$$

The probabilities for each individual combination can then be calculated. Finally, the number of combinations and the individual probabilities can be multiplied together (Table 4.2) to provide the total probability for each combination (Step 4). The resulting probability mass function is graphed in Figure 4.8.

By combining the equation for the number of combinations (4.17) with that for the probability of the individual combination (4.18), we derive the *binomial probability mass function*. For n experiments and with P(an event, X) equal to p,

$$P(X = x) = y = f(x) = \frac{n!}{x!(n - x)!} p^x(1 - p)^{n-x} \text{ for } x = 0,1, \ldots, n \tag{4.19}$$

We thus have a probability function that describes the distribution of successes from a series of experiments (selections) from a binary population variable.

We have spent considerable time deriving the simple point binomial probability function and its extension to the general binomial probability func-

TABLE 4.2 Calculation of Probabilities for Binomial Function

Number of Successes x	Number of Possible Occurrences $\frac{4!}{x!(4-x)!}$	Probability of Individual Intersection $p^x(1-p)^{4-x}$	Probability of Combination $P(X = x)$
0	1	$(.4)^0\ (0.6)^4 = .1296$	.1296
1	4	$(.4)^1\ (0.6)^3 = .0864$	.3456
2	6	$(.4)^2\ (0.6)^2 = .0576$	.3456
3	4	$(.4)^3\ (0.6)^1 = .0384$	.1536
4	1	$(.4)^4\ (0.6)^0 = .0256$	.0256
TOTAL	16		1.0000

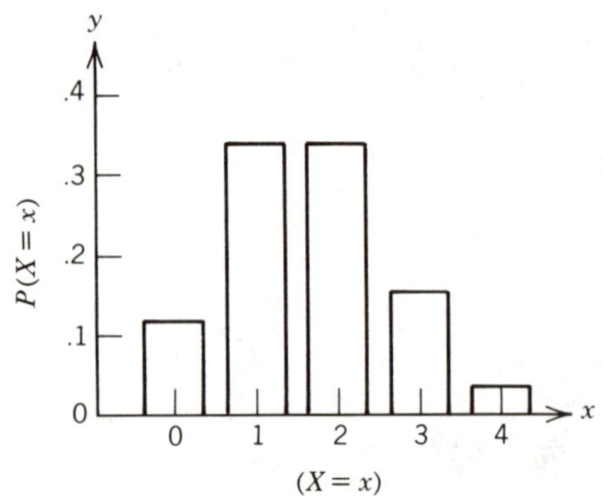

Figure 4.8 Probability mass function for binomial ($p = .4$, $n = 4$).

tion, and we have done so not because these are the most important discrete probability functions in geographic research but primarily because they provide the simplest examples to illustrate the derivation of probability functions in general. In addition, some of the more important probability functions that are used are extensions of the binomial probability function. These will be examined in Section 4.4.

Probability mass functions have two important properties, which can be seen from their graphed functions (Figures 4.6 and 4.8).

$$\text{I } f(x_j) \geq 0 \qquad \text{for all } x_j \tag{4.20}$$

$$\text{II } \Sigma f(x_j) = 1.0 \qquad \text{for all } x_j \tag{4.21}$$

An alternative form of expressing the probability functions we have been describing is to present them in cumulative form. In certain situations, as we will see later, this modified form will be easier to utilize. The graphs in cumulative form for the two examples that have been cited are illustrated in Figure 4.9. These functions are designated *probability distribution functions,* and, using $F(x)$ to describe the function, they are defined as

$$y = F(x) = P(X \leq x) \tag{4.22}$$

These distribution functions have the following properties:

$$\text{I } F(-\infty) = 0 \tag{4.23}$$

$$\text{II } F(\infty) = 1.0 \tag{4.24}$$

$$\text{III } F(x_1) \leq F(x_2) \text{ if } x_1 \leq x_2 \tag{4.25}$$

$$\text{IV } P(a < X < b) = \Sigma f(x_j) \text{ for all } x_j \text{ where } (a < x_j < b) \tag{4.26}$$

$$= F(b) - F(a) \tag{4.27}$$

This last property especially will prove of immense value.

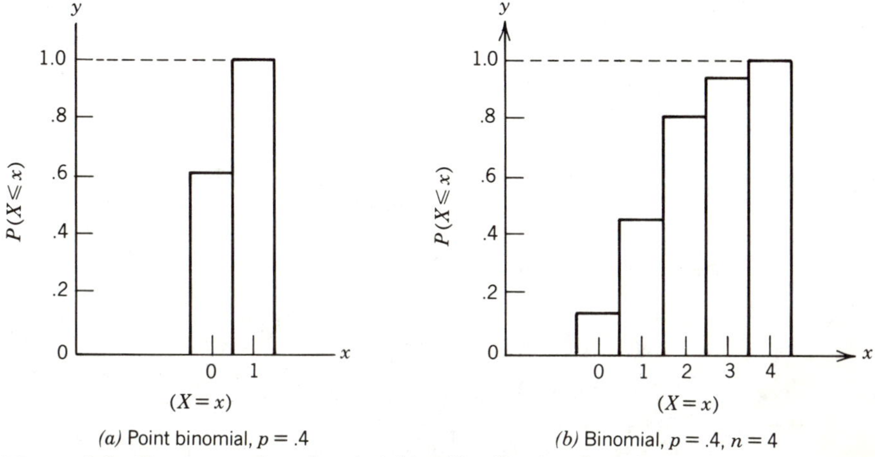

Figure 4.9 Two examples of probability distribution functions.

Probability Functions for Continuous Random Variables

The same principles developed for discrete random variables holds true for continuous random variables, but the interpretation and derivation is slightly different and perhaps a little more complicated. As an example of a continuous random variable, let us take the variable Land Value per Square Meter from the Urban Land Value Survey as used in Chapters 2 and 3, where we had scores for 86 property lots, each score taking on a numeric value within the range \$32 to \$274. If we reconstruct the histogram (Figure 2.5) in the form of a Venn diagram with 86 cells (Figure 4.10), we can use this diagram to illustrate some of the associated probabilities. If we were to take one selection from our population of 86 property lots, the probability of the selection of *any* one property lot must be 1/86 or approximately .0116. Because selections will be mutually exclusive, the probability of selecting a property lot with a value within the range \$65 to \$99 (the second class in the distribution) will be 26/86 or .3023; that is, the sum of the individual probabilities of selection of each of the 26 lots in that class ($26 \times .0116 =$ approximately .30). Thus, we can use the observed *relative* frequencies and the histogram in the same way as we did for discrete random variables—to determine the probabilities associated with each class.

However, there is one important distinction. For the discrete random variable the class was a single score and the width of the bar in the histogram had no numeric value and was used to illustrate the distribution, whereas for the continuous random variable the width of the bar is scaled and represents a range of possible scores, and a particular score could fall anywhere within that range. Two questions arise. First, what is the probability of selecting a property lot with a value of, say, \$74? A little thought should indicate that its probability of selection would be zero for all practical pur-

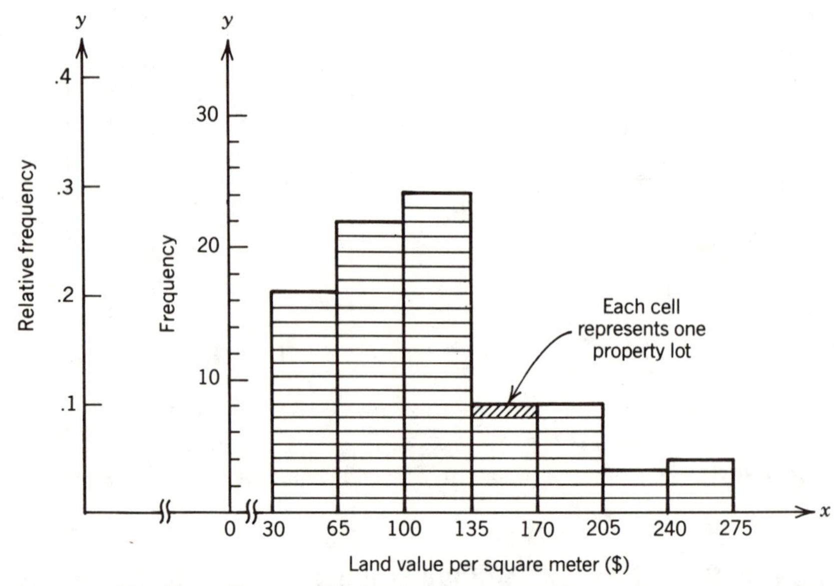

Figure 4.10 Venn diagram for land value per square meter.

poses. Because of the measurement process there is no chance that a property lot would have a value of exactly $74, even if one has been listed approximately at that value. Second, what is the probability of selecting a lot with a value between $65 and $82? Referring back to our original set of 86 scores, we could soon count up the number of lots within that range and then determine the probability. But if this detailed information were not available (which would be the case if our data had been sampled), we could use the information obtained from the graphed observed distribution. Because the range $65 to $82 covers approximately one-half the range of the class $65 and $99, then a reasonable estimate of the probability might be .1512, half of the probability of that class. Better estimates can be made, however. Classes lying between a class with a higher frequency and a class with a lower frequency would probably have more individuals in the half of the class interval lying adjacent to the larger class frequency (Figure 4.11).

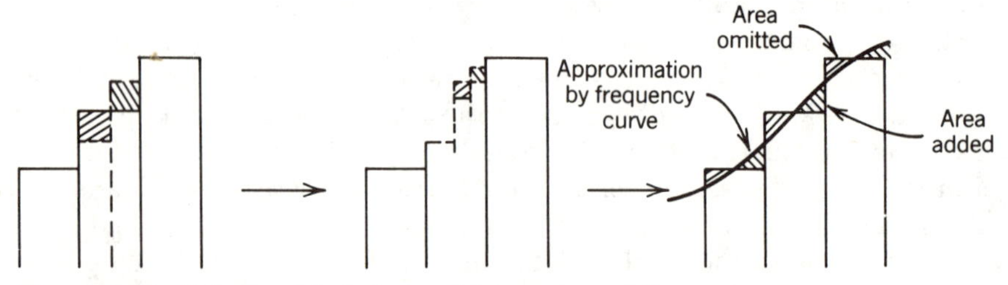

Figure 4.11 Relationship between histogram and frequency curve.

Within that half (adjacent to the larger class frequency), there are probably more individuals in the fourth adjacent to the higher class and so on for as many subdivisions as we care to pursue (assuming that the total number of individuals was large). This means that the approximation of a histogram by a frequency curve would usually provide a more realistic picture of the distribution by averaging out the sharp breaks suggested by the class boundaries. The accuracy of this approximation increases as the sample size gets larger and as the size of the class interval gets smaller. A frequency curve, then, provides a means of estimating probabilities for any interval of scores within the distribution.

Two important corollaries follow. Because the histogram is made up of columns, the width of which represent a range of scores and not a single score as for discrete random variables, the simple types of mathematical functions described in the preceding section cannot be applied. However, for many frequency curves, mathematical functions can be derived to describe the curve. Such functions (the equivalent of the probability mass function of discrete random variables) are called *probability density functions.* A second corollary concerns the area of the sample space. Looking at the Venn diagram (Figure 4.10) with 86 cells of equal area, each representing a probability of selection of .0116 for an individual lot, the total *area* of the sample space will be 1.0, and the probability of selecting a lot within a class will simply be the area of the sample space falling into that class as a proportion of the total area. If the Venn diagram (histogram) is in relative frequency form, then this proportioning has already been done. Thus, probabilities are given by the proportion of the total area within the desired range of numeric values. The frequency curve approximation to the histogram will also have this property, because the areas of the histogram lying above the curve will usually be compensated for by including areas below the curve that do not fall into the bar of the histogram (Figure 4.11). Thus, the total area beneath the curve will be approximately 1.0. Probabilities can then be calculated for any interval by measuring the area lying beneath the curve in that interval and expressing it as a proportion of the total area (Figure 4.12). Graphically, this would be a tedious process, but the use of the *integration* of a curve provides a mathematical means of carrying out this estimation.

We do not attempt here to describe the principles and methods of integration. To the reader not familiar with calculus, we note that the integration of a mathematical function between two numeric values can be described simply as the calculation of the area beneath the curve between the two values. The integration sign $\int$ is used to signify the process, with the two limiting values placed below and above it, and what is being integrated (the mathematical function) placed to the right, as follows.

$$\int_a^b f(x)dx$$

The foregoing expression would indicate that the integration process is

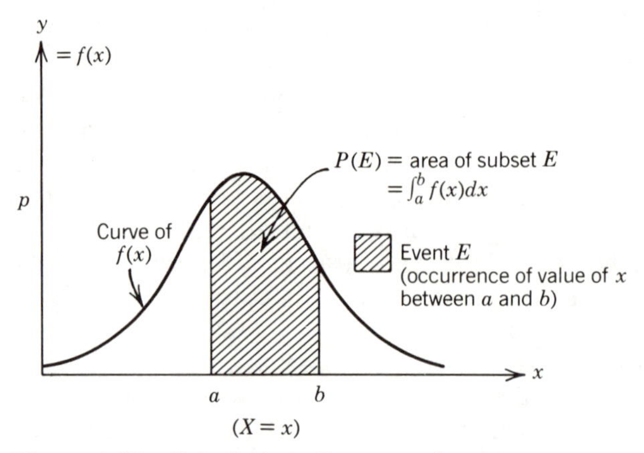

Figure 4.12 Calculation of areas under a curve.

to be used to calculate the area under the curve described by the mathematical function $f(x)$, along the x axis (for all values of x denoted by dx) from a numeric value labeled a to a numeric value labeled b (Figure 4.12). However, because most of the probability density functions with which we are concerned have been tabled, calculations of this type are rarely necessary.

There is a wide variety of different probability density functions and no attempt will be made to show how these are derived. However, there is one probability density function that, because of its critical importance in statistical analysis, deserves at least an illustration of its derivation. This is the *normal* (or *Gaussian*) *probability density function.* Originally conceived as a means of describing the outcomes in games of chance, and later as a method of deriving the probability of accidental errors in the measurement process, in modern statistical theory, this function forms the central concept of many of the techniques in the analysis of data distributions. It can best be illustrated by referring back to probabilities associated with discrete random variables and particularly to the binomial probability function.

Let us take a simple example of an experiment utilizing the binomial probability function. We make three basic assumptions: that we are dealing with a binary variable, that the probability of the event and its complement are equal, and that successive selections (experiments) are independent and mutually exclusive. A commonly used example is tossing a coin (with the event E being the result "head," and its complement $\tilde{E}$ a "tail"). Using the terminology developed in the last section, the event E has a probability of $P(E) = p = .5$, and $P(\tilde{E}) = 1 - p = q = .5$. If we carry out a series of experiments (tossing of the coin), our distribution of possible combinations of results is given by the binomial probability mass function or the expansion of the binomial term $(p + q)^n$. Graphs for $n = 2$, $n = 10$, and $n = 20$ are shown in Figure 4.13. Each of these density functions is symmetrical and as n increases, the histogram becomes more smoothed and bell shaped, and

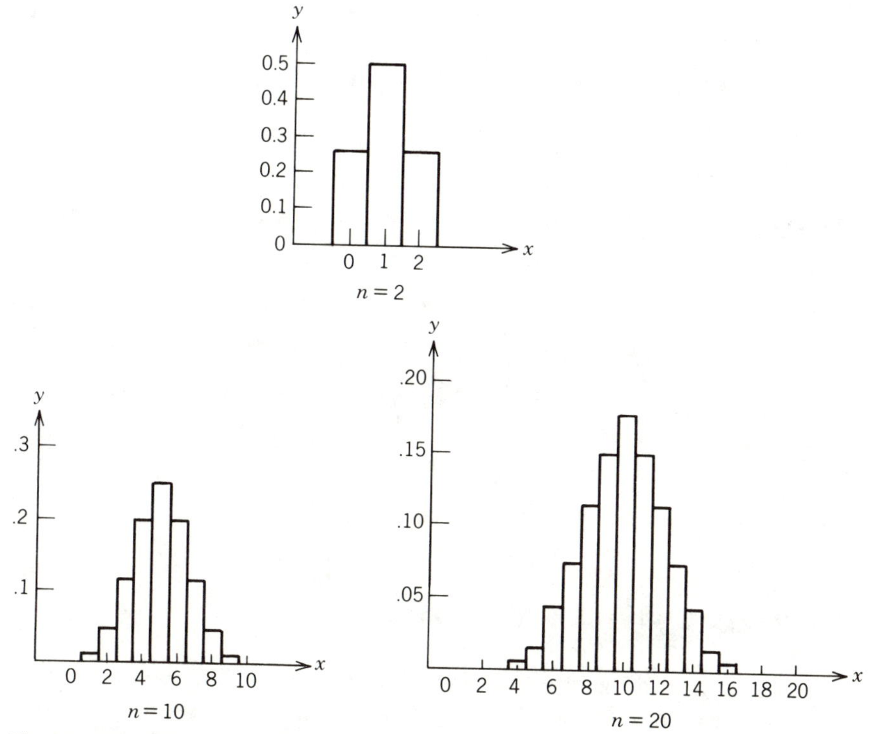

Figure 4.13 Three examples for binomial probability function with $p = .5$.

can be represented by a frequency curve of increasing accuracy. We can show that as n approaches infinity, the curve of $(.5 + .5)^n$ approaches as a limit—the curve of the normal probability density function. In fact, even if we relax the first two assumptions and make $p \neq q$, or increase the number of possible outcomes at each experiment to greater than two (multinomial rather than binomial), the limit as n approaches infinity of both these distributions is also the normal probability density function. The importance of this function is that it approximates the frequency distributions derived from a wide variety of different practical situations, whenever events making up the possible outcomes can be described as independent and mutually exclusive. Its place in statistical methods should become apparent in Chapter 6.

The definition of any *probability density function* is

$$y = P(E) = f(x) = \int f(x)dx \text{ for all } x \text{ contained in the event } E \quad (4.28)$$

Probability density functions have the following properties

$$\text{I } f(x) \geq 0 \text{ for all } x \quad (4.29)$$

$$\text{II } \int_{-\infty}^{\infty} f(x)dx = 1.0 \quad (4.30)$$

In the same way as for discrete random variables, the probability density function can be rewritten and redrawn in cumulative form to define a *probability distribution function*. By definition, this will be

$$y = F(x) = \int_{-\infty}^{b} f(x)dx \text{ for all } x \leq b \tag{4.31}$$

The properties of the probability distribution function for continuous random variables are

$$\text{I } F(-\infty) = 0 \tag{4.32}$$

$$\text{II } F(\infty) = 1.0 \tag{4.33}$$

$$\text{III } F(x_1) \leq F(x_2) \text{ if } x_1 \leq x_2 \tag{4.34}$$

$$\text{IV } P(a < x < b) = \int_{a}^{b} f(x)dx \tag{4.35}$$

$$= F(b) - F(a) \tag{4.36}$$

4.4 UNIVARIATE PROBABILITY FUNCTIONS

One of the major aims of inferential statistics is to obtain a random sample from a population and to use this sample to make probabilistic conclusions about the population as a whole. To do this, we need to know something about the characteristics of the population from which the sample was drawn—what probability function best describes the population and therefore what probabilities are associated with each possible outcome of selection.

The mathematical functions that describe these probability distributions are simply functional equations describing the relationship between possible outcomes (numeric values of X) and their probability. As a functional equation, they consist of a set of parameters with the parameters varying from one example of the probability function to another. There is, then, a variety of probability functions, and for each probability function there is a variety of forms of the function depending on the values of the parameters. For example, one simple probability function was the point binomial with a single parameter p that represented the probability of selection of an individual belonging to a particular event. The form of the function, both graphically and by equation, will vary as the value of the parameter p varies. Some authors call the general function, such as the point binomial, a *probability family* or a *probability law* and particular forms of the probability law, such as the point binomial when $p = .4$ as a *probability function*. We will simply retain the use of the general term *probability function*. It is also common to find probability functions referred to as *probability distributions*, even when it is the mass or density function that is being discussed.

Probability Functions and Statistical Methods

There is a large number of different probability functions, but fortunately only a few are of major practical importance. There are, basically, two reasons for describing population distributions by probability functions: (1) by assuming that a population variable is distributed according to a particular probability function we can use the probability function to make inferences about the characteristics of the population; (2) by hypothesizing that a particular population distribution results from the operation of one or more processes that can be described by a probability function, this can then be tested and either retained or rejected. These are the aims of inferential statistics and probability model building, respectively.

To some extent, geographers have been forced to consider a wide variety of probability functions in an endeavor to find the probability function that best approximates the population distribution in which they are interested, or from which they assume the sample has been selected. This has been especially true with spatial samples. However, there are few probability functions that are frequently used, and these have a wide variety of applications simply because most phenomena—whether they be biological, geological, psychological, etc.—tend to distribute approximately according to these functions. That is, phenomena distribute according to a few "natural laws" that span a wide variety of disciplines, and there is no reason to assume that most geographic phenomena (of whatever kind) will not follow similar distributions, even if this has not been determined as yet. Moreover, some functions can be approximated by the normal probability function under certain conditions (usually large sample size), and thus the normal distribution can be used in their place. And, perhaps most important of all, certain functions (again especially the normal probability function) have a number of properties that allow their use regardless of the underlying population distribution. This will be taken up in Chapter 6. Finally, if certain information is available, we may use probability ideas without reference to the underlying probability function (the distribution-free approach).

Expected Values and Moments of Probability Functions

To determine the form of a particular probability function from experimental data requires the estimation of its parameter(s). As an aid in this determination or estimation, the summarizing properties of the moments of a statistical distribution (Section 3.5) can be extended to include the characteristics of probability functions. These properties provide greater assistance when they are redefined in terms of mathematical expectation, a concept that will have application in later chapters. The *expected value* of any function of a random variable, denoted by $E[\ ]$, is defined as the weighted average (weighted by the probability of occurrence) of the function over all possible values of the variable; or expressing it in a different way, it is the average value obtained by infinitely repeated sampling from the population distribution.

When applying the system of moments to probability functions we can make one important simplification. Probability functions have been defined in terms of *relative frequencies,* thus the size of the population (N) is not required in deriving the "average" moment, and the moment will simply be the sum of the "distance" measurement times the relative frequency. For any random variable X, the expected value for X^n is given by

$$E[X^n] = \alpha^n$$

where

α^n is the moment around zero

In particular,

$$E[X] = \alpha^1 = \text{the arithmetic mean} = \mu \tag{4.37}$$

That is, the expected value of a set of scores for the random variable X is the first moment about zero, the arithmetic mean. The symbol μ (Greek lowercase mu) has been given to this summary measure, defined in this manner.

A similar relationship can be constructed for the central moments about the arithmetic mean. For a random variable X and its arithmetic mean μ, the expected value for $(X - \mu)^n$ is given by

$$E[(X - \mu)^n] = \mu_n$$

where

μ_n is the central moment about μ

In particular,

$$E[(X - \mu)^2] = \mu_2 = \text{the variance} = \sigma^2 \tag{4.38}$$

where σ is Greek lower case sigma.

The two measures—arithmetic mean and variance—can be used to assist in the process of fitting probability functions to observed data, a topic to be examined in Chapter 6. It can be shown that the arithmetic mean and the variance can be found for almost all probability functions by using the method of moments.

For discrete random variables:

$$\mu = E[X] = \sum_{j=1}^{n} x_j f(x_j) \tag{4.39}$$

where

$$f(x_j) = P(X = x_j) \text{ for all } x_j = 1, 2, \ldots n$$

Note that the $x = 0$ term drops out as x_0 will be zero

and

$$\sigma^2 = E[(X - \mu)^2] = E[X^2] - (E[X])^2$$
$$= \sum_{j=1}^{n} x_j^2 f(x_j) - \left(\sum_{j=1}^{n} x_j f(x_j)\right)^2 \quad (4.40)$$

For continuous random variables:

$$\mu = E[X] = \int_{-\infty}^{\infty} x f(x)dx \quad (4.41)$$

and

$$\sigma^2 = \int_{-\infty}^{\infty} x^2 f(x)dx - \left(\int_{\infty}^{\infty} xf(x)dx\right)^2 \quad (4.42)$$

To illustrate the derivation of these moments, we limit our discussion to a simple example—that of the moments of the binomial probability function as discussed in Section 4.3. The same principles apply to the derivation of moments for other discrete and continuous random variables. Using the example where four experiments (selections) were taken (Table 4.2), the values of x_j and $f(x_j)$ and some derived values are calculated (Table 4.3). The mean and variance can then be determined. From Equation 4.39,

$$\mu = \sum_{j=1}^{n} x_j f(x_j) = 1.6$$

From Equation 4.40,

$$\sigma^2 = \sum_{j=1}^{n} x_j^2 f(x_j) - \left(\sum_{j=1}^{n} x_j f(x_j)\right)^2$$
$$= 3.52 - (1.6)^2 = .96$$

The calculation of means and variances in this manner can be tedious, and for most probability functions, the equations for the moments of the distribution can be simplified by replacing the $f(x)$ portion of the equations by the mathematical function itself. Thus, in the case of the binomial function for the arithmetic mean μ, this becomes

$$\mu = \sum_{j=1}^{n} \left((x_j)\left(\frac{n!}{x!(n - x)!} p^x (1 - p)^{n-x}\right)\right)$$

which, by summing for all n of the x values, reduces to

$$\mu = np$$

TABLE 4.3 Mean and Variance Calculation—Binomial Function (p = .4, n = 4)

Number of Successes x_j	Probability of Combination $P(X = x_j) = f(x_j)$	x_j^2	$x_j\, f(x_j)$	$x_j^2\, f(x_j)$
0	.1296	0	0.0	0.0
1	.3456	1	.3456	.3456
2	.3456	4	.6912	1.3824
3	.1536	9	.4608	1.3824
4	.0256	16	.1024	.4096
TOTAL	1.0000		1.6000	3.5200

and, similarly,

$$\sigma^2 = np\,(1 - p)$$

Thus, in our example,

$$\mu = 4(.4) = 1.6$$

and

$$\sigma^2 = 4(.4)(.6) = .96$$

Basic Probability Functions

No attempt will be made to provide a comprehensive account of the various types of probability functions that could be employed by geographers. Instead, we will limit discussion to the binomial and normal functions and some modifications of these basic probability functions. The functional form of these probability mass functions or probability density functions is outlined in Table 4.4 and some examples are illustrated in Figure 4.14.

The *point binomial probability function* is simply the limiting case of the binomial probability function with $n = 1$. Thus, it has a limited application in the description of the *proportions* of results that fall into one of two categories.

The *binomial probability function* is the basic discrete probability function and was outlined in detail in Section 4.3. Its use is limited to situations where there are repeated random samples of a specified size from a discrete population where each selection results in a binary (binomial) decision, and the events are independent of each other—a situation that is only infrequently encountered in geographic analyses. Where its use in geographic analyses would be suggested, we are usually dealing with large sample sizes, and other distributions (the Poisson and the normal) that can approximate the binomial can handle these large frequencies much more easily.

TABLE 4.4 The Important Probability Functions

Function	Functional Form	Restrictions	Moments	
			Mean	Variance
DISCRETE				
Point binomial $X_i \sim PB(p)$	$y = f(x) = p^x (1 - p)^{1-x}$	$x = 0, 1$; $0 < p < 1$	p	$p(1 - p)$
Binomial $X_i \sim B(p, n)$	$y = f(x) = \frac{n!}{x!(n - x)!} p^x(1 - p)^{n-x}$	$x = 0, 1, 2, \ldots, n$; $n = 1, 2, 3, \ldots$; $0 < p < 1$	$n \cdot p$	$n \cdot p(1 - p)$
Poisson $X_i \sim P(\lambda)$	$y = f(x) = \frac{e^{-\lambda} \lambda^x}{x!}$	$x = 0, 1, 2, \ldots$; $\lambda > 0$	λ	λ
CONTINUOUS				
Normal $X_i \sim N(a, b)$	$y = f(x) = \frac{1}{b\sqrt{2\pi}} \exp(-(x - a)^2/2b^2)$	$-\infty < x < \infty$; $b > 0$	a	b^2
Standard normal $Z_i \sim SN(0, 1)$	$y = f(z) = \frac{1}{2\pi} e^{-z^2/2}$ where $z = \frac{x - \mu}{\sigma}$	$-\infty < z < \infty$	0	1
Lognormal $\log X_i \sim LN(a_L, b_L)$	$y = f(\log x) = \frac{1}{b_L\sqrt{2\pi}} \exp(-(\log x - a_L)^2/2b_L^2)$	$\log x > 0$	a_L	b_L^2

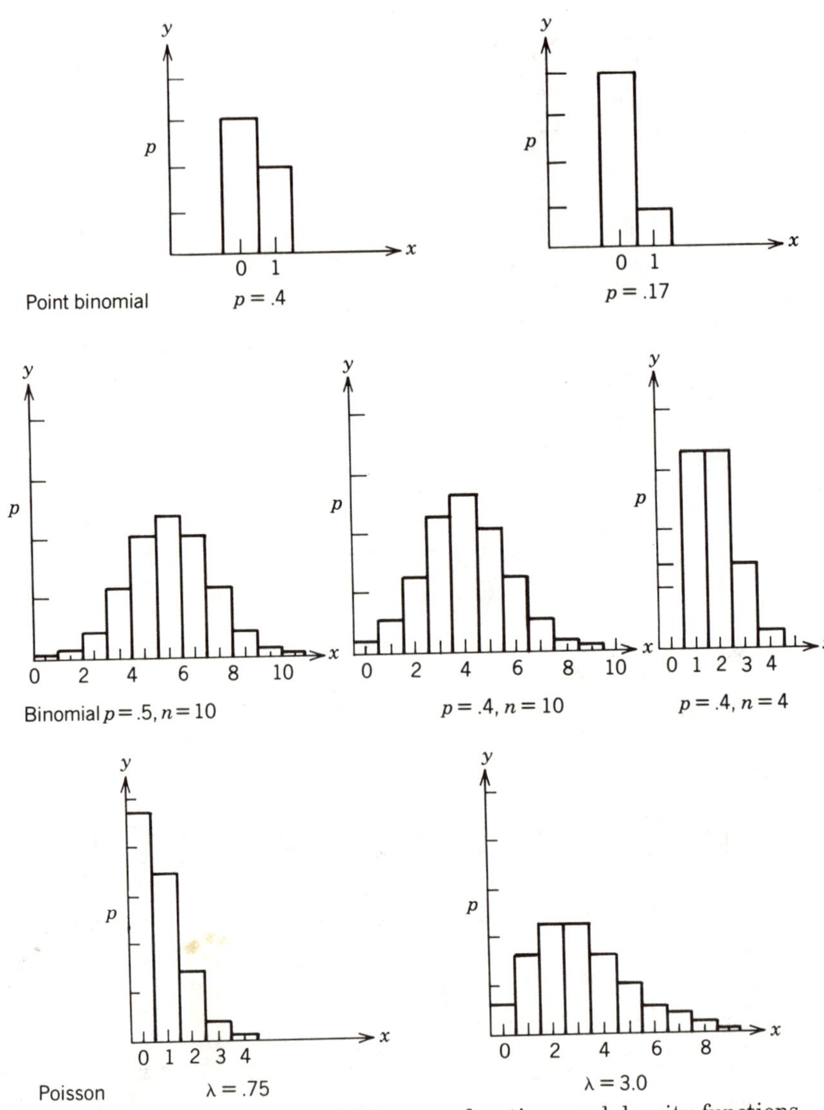

Figure 4.14 Examples of probability mass functions and density functions.

The *Poisson probability function* (named after the French mathematician Poisson who described it in 1837), which is defined as the limit of the binomial probability function when p (or $1 - p$) is very small and n approaches infinity, has important uses in geographic analysis. In the same way as the binomial probability function, the Poisson function describes the probabilities associated with the possible frequencies of occurrence of a particular event. It is characterized by two important properties. First, the mean occurrence of the event must be small relative to the total possible range of

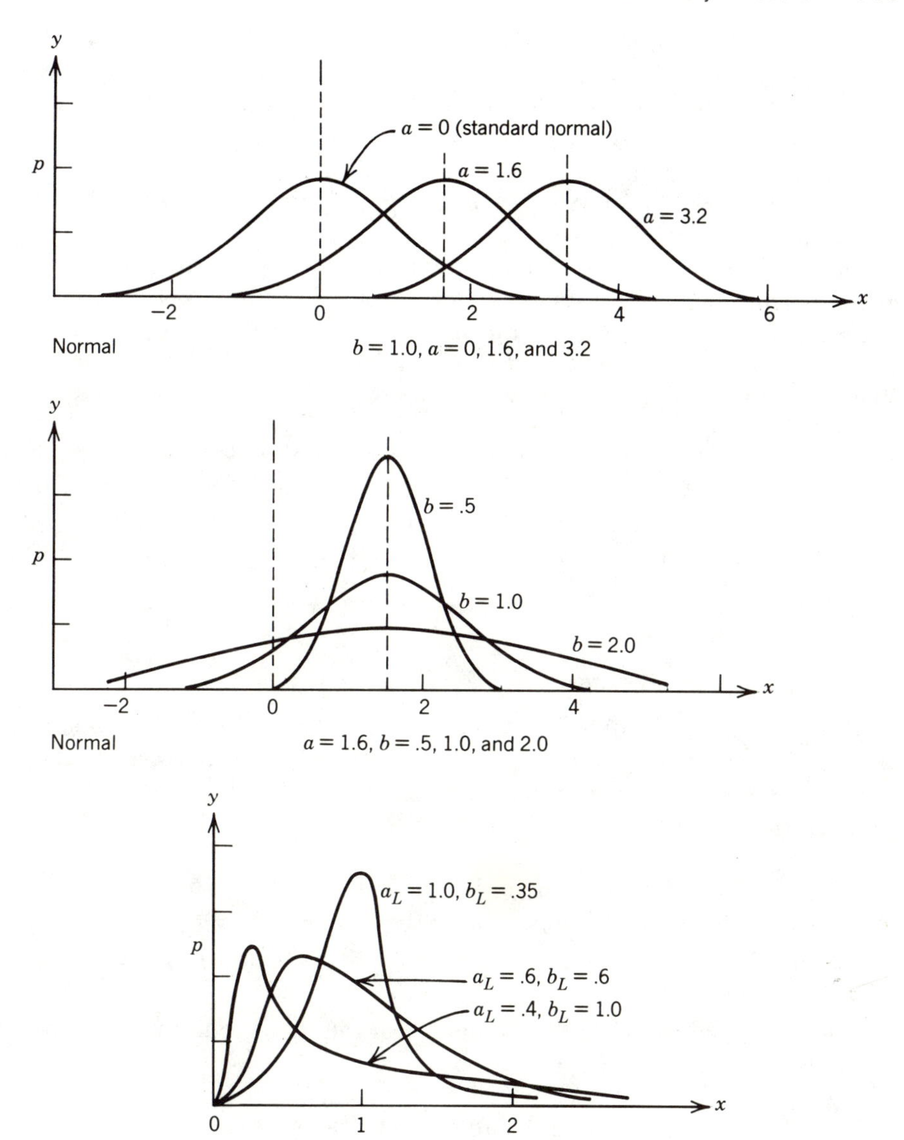

Figure 4.14 (*Continued*)

occurrences (a *rare* event), which is exemplified by the fact that the function's variance equals its mean. Second, the occurrences of events must be independent; that is, they occur randomly. It is this property that makes the use of the Poisson so important in geographic research, especially in the description of the random distribution of spatial phenomena, or rare events over time.

The *normal probability function* is the most important of all the probability functions, not so much because many population variables are distributed normally, but more because the normal probability function describes the pattern of *random errors* around a mean value. This property allows its use to assess the probability of errors in fitting a variety of statistical models to observed data. The importance of this probability function will become apparent in later chapters.

The *standard normal probability function* is simply the operational form of the normal probability function. Any given combination of the two parameters a and b of the normal probability function defines a separate probability function, but by standardizing the data using standard scores (outlined in Section 3.4), the standard normal probability function describes any variable that is distributed normally.

The *lognormal probability function* is another form of the normal probability function where the data have been transformed, in this case by logarithms, so that the data previously positively skewed may fit the normal probability function. The wide occurrence of population distributions in this form justifies its inclusion for treatment in some detail.

Other Probability Functions

Although it is not possible to discuss all of the other probability functions used by geographers, some of the most commonly encountered should be mentioned. Three of the most important discrete probability functions (Table 4.5) are as follows.

Negative binomial similar to the Poisson, but where the distribution is more "clumped" than the random Poisson.

Double Poisson where two independent Poisson (random) processes are hypothesized as operating.

Geometric where declining frequencies from a lower limit peak, produce a reverse J-shaped curve.

Of the many continuous probability functions, three deserve mention (Table 4.6).

Uniform where the frequencies are evenly distributed over an interval.

Gamma where the distribution is characterized by a very marked asymmetry (greater than the lognormal).

Exponential a special case of gamma, where the distribution is even further skewed with its peak at the lower limit (reverse J shaped).

TABLE 4.5 Other Discrete Probability Functions

Function	Functional Form	Restrictions	Moments: Mean	Moments: Variance
Negative binomial $X_i \sim NB(r, p)$	$y = f(x) = \frac{(r + x - 1)!}{x!(r - 1)!} p^r(1 - p)^x$	$x = 0, 1, 2, \ldots$ $r = 0, 1, 2, \ldots$ $0 < p < 1$	$\frac{r(1 - p)}{p}$	$\frac{r(1 - p)}{p^2}$
Double Poisson $X_i \sim DP(\lambda_1, \lambda_2, k_1, k_2)$	$y = f(x) = k_1 \left(\frac{\lambda_1^x e^{-\lambda_1}}{x!}\right) + k_2 \left(\frac{\lambda_2^x e^{-\lambda_2}}{x!}\right)$	$\lambda_1 > 0$ $\lambda_2 > 0$ $k_1 + k_2 = 1$ $x = 0, 1, 2, \ldots$	$k_1\lambda_1 + k_2\lambda_2$	
Geometric $X_i \sim Ge\ (p)$	$g = f(x) = p(1 - p)^x$	$x = 0, 1, 2, \ldots$ $0 < p < 1$	$\frac{1 - p}{p}$	$\frac{1 - p}{p^2}$

TABLE 4.6 Other Continuous Probability Functions

			Moments	
Function	**Functional Form**	**Restrictions**	**Mean**	**Variance**
Uniform $X_i \sim R(p)$	$y = f(x) = p$	$0 < p < 1$	p	0
Gamma $X_i \sim Ga(a, b)$	$y = f(x) = \dfrac{x^a e^{-x/b}}{a!b^{a+1}}$	$x \geq 0$ $a > -1$ $b > 0$	$b(a + 1)$	$b^2(a + 1)$
Exponential $X_i \sim E(\theta)$	$y = f(x) = \dfrac{e^{-x/\theta}}{\theta}$	$x > 0$ $\theta > 0$	θ	θ^2

Many of the functional equations outlined (Table 4.4) involve tedious calculation when the probabilities have to be derived for particular values of the variable X. Thus, for the more important functions, tables have been constructed listing the values for $f(x)$ or $F(x)$ based on particular values for the parameters. However, for each value of the parameter, a separate table would be required; and where two parameters are involved, the number of possible tables that might be needed would be considerable. The derivation and uses of these tables are discussed in Chapter 6.

For the moment, we note the functional form of the various probability functions and their summary parameters and introduce a simple nomenclature to describe the distribution of scores of a variable X when its probability functional form is known. For example, for the binomial probability function, we can use

$$X_i \sim B(p, n)$$

to signify that the X_i scores are distributed according to the binomial probability function with parameters p and n. For the normal probability function, the abbreviation is

$$X_i \sim N(a, b)$$

but because the parameter a is also defined as the arithmetic mean, and b as the standard deviation (and b^2 is, of course, the variance), we can rewrite this expression in a more convenient form as

$$X_i \sim N(\mu, \sigma)$$

References and Readings

1. Probability Concepts

Bulmer, M. G. (1967) *Principles of Statistics,* Second Edition, Oliver and Boyd: Edinburgh, UK.

Feller, W. (1957) *An Introduction to Probability Theory and Its Applications,* Wiley: New York.

Hogben, L. (1957) *Statistical Theory,* Norton: London.

Pielou, E. C. (1969) *An Introduction to Mathematical Ecology,* Wiley: New York.

2. Probability Models

Cliff, A. D. and J. K. Ord (1981) *Spatial Processes,* Pion: London.

Gudgin, G. and J. B. Thornes (1974) "Probability in geographic research: Applications and problems," *Statistician* 123:157–178.

Haggett, P., A. D. Cliff, and A. E. Frey (1977) *Locational Analysis in Human Geography,* Second Edition, Volume Two, *Locational Methods,* Arnold: London.

McConnell, H. and J. M. Horn (1972) "Probabilities of surface karst," in R. J. Chorley (ed.), *Spatial Analysis in Geomorphology,* Methuen: London.

CHAPTER 5
Sampling Designs and Sampling Methods

Populations with an infinite number of elements have to be sampled before the nature of the population can be ascertained. Even populations with a finite and known population size may be sampled for reasons of time, cost, or accessibility. In fact, in many situations the powerful concepts of probability make it unnecessary to examine all the elements of a population. With a carefully designed sample, the methods of inferential statistics can be employed to draw meaningful conclusions about the population from the sample. After examining the relationships between populations, samples and inferential statistical methods (Section 5.1), we outline the major sampling designs and methods—first for statistical distributions (Section 5.2) and then extend them to handle geographic distributions (Section 5.3). To use a sample to make inferences about a population requires knowledge of the form of the population being sampled, but it also requires knowledge of the characteristics of the statistics obtained from the sample. The most important of these characteristics is the form of the *sampling distribution* for a particular statistic. This provides the topic for the final section (5.4).

5.1 SAMPLING AND INFERENTIAL STATISTICS

Before we outline the various sampling designs that are available, there are a number of concepts (and problems) that we need to consider. These concepts and problems center on the role of sampling and inferential statistical methods within the whole range of methods that can be used for analyzing real-world data.

Some Preliminary Definitions

Occasionally, there is confusion over the distinction between populations and samples, and it may be of benefit to define (or redefine) these and some other related terms. Note that not all the definitions used here will agree with many statistical texts, but they follow the concepts outlined in Chapter 1 and they provide a clear terminology as well as a base for discussing inferential methods. We can illustrate these definitions with reference to the examples used in these introductory chapters (Table 5.1).

Recalling Section 1.1, an *element* is an individual unit being examined, capable of being measured in some form or other. Each element has a particular *score* for any variable, obtained by the measurement process. Occasionally, textbooks distinguish between *universe*—the totality of all elements that could be examined—and *population*—the totality of all scores for a particular variable. With this distinction, a universe could have several populations. However, following generally accepted terminology, we will not differentiate between the two terms, and we will simply define a population as a set of elements that has particular scores for the variables with which we are concerned. The population, then, can be defined as a *sample space* (Section 4.1) made up of a finite or an infinite set of elements or set of scores.

It is often convenient to distinguish between the target population and the sampled population. The *target population* is the population of elements or scores for which we hope to obtain information by selecting a sample. The *sampled population* is the set of elements or scores from which the sample is actually obtained (Table 5.1).

Inferential or *inductive statistics* form a group of methods that enables inferences to be made about the sampled population, based on the laws of probability and information provided by the sample. It is obvious that the degree to which the conclusions can be extended to the target population depends on how representative the sampled population is of the target population. The inferential process proceeds by using summary measures calculated from the sample data—which we will call *sample statistics*—and using these to estimate their equivalent *population parameters*. To some degree, we describe this procedure in this chapter, but its full discussion is deferred until Chapter 6.

TABLE 5.1 Examples of Population/Sample Relationships

	Urban Land Value Survey	Cumberland County Farm Survey	Northland Drainage Basin Survey
Element	Property lot	Farm	Drainage basin
Universe/population/sample space	City of Woodford, CBD made up of 86 elements	Cumberland County, consisting of 94 farms	All drainage basins in north (unknown number)
Target population	Not applicable	As above	As above
Sampled population	Not applicable	As above	All basins for which records are kept by catchment authorities ($N = 62$)
Sample	Not applicable	A sample of 37 farms	A sample of 26 basins

Population Surveys and Inferential Methods

Before we proceed, there is a question that needs to be examined concerning the appropriateness of inferential methods when dealing with a complete enumeration of the population. This is not an unusual situation, but it is one that requires a little elaboration. With complete information on a population that we have specified, the parameters and other characteristics of the population are then defined and inferential methods appear to be unnecessary. However, when we examine certain situations a little more closely, this conclusion seems partially unsatisfactory.

Let us look at the example of the Urban Land Value Survey. Our population was clearly defined as made up of 86 elements (the individual property lots), and our aim was to provide summary information about these property lots. One of the results of the survey was that the mean Land Value per Square Meter was $115.20. This is a population mean (μ) and it would be incorrect to assess this value using the methods to be outlined in Chapters 6 and 7. However, it is possible to speculate on whether another city of similar character would also have a mean of approximately the same value or, whether in a few years time with some changes to the lots and to land values, the mean would be about the same. Such inferential questions are feasible, but the inferential methods and sampling design that would be necessary to make interpretations of this sort are not possible with the classical statistical methods of this book.

The problem becomes even more complex when we look at bivariate and multivariate relationships among the variables. Suppose that the mean Land Value of the six warehouses is $101.40 and the mean Land Value of the five hotels is $114.60. Are these two means significantly different? Obviously, the answer is not a matter of significance at all—there just *is* a difference—hotels on average have land values of $13.20 per square meter greater than warehouses. Thus, one of the most frequently used inferential statistical tests—the difference of means test—is not applicable. But, on the other hand, the *type* of inferential question just asked might be valid. If we had taken two random samples of size 5 and 6 from the 86 property lots, isn't it feasible that we might have obtained a difference as great as this due to random chance?

The problem can be confusing, and as a general conclusion we make the following observations. First, the inferential methods outlined in the remainder of this book are designed for random samples from a specified (finite or infinite) population. Second, when we are dealing with a completely measured population, then these inferential methods should not be used except when we are testing the fit of a set of population scores to a statistical model or examining ecological relationships. These topics will be taken up later. Third, when answers to certain inferential questions are attempted using population data, the statistical methods should be examined carefully to see that the assumptions involved are justified, and the inferential process should be explained fully and explicitly.

Sample Size

The second major question that we have delayed examining until now concerns sample size. In the examples we have mentioned so far the size was already stated, but this is usually a question that would have to be answered at the beginning of the survey. No rule can be provided. However, the size of the sample could be governed by one or more of the following principles.

First, if the desired confidence interval (range) about an estimate of a certain parameter (usually the mean) for a particular variable is regarded as critical, the required sample size can be found if we have reasonable estimates of the population variance. The method is discussed in Section 7.2. This is the standard method, outlined in all books on statistics and would usually involve a preliminary pilot survey to obtain the necessary estimates. In geographic problems we are rarely interested in just the one parameter for one particular variable, and we also might have no idea of what would be a satisfactory confidence interval.

Second, following similar surveys, a sample of a certain proportion of the total population might be desirable. Thus a 3 percent sample may be the standard size used for demographic studies of urban populations. There is often no statistical basis for such a figure, and there may even be no reason to justify its use in comparable surveys. It is quite possible that the first study of that type employed a 3 percent sample, and all other studies since then have attempted to duplicate it.

Third, certain statistical tests require a reasonable sample size, especially when there is little information about the characteristics of the distribution of the population scores. Again, no fixed minimum can be stated, but in most situations a sample size of at least 30 is employed.

Fourth, finance and time may dictate a certain maximum sample size. Thus, if it were found from a pilot survey that it took x minutes to administer a questionnaire (or x hours to obtain the necessary measurements for a single element), and therefore we might obtain data for y elements in a week of work, but only z weeks are available for the survey, then the maximum size of a *practical* sample is obtained.

All the above principles point to a simple conclusion—that realistically, time and money dictate sample size. A sample size thus determined is either adequate, too small, or too large. With a carefully designed problem—both in its aims and its operational definitions—the sample size would rarely be too small. If as the sample is being obtained it is obvious that the amount of variability is much greater than expected, or if the time necessary for the collection of the data has been underestimated, then a major revision of the aims and methods should be undertaken. The possibility of major revision emphasizes the importance of the pilot survey. The other possibility—that the sample size is too large (similar results could have been obtained from a smaller sample and therefore time and money have been wasted)—is also rare.

Ideally, then, the sampling situation is as follows. Through a small pilot survey, we obtain some idea of the variability of the data and the time it takes to obtain the necessary information. Then we either decide on a sample size and a sampling design and collect the information, or we decide on an absolute minimum sample size and select a sampling design that allows a certain flexibility of sample size. Suppose we are using a simple random sampling design (Section 5.2) and we decide to examine at least 40 elements, but we expect that it is desirable and possible to examine more. We randomly select 100 elements, recording them in order, and examine the first 40. We analyze our results and decide that another 20 elements would be desirable—we then examine the 20 elements that were randomly selected in the forty-first to sixtieth selections. The process can be repeated until we feel we have obtained a satisfactory sample—as satisfactory as time and money will allow—in addition to being statistically sound.

Although the preceding discussion may seem to relegate the importance of obtaining a statistically satisfactory sample to a minor position, it outlines a realistic approach to sampling—an approach that is almost invariably used. However, the reliability of the sample survey depends to such a large degree on the sampling design and size that their importance should not be underestimated. Every survey in which a sample is employed should state clearly the size, proportion (if known), and design used, and some estimate should be made of the reliability of the sample statistics calculated. The topic of the structuring of confidence intervals is discussed in Chapter 7.

Sample Design and Experimental Design

In Section 5.2 we examine the major sampling designs that are used, and we emphasize the use of one—the random sample design. We want to concentrate on this design because the inferential methods we examine in Chapters 6 and 7 are based on the analysis of distributions derived through this design. To answer the same sort of inferential questions about distributions derived using other sample designs requires some modification of the techniques outlined. That is, the methods of analysis are dependent on the design used to obtain the sample. This emphasizes the need for a careful design of the whole statistical project before the sample is obtained so that the answers that are required from the analysis will be forthcoming.

Sample design methods can be considered as part of the broader concept of *experimental design*. Here, we are using the term "experiment" in the same way as we did in discussing empirical probability concepts (Section 4.1). The process involves selecting elements from the sample space in such a way that we can ascribe certain probabilities to the outcomes of those selections. The experiment can range from selecting a sample from a population (as we are describing) to recording scores from an experimental instrument. Our aim in this book is to present the most common experimental designs and the methods for their analysis. Departures from these designs into other designs require modifications of these techniques. Multivariate analysis techniques in particular are very much dependent on the design used, and the designs available become much more numerous and their analyses more varied.

5.2 METHODS OF SAMPLING

Although our aim in sampling is usually to make inferences about specified characteristics of the population from which the sample was obtained, this is not always the case. Samples may be obtained simply to seek information about the sampled elements themselves. Such *search sampling* and *purposeful sampling* methods (broadly speaking, *subjective sampling*) are not discussed here. We are interested in *probability sampling*, a method in which each element of the population has a known probability of being selected. This approach ensures that we can draw conclusions about the characteristics of the population from the sample and that we can defend our conclusions by recourse to the laws of probability. More particularly, we are interested in *random sampling* methods, where each element in the population has an *equal* chance of being sampled. We will confine our discussion primarily to random sampling methods throughout the remainder of this book, but we note that if individual elements have known probabilities of selection that are other than equal, the same methods can be applied by weighting the results of the sample according to the probability of selection.

Random Sampling

In most statistical analyses, we want to obtain a *random sample,* although this simple aim may be difficult to achieve in practice. Resort then may have to be made to some approximately random sampling method or a variation of a random design. As stated, here we emphasize the methods of obtaining, or approximating, a random sample. However, we differentiate the following four, frequently used, basically random designs.

A random sampling design, in which each element in the population has an equal chance of being selected. Strictly speaking, this is the only truly random sampling method, and it is this design with which we are predominantly concerned.

A stratified random sampling design, in which the elements of the population are allocated into subpopulations or strata before the sample is taken, and then each stratum is randomly sampled.

A systematic random sampling design, in which the population elements are arranged in some order, a single element is randomly selected, and then the remaining elements are selected systematically from around that single sampled element using a fixed interval.

A clustered random sampling design, in which the elements of the population are allocated into subpopulations or clusters before the sample is taken, and then these clusters are randomly sampled. All the elements within the sampled clusters may then constitute the sample or the sample clusters themselves may be randomly sampled to obtain the sample elements.

The Simple Random Sample Design

With reference to inferential statistical methods, the aim of sampling is to obtain a sample that will produce sample statistics that are *unbiased* and *efficient* estimates of the relevant population parameters (see Section 6.2). In its simplest form, it would seem that the statistical requirements of unbiasedness and efficiency would necessitate as large a sample as possible, whereas time and money would dictate the use of as small a sample as possible. This is not strictly correct, however, because in certain situations, an increase in sample size will not increase the degree of unbiasedness and efficiency. For the moment, we will consider the problem of obtaining a sample when the sample size has already been determined.

We defined a random sample as one in which each element in the population has an equal chance of being selected. Random sampling methods enable the use of the two important probability concepts of mutually exclusive events and independent events. Equating the *sample space* to our population of elements, and an *event* to the selection of an *element* in that sample space, random sampling requires that the individual elements and their selection be mutually exclusive and independent. The property of mutual exclusiveness requires that population elements be defined exactly so that there is no possibility of overlap—this is rarely a problem, although the property should not be dismissed. However, the property of independence poses more of a problem. In Chapter 4, we noted that independence occurred only when sampling was performed with replacement. The relationship between independence of selection of elements and sampling with and without replacement can be summarized in three statements.

1. If the population being sampled is described by a continuous probability density function, then sampling with or without replacement does not affect the independence of selection.

2. If the population being sampled is described by a discrete probability mass function, and the population size is infinite, then sampling with or without replacement does not affect the independence of selection.

3. If the population being sampled is described by a discrete probability mass function, and the population size is finite, then sampling with replacement will constitute a random sample but sampling without replacement will not.

That is, if the population of scores from which we are sampling is for a variable assumed to be described by a discrete probability mass function, and the population size is finite, then sampling without replacement will not constitute a random sample. More commonly, when we are sampling elements from a finite population and we are interested in a number of variables, we cannot consider the sample to be a random sample of scores for those variables that are discrete. Referring back to our simple example in Chapter 4, the variable we were interested in was Farm Type (binary—dairy farm/nondairy farm), our population was finite ($N = 10$), and samples of sizes 1,2,3, etc. were taken. We obtained different results for the probabilities of intersection for sampling with replacement and without replacement. Only sampling with replacement constituted a random sample as just defined.

This is an important point that must be evaluated carefully. More frequently than not, geographers are concerned with finite populations, and often with discretely measured scores. Does this mean that we should obtain a sample from a population using some replacement technique? Common sense seems to dictate that we should not—the reason for sampling was to save time and money, but also to evaluate as many elements as possible. To evaluate the same element twice seems to contradict the aims of sampling. Also, if the size of the finite population were large, what would be the chances of selecting the same element twice? Reviewing the principles outlined in Chapter 4, we can see that it would be very small.

Thus, selection of elements without replacement from a finite population whose scores are discretely measured will modify the probabilities of selection of the remaining elements at each step in the selection process. Despite this modification, it is common practice to retain the sampling method without replacement. The design is called a *simple random sample*. It has two major requirements: (1) all individual elements initially have equal chances of selection, and (2) any set of elements of size *n*, which we call the sample size, has an essentially equal chance of selection. We can consider this as a less restrictive definition of independence of selection of elements.

However, the use of the simple random sampling design, rather than the random sampling design, does have repercussions when we deal with discrete variables and finite population size. The relationship between the values of sampling statistics and their corresponding population parameter values is different for these two sampling designs (in fact, they are different for all sampling designs), and the bulk of the inductive statistical methods requires sample statistics assumed to be derived from a random sample. Nevertheless, the differences are small and can be overlooked when dealing with small (relative to population size) samples. As the relative sample size increases, the discrepancy increases, and when the relative sample size is about 50 percent or greater ($2n > N$), modified definitions of sample statistics should be used. These are not discussed here, and the reader is referred to one of the standard texts such as Cochran and Cox (1957).

The simple random sample design, then, involves the selection without replacement of a set of population elements. We illustrate its use with reference to the sample needed for the Cumberland County Farm Survey—the example to be used extensively in the next two chapters. We should note that while this survey is geographic in that our farms are distributed spatially over the county, our analysis of the data is not spatial. That is, we are not interested in where within the county the farms are located. Thus, we can use this example as a nonspatial sampling problem. If we had wanted a spatial random sample we would have had to apply the methods outlined in Section 5.3.

Time dictated the selection of sample size, and it was decided to obtain a sample of approximately 40 farms. A map of the county was available, with the farms clearly demarcated. To sample from the total of 94 farms located in the county, each farm needed to be identified in some manner. The farms were arbitrarily identified by a pair of digits (numbered 1 to 94) starting in the northeast corner (Figure 5.1). Our task is simple—to randomly select 40 identified elements from a total of 94. Any random selection process could accomplish this, but the easiest is to use the table of random digits (Table A at the back of the book). The table consists of sets of digits, in this instance arranged in pairs, which are randomly positioned in any direction in the table. We enter the table by randomly selecting a starting position, and we assume that we have selected the random pair of digits 74 located near the center of the first page of the table. We proceed from that point in any direction (up, down, across, diagonal) making up a rule as to what to do at the edge of the table. Each pair of digits is recorded (where our population exceeded 100, we would have recorded every set of three digits ignoring the column breaks). In our example, we proceed to the right and then return across the table on the next row down. We reject numbers greater than 94 (outside the range) and omit any pair of digits repeated (already selected). Selection continues until we have the required number of elements (in our case, 40). Thus the selection was:

74	41	56	23	82	19	95*	38	04	71	36
69	94	55	64	20	21	61	39	21*	35	53
05	95*	06	52	17	59	14	98*	41*	95*	07
41*	34	88	52*	60	83	59*	63	56*	55*	06*
95*	89	29	83*	05*	12	80	97*	19*	77	43
35*	37	83*	92	(*rejected)						

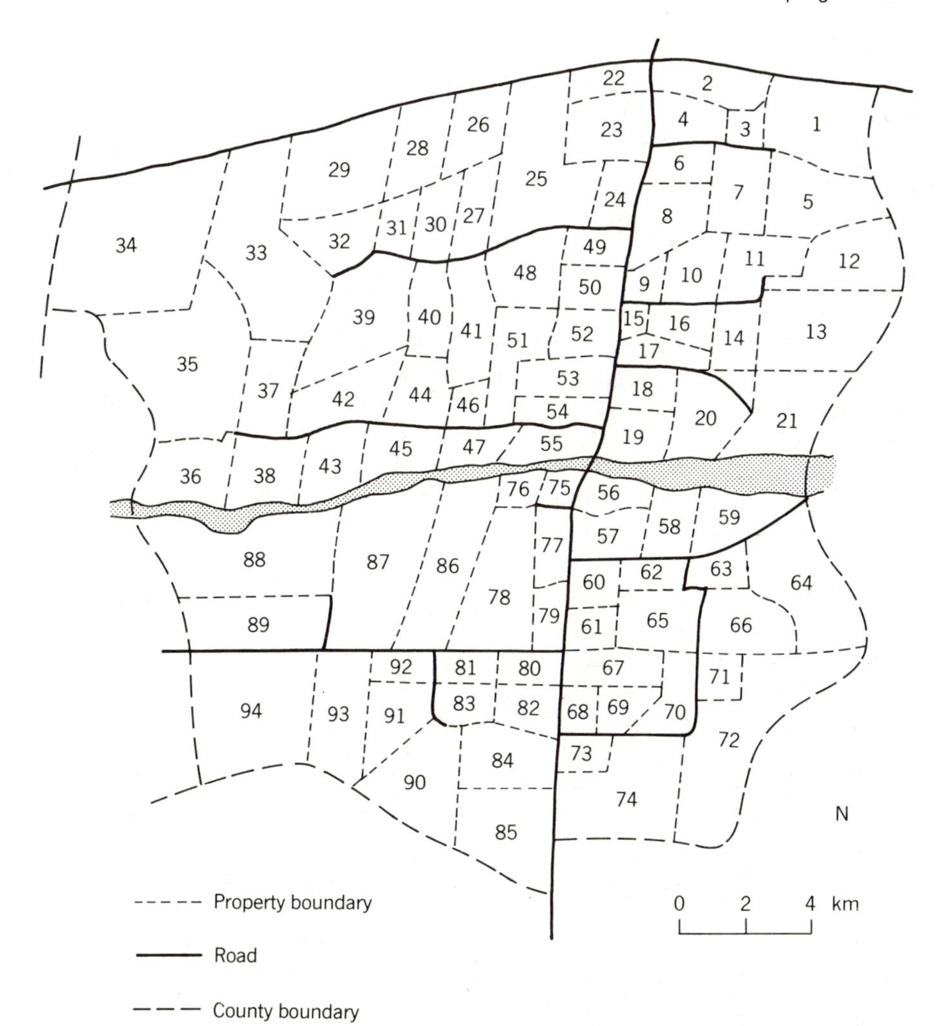

Figure 5.1 Location of farms in Cumberland County.

Our 40 farms selected would then be (in numeric order):

4 5 6 7 12 14 17 19 20 21 23
29 34 35 36 37 38 39 41 43 52 53
55 56 59 60 61 63 64 69 71 74 77
80 82 83 88 89 92 94

The selected farms constitute our simple random sample of 40 farms, and the survey then began. Note that in carrying out the survey, 3 farmers failed to respond, thus the eventual sample size was $n = 37$. The response rate was 92 percent (37/40) and the final relative sample size was 39 percent (37/94).

Extensions to the Simple Random Sample Design

The example of the selection of a sample of 40 farms from a tabulated set of 94 is a sampling situation that, unfortunately, is only rarely encountered. More frequently, we find that the population size is unknown, or the population elements are not tabulated, or they may not be identifiable, or even a combination of all of these. The sampling designs that might be appropriate in a variety of situations are outlined in Table 5.2, with some illustrative examples. The key definitions of the types of elements are as follows.

A *defined element* is one that, given complete information, can be identified as a point in space or time—such as a farm, a person, etc. The location of points and areas in a continuous spatial surface are, of course, undefined—even with complete information they could be located anywhere. Basically, the distinction between defined and undefined elements is whether we are concerned with a finite number of elements (even if large or unknown) or an infinite number. This ultimately reduces to an examination of nonspatial and spatial populations and the consideration of spatial sampling designs (Section 5.3).

A *tabulated element* is one that is clearly defined and already has some form of identification. Populations in which the elements are already tabulated would usually be restricted to very small sizes. An example would be the listing of the cities in a census volume.

An *identifiable element* is one that, even if it is not tabulated, can be identified through some simple search procedure. Thus, if we know that there are 94 farms in a county, or 212 workers in a factory, and if a map of farms or list of workers is available, we could identify each of the elements and allocate some identifying set of digits to each.

An *aggregated population* is one that can be subdivided into some set of component parts—preexisting or arbitrary—which may assist in the sampling process. Thus, people in an urban area are located in census tracts, farms in a country are located in counties, etc. The size of each of the aggregates may be known or unknown.

In all the examples we examine, we want to provide a sampling design that is as similar to a pure random sample as possible, considering the nature of the population elements.

In the example in which 40 farms were sampled from the 94 in Cumberland County, the number of population elements was small enough to enable us to identify each element and then obtain a simple random sample. If, however, the number of population elements is very large, or even unknown, then individual identification becomes too tedious and time consuming. In certain situations it may not be possible to identify the individual population elements without complete evaluation of the population. Nevertheless, it is not too difficult to envisage some simple method for approximating a simple random sample.

A frequently encountered design is the *systematic random sample,* where the elements can be considered to be in some sort of order (over space or time or simply a listing). A single element is randomly sampled and then the other elements are added to the sample, using some sort of constant interval or "distance." Thus, houses could be sampled by randomly selecting a house within the first interval of 10 houses and then taking every tenth house along a street, farms by every fifth farm along a road, shoppers by every twentieth shopper entering a supermarket, etc. The method rests on the assumption that there is no periodic (cyclic) trend in the ordering of the individual elements. The design would have to be carefully structured, and perhaps a pilot sample taken to determine any pitfalls in the method. Before the main survey begins, the operational definitions for the sampling design should be set up exactly—how the ordering is to be accomplished, how the initial element is to be randomly chosen, what interval to use, what to do when a selected individual element fails to respond; and then it must be followed exactly. If the design is structured carefully—and intuition and experience will usually see that this occurs—the inferential methods that assume random sampling can be used.

Another commonly encountered situation is where the elements, although they may be untabulated, are located in some aggregated unit that may assist in the identifying and selection of the sample elements. Two basic sampling designs are available. The *stratified random sampling design* can use these aggregates to decide how many, or approximately how many, elements should be selected from each aggregate or *stratum*. Elements within each stratum would then be randomly selected as far as possible. The proportion of the sample drawn from each stratum would be based on its known or estimated total number of elements as a proportion of the total population. The method, then, is simply a means of making sure that the sample contains a similar proportion of elements from the different strata that occur in the population. Such a design is a *proportional* stratified random sample, and most of the standard inferential methods would apply. It is assumed that a large simple random sample from a large population would also have proportional stratified representation in each of the strata, and thus the proportional stratified random sample attempts to approximate this. *Disproportional* stratified random samples would be used when we are more interested in the strata themselves—and how the population scores vary from one stratum to another. This practice is sometimes known as "oversampling," and proper statistical analysis of data collected according to this design requires weighting to compensate for disproportional probabilities of selection.

As an alternative to the stratified random sample, the *clustered random sampling design* is used when there is a large number of aggregates (which we now call *clusters*) and time, money, or practicality prohibits the selection of subsamples from each of the clusters. Instead, the clusters themselves are randomly sampled, and, in its simplest form, all elements within the selected clusters are examined. Two-stage cluster sampling involves a further sampling (simple random, stratified, or systematic) within each of the clusters.

TABLE 5.2 Examples of Modified Random Sampling (RS) Designs

Type of Element		Sampling Design (Population Size)				Examples		
		Finite (N Small)	Finite (N Large)	Finite (N Unknown)	Infinite	Element	Aggregate	Population
Undefined	—spatial point	—	—	—	Spatial RS	A slope unit	—	All slopes in an area
	—spatial area	—	—	—	Spatial RS	A sample soil plot	—	All soil surfaces in an area
Defined	—tabulated	Simple RS	Simple RS	—	—	A city	—	All cities listed in a census
Defined/ untabulated/ unaggregated	—identifiable	Simple RS	Approx. simple RS Systematic RS	Approx. simple RS Systematic RS	—	A farm A building (or lot)	— —	All farms in a county All bldgs. (or lots in urban county)
	—unidentifiable		Approx. simple RS Systematic RS		—	A person A shopper	—	All people in a census tract All shoppers visiting shopping centers

Defined/ untabulated/ aggregated/ (size known)	—identifiable	Simple RS	Strat. RS Cluster RS	—	—	A worker	A factory	All workers in all factories in an urban area
						A farm	A county	All farms in all counties in a country
	—unidentifiable		Approx. strat. RS Approx. cluster RS	—	—	A person	A census tract	All people in all census tracts in an urban area
Defined/ untabulated/ aggregated (size unknown)	—identifiable	—	—	Approx. strat. RS Approx. cluster RS	—	A person	A city block	All people in all blocks in an urban area
						A telephone subscriber	Page(s) in the directory	All people listed in the directory

Cluster sampling involves the use of large numbers of clusters and a heterogeneous composition (with respect to the population of scores being sampled) within each cluster. In many situations, cluster sampling provides the only economical means of obtaining a sample—but it should be emphasized that in almost all of these situations it is impossible to obtain a sample that approximates a strictly random sample and, consequently, few of the inferential methods outlined in this text are applicable without considerable modification.

The foregoing brief review of sampling designs does not do justice to the enormous number of designs, combinations, and extensions of the four basic designs that can be used. However, the review does emphasize the following points.

1. The design of the sampling method can be one of the most critical steps in geographic analyses.

2. Most of the inferential methods that have been developed and the ones to be discussed in the remaining chapters assume that a random sample of elements (defined in the introduction to Section 5.1) has been obtained.

3. A simple random sampling design, or as near an approximation as is physically and economically possible, should be used.

4. If an alternative design must be considered, then the literature should be examined carefully, or a professional statistician should be consulted *before* the sampling beings. All too frequently in geography, a sample is obtained with little regard to the ensuing methods of analysis.

5.3 SAMPLING GEOGRAPHIC DISTRIBUTIONS

When our interest is in the spatial distribution of scores as well as their statistical distribution, then our sampling design needs to be spatially random as well as statistically random. In addition, if we are dealing with a variable that is spatially continuous, then we need a spatial sampling design simply to obtain sample scores for that variable. The methods needed to handle sampling geographic distributions are modifications of those already outlined. We want to obtain a random selection of points, areas, or sample points from a continuous spatial distribution. All three situations revert to the requirement that a set of points be randomly located in an area. In the case of the area elements, an internal point within an area would represent that area. The sampling procedure might be based on a map, or it might be necessary to carry out the sample in the field. The former method is obviously more straightforward, although there would be associated problems of establishing the precise location of the sampled points. With a map base and some arbitrary or otherwise coordinate system, sampling involves the random selection of x and y coordinates (Figure 5.2).

Simple random, systematic, stratified, and clustered sampling designs can

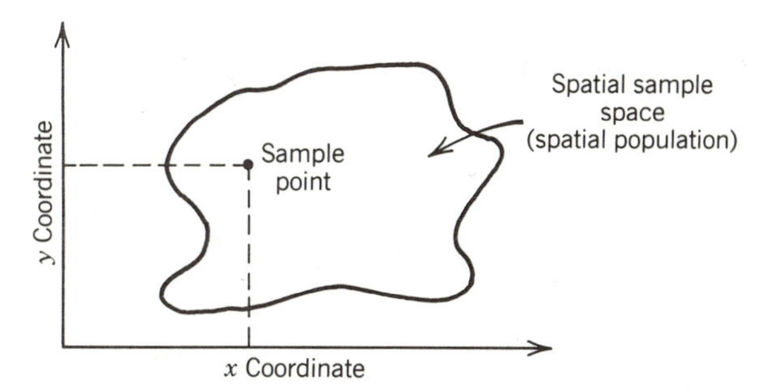

Figure 5.2 Location of sample point by random x and y coordinates.

all be applied in a spatial context. Figure 5.3 illustrates the differences among these four sampling procedures using spatial units. The simple random sample is derived by establishing a coordinate system with some specified resolution, and taking two random numbers to establish the location of each point. The first random number might apply to the north–south direction, the second random number to the east–west direction. Because the two selections are independent, the resulting intersection provides a simple random sample of points within the grid structure. Assuming a very fine grid over the area, every intersection within the grid has an equal chance of selection, and therefore, every point within the area has an equal chance of selection.

The spatial systematic sample, on the other hand (as in the case for the nonspatial systematic sample), establishes an initial location and a fixed interval, and points are selected at that fixed interval in both directions across the area to be sampled. As in the nonspatial systematic sample, once the initial point is selected, with random x and y coordinates, all the other points are determined. The systematic sample overcomes one of the perceived problems of the simple random sample—that it may not give good coverage within the spatial sampling space. In a simple random sample, the selected points are often bunched in particular sections of the map. On the other hand, if there is some regular periodicity in the population, systematic sampling is likely to reproduce only a part of that periodicity, or alternatively, to bias the sample by being overrepresented by the periodicity. Single-dimensional systematic sampling has important applications particularly in physical geography, where samples are taken at regular intervals along a stream or down a slope.

A stratified spatial sample can be designed in a number of ways. In simplest form, a cellular grid is placed over the map and a fixed number of sample points is randomly obtained from each cell (Figure 5.3). This design is the spatial equivalent of the stratified random sampling design outlined in the last section. Alternatively, areal units other than arbitrary grid cells

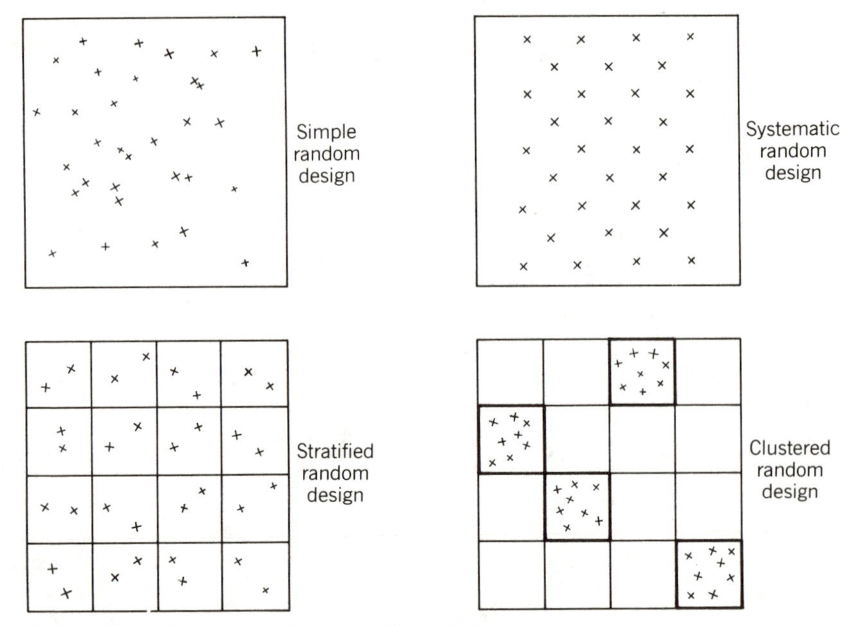

Figure 5.3 Random spatial sampling designs.

could be used to stratify the spatial population—the strata could be weighted by area, population, income, etc.

A clustered sampling design consists of the random selection of a set of the arbitrary or fixed areal cells (Figure 5.3), and then a total enumeration, simple random selection, or systematic selection within the selected cells. In a spatial context, there are few practical situations in which this design can be used with advantage.

Simple random and systematic sampling designs appear to be the obvious choices in most sampling situations, although the restricted extent of the former, and the possible built-in error of the latter, are still limitations. Berry and Baker (1968) suggested a new design—a stratified systematic unaligned design—that overcomes these two problems by combining the two methods. In this procedure (Figure 5.4), a grid structure is laid over an area to define cells of specified size, with the number of grid cells equal to the final sample size. Within each of these cells, a point is chosen based on a set of rules. The procedure is designed so that there will be one point selected per cell, but the location of the point within the cell will vary randomly between cells. Berry and Baker suggested that the results of a stratified systematic sample are less biased and have greater precision than any of the other methods mentioned.

The procedures for choosing a spatial stratified systematic unaligned sample are as follows.

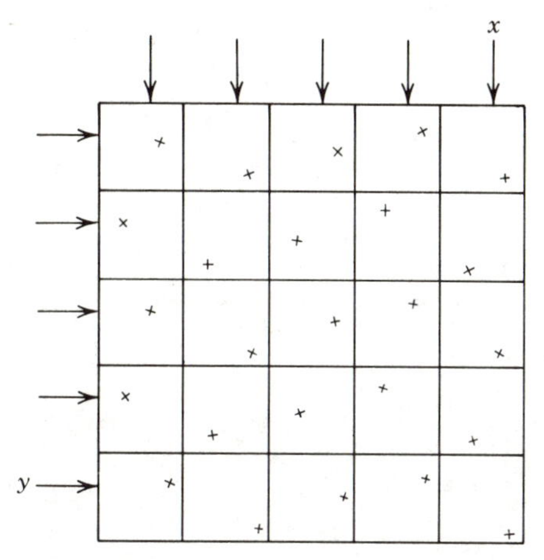

Figure 5.4 Stratified systematic unaligned design.

1. Grid the area, recognizing that the number of cells chosen must exhaust the spatial population units and be the same size.

2. Establish a random x and y coordinate within the uppermost left cell.

3. Keeping the random x value from the first cell, choose new random values for the y coordinate for each cell in the first row.

4. Keeping the y value from the first cell, select new random x values for each of the succeeding cells in the first column.

5. For all the succeeding rows, take the value from the leftmost cell, maintain that as a constant, and take the y value from the topmost cell.

The advantages of this approach are that it avoids the periodicities of the systematic approach, it gives good coverage over an area, it is efficient, and it deals with most distributions.

5.4 SAMPLE DISTRIBUTIONS AND SAMPLING DISTRIBUTIONS

In the Cumberland County Farm Survey, a sample of 40 farms was selected from the population of 94 farms, using a simple random sampling design. Inadequate responses led to the elimination of three farms. The effective sampling size was $n = 37$. Using a variety of measurement processes, scores were obtained for a number of variables for each farm, and the data were entered as a SAS data set (Table 5.3). The scores were printed (Table 5.4) and checked. For each variable there is a distribution of scores that we will call a *sample distribution.*

All the methods outlined in Chapters 2 and 3 for portraying and sum-

TABLE 5.3 The Cumberland County Farm Survey (SAS Setup)

```
FILE: FARMS    SAS       A    UNIVERSITY OF AUCKLAND

DATA FARMS;
   INPUT FARMSIZE 1-3 RUNOFF 5 HERDSIZE 7-9 BREED 11 OTHERLIV 13-17 MLKSHED 19
         MILKFAT 21-25 NONDAIRY 27-28 HAYPROD 30-33 FEEDPRCH 35 OWNERSHP 37
         AGEFARMR 39-40 FARMEXP 42-43 LABOR 45-48 EQUIPMNT 50 PASTURE 52-53
         PAST_DIV 55 PLAN_DIV 57 CODENO 59-60;
   LABEL FARMSIZE=EFFECTIVE AREA OF FARM IN HA
         RUNOFF=BINARY: 1 FOR USE OF NON-CONTIGUOUS LAND
         HERDSIZE=NUMBER OF COWS IN MILK (1/1/84)
         BREED=DOMINANT BREED DAIRY COWS (8 CLASSES)
         OTHERLIV=TOTAL BEEF AND SHEEP IN LIVESTOCK UNITS
         MLKSHED=TYPE OF MILKING FACILITY (5 CLASSES)
         MILKFAT=TOTAL MILKFAT PRODUCTION 1983 (KG)
         NONDAIRY=ESTIMATED PER CENT INCOME NONDAIRY
         HAYPROD=NO. BALES OF HAY PRODUCED 1983 SEASON
         FEEDPRCH=NO. OF YEARS IN LAST 10 FEED PURCHASED
         OWNERSHP=TYPE OF FARM OWNERSHIP (8 CLASSES)
         AGEFARMR=AGE OF FARMER (YEARS)
         FARMEXP=NO. OF YEARS FARMER ENGAGED IN FARMING
         LABOR=TOTAL FULL-TIME EQUIVALENT LABOR
         EQUIPMNT=STATUS OF EQUIPMENT (3 ORDINAL CLASSES)
         PASTURE=NO. OF FIELDS (OUT OF 15) WEED INFESTED
         PAST_DIV=PAST DIVERSIFICATION (4 ORDINAL CLASSES)
         PLAN_DIV=PLANS FOR CHANGES (4 ORDINAL CLASSES)
         CODENO=SAMPLE FARM CODE NUMBER;
   MILK_HA=ROUND(MILKFAT/FARMSIZE,.1);
   MILK_COW=ROUND(MILKFAT/HERDSIZE,.1);
   LABEL MILK_HA=MILKFAT PRODUCED PER EFFECTIVE HECTARE
         MILK_COW=MILKFAT PRODUCED PER COW;
   CARDS;
```

marizing scores for *population distributions* can be applied to these sample distributions, with the exception of some of the techniques for spatial distributions. However, we are now calculating sample statistics, whereas in Chapter 3 we were calculating values for populations. We can still use the conventional symbols $\overline{X}$ for the arithmetic mean and s^2 for the variance, as in Chapter 3, but there is an important distinction between sample statistics and the population values that the sample statistics are estimating. In Chapter 4, we introduced the use of μ for the population mean and σ^2 for the population variance, and our interest lies in these population values. Thus, we can retain the use of $\overline{X}$ to symbolize the sample mean, noting that it will be our estimate of the population mean μ, but it is worthwhile at this stage to use the symbol $\hat{\sigma}^2$ (sigma square "hat") for the sample variance when estimating the population variance σ^2. We discuss the process of estimation in Chapter 6.

There is one change to the calculation procedure that we need to note now, although we cannot really justify it without further developing our statistical methodology. In calculating the variance for a population distribution, we divide the total variation ($\sum_{i=1}^{N} (X_i - \mu)^2$) by a denominator of N (total population). The definition of the variance for a sample distribution is the same using n, the sample size, to replace N, except when these summary measures are considered as *sample statistics* that are to be used to estimate their equivalent *population parameters*. In this situation, the variance just defined tends to underestimate the population variance. By dividing by $(n - 1)$ rather than n, we obtain a "better" estimate. For the moment, we note this difference and redefine the variance and standard deviation to incorporate the change. Other summary measures based on the variance—notably the statistical moments describing skewness and kurto-

TABLE 5.4 Sample Data for the Cumberland County Farm Survey

CUMBERLAND COUNTY DAIRY FARM SURVEY

OBS	FARMSIZE	RUNOFF	HERDSIZE	BREED	OTHERLIV	MLKSHED	MILKFAT	NONDAIRY	HAYPROD	FEEDPRCH	OWNERSHP	AGEFARMR	FARMEXP	LABOR	EQUIPMNT	PASTURE	PASTDIV	PLANDIV	CODENO	MILKHA	MILKCOW
1	34	0	61	1	0.0	5	8600	10	700	4	1	41	26	1.03	3	6	2	2	4	252.9	141.0
2	53	0	93	2	0.0	2	12900	0	1500	1	6	63	45	2.40	1	5	2	2	5	243.4	138.7
3	53	0	80	2	40.4	5	10000	0	1500	0	1	50	35	1.60	2	7	2	3	9	188.7	125.0
4	71	0	155	5	20.0	1	16300	5	2000	0	3	22	6	1.02	1	7	2	2	10	229.6	105.2
5	49	0	107	5	0.0	2	11300	0	1500	1	1	33	15	2.01	1	6	2	2	12	230.6	105.6
6	69	0	159	8	0.0	2	15900	0	1600	2	3	37	22	2.54	2	10	2	4	14	230.4	100.0
7	42	0	67	2	16.2	2	8800	5	750	0	1	35	19	1.00	2	4	2	2	15	209.5	131.3
8	81	0	103	2	80.0	4	13600	1	500	0	1	50	35	1.04	1	4	2	4	19	167.9	132.0
9	111	0	164	1	20.0	4	18100	0	1750	1	2	41	25	3.02	2	5	2	2	20	163.1	110.4
10	74	1	65	2	475.6	3	6800	55	500	1	1	47	15	1.01	2	6	4	4	21	91.9	104.6
11	45	1	48	4	202.0	5	3400	70	750	0	2	27	8	2.03	2	12	4	1	23	75.6	70.8
12	55	0	69	1	118.4	2	5000	50	2000	0	1	45	30	1.03	1	9	2	1	29	90.9	72.5
13	72	0	51	4	152.6	5	7300	60	1000	0	1	56	41	2.01	2	9	2	2	35	101.4	143.1
14	61	0	84	2	116.0	5	11300	33	1800	0	1	63	35	1.50	2	7	1	3	36	185.2	134.5
15	57	0	105	5	67.0	4	14300	10	1200	0	6	25	7	1.00	2	6	2	2	37	250.9	136.2
16	26	0	48	2	0.0	5	6300	0	500	2	1	59	32	2.00	3	7	2	4	38	242.3	131.3
17	40	0	83	2	0.0	3	8800	20	600	0	1	33	15	2.00	2	8	2	2	33	220.0	106.0
18	45	0	90	7	24.0	3	9500	5	1400	1	1	44	27	1.50	2	7	2	4	44	211.1	105.6
19	61	0	103	3	502.4	4	11800	25	5000	0	1	58	43	2.00	1	6	3	1	52	193.4	114.6
20	53	0	58	2	96.0	4	6600	25	1400	0	1	54	39	1.00	2	8	2	2	53	124.5	113.8
21	57	0	83	2	227.0	2	8600	50	1628	0	1	37	22	1.50	2	7	4	2	54	150.9	103.6
22	55	0	75	2	88.4	2	10400	20	1700	0	1	64	30	1.05	2	5	4	2	55	189.1	138.7
23	89	1	120	2	520.0	2	16300	25	2500	0	2	49	27	4.00	1	5	4	4	51	183.1	135.8
24	69	0	90	2	255.0	5	9500	30	1300	0	1	55	38	2.00	1	9	4	4	60	137.7	105.6
25	58	1	87	7	131.0	5	9500	10	200	1	1	54	22	1.00	2	6	3	4	61	163.8	109.2
26	78	0	117	2	155.0	2	13600	0	900	2	7	26	4	2.02	2	7	2	3	63	174.4	116.2
27	34	0	82	2	85.6	4	9100	30	1700	1	1	55	39	1.00	2	9	2	2	64	267.6	111.0
28	44	0	70	8	130.0	5	8200	50	2000	0	1	33	5	0.50	3	7	2	2	69	186.4	117.1
29	40	0	67	2	0.0	2	9100	0	1000	0	1	42	19	2.00	1	3	2	2	71	227.5	135.8
30	107	1	100	2	224.0	3	12200	25	3000	0	1	43	28	2.10	2	4	2	2	62	114.0	122.0
31	45	0	85	5	5.0	3	9100	0	800	1	1	38	10	2.00	2	8	2	4	66	202.2	107.1
32	19	0	21	7	249.0	5	2300	90	500	1	1	58	30	1.00	2	3	4	4	58	121.1	109.5
33	97	1	120	2	204.0	2	13200	40	1100	1	1	38	20	2.50	2	8	4	3	82	136.1	110.0
34	81	0	100	7	308.8	2	10900	30	2000	3	2	42	26	2.00	2	8	3	4	88	134.6	109.0
35	103	1	138	5	500.0	2	17700	5	3200	0	2	44	25	1.50	1	6	3	4	89	171.8	128.3
36	38	0	78	2	15.0	5	10400	5	2000	1	1	28	3	1.50	2	7	3	1	92	273.7	133.3
37	130	0	110	7	415.0	2	13600	40	3000	0	1	32	15	1.55	1	7	4	4	94	104.6	123.6

sis—will also be affected. Thus, we can define the sample variance, an estimate of the population variance, as

$$\hat{\sigma}^2 = \frac{\sum_{i=1}^{n} (X_i - \overline{X})^2}{n - 1} \tag{5.1}$$

$$\hat{\sigma}^2 = \frac{\sum_{i=1}^{n} X_i^2 - \dfrac{\left(\sum_{i=1}^{n} X_i\right)^2}{n}}{n - 1} \tag{5.2}$$

Note that almost all computer statistical packages assume that the data being processed are sample data, and that the sample data will be used to make inferences about a population. Therefore, the statistics that are calculated (5.1, 5.2) incorporate the modification of $n - 1$ as the denominator in the derivation of the variance.

If our aim in taking the sample is simply to obtain information about the sampled elements themselves, then our task is completed. We have obtained information about a certain number of variables for a collection of sample elements and summarized the distribution of scores for each variable graphically, by summary measures (Table 5.5) and perhaps by maps. More usually, however, our main interest is not in the sample but in the population from which the sample was obtained. In this situation, our task has barely begun—we have to find methods for making inferences about the population from the sample data.

Sampling Distributions

The key to the inferential process lies in the concept of sampling distributions. We have already mentioned a number of different types of distributions; we now introduce sampling distributions by reviewing the various statistical distributions we have encountered.

First, for a particular variable, sampling was carried out and a certain set of scores resulted—we called the distribution of sample scores the *sample distribution.* The individual scores were denoted by X_i, the number of scores making up the distribution (the size of the sample) was denoted by n, the mean of the distribution was denoted by $\overline{X}$, and the estimate of the population variance by $\hat{\sigma}^2$.

Second, this sample distribution was obtained from a *population distribution,* either finite or infinite. In most sampling situations, we would not know the characteristics of the population distribution in any detail; however, we would still denote its scores by X_i, the number of scores making up the population, if finite, was denoted by N (known or unknown), the mean of the distribution was denoted by μ (usually unknown), and the variance by σ^2 (usually unknown).

Third, in certain situations, we would have assumed that this population

TABLE 5.5 Summary Statistics for Cumberland County Farm Survey

CUMBERLAND COUNTY DAIRY FARM SURVEY

VARIABLE	LABEL	N	MEAN	STANDARD DEVIATION	MINIMUM VALUE	MAXIMUM VALUE	STD ERROR OF MEAN	SUM
FARMSIZE	EFFECTIVE AREA OF FARM IN HA	37	62.05	25.01	19.00	130.00	4.11	2296.00
RUNOFF	BINARY: 1 FOR USE OF NON-CONTIGUOUS LAND	37	0.19	0.40	0.00	1.00	0.07	7.00
HERDSIZE	NUMBER OF COWS IN MILK (1/1/84)	37	90.16	31.47	21.00	164.00	5.17	3336.00
OTHERLIV	TOTAL BEEF AND SHEEP IN LIVESTOCK UNITS	37	147.12	160.25	0.00	520.00	26.35	5443.40
MILKFAT	TOTAL MILKFAT PRODUCTION 1983 (KG)	37	10548.65	3728.32	2300.00	18100.00	612.93	390300.00
NONDAIRY	ESTIMATED PER CENT INCOME NONDAIRY	37	22.27	23.27	0.00	90.00	3.83	824.00
HAYPROD	NO. BALES OF HAY PRODUCED 1983 SEASON	37	1526.43	943.52	200.00	5000.00	155.11	56478.00
FEEDPRCH	NO. OF YEARS IN LAST 10 FEED PURCHASED	37	0.65	0.95	0.00	4.00	0.16	24.00
AGEFARMR	AGE OF FARMER (YEARS)	37	43.81	11.70	22.00	64.00	1.92	1621.00
FARMEXP	NO. OF YEARS FARMER ENGAGED IN FARMING	37	23.86	11.77	3.00	45.00	1.93	883.00
LABOR	TOTAL FULL-TIME EQUIVALENT LABOR	37	1.67	0.69	0.50	4.00	0.11	61.96
PASTURE	NO. OF FIELDS (OUT OF 15) WEED INFESTED	37	6.70	1.93	3.00	12.00	0.32	248.00
MILK_HA	MILKFAT PRODUCED PER EFFECTIVE HECTARE	37	179.51	54.17	75.60	273.70	8.91	6641.90
MILK_COW	MILKFAT PRODUCED PER COW	37	117.24	17.12	70.80	143.10	2.81	4338.00

distribution followed a particular probability function, and for any particular value (or range of values) of X_i there would be a certain probability value $y_i = f(X_i)$ and these probability values formed another distribution, although strictly speaking, we had confined the term "distribution" only to the cumulative form. The nomenclature for scores, the number of scores, the mean, and the variance would depend on what particular function we were referring to. Thus, if it were the normal probability function, the scores would be indicated by y_i, the number of scores would be infinite, the mean would be denoted by a $(=\mu)$, and the variance by b^2 $(=\sigma^2)$. If it were the point binomial function, the scores would be denoted by y_i, the number of scores would be N, the mean would be defined as $p(=\mu)$ and the variance as $p(1-p)$ $(=\sigma^2)$. The relationship between the population distribution and the assumed probability function for these two examples would be summarized by $X_i \sim N(a,b)$ and $X_i \sim PB(p)$, respectively.

Fourth, we are now in a position to define *sampling distributions*. We are interested in the distribution of a set of scores where each score is an estimate of a parameter of a function (i.e., a sample statistic) taken from a single sample. A sampling distribution is, then, a theoretical probability function describing the distribution of some property obtained from a sample. It applies to each of an infinite number of samples drawn from the same population and employs identical selection procedures. For example, the property may be the sample mean or variance, or some function of them, or some specified score (say the largest value). We rarely are able to derive the sampling distribution because we usually take only a single sample from a population, and from that sample we calculate sample statistics for each of the population parameters in which we are interested. The importance of sampling distributions is that we can compare the sample statistics we have obtained from the single sample with their relevant known sampling distributions, which will enable us to make conclusions about the unknown population parameter. The methods for doing so are discussed in Chapters 6 and 7.

The Four Key Probability Functions

What we need to know, then, is the probability form of the sampling distributions for the statistics in which we are interested. As we will see, if the population from which the sample was obtained is distributed normally (or approximately normally), then our discussion of sampling distributions for various parameters (bivariate and multivariate as well as univariate) centers on only four probability functions describing these sampling functions: the standard normal (z), chi square (χ^2), Student's (t), and Snedecor's (or Fisher's) (F) distributions. The functional form, parameters, mean, and variance for these functions are given in Table 5.6.

All four of these probability functions are continuous probability functions, but we will define them for integer values only. Note that the four

TABLE 5.6 Functional Form of Four Sampling Distributions

Function	Functional Form	Restrictions	Parameters	Moments: Mean	Moments: Variance
Standard normal	$y = f(z) = \frac{1}{\sqrt{2\pi}} e^{-z^2/2}$	$-\infty < z < \infty$		0	1
Chi square	$y = f(x) = \frac{x^{(V-2)/2} e^{-x/2}}{2^{V/2}\left(\frac{V-2}{2}\right)!}$	$x > 0$ $V = 1, 2, \ldots$	V	V	$2V$
Student's t	$y = f(x) = \frac{\left(\frac{V-1}{2}\right)!}{\sqrt{V\pi}\left(\frac{V-2}{2}\right)!}\left(1 + \frac{x^2}{V}\right)^{-(V+1)/2}$	$-\infty < x < \infty$ $V = 1, 2 \ldots$	V	0	$\frac{V}{V-2}$ for $V > 2$
Snedecor's F	$y = f(x) = \frac{\left(\frac{V_1+V_2-2}{2}\right)!\left(\frac{V_1}{V_2}\right)^{V_1/2}(x)^{(V_1-2)/2}}{\left(\frac{V_1-2}{2}\right)!\left(\frac{V_2-2}{2}\right)!\left(1+\frac{V_1x}{V_2}\right)^{(V_1+V_2)/2}}$	$x > 0$ $V_1 = 1, 2 \ldots$ $V_2 = 1, 2 \ldots$	V_1, V_2	$\frac{V_1}{V_2 - 2}$	$\frac{V_2^2(V_1+2)}{V_1(V_2-2)(V_2-4)}$

TABLE 5.7 Equivalence of Values of F (with Parameters V_1 and V_2) to Certain Values of t, χ^2, and z

		Value of Parameter V_1			
		1	V_1	∞	
Value of parameter V_2	1	$(t_{1-p/2}(1))^2$	$\frac{1}{(t_{(1-p)/2}(V_1))^2}$	$\frac{1}{(z_{(1-p)/2})^2}$	
	V_2	$(t_{1-p/2}(V_2))^2$	$F_{1-p}(V_1,V_2)$	$\frac{V_2}{\chi_p^2(V_2)}$	Table of Values of $F_{1-p}(V_1,V_2)$
	∞	$(Z_{1-p/2})^2$	$\frac{\chi^2_{1-p(V_1)}}{V_1}$	1	

functions are related (Table 5.7), and although we will not examine this topic, we will note the similarity of certain inferential tests using various of these probability functions in later chapters. The definition and use of p (a probability value) is delayed until Chapter 6. The cumulative form of these four probability functions is given in Tables B, C, D, E, and G at the end of the book. We will examine the structure and use of these tables in later chapters. The cumulative form (distribution function), rather than the density form, enables a more flexible application of the tables.

For all but the most advanced statistical analysis, we need not concern ourselves with the derivation of sampling distributions. The sampling distributions for all the commonly used statistics are well known and frequently tabled. To apply an inferential test to a sample statistic derived from a sample, all we need to know are the following. (1) What sampling design was used to obtain the sample? (2) What probability function does the population follow? (3) What probability function does the sampling distribution of the sample statistic follow? For most statistical work, these three questions reduce to two assumptions: (1) the sample is a random sample from the population, and (2) the population follows a normal distribution.

The Central Limit Theorem

The reason for the restriction of our discussion to sample statistics derived from samples from normally distributed population variables can now be examined. It derives not just from the fact that many variables are distributed normally but also because of a very remarkable theorem that will be stated here without proof. The *central limit theorem* can be presented in a number of ways but is stated here in its simplest form.

> **If a population has a mean μ and a finite variance σ^2, the sampling distribution of the sample mean approaches a normal distribution with mean μ and variance σ^2/n, as the sample size n increases.**

The importance of the normal distribution can thus be seen. No matter what probability function best describes the distribution of scores for a population, as long as the sample size n is large enough, the sample mean has a sampling distribution that is approximately normal. No minimum size of n can be quoted, because this would vary from one situation to another, but in most problems $n \geq 30$ is sufficient.

References and Readings

1. General Sampling Methods

Babbi, E. S. (1973) *Survey Research Methods,* Belmont: California.

Cochran, W. G. (1953) *Sampling Techniques,* Third Edition, Wiley: New York.

Cochran, W. G. and G. M. Cox (1957) *Experimental Designs,* Second Edition, Wiley: New York.

Kish, L. (1965) *Survey Sampling,* Wiley: New York.

Yates, F. (1953) *Sampling for Censuses and Surveys,* Griffen: London.

2. Spatial Sampling Methods

Berry, B. J. L. and A. M. Baker (1968) "Geographic sampling" in B. J. L. Berry and D. F. Marble (eds.), *Spatial Analysis: A Reader in Statistical Geography,* Prentice Hall: Englewood Cliffs: N.J.

Haggett, P., A. D. Cliff, and A. E. Frey (1977) *Locational Models,* Volumes 1 and 2, Arnold: London.

Holmes, J. (1967) "Problems in location sampling," *Annals, Association of American Geographers* 57:757–780.

Holmes, J. (1970) "The theory of plane sampling and its application in geographic research," *Economic Geography* 46:379–392.

3. Sampling and Inferential Methods

Cliff, A. D. and J. K. Ord (1973) *Spatial Autocorrelation,* Pion: London.

Court, A. (1972) "All statistical populations are estimated from samples," *Professional Geographer* 24:160–161.

Gould, P. (1970) "Is 'statistix inferens' the geographical name for a wild goose?" *Economic Geography* 46:439–448.

Meyer, D. R. (1972) "Geographical population data: Statistical description not statistical inference," *Professional Geographer* 24:26–28.

CHAPTER 6
Statistical Inference: Fitting Probability Functions

6.1 Selection of the Probability Function:
The fundamental use of the normal probability function; The problem of nonrecognizable distributions.

6.2 Parameter Estimation:
The principles of parameter estimation; Point estimation of the summary parameters; Point estimation of the parameters of probability functions.

6.3 The Normal Probability Function:
Preliminary processes; Fitting the normal probability function; Deriving the expected distribution; Fitting the curve graphically; Rapid methods of assessing normality.

6.4 The Lognormal Probability Function:
Arithmetic and logarithmic scales; Fitting the lognormal probability function.

6.5 Other Probability Functions:
The Poisson probability function; The binomial probability function; Approximations for the binomial and Poisson.

The methods used to estimate characteristics of populations from samples can seem difficult and confusing. This arises from the fact that, with some exceptions, to infer characteristics of a population from a sample not only implies that the sample was carefully chosen, but also that we know something about what probability function is followed by the population distribution. The latter may be a difficult assumption to justify unless we are dealing with a variable whose population has been examined repeatedly and for which a particular probability function has been shown to fit adequately, or unless we can resort to the central limit theorem.

The type of estimation problem may be the simple estimation of a single parameter of a population, or it may be a complete evaluation of how closely a sample distribution approximates some probability function. Because the latter includes the former, we examine in this and in Chapter 7 a complete step-by-step evaluation of the fitting of probability functions of various types to sample data. The simpler estimation problems will require only a small part of the total method. But we note here that estimation is usually involved in one of the following three situations.

First, we want to estimate the value of a single parameter—a summary parameter of the population such as the mean (μ) or a functional parameter such as (p) in the binomial probability function—from sample data.

Second, for later statistical analysis, we want to know whether the population distribution from which we have obtained a sample follows a normal probability function. Many parametric statistical tests require this assumption. This involves the estimation of the parameters a $(=\mu)$ and b $(=\sigma)$ of the normal probability function, calculation of a normal density or distribution function using these parameters, and assessing how closely the two distributions correspond.

Third, we want to test whether a population distribution or a sample distribution derived from a population distribution approximates a particular probability function. The methods are the same as those described immediately above, although the aim is slightly different. This is a theoretical problem. A certain theory suggests that the individuals of a population would be distributed according to a particular probability function; data for the population (or a sample from it) have been obtained. How well do the data fit the assumed function? Alternatively, we may be interested in the reverse of this situation: What probability function best describes the frequency distribution we have recorded?

With these three aims in mind, we can set out six (not necessarily sequential) steps in the complete evaluation of probability functions and sampled data. Three of these are examined in this chapter and three in the next. In this chapter, we examine the selection of the appropriate probability function (Section 6.1), the principles of estimating the functional parameters of the probability function (Section 6.2), and the construction of an expected distribution based on the chosen probability function and the point estimates of its parameters—*fitting the curve* (Sections 6.3 to 6.5).

Chapter 7 examines (1) tests of how well the expected distribution corresponds to the sample distribution—*testing the goodness of fit*, (2) providing an error margin around the point estimate based on the chosen probability function—*interval estimation*, and (3) testing the value of the point estimate against some specified theoretical value based on the assumed probability function—*hypothesis testing of parameters*. Testing for the equality of two sample means is deferred until Chapter 8.

If our aim is simply to estimate the value of some summary or functional parameter, then we would only be concerned with the first two and last two of these steps. On the other hand, if we were interested in matching up our recorded data with a specific probability function, then we would probably wish to employ all six steps.

6.1 SELECTION OF THE PROBABILITY FUNCTION

One of the most difficult problems in inferential statistics is deciding which probability function is approximated by the population being examined. Frequently, a variable is assumed to be *distributed normally* (i.e., the variable follows a normal probability function), without any examination or investigation to see whether this assumption is justified or not. In some situations, violating this assumption may be of little consequence, but in others, it may completely invalidate conclusions drawn from later statistical tests. This violation is especially critical when the sample size is small.

Selecting the probability function is a trial-and-error process. The process involves comparing the distribution of scores obtained from the sample with distributions based on various probability functions. However, three lines of evidence might simplify the process. First, and fairly obvious, the characteristics of the variable itself may suggest its probability relationships. From previous research it may be suggested that a particular probability function provides the best fit for the data, or the theory behind the structure of the variable may be based on a particular probability rule. Second, the calculation of summary measures for the sample may suggest the type of probability function. For example, in the normal probability function, the mean, mode, and median should approximately coincide; or in the Poisson probability function, the mean and the variance should be approximately equal. Third, some probability functions have fixed *domains* or range of values—for example, the positive integers only—and the characteristics of the variable can be compared to these.

More usually, however, it is necessary to carry out some form of comparison of the sample data with a set of probability functions. In fact, this should be standard practice even when the selection is supported by other evidence. The easiest method (although time consuming) for doing this is to graph the distribution in the form of a frequency polygon or curve and compare the distribution with known probability density or mass functions. However, for any probability function there are a wide variety of curves, depending on the value of the parameter or parameters, and many different probability functions have similar curves. But graphed examples give some idea of the type and general form of the population distribution and what probability function might be appropriate. Using the frequency curve in this manner was suggested in Section 3.1, where we outlined some of the more important curve shapes. For certain probability functions, special graph paper has been constructed that provides linear (straight line) plots for cumulated data that follow a particular function. The use of these graph papers is examined later in the chapter.

In most instances, we made assumptions about the form of the population distribution, and it remains to be seen whether the assumption was justified or not. The assumption most commonly made is that the population is distributed normally.

The Fundamental Use of the Normal Probability Function

Throughout the previous chapters, we stressed the importance of the normal probability function in statistical analysis. The bulk of this book is, in fact, concerned with the analysis of variables assumed to be distributed normally. The importance of the normal probability function arises from a number of reasons as follows.

First, the properties of the central limit theorem and its extensions (Section 5.4) stated that the sampling distribution of the sample means approaches a normal distribution as n (the size of the sample) increases, even when the population from which the sample means were obtained is *not* distributed normally. This allows us to use inferential methods based on the normal probability function when testing parameters related to the mean, even when we cannot assume that the population from which the data were sampled follows the normal probability function, *as long as the sample size is large enough*. How large must the sample size be? This depends on the variability of the data: the greater the variability the larger the necessary sample size. A general rule of thumb sets the minimum size at $n = 30$, but this value should not be regarded as absolute.

Second, inferential tests can either be described as *parametric* (distribution-assumed methods) or *nonparametric* (distribution-free methods); that is, the method either implies that the variable(s) is distributed according to a certain probability function (distribution) or no assumption is made about the form of the population. Almost all parametric methods assume the normal probability function. Distribution-free methods, although they relax the assumption about the form of the distribution, are often less efficient (see Section 6.2) and may provide less satisfactory statistical conclusions. As more statistical methods are available using the normal probability function than any other function, its use is preferred.

Third, a large number of variables that are characterized by a distribution of scores around some central score follow a normal probability function. The classic measurement-error model (random errors around an unknown "true" value) introduced in Section 1.2 provides the most common example. However, not all symmetric distributions are normal (Figure 6.1).

Fourth, the normal probability function has two parameters (a and b) that correspond to the mean and standard deviation, respectively, and these parameters can be estimated with a minimal amount of calculation.

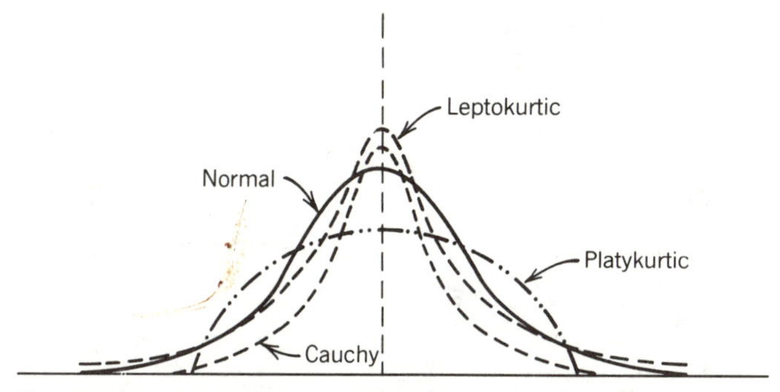

Figure 6.1 Symmetric distributions.

Fifth, by transforming a normally distributed variable into standard normal form using the Z-score transformation $Z_i = (X_i - \overline{X})/s$, with mean 0 and standard deviation 1, the probability properties of the distribution can be contained in a single universal equation or table. As an illustration, Figure 6.2 shows how three normal curves with different parameters can be transformed into a standard normal curve.

Finally, many variables that are not normally distributed can be transformed into an approximately normal distribution by simple manipulation. There is a variety of transformations that could be used, but the most frequently encountered is the logarithmic transformation for skewed distributions. This transformation and the lognormal probability function will be discussed later in the chapter.

The Problem of Nonrecognizable Distributions

Inevitably, in searching for a probability function to describe a frequency distribution, we may find that none of the functions outlined here fit the data. It is important to recognize that not only are there many probability functions other than the ones included here but that in a research situation it may be necessary to create "new" probability functions. It is not possible to discuss this in detail here, but we can illustrate this situation by examining two of the most frequently encountered departures from common distributions. These concern *truncated* distributions (Figure 6.3) and *bimodal* distributions (Figure 6.4).

Truncated distributions can occur quite frequently in geographic research, especially when dealing with positively scored data only, where the distribution may be truncated at zero. Fitting a normal probability function to these data could be a problem, for example. If the truncation is near the extreme end of the left-hand tail, then the departure from normality is probably small, but if a major truncation occurs, then some form of transformation may assist. Alternatively, with a little manipulation, a *truncated*

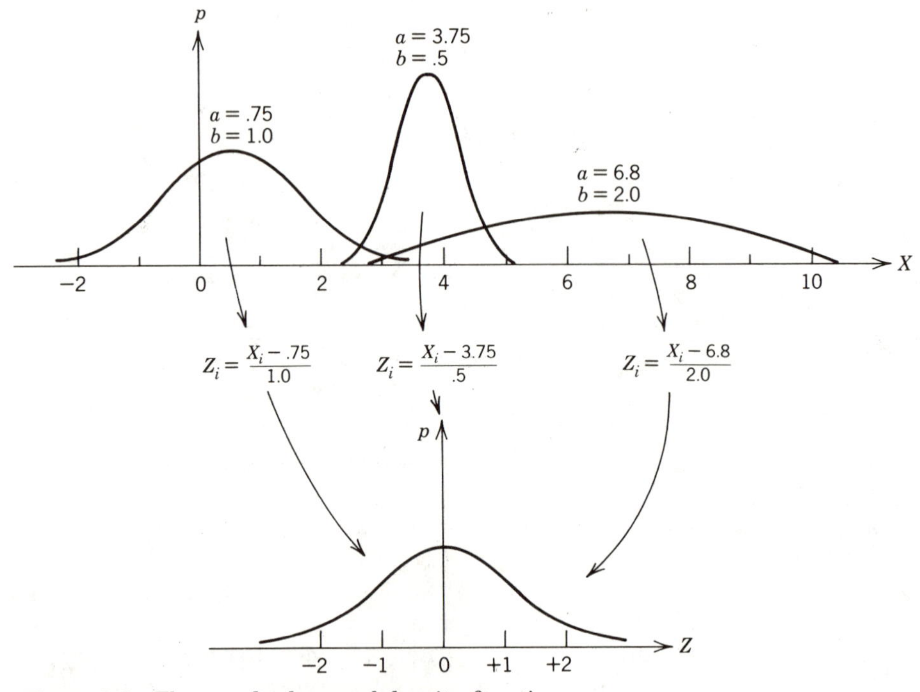

Figure 6.2 The standard normal density function.

normal probability function can be created, although this would not allow the use of standard parametric (normality-based) methods. Other truncations may occur at some natural or artificial boundary in the data—and again, functions can be created to fit the data, but alternative inferential testing methods would have to be found.

The problem of bimodal distributions is usually considered part of the general problem of *mixed* distributions. That is, it is assumed that the fre-

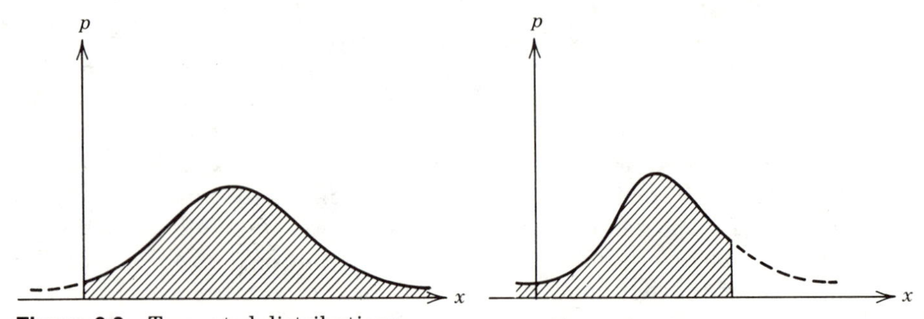

Figure 6.3 Truncated distributions.

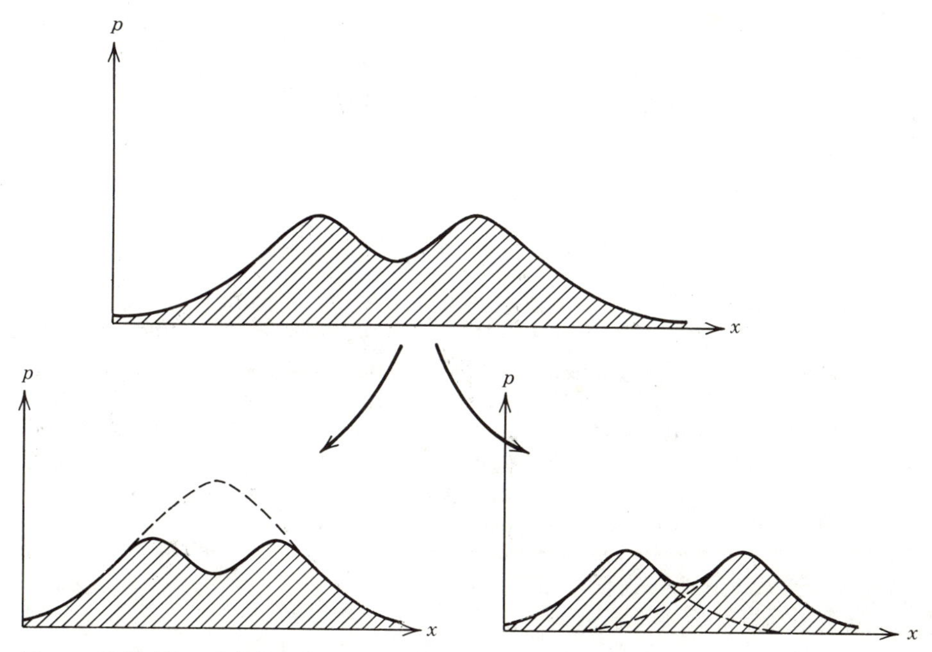

Figure 6.4 Bimodal distributions.

quency distribution contains elements from two or more populations. A common illustrative example is in the analysis of sediment particle size from a deposit with a bimodal distribution. Because deposition of a sediment results from a particular process (e.g., by running water) and the size of the sediment being deposited is governed by the magnitude of the force of the process, when a bimodal distribution is encountered one of three things has occurred.

1. The particular process has operated at two different magnitudes to provide two modal sizes.

2. Some other process has operated to differentially remove sediments between the two modal sizes.

3. Some other process (e.g., wind deposition) may be operating in addition to the assumed process (deposition by running water), resulting in two separate populations being sampled.

Bimodal populations cannot be handled using the standard statistical procedures; thus, variables with this type of distribution must be examined carefully to see if they are mixed distributions and, if they are, if they can be modified in some way. The modification would mainly take the form of some sort of separation into two component distributions or the selection of some additional variable that enables sampling from a more uniform population. Alternatively, the distribution could be fitted to a contagious probability function that is simply the combination of two or more separate probability functions. These functions are examined in Pielou (1969).

6.2 PARAMETER ESTIMATION

We defined parameter or point estimation as the process of obtaining an estimate of a particular characteristic of a population. The estimates of the *parameters* of a population are referred to as *sample statistics* or *estimators*.[1] The aim of parameter estimation is simple—to obtain the most accurate, or "best" estimate available. But the process is very difficult in practice, because there are a number of ways to define a best estimate; and also, what may be best for one purpose may be inadequate for other purposes. We will examine the principles of parameter estimation and then apply these principles to estimating the summary parameters for a population and the functional parameters for a probability function.

The Principles of Parameter Estimation

It is outside the scope of this book to do more than briefly examine the principles behind estimation. The topic, however, is extremely important and the reader who is unwilling to take a lot for granted in the remainder of this (and later) chapters should examine one of the many books that discuss estimation in more detail (Afifi and Azen, 1979; Mood and Graybill, 1963; or a textbook in econometrics). We first discuss a number of ways in which we could assess the "best" estimator, and then briefly outline the major methods of deriving estimators.

Let us take a simple example of a probability function ($y = f(x)$), with the values of y (the probabilities) distributed in some known form (i.e., we might know that it was distributed according to the binomial probability function, or the normal probability function, etc.). We will assume that this function is defined by a single parameter θ (Greek lowercase theta). From the population defined by the probability function $f(x)$, we take a sample, and compute from the sample a *statistic* that is an *estimate* of the *parameter*, and we designate the estimate by $\hat{\theta}$ (theta-hat). For any θ, there is a variety of $\hat{\theta}$, and of these we wish to decide which is best.

The determination of the "best" estimator depends on what properties of the estimate we consider to be important. A number of criteria are often used, but four of these are considered to be of special importance:

[1]Generally the term *estimator* is the term reserved for the mathematical/statistical formula used to calculate the estimate or sample statistic.

1. An estimator is said to be *unbiased* if the expected value of the estimate is equal to the population parameter (for a review of expected values, see Section 4.4).

$$E[\hat{\theta}] = \theta$$

That is, if we took a series of random samples (of any size n), calculated the $\hat{\theta}$ for each sample, then averaged the values of $\hat{\theta}$, we would derive the population parameter. That is,

$$\bar{\hat{\theta}} = \theta$$

Thus, although an unbiased estimator may not closely estimate the value of θ in a single sample, the average of $\hat{\theta}$ from a number of samples would be very close to θ. An estimator without this property is said to be *biased.*

2. An estimator is said to be a *consistent* estimator of θ if $\hat{\theta}$ converges to θ as n (the sample size) approaches N (the population size). In other words, as larger and larger samples are taken from the population, the probability gets higher and higher that the difference between $\hat{\theta}$ and θ is very small.

3. Of all possible estimators, the one with the smallest variance is called an *efficient* estimator. That is, in the long run, the efficient estimator provides the closest estimate (least variance) of the true value of the parameter.

4. An estimator is said to be *sufficient* if it contains all the information about the parameter θ through a sample of size n. This means that all the characteristics of parameter θ can be derived from the characteristics of the estimator.

Ideally the "best" estimator would be one that is unbiased, consistent, efficient, and sufficient. This estimator rarely exists, and thus, we resort to using an estimator that does not fulfill one or more of these criteria.

A number of different methods for determining estimators has been devised. They vary according to which of the above properties are considered to be of most importance. Without deriving these methods, we can briefly comment on the four most common approaches.

The method of moments is the oldest method of estimation and is perhaps the easiest to visualize. In Chapter 4, several important probability functions were introduced, and for each function, the first moment and second central moment (the mean and the variance, respectively) were calculated. These moments were shown to be derived from the unknown parameter or parameters. Thus, in reverse fashion, the sample moments can be used to set up equations to estimate the unknown functional parameters. Under suitable conditions, this method generates consistent and sufficient estimators that are frequently unbiased.

The method of maximum likelihood is probably the most frequently used method to estimate parameters of univariate distributions. It is based on the principle of selecting values for the parameters that maximize the probability of the observed data. The estimators are usually consistent and sufficient, but can be biased.

The method of least squares, under certain conditions, gives results that are identical to the maximum likelihood method, and is often used for the estimation of linear functions. We will see its fundamental use in the estimation of regression parameters in Chapter 9. The method involves minimizing the sum of the squares of the deviations between the observed and expected values for the variable X.

The minimum chi-square method is used when the random variable X is made up of frequencies, and is therefore concerned mainly with discrete probability functions. It derives an estimator that minimizes the chi-square value determined by comparing the observed frequencies with the expected frequencies (using the estimator).

One of the four methods briefly outlined can be used to produce an estimator for the particular parameter or parameters. The degree to which the estimator departs from the properties of unbiasedness, consistency, efficiency, and sufficiency can usually be measured. The estimator may then be accepted, rejected, or modified to conform to the desired property or properties. In most circumstances, an estimator that is consistent and unbiased (or with only small bias) is desired, and, in addition, it is hoped that it is efficient and sufficient. However, an estimator with all these properties is rarely possible.

Point Estimation of the Summary Parameters

It can be shown that the sample mean ($\overline{X}$) is an unbiased, consistent, sufficient, and highly efficient estimate of the population mean (μ). Therefore, we retain the definition of the sample mean as being the best estimator of the population mean. We repeat the definition. Estimator of the population mean is:

$$\hat{\mu} = \overline{X} = \frac{\sum_{i=1}^{n} X_i}{n} \tag{6.1}$$

However, it can be shown that the sample variance in its original form is not an unbiased estimator of the population variance. We defined an unbiased estimator as one where the expected value equals the true value of the parameter, that is, that the mean of the *sampling distribution* of the estimates (a distribution made up of the variances of a number of samples) equals the population parameter. It can be shown that the mean of the sampling distribution of the s^2 is:

$$\overline{s}^2 = \frac{n-1}{n}\sigma^2$$

That is, the estimator gives values slightly too small in the long run, although for large samples, the difference would obviously be minor. Thus, s^2 with n in the denominator, which is the estimator determined by both the methods of moments and maximum likelihood, is rarely used. Manipulation of the relationship produces an unbiased estimator

$$\hat{\sigma}^2 = \frac{\sum_{i=1}^{n} (X_i - \overline{X})^2}{n - 1} \tag{6.2}$$

This is the estimator we introduced indirectly in Section 5.4. Although many texts use Equation 6.3, to avoid confusion we emphasize $\hat{\sigma}^2$ as an estimate of the population variance. It is clear, however, that when n is large, the difference between dividing by n and $n - 1$ will be minimal.

$$s^2 = \frac{\sum_{i=1}^{n} (X_i - \overline{X})^2}{n - 1} \tag{6.3}$$

Not only is the estimate $\hat{\sigma}^2$ unbiased, it is also consistent and sufficient, with a high degree of efficiency. However, the positive square root of the estimator s—the standard deviation of the sample—is slightly biased, but because the amount of bias is very small, and because of its other properties, s is retained as the "best" estimator of σ.

There is another way to look at the modification of the denominator to produce an unbiased estimator of the variance. The reason why s^2, derived using a denominator of n rather than $n - 1$, was biased was because in its calculation, we had to use an estimate of the population mean μ (i.e., $\overline{X}$) to estimate the variance. In calculating the value of an estimator that requires the use of another estimator, we use the concept of *degrees of freedom*. The degrees of freedom will be the value we divide a total by (the denominator) to obtain an average value. In calculating the estimate of the mean, the degrees of freedom are at a maximum and equal the size of the sample n. That is, if we selected *any* n values from the population, and averaged them, this would be our estimate of the population mean. However, in the case of the calculation of the estimate of the variance, a different situation exists. Looking at the situation as if we were taking a sample of size n one at a time from the population, we could select any element of the population for all but one ($n - 1$) of the sample. In the last selection (the nth element of the sample), a value must be selected that provides us with the correct estimate for the value of $\overline{X}$. Thus, because we used an estimate for the value of the mean, we lost one degree of freedom for our estimate of the variance. Degrees of freedom are then $n - 1$.

The number of degrees of freedom can then be defined as the number of elements being considered (the sample size, n, the values of which will be unknown until the sample has been selected) *minus* the number of estimates already made in the calculation (i.e., the number of equations already linking the elements). In later estimates, more than one estimate will have already been made, and degrees of freedom of $n - 2$, $n - 3$, etc. will be encountered. This initial explanation of degrees of freedom may seem a little confusing, but its use should become clearer as the concept is used in later chapters. The concept of degrees of freedom also has practical advantages in that it enables the construction of tables for such statistics as t, F, and chi square, using degrees of freedom rather than sample size—allowing one table to be used in a number of different situations.

Alternative forms for the calculation of the variance of a sample are now presented.
From

$$\hat{\sigma}^2 = \frac{\sum_{i=1}^{n} (X_i - \overline{X})^2}{n - 1}$$

can be derived

$$\hat{\sigma}^2 = \frac{\sum_{i=1}^{n} X_i^2 - \frac{\left(\sum_{i=1}^{n} X_i\right)^2}{n}}{n - 1} \tag{6.4}$$

and for grouped data

$$\hat{\sigma}^2 = \frac{\sum_{j=1}^{k} f_j(M_j)^2 - \frac{\left(\sum_{j=1}^{k} f_j M_j\right)^2}{n}}{n - 1} \tag{6.5}$$

where f_j and M_j are the frequency and midpoint of class j respectively, and k is the number of classes.

Calculating s^2 from grouped data produces a more biased estimate of σ^2 than calculating s^2 from ungrouped data, and Sheppard's correction is often used. This involves subtracting from the calculated s^2, a value equal to the class interval squared, divided by 12. That is,

$$\hat{\sigma}^2 = s^2 - \frac{I^2}{12} \tag{6.6}$$

The amount of correction is usually small and because it results in problems in using the resulting value for $\hat{\sigma}^2$ in later tests, it will not be used here.

We have discussed the estimation of the two most important summary parameters—the mean and the variance (and the standard deviation). Equivalent methods are available for the other summary measures introduced in

Chapter 4. However, we need not go into their derivation here, and we simply note that in almost all cases the estimate of the population parameter follows the equivalent definition applied to the sample, with $n - 1$ used in the denominator when our parameters involve deviation around a measure of central tendency (including skewness and kurtosis). One additional point should be emphasized. When samples are very small (e.g., 10 or fewer), the estimation methods just described may provide poor estimates of their equivalent population values. In these circumstances, other estimation procedures have been suggested, but because these are usually fairly specific to individual situations, they are not discussed here.

Point Estimation of the Parameters of Probability Functions

As might be expected, for any particular probability function, a variety of point estimators have been suggested. Table 6.1 outlines the most frequently encountered. For the most part, they are the estimators produced by the method of maximum likelihood and/or the method of moments. Note that for the normal probability function, a more accurate, if tedious, method is to use the central limit theorem and obtain estimates of a and b by taking a large number of samples and calculating the mean for each. The estimate of a would then be the mean of the sample means, and the estimate of b would be equal to the standard deviation of the sample means multiplied by the square root of n, the sample size. For the lognormal distribution, the estimators $\overline{X}_L s_L$ are the estimators of the mean and standard deviation of the data in logarithmic form, and therefore, their antilogarithm is *not* the mean and standard deviation of the untransformed data. This point will be taken up later in this chapter. Examples of the computation of four of these various estimators (for the binomial, Poisson, normal, and lognormal probability functions) are given in the following sections. Discussion of the remaining probability functions is outside the scope of this book and the interested reader is referred to the references at the end of the chapter.

6.3 THE NORMAL PROBABILITY FUNCTION

We begin the process of fitting probability functions to sample data by examining the normal probability function. The process is often described as "fitting a curve to a frequency distribution" and, indeed, graphical methods can be used, and in some situations provide the quickest methods. However, the usual method is to construct an *expected frequency distribution* based on the normal probability function to compare with the *observed frequency distribution*. We will examine both tabular and graphical methods, but will delay any discussion of assessing the degree of fit until the next chapter. The methods outlined in this section can be applied to the other probability functions discussed in the remainder of the chapter.

TABLE 6.1 Point Estimators of Some Functional Parameters

Function	Estimators
Point binomial	$\hat{p} = \overline{X}$
Binomial	$\hat{p} = \dfrac{\overline{X}}{n_s}$ where n_s = sample size
Poisson	$\lambda = \overline{X}$
Negative binomial	$\hat{r} = \dfrac{(\overline{X})^2}{s^2 - \overline{X}}$
	$\hat{p} = \dfrac{\overline{X}}{s^2}$
Geometric	$\hat{p} = \dfrac{1}{\overline{X} + 1}$
Normal	$\hat{a} = \overline{X}$
	$\hat{b} = s$
Lognormal	$\hat{a}_L = \overline{X}_L$
	$\hat{b}_L = s_L$
Exponential	$\hat{\theta} = \overline{X}$
Gamma	$\hat{a} = \dfrac{(\overline{X})^2 - s^2}{s^2}$
	$\hat{b} = \dfrac{s^2}{\overline{X}}$

Preliminary Processes

Examples will be drawn from the 37-farm sample of the Cumberland County Dairy Farm Survey (Section 5.2). One of the variables measured was Herd Size—the number of cows on the farm used for milk production. The frequency distribution based on a 20-cow constant class interval (Table 6.2) was plotted as a histogram (Figure 6.5). The distribution appears normal both in the histogram and as a cumulative frequency distribution (Figure 6.6). The statement that the distribution appears to be similar to the distribution of a normal probability function is the first step in the inferential process.

Note also that Herd Size is an enumerated variable and the scores are discrete. However, we have selected a continuous probability function to approximate the distribution, a common occurrence when the magnitude of the scores is high relative to the magnitude of the discrete measurement units (integers in this instance).

We now require point estimates of the functions parameters a $(= \mu)$ and

TABLE 6.2 Frequency Distributions—Herd Size

Class	Frequency	Class	Cumulative Frequency	Relative Cumulative Frequency
		<20	0	.00
20–39	1			
		<40	1	.03
40–59	4			
		<60	5	.14
60–79	8			
		<80	13	.35
80–99	10			
		<100	23	.62
100–119	8			
		<120	31	.84
120–139	3			
		<140	34	.92
140–159	2			
		<160	36	.97
160–179	1			
		<180	37	1.00
TOTAL	37			

Source: Cumberland County Dairy Farm Survey.

b $(=\sigma)$. These point estimates are $\overline{X}$ and $\hat{\sigma}$, respectively.

$$\overline{X} = \frac{\sum_{i=1}^{n} X_i}{n}$$

$$= 90.2 \text{ cows}$$

$$\hat{\sigma} = \sqrt{\frac{\sum_{i=1}^{n} X_i^2 - \frac{\left(\sum_{i=1}^{n} X_i\right)^2}{n}}{n - 1}}$$

$$= 31.5 \text{ cows}$$

These estimates were calculated using all 37 scores (not from the 20-cow interval frequency table). Estimates derived in this manner are presumably more accurate, and when complete data are available then the ungrouped

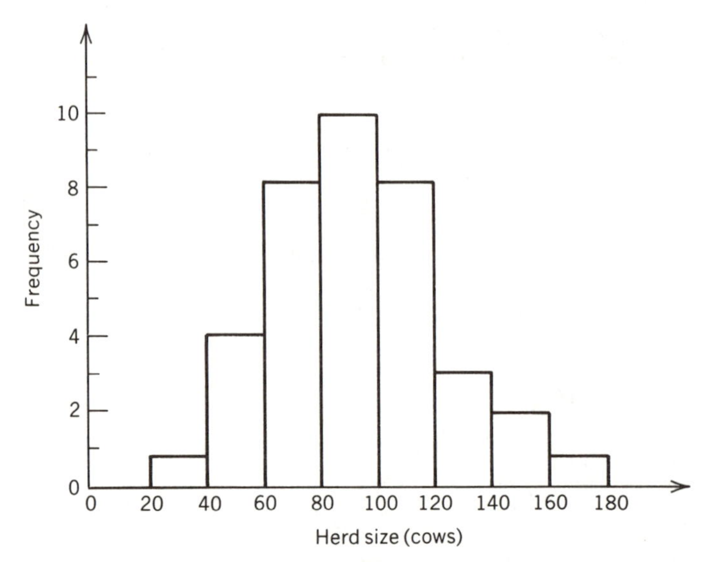

Figure 6.5 Histogram—Herd Size variable. *Source:* Cumberland County Dairy Farm Survey.

data should be used in calculating point estimates, unless we wish to deliberately smooth the data.

Fitting the Normal Probability Function

The equation describing the normal probability function is given by (from Table 4.4):

$$y = f(x) = \frac{1}{b\sqrt{2\pi}} e^{-(x-a)^2/2b^2} \tag{6.7}$$

Using $\overline{X}$ as the estimator of a and $\hat{\sigma}$ as the estimator of b:

$$y = f(x) = \frac{1}{\hat{\sigma}\sqrt{2\pi}} e^{-(x-\bar{X})^2/2\hat{\sigma}^2} \tag{6.8}$$

The expression gives the height of the curve at any point x, where the height is expressed in probability terms (proportions of 1.0). The total area under the curve can also be considered to have the value of 1.0. However, with histograms and frequency polygons, the area under the curve of the frequency polygon will be equal to the sum, across all classes, of the frequency of the class multiplied by its magnitude. When the class interval is constant, the total area under the polygon will be equal to the number of elements in the sample (n) multiplied by the magnitude of the class interval (I). Thus,

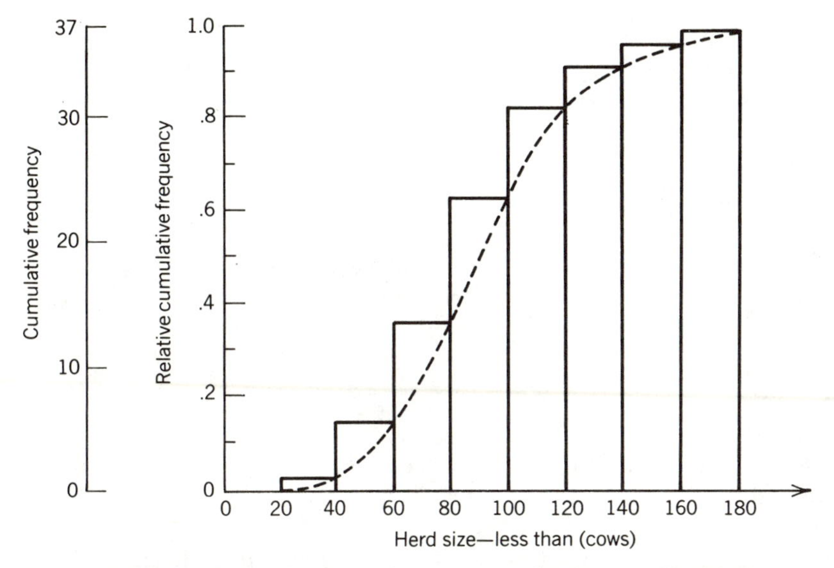

Figure 6.6 Cumulative frequency histogram and polygon—Herd Size variable. *Source:* Cumberland County Dairy Farm Survey.

an equivalent form for a density function equation that provides an approximation for a frequency polygon is

$$y = f(x) = \frac{nI}{\hat{\sigma}\sqrt{2\pi}} e^{-(x-\bar{X})/2\hat{\sigma}^2} \tag{6.9}$$

In our example, for any value ($X_i = x$) the Y_i value (the height of the curve) at that point is

$$Y_i = f(X_i) = \frac{37\ (20.0)}{31.5\ \sqrt{2\pi}} e^{-(X_i-90.2)^2/2(31.5)^2}$$

where the constants

$$e = 2.7183$$

$$\pi = 3.1416$$

and

$$\sqrt{2\pi} = 2.5066$$

This equation can be used to calculate the height of the curve at the boundaries between the classes, thus enabling the construction of an expected frequency curve or the calculation of estimated numbers in each class. However, the calculations involved are very tedious and the process can be simplified by using the properties of the *standard normal probability function* and its tabled values. By converting the individual X_i scores or any selected X_j scores (such as the class boundaries) into z scores using the standardizing equation

$$Z_i = \frac{X_i - \overline{X}}{\hat{\sigma}} \tag{6.10}$$

the standard normal probability function can be used. With a mean of 0 and a standard deviation of 1, the normal function expressed in Equation 6.7 simplifies to

$$y = f(z) = \frac{1}{\sqrt{2\pi}} e^{-z^2/2} \tag{6.11}$$

Table B at the end of the book is constructed to yield values for the height of the ordinate of the standard normal density function based on Equation 6.11.

In many instances, however, it is more convenient to focus on the areas between the class boundaries rather than the coordinates at those points and to use the cumulative areas between the class boundaries. This can be done with the cumulative standard normal distribution based on the distribution function.

$$y = F(z) = \int_{-\infty}^{z} \frac{1}{\sqrt{2\pi}} e^{-z^2/2} \tag{6.12}$$

The data are given in Table C at the end of the book.

Deriving the Expected Distribution

We will first illustrate fitting a normal distribution to an observed frequency distribution. In essence, we are comparing an expected (or theoretical) normal to an actual distribution. Table 6.3 lists the values of the classes (column 1) and the class boundaries (column 2). All the possible classes are included to accommodate a normal distribution (infinite in both directions), so that the table has open classes at each end. Standard scores (Z values) are then calculated for each of the boundaries (column 3). For example, the boundary value of $X_1 = 20$ would convert to

$$Z_1 = \frac{20.0 - 90.2}{31.5}$$

$$= -2.23$$

For each Z value, its $F(z)$ value (column 4) is taken from the table of the cumulative standard normal distribution (Table C). This $F(z)$ value gives the probability of a value of z being located between minus infinity and that

TABLE 6.3 Calculation of Expected Class Frequencies—Herd Size

Class	Approximate Class Boundary	Z Value	Cumulative Area, $F(z)$	Class Probability	Expected Frequency
<20				.0129	.48
	20	−2.23	.0129		
20–39				.0430	1.59
	40	−1.59	.0559		
40–59				.1126	4.17
	60	−.96	.1685		
60–79				.2060	7.62
	80	−.32	.3745		
80–99				.2472	9.15
	100	.31	.6217		
100–119				.2072	7.67
	120	.95	.8289		
120–139				.1140	4.22
	140	1.58	.9429		
140–159				.0439	1.62
	160	2.22	.9868		
160–179				.0110	0.41
	180	2.85	.9978		
≥180				.0022	0.08
TOTAL				1.0000	37.01

Source: Cumberland County Dairy Farm Survey.

specified value of z. By calculating the difference between each adjacent pair of $F(z)$ values, we obtain the probabilities associated with each class (column 5). The open classes at each end will have probabilities determined by finding the difference between the $F(z)$ value of the lowest boundary and an $F(z)$ of 0.00 (the probability at $-\infty$), and the difference between the $F(z)$ value of the highest boundary and an $F(z)$ of 1.00 (the probability at $+\infty$). By doing this we also have a check on this step of our method because the class probabilities should sum to 1.0. With these class probabilities and a sample size of 37, the expected number in each class will be the relative proportion of 37 (column 6). Thus, for example, the expected number of farms in the class 40 to 59 cows is

$$.1126\ (37) = 4.17$$

The expected numbers in the classes should sum approximately (allowing for some rounding error) to the sample size (in this instance 37). The observed and expected frequency distributions can then be compared (Table 6.4). A reasonable degree of correspondence can be seen, but we delay any statistical

TABLE 6.4 Observed and Expected Frequencies—Herd Size

	Class										
Frequency	**<20**	**20–39**	**40–59**	**60–79**	**80–99**	**100–119**	**120–139**	**140–159**	**160–179**	**≥180**	**Total**
Observed	—	1	4	8	10	8	3	2	1	—	37
Expected	.48	1.59	4.17	7.62	9.15	7.67	4.22	1.62	.41	.08	37.01

evaluation of this correspondence until we examine hypothesis testing in Chapter 7.

Fitting the Curve Graphically

There are a number of ways that the normal curve could be fitted graphically. First, we could plot the expected distribution by classes as calculated in Table 6.3. However, in trying to produce a smooth curve from this grouped data, the peak and points of inflection are not easily determined. For a more accurate portrayal, the height of the normal curve at the mean can be calculated and then used either to calculate the height of the class boundaries or to construct an $f(z)$ axis parallel to the y axis, which can then be used to plot the curve.

Because the curve of the normal probability density function will peak at the mean, the highest point on the expected curve will have its highest point above the estimated mean. To calculate the height of the curve above the point of the mean, we replace x by $\overline{X}$ in Equation 6.9, and we can see that

$$e^{-(\bar{X}-\bar{X})^2/2\hat{\sigma}^2}$$

simplifies to 1.0.

The equation for y at the point of the mean then becomes

$$y = \frac{nI}{\hat{\sigma}\sqrt{2\pi}} \tag{6.13}$$

In our example

$$y = \frac{37\ (20.0)}{(31.5)(2.5066)}$$

$$= 9.37$$

This value can then be plotted on the graph (Figure 6.7) as the peak of the normal approximation. Calculation of the height of the curve for other values of X can use the full equation (6.9) or, again, we can use the standard normal function. Class boundaries are converted to Z values, the height of the curve ($f(z)$) at these Z values is obtained from Table B. These are next converted to $f(x)$ values by proportioning them to the $f(x)$ value at the mean (Table 6.5). The points can then be plotted on the graph (Figure 6.7).

Rapid Methods of Assessing Normality

In many circumstances, rather than proceed with the complete construction of an expected distribution based on a normal approximation plus any subsequent testing of the goodness of fit, all that is required is a rapid assessment of whether we can reasonably assume that the sample frequency distribution comes from a population whose scores are distributed normally. If any major departure is noted, then the full testing procedure can be implemented. The observed frequency distribution itself, especially when drawn as a histogram or frequency curve, and its summary statistics will generally indicate any strong departure from normality. Several questions might be asked.

Do the mean and median (and mode) roughly coincide? In the sample of the Herd Size variable, we have

$$\left.\begin{aligned}\text{Mean} &= 90.2 \text{ cows}\\ \text{Median} &= 85 \text{ cows}\end{aligned}\right\} \text{based on the full distribution of 37 scores}$$

$$\text{Middle of the modal class} = 90 \text{ cows}$$

Does the shape of the frequency curve follow the properties of the normal curve—symmetry, peaks at the mean, inflection points approximately one standard deviation on either side of the mean? More specifically, is the coefficient of skewness approximately equal to zero? The Herd Size sample had a coefficient of skewness of .48 indicating a slight degree of positive skewness. Does the observed distribution have excessive peakedness (leptokurtosis) or flatness (platykurtosis)? Or is the coefficient of kurtosis approximately equal to zero? In the Herd Size sample, the coefficient is .19.

A more exact method, one that is capable of being extended to provide an agreement/disagreement decision on whether the normal fit is adequate, is

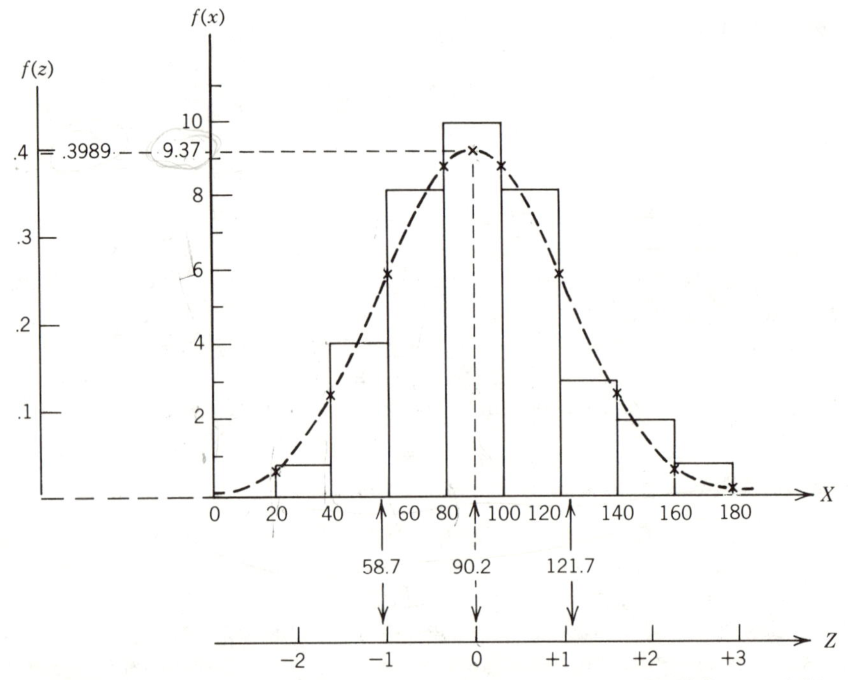

Figure 6.7 Observed histogram and fitted normal curve—Herd Size variable.

TABLE 6.5 Calculation of Ordinates—Herd Size

Class	Approximate Class Boundary	z Value	Height of Curve, $f(z)$	Height of Curve, $f(x)$
<20				
	20	−2.23	.0332	.80
20–39				
	40	−1.59	.1127	2.65
40–59				
	60	−.96	.2516	5.91
60–79				
	80	−.33	.3778	8.87
80–99				
	$\bar{x}$ = 90.2	.00	.3989	9.37
	100	.31	.3802	8.93
110–119				
	120	.95	.2541	5.97
120–139				
	140	1.58	.1145	2.69
140–159				
	160	2.22	.0339	.80
160–179				
	180	2.85	.0069	.16
≥180				

to use *a fractile diagram*. Only an outline of the method is presented here, the interested reader is referred to Hald (1952) for a more elaborate discussion. The method involves the use of normal probability paper (Figure 6.8), which has been constructed so that when relative cumulative frequencies are plotted against their X values for a normally distributed variable, a straight line results. Illustrating the procedure with reference to the Herd Size variable the x axis is scaled according to the range of herd-size values. Cumulative frequencies at each of the boundaries between classes (from Table 6.2) are then plotted on the graph (Figure 6.8).

While it is convenient to plot data in relative cumulative frequency form based on classes, a similar method can be applied to the raw scores. In this situation the X values are ranked from smallest, where the *rank* value (which we designate by ro_i) equals 1, to the largest, where ro_i will equal n, the size of the sample. We can then define points Po_i to plot on the graph by

$$Po_i = \frac{ro_i - .5}{n} \tag{6.14}$$

The .5 is subtracted to approximate the midpoint of an actual range of X values (ranked) represented by a single rank–score. A variety of other methods to determine the plotting point Po_i have been suggested, but the above is the simplest. Unless the sample size n is very small, all points need not be plotted, but only a selection of say 5 to 10 representative scores.

For convenience, we refer to the class boundary values as the Po_j values (i.e., the observed probability for the jth case). Because the normal probability function is infinite in both directions, we do not plot the two extreme values of class boundaries (20 and 180); with Po_j values of 0.0 and 1.0 these would not be defined on the graph.

A straight line representing the normal curve can then be drawn by eye or the tabled values for the cumulative standard normal distribution can be used (Table C). Three or four X scores are converted to Z scores, and their appropriate $F(z)$ value obtained from Table C. The X score and the $F(z)$ score then define the coordinates for the location of a point on the fitted normal curve. These points are then joined to form a straight line. The points selected (Figure 6.8) were as follows:

X Score	Z Score	$F(z)$ = Relative Cumulative Frequency
60	−.96	.1685
90.2	.00	.5000
120	.95	.8289
160	2.22	.9868

We will refer to the points along the fitted line as the Pe_j values (the "expected probability" of the jth case).

From the graph (Figure 6.8), we see that the observed points lie fairly close to the normal straight line, with a slight departure beneath the line at either end of the distribution. This indicates a frequency distribution with a small positive skew. In fact, fractile diagrams provide a rapid means of assessing the type of departure from normality simply by the shape of the pattern of points. Some more common departures are sketched in Figure 6.9. If desired, the fractile diagram can be used to provide an evaluation of the degree of correspondence of the observed frequency distribution and the normal curve—by assessing whether the departure of the Po values from the Pe values can be ascribed to random fluctuations. The method involves the fitting of a confidence interval about each of the Pe values.

6.4 THE LOGNORMAL PROBABILITY FUNCTION

The methods that have been outlined in the last section with reference to the normal probability function can be applied to the fitting of any probability function. There are only small differences from one function to another.

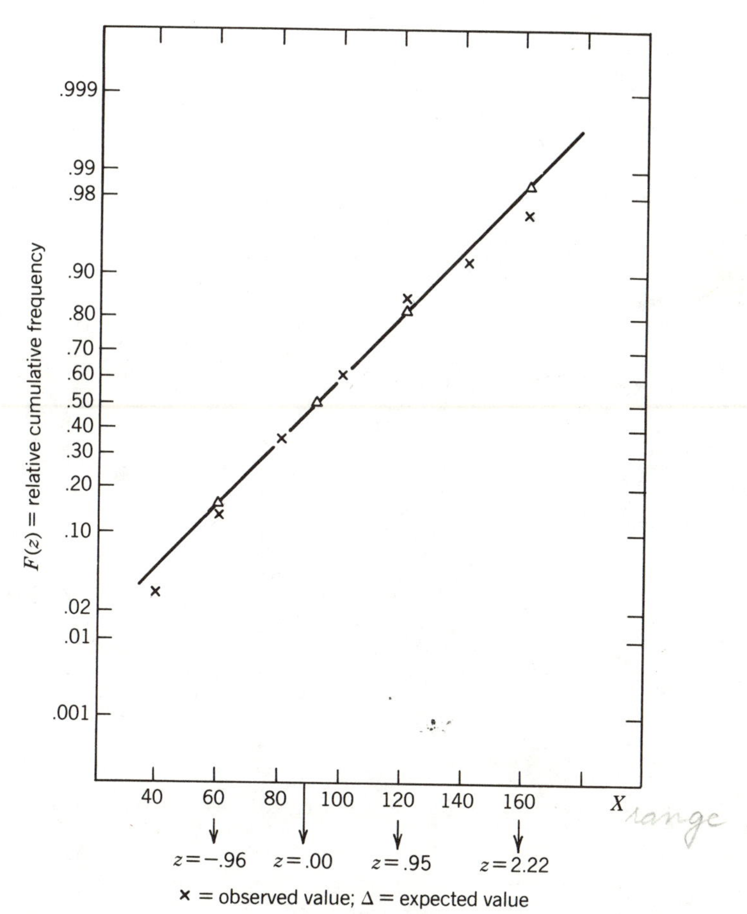

Figure 6.8 Fractile diagram—Herd Size variable.

However, it is worthwhile considering three further probability distributions that occur quite frequently: the lognormal, the Poisson, and the binomial.

A great number of variables of interest to geographers have a distribution with a marked skew to the right. It is assumed that many of these are lognormal, although there are a variety of probability functions with skewed distributions other than the lognormal. Nevertheless, the lognormal distribution has proved to be of some importance in geographic research. Before we examine this distribution, however, we digress slightly to examine the nature of logarithms.

Arithmetic and Logarithmic Scales

In the introduction to the use of graphs of Chapter 2 we commented on the dominant use of the arithmetic scale to portray distributions of scores. This

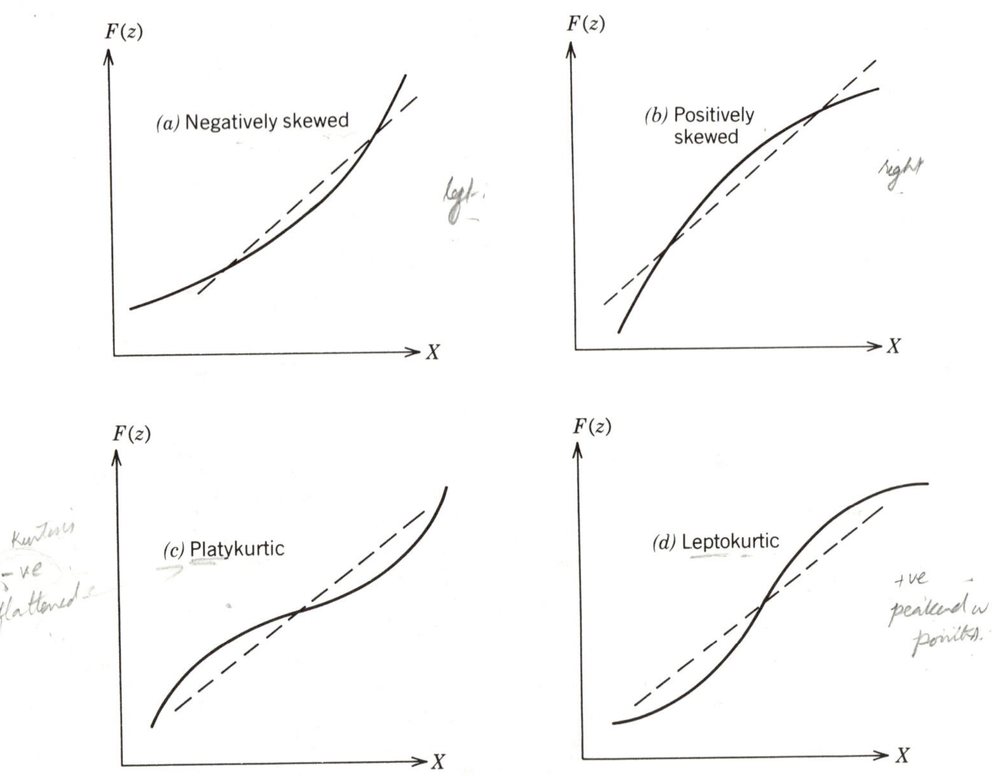

Figure 6.9 Fractile diagrams illustrating departures from normality.

is the scale used on the x axis of a histogram where the range of scores was divided using regular arithmetic increments. The normal probability function describes a frequency distribution of scores that produces the characteristic bell-shaped curve when the x scores are arranged arithmetically along the x axis. It should be obvious, however, that if we transform the x scores into some nonarithmetic form then the shape of the frequency curve will change. Conversely, it might be possible to transform some distributions that do not follow a normal probability function into some nonarithmetic form that does (Figure 6.10). Logarithmic transformations are one of a number of transformations that are used for this purpose.

One of the most common departures from normality is the distribution that is skewed to the right—the right-hand tail of the frequency curve is stretched in the positive direction. Many (but not all) of these skewed distributions can be transformed into near-normal distributions by converting the x scores into logarithmic form. There are a variety of logarithmic transformations depending on the *base* of the logarithm, but all of these logarithms transform the distribution in the same manner.

Using $\log_a$ to symbolize a logarithmic transform to the base a we can define a logarithm simply as being the number that satisfies:

$$\log_a x = b \tag{6.15}$$

That is, *the logarithm* of a number x is the amount we have to *raise* (power) the base to get to that number x. This is most easily illustrated by using *common logarithms*—the logarithms to the base 10 where

$$\log_{10}(100) = 2 \text{ as } 10^2 = 100$$

$$\log_{10}(1000) = 3 \text{ as } 10^3 = 1000$$

$$\log_{10}(11.5) = 1.0607 \text{ as } 10^{1.0607} = 11.5 \text{ (approximately)}$$

Logarithms can be constructed to any base, but almost all logarithms used are to the base 10 (the common logarithm) or the base e (the *natural logarithm*). The letter e (named after the mathematician Euler) has a value of approximately 2.7183. The latter logarithms are often designated by *ln* to differentiate them from common logarithms, designated by $\log_{10}$ or simply log.

Logarithm values have some important properties. The logarithm of x is not defined for $x \leq 0$. For x from 0 to 1, $\log_a x$ ranges from $-\infty$ to 0; for x from 1 to $+\infty$, $\log_a x$ ranges from 0 to $+\infty$.

$$\log_a a = 1 \tag{6.16}$$

$$\log_a x^n = n \log_a x \tag{6.17}$$

$$\log_a (x_1 x_2) = \log_a x_1 + \log_a x_2 \tag{6.18}$$

$$\log_a (x_1/x_2) = \log_a x_1 - \log_a x_2 \tag{6.19}$$

We can convert from a logarithm in one base to one in another base by a simple linear transformation. Thus, we can use any base for a logarithmic transformation (in practice this will mean either common or natural logarithms). The transformed scores differ by some simple ratio (Figure 6.10). For example, to convert from base e (natural) logarithms to base 10 (common) logarithms:

$$\log_{10} x = \frac{\ln x}{\ln 10} = \frac{\ln x}{2.302585} \tag{6.20}$$

Fitting the Lognormal Probability Function

In regard to the fitting of distributions to data, we can consider logarithmic transformations as transformations to allow the fitting of a normal distribution, or we can attempt to fit the lognormal probability function itself to

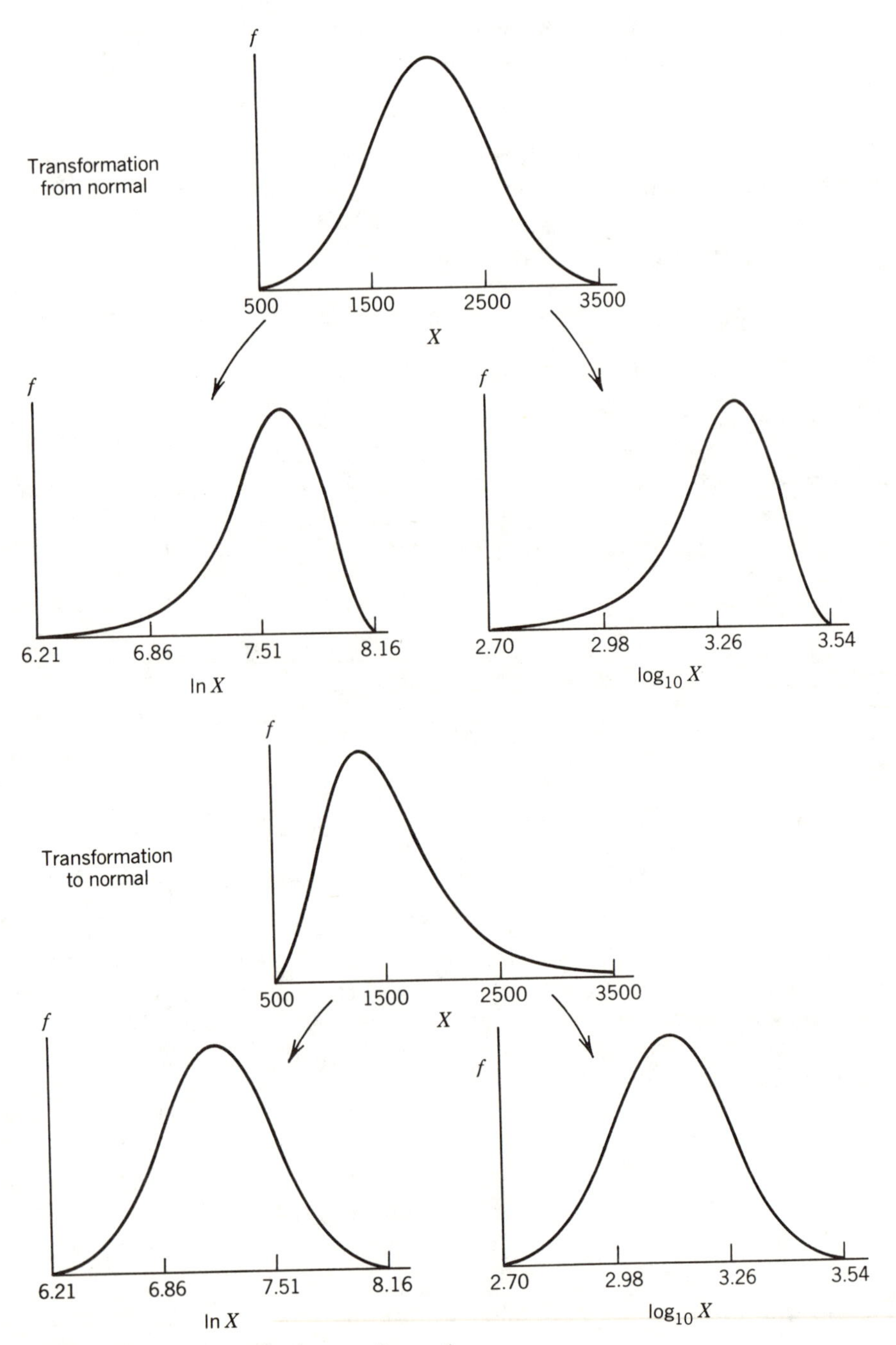

Figure 6.10 Logarithmic transformations.

the raw scores. The lognormal probability density function is given by

$$y = f(x) = \frac{1}{\theta\sqrt{2\pi}} e^{-(\ln x - \gamma)^2/2\theta^2} \tag{6.21}$$

where ln = logarithms to the base e (natural logarithms).

The estimation of the parameters (γ, θ) of the lognormal distribution and the fitting of the distribution is quite a complex procedure and will not be examined here. The interested reader is referred to the work of Aitchison and Brown (1957). Instead, we note that the lognormal probability function has the property that when the X values of a variable with a lognormal distribution are converted to logarithms, of any base, these new L values are distributed as a normal probability function with parameters a_L and b_L (equals μ_L and σ_L). There are three differences between the lognormal distribution of X values and the normal distribution of L values. First, the $\bar{X}$ and s calculated from a distribution of X scores are obviously not estimates for the normal distribution of L scores, but they also are not good estimates of the two parameters (γ, θ) of the lognormal probability function. Second, $\bar{X}_L$ and s_L calculated from the distribution of L scores (the X scores in logarithmic form) are the point estimates for $a_L(\mu_L)$ and $b_L(\mu_L)$ for the normal distribution, but again they are not the best point estimates for the lognormal probability function. Three, the antilogarithms of $\bar{X}_L$ and s_L are also not good point estimates for the lognormal probability function.

Thus, we are not really considering the lognormal probability function, but simply an extension of the normal probability function, where the X values have been transformed using logarithms to remove the positive skew. The form of this probability function is

$$y = f(L) = \frac{1}{s_L\sqrt{2\pi}} e^{-(L - \bar{X}_L)^2/2s_L^2} \tag{6.22}$$

where $\bar{X}_L$ and s_L are the point estimates of a_L and b_L. If a good fit to this function is found for the sample distribution, then the sample distribution basically follows the lognormal distribution, but the form of this latter function has not been fully determined.

All the methods outlined in the previous section can be applied to the transformed data. An example is provided by the variable Farm Area from the Cumberland County Dairy Farm Survey. From the frequency distribution (Table 6.6), a histogram was constructed (Figure 6.11) and the form suggested a lognormal distribution. A rapid check was provided for this by plotting the class boundaries of the relative cumulative frequency distribution on lognormal probability paper (where the x axis is constructed using a logarithmic scale—in this instance using a base of 10) or by converting the boundary values into logarithmic form and plotting these on normal probability paper. For comparison, the boundary values are plotted as X scores on normal probability paper (Figure 6.12*a*); as L scores on normal probability paper (Figure 6.12*b*); and as X scores on lognormal probability

TABLE 6.6 Frequency Distributions—Farm Area

Class	Frequency	Approximate Class Boundary	Log_{10} of Boundary	Cumulative Frequency (Less Than)	Relative Cumulative Frequency
		15	1.1761	0	.00
15–29	2				
		30	1.4771	2	.05
30–44	7				
		45	1.6532	9	.24
45–59	12				
		60	1.7782	21	.57
60–74	7				
		75	1.8751	28	.76
75–89	4				
		90	1.9542	32	.86
90–104	2				
		105	2.0212	34	.92
105–119	2				
		120	2.0792	36	.97
120–134	1				
		135	2.1303	37	1.00
TOTAL	37				

Source: Cumberland County Dairy Farm Survey.

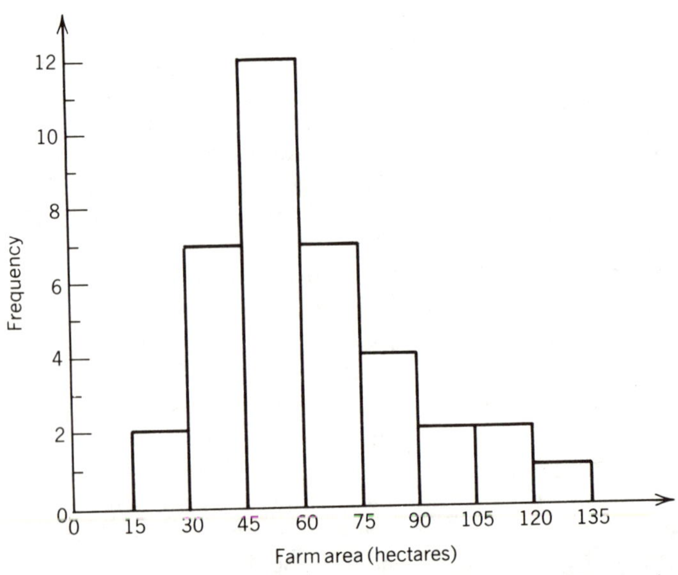

Figure 6.11 Histogram (arithmetic scale)—Farm Area variable. *Source:* Cumberland County Dairy Farm Survey.

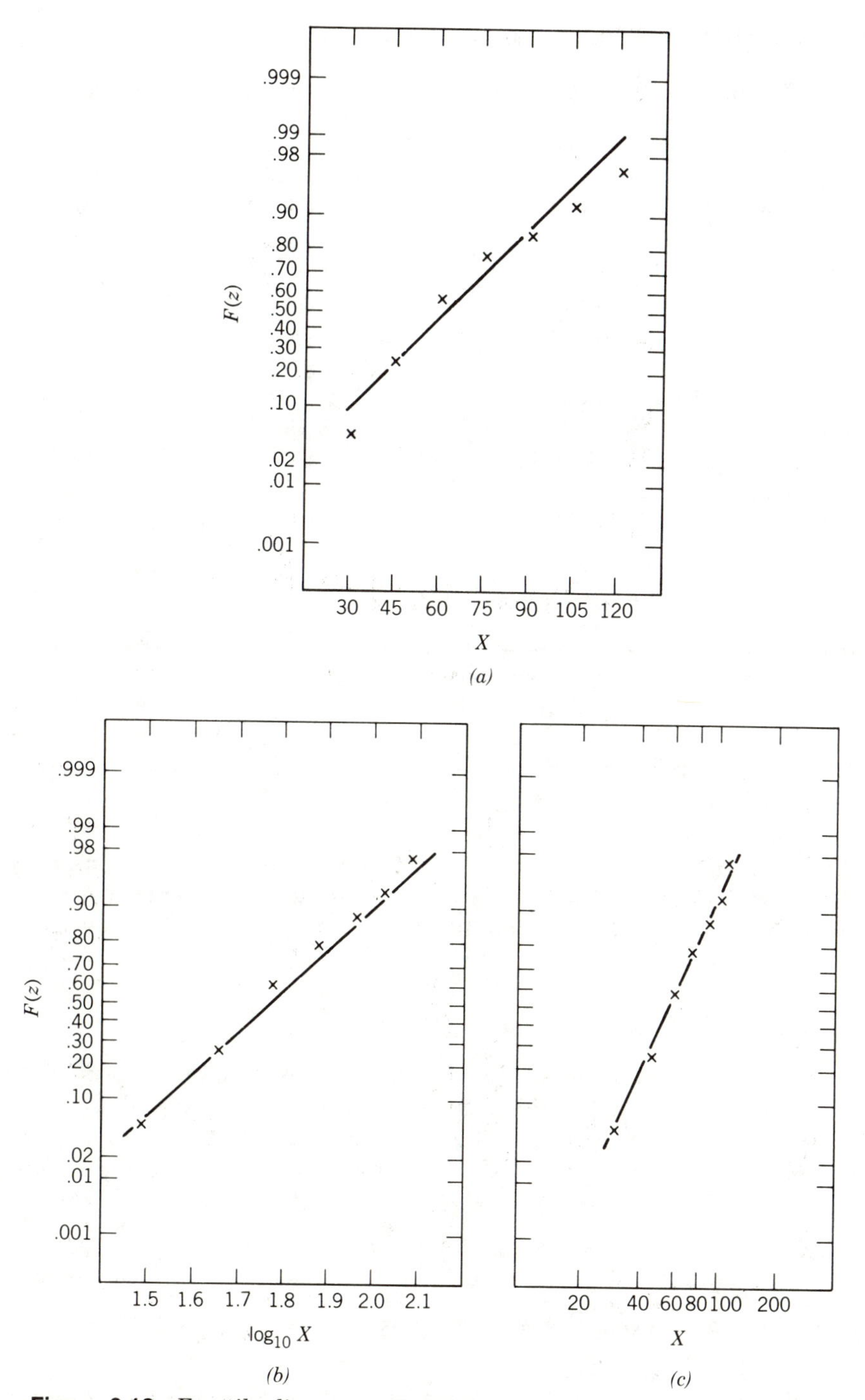

Figure 6.12 Fractile diagrams—Farm Area. (*a*) X scores on normal probability paper. (*b*) L scores on normal probability paper. (*c*) X scores on lognormal probability paper.

TABLE 6.7 Observed and Expected Frequencies—Farm Area (log)

Class (L Scores)	Approximate X Score Classes	Observed Frequency	Expected Frequency
<1.20	<16	—	.03
1.20–1.29	16–19	1	.16
1.30–1.39	20–25	0	.65
1.40–1.49	26–31	1	1.93
1.50–1.59	32–39	3	4.14
1.60–1.69	40–50	8	6.81
1.70–1.79	51–63	10	8.01
1.80–1.89	64–79	6	7.22
1.90–1.99	80–99	4	4.66
2.00–2.09	100–125	3	2.31
2.10–2.19	126–158	1	.82
≥2.20	>158	—	.27
TOTAL		37	37.01

Source: Cumberland County Dairy Farm Survey.

paper (Figure 6.12*c*). The straightening of the curve in Figures 6.12*b* and 6.12*c* suggests that the distribution is basically lognormal, although the fit is by no means perfect.

The Farm Area scores were converted to base 10 logarithms and a frequency distribution based on a new class interval was constructed (Table 6.7) for the 37 sampled farms. Point estimates for the parameters were:

$$\bar{X}_L = 1.7583 \text{ (log hectares)}$$

$$s_L = 0.1786 \text{ (log hectares)}$$

Using the methods outlined in Section 6.3, an expected distribution was calculated (Table 6.7) and a histogram drawn based on the L_i data plotted with the normal curve overlaid (Figure 6.13). The fitted curve has a reasonable degree of correspondence, but there still remains some departure from normality. Tests for the goodness of fit will be taken up in the next chapter.

6.5 OTHER PROBABILITY FUNCTIONS

The last two probability functions to be examined are the Poisson and binomial. Both of these examples involve fitting discrete probability functions to sample data.

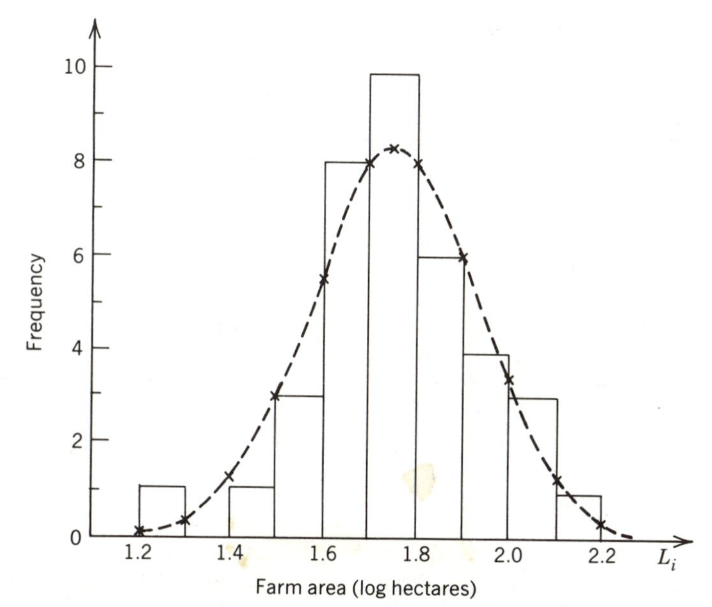

Figure 6.13 Observed histogram and fitted normal (logarithmic scale)—Farm Area variable. *Source:* Cumberland County Dairy Farm Survey.

The Poisson Probability Function

The Poisson probability function was one of the first functions, apart from the normal and lognormal, to be used by geographers. Its use was initiated because of two basic properties. (1) It is especially suitable for describing frequencies recorded by areal units, more particularly equal-sized areal units. (2) More importantly, it describes the actual distribution of the number of occurrences (enumerated) for each element (areal or otherwise) of a phenomenon where the phenomenon is distributed randomly. Thus, it has frequently been used to assess whether the distribution of a phenomenon (towns, plants, shopping centers, etc.) measured by the frequency per unit area is distributed randomly or not. The example we will use from the Cumberland County Dairy Farm Survey is the number of years (in the last 10 years) in which the farmer had to purchase feed to supplement the production of grass, hay, and silage grown on the farm. The assumption is that each farmer attempted to be self-sufficient in feed production, but in an occasional year through miscalculation, mismanagement, or a "bad year" he had to supplement his own production—such a supplement is assumed to have occurred randomly over the 37 farms. The frequency distribution of the variable (Table 6.8) shows that most farms did not supplement their own feed, and a few had to in one or more years.

TABLE 6.8 Frequency Distribution—Years Feed Purchased

Number of Years Feed Purchased	Frequency	Number of Occurrences (Frequency · Years)
0	21	0
1	11	11
2	3	6
3	1	3
4	1	4
TOTAL	37	24

Source: Cumberland County Dairy Farm Survey.

The Poisson density function is given by

$$y = f(x) = \frac{e^{-\lambda}\lambda^{x}}{x!} \tag{6.23}$$

where

$$e = 2.7183$$

The estimate of the parameter λ (Greek lowercase lambda) is found by

$$\hat{\lambda} = \overline{X}$$

$$= 24/37 = .65$$

Using this point estimate, the probability function then becomes

$$f(x) = \frac{(2.7183)^{-.65}\,(.65)^{x}}{x!} \qquad \text{for } x = 0, 1, 2, 3, 4 \ldots \tag{6.24}$$

The calculation of the probabilities simply involves the substituting of different values for x in Equation 6.24, which can be simplified by calculating the value of $(2.7183)^{-.65}$.

$$(2.7183)^{-.65} = \frac{1}{(2.7183)^{.65}}$$

$$= .5220$$

The steps in the calculation are outlined in Table 6.9. The recorded values of x are entered (column 1) with an additional open-ended class to include the remaining values of x (to plus infinity); the functional form is specified (column 2) and the resulting probabilities listed (column 3). The probability for the open-ended class is found by subtracting the sum of the probabilities calculated from the total probability of 1.0. The expected frequency can then be derived by using the probabilities as a proportion of the total sample size

TABLE 6.9 Expected Frequencies for Poisson Function—Years Feed Purchased

x	$P(X = x)$	p	Expected Frequency
0	$\frac{(.5220)(.65)^0}{0!} = \frac{.5220}{1}$	.5220	19.31
1	$\frac{(.5220)(.65)^1}{1!} = \frac{.3393}{1}$	.3393	12.55
2	$\frac{(.5220)(.65)^2}{2!} = \frac{.2205}{2}$	.1103	4.08
3	$\frac{(.5220)(.65)^3}{3!} = \frac{.1434}{6}$	.0239	.88
4	$\frac{(.5220)(.65)^4}{4!} = \frac{.0932}{24}$	.0039	.14
>4		.0006	.02
TOTAL		1.0000	36.98

(i.e., n times p). A comparison of the observed and expected frequencies shows a reasonable degree of correspondence, but evaluation of the goodness of fit is taken up in Chapter 7.

The Poisson probability function is most frequently used when the value of lambda is small (i.e., the probability of an individual occurrence of the phenomenon being examined is low—a rare event), but the distribution can be fitted for any value of lambda. As lambda increases, the number of realistic X values also increases and the calculations become very tedious. However, examples of graphs of the Poisson function (Figure 4.14) show that as the parameter lambda increases the distribution becomes similar in form to the normal probability function. Thus, when lambda is say greater than 5.0, even though the Poisson is a discrete distribution, a *normal approximation to the Poisson* can be used—and a normal curve is fitted in the same manner as in Section 6.3. The estimate for a and b of the normal curve would be $\hat{\lambda}$ and $\sqrt{\hat{\lambda}}$, respectively.

The Binomial Probability Function

Although the binomial probability function is usually used to illustrate the properties of probabilities (see Chapter 4) it has been found of little practical value in geographic research. However, various modifications of the binomial have found limited use. In addition, the binomial situation in which an element of a population or sample either has or does not have a certain characteristic is encountered quite frequently. A point binomial probability function can be fitted to a variable measured in this manner, but the fitted distribution using the estimate of p is simply a description of the sample since the function is fully described by the single parameter p—the proportion of the population with the particular characteristic being examined. The estimate of p, however, can be tested to see whether it corresponds to an assumed value (Chapter 7). An example of the point binomial function was for the variable Presence of a Runoff, where 7 of the 37 sampled farms had runoffs, the remaining 30 farms did not. Using $\overline{X} = 7/37 = .19$ as the estimate of p, the point binomial probability function becomes

$$y = f(x) = (.19)^x(.81)^{1-x} \quad \text{for } x = 0, 1 \tag{6.25}$$

Expanding the point binomial to form the binomial probability function is useful when dealing with a situation where for each of our n sampled elements of the population, we obtain a count of the results of a series of "experiments." That is, for each of the sampled elements we are examining, we have recorded a certain number of experiments, or the results of a certain number of events. We designate this number n_e. Each experiment or event had a binomial response—a yes/no, presence/absence, 1/0 result. Our score, then, for each of the sampled elements of the population is the number of successful events (or values of 1) out of n_e possible successes. Although variables structured in this manner are by no means inconceivable in geographic problems, they are rarely encountered. However, three examples might be a sample of shopping centers, where within each shopping center 10 stores are randomly selected and described in terms of self-service/non-self-service, a sample of sediments, where for each sediment sample 25 pebbles were described as rounded/nonrounded, or a sample of drainage basins, where within each drainage basin a random sample of 20 m^2 plots are described as vegetated/nonvegetated. The X_i scores for these three variables will be the set of integers from 0 through to a maximum of 10, 25, or 20, respectively.

One of the variables examined in the Cumberland County Dairy Farm Survey was of this type—the Pasture Assessment Index. To try and obtain some estimate of the extent of pasture deterioration on the farms, 15 fields were randomly selected for each farm and for each field a decision was made as to whether weed infestation was greater than some critical amount. Each of the 37 sampled farms had either 0, 1, 2, 3, . . . , or 15 of the fields classified as weed infested. The frequency of the resulting values, the X_i scores (Table 6.10) shows that a wide variety of values occurred, with the center of the distribution around a value of 7.

The binomial probability function is given by

$$y = f(x) = \frac{n!}{x!(n - x)!} p^x (1 - p)^{n-x} \tag{6.26}$$

where n is the n_e just defined, which equals 15 in our example. Our estimate of p, the parameter of the binomial function was determined from

$$\hat{p} = \frac{\overline{X}}{n_e} \tag{6.27}$$

TABLE 6.10 Frequency Distribution—Pasture Assessment Index

Number of Weed Infested Fields (Out of 15)	Frequency	Number of Occurrences
0–2	0	0
3	2	6
4	3	12
5	4	20
6	7	42
7	10	70
8	5	40
9	4	36
10	1	10
11	0	0
12	1	12
13–15	0	0
TOTAL	37	248

Source: Cumberland County Dairy Farm Survey.

The mean number of weed-infested fields (out of 15) per farm is given from column 3 of Table 6.10 as

$$\overline{X} = \frac{248}{37} = 6.705 \text{ fields}$$

Our estimate of p is, therefore,

$$\hat{p} = \frac{6.705}{15} = .447$$

that is, our estimate of the probability of a single field being classified as weed infested is .447.

The calculation proceeds by the method outlined in Chapter 4. The probability function for this example is

$$y = f(x) = \frac{15!}{x!\,(15 - x)\,!}\,(.447)^x\,(.553)^{15-x} \quad \text{for } x = 0, 1, 2, 3 \ldots 15 \qquad (6.28)$$

Table 6.11 illustrates the method of calculation and the results. The probability of occurrence of each of the X values (column 4) is given by the multiplication of columns 2 and 3, and as a check on the calculations, these should sum to approximately 1.0. The expected frequency (column 5) would be derived, as before, by multiplying the probabilities by the sample size ($n = 37$). A comparison of the observed and expected distributions shows a basic similarity.

TABLE 6.11 Calculation of Expected Frequency for Binomial Distribution—Pasture Assessment Index

Number of Weed Infested Fields (x)	Number of Combinations $\frac{15!}{x!\,(15-x)!}$	Probability of Individual Outcome $(.447)^x(.553)^{15-x}$	Probability of Combination $P(X = x)$	Expected Frequency (np)
0	1	.0001383	.0001383	.005
1	15	.0001118	.0016771	.062
2	105	.0000904	.0094892	.351
3	455	.0000731	.0332381	1.230
4	1365	.0000590	.0806008	2.982
5	3003	.0000477	.1433324	5.303
6	5005	.0000386	.1930969	7.145
7	6435	.0000312	.2006792	7.425
8	6435	.0000252	.1622126	6.002
9	5005	.0000204	.1019818	3.773
10	3003	.0000165	.0494602	1.830
11	1365	.0000133	.0181726	.672
12	455	.0000108	.0048964	.181
13	105	.0000087	.0009134	.034
14	15	.0000070	.0001055	.004
15	1	.0000057	.0000057	.000
TOTAL			1.0000002 (rounding errors)	36.999

Approximations for the Binomial and Poisson

The calculations for the binomial probability function are obviously tedious, and while tables are available for some values of n_e and p, these are only infrequently applicable. We have already noted that when the Poisson parameter lambda is sufficiently large (greater than 5.0) then the discrete Poisson distribution can be approximated by the normal probability function. Similarly, when the probability (p) and the sample size n_e are sufficiently large, the binomial function can be approximated by the Poisson or the normal probability functions. The latter approximation, especially, will save a lot of calculation. No hard-and-fast rules can be quoted but it is frequently recommended that the approximations can be used in the following circumstances:

If $n_e(p)$ *and* $n_e(1 - p) \geq 5.0$, the normal approximation can be used.

If $n_e(p)$ *or* $n_e(1 - p) < 5.0$ and $n_e \geq 30$, the Poisson approximation can be used.

If $n_e(p)$ *or* $n_e(1 - p) < 5.0$ and $n_e < 30$, the binomial function must be used.

TABLE 6.12 Normal Approximation to the Binomial—Pasture Assessment Index

Value of x	Class Boundary	z Value	Cumulative Area $F(z)$	Class Probability	Expected Frequency
<0				.0001	.004
	−.5	−3.74	.0001		
0				.0005	.019
	.5	−3.22	.0006		
1				.0029	.107
	1.5	−2.70	.0035		
2				.0111	.411
	2.5	−2.18	.0146		
3				.0339	1.254
	3.5	−1.66	.0485		
4				.0786	2.908
	4.5	−1.14	.1271		
5				.1372	5.076
	5.5	−.63	.2643		
6				.1919	7.100
	6.5	−.11	.4562		
7				.2029	7.507
	7.5	.41	.6591		
8				.1647	6.094
	8.5	.93	.8238		
9				.1027	3.800
	9.5	1.45	.9265		
10				.0491	1.817
	10.5	1.97	.9756		
11				.0180	.666
	11.5	2.49	.9936		
12				.0051	.189
	12.5	3.01	.9987		
13				.0011	.041
	13.5	3.53	.9998		
14				.0001	.004
	14.5	4.05	.9999		
15				.0000	.000
	15.5	4.57	.9999		
>15				.0000	.000
TOTAL				.9999	36.997

In our example $n_e = 15$ and $p = .447$ and, therefore,

$$n_e(p) = 6.705$$

$$n_e(1 - p) = 8.295$$

we are justified in using the normal approximation, and we illustrate its use and compare the results with the full binomial calculations. For the normal approximation, our estimates of the parameters a and b are

$$\hat{a} = n_e p = 6.705$$

$$\hat{b} = \sqrt{n_e p(1 - p)} = \sqrt{3.708} = 1.926$$

We use these estimates in the approximation, constructing an expected distribution in exactly the same way as in Section 6.3 (Table 6.12). Because we are using a continuous distribution to approximate a discrete distribution, we need to make one adjustment. Each X value in the binomial probability function must be assumed to be the midpoint of a class covering values of X from halfway between the discrete values on either side. Thus an X value of 4 is assumed to be the midpoint of a class ranging from 3.5 to 4.5. Thus, .5 is added to or subtracted from each X value to define its class boundaries—such an adjustment is called a correction for continuity. The approximation produces expected frequencies very similar to those derived using the binomial probability function (Table 6.13). The normal approximation obviously involves far less calculation.

The foregoing is an example of using the normal probability function to approximate the binomial probability function. Its use is based on the fact

TABLE 6.13 Comparison of Frequency Distributions—Pasture Assessment Index

X Score	Observed	Expected Binomial	Expected Normal Approximation
<0	—	—	.004
0	0	.005	.019
1	0	.062	.107
2	0	.351	.411
3	2	1.230	1.254
4	3	2.982	2.908
5	4	5.303	5.076
6	7	7.145	7.100
7	10	7.425	7.507
8	5	6.002	6.094
9	4	3.773	3.800
10	1	1.830	1.817
11	0	.672	.666
12	1	.181	.189
13	0	.034	.041
14	0	.004	.004
15	0	.000	.000
>15	—	—	.000
TOTAL	37	36.999	36.997

that the binomial (and the Poisson) become closer and closer to a normal distribution as the sample size increases or alternatively, when the probabilities give modal values near the center of the distribution. The result of the process outlined was an expected distribution based on the binomial probability function but approximated by the normal distribution. We can also look at a similar process from the opposite direction. This occurs when our interest is not in the underlying probability function, but simply in finding a probability function (especially the normal) that will fit the data, enabling us to make use of a variety of inferential tests. Such was the situation where we looked at the example of the variable Herd Size in Section 6.3. We showed that the variable closely followed a normal distribution—despite the fact that it was obviously a discretely measured variable. These results enable us to use inferential methods based on the normal probability function on this variable. This form of approximation is common.

References and Readings

Testing Probability Models

Afifi, A. and S. P. Azen (1979) *Statistical Analysis: A Computer Oriented Approach*, Academic Press: New York.

Aitchison, J. and J. A. C. Brown (1957) *The Lognormal Distribution*, Cambridge University Press: London.

Feller, W. (1957) *An Introduction to Probability Theory and Its Applications*, Wiley: New York.

Hald, A. (1952) *Statistical Theory with Engineering Applications*, Wiley: New York.

Johnson, N.L. and S. Kotz (1972) *Distributions in Statistics*, Wiley: New York.

King, L. J. (1969) *Statistical Analysis in Geography*, Prentice-Hall: Englewood Cliffs, N.J.

Krumbein, W. and F. A. Graybill (1965) *An Introduction to Statistical Models in Geology*, McGraw-Hill: New York.

Mood, A. M. and F. A. Graybill (1963) *Introduction to the Theory of Statistics*, McGraw-Hill: New York.

Moser, C. A. and G. Kalton (1971) *Survey Methods in Social Investigation*, Second Edition, Heinemann: London.

Pielou, E. C. (1969) *An Introduction to Mathematical Ecology*, Wiley: New York.

CHAPTER 7

Statistical Inference: Interval Estimation and Hypothesis Testing

A point estimate of a population parameter provides the "best" estimate for that parameter, but it gives no indication as to the error margin of the estimate. It is standard practice to provide with the point estimate an additional estimate of the possible range of error. For example, we might say that a certain item weighs 107 kg $\pm$ 5 kg. When we want to provide a statement of how close we think our sample estimate is to the true population value, we can either (1) provide a range of scores about the parameter estimate, *a confidence interval estimate,* and state the probability of the true value of the parameter lying within that interval or (2) set up a *hypothesis test* stating that the true value of the parameter equals some specified value and then reject or fail to reject the hypothesis with a certain probability. In fact, the two approaches are similar and yield the same results.

Setting up a confidence interval about an estimate of a population mean, for example, implies testing an infinite number of hypotheses that state that the true population mean takes on every possible value between the upper and lower limits of the confidence interval. Although the confidence interval uses the estimate and an assumed sampling distribution to define limits within which we have a certain confidence (expressed as a probability) that the true population value lies, hypothesis testing examines how close the sample statistic is to an assumed population parameter. The latter approach uses the sampling distribution of a test statistic and rejects or fails to reject the hypothesis that the estimate and the population parameter are the same. We will find interval estimates of more use in analyzing the reliability of parameter estimates from a single sample distribution, and hypothesis testing more applicable in testing the differences between the estimated values of a parameter from two or more samples.

The discussion is best initiated with an examination of interval estimation in general (Section 7.1). We then extend these methods to handle interval estimation for some of the more important summary and functional parameters of a population (Section 7.2). Hypothesis testing is introduced (Section 7.3) and applied to testing hypotheses about some summary and functional parameters (Section 7.4). Specific hypothesis tests, usually referred to as goodness-of-fit tests (Section 7.5), provide a means for assessing the correspondence between observed frequency distributions and an expected frequency based on the fitting of a probability function. The last section (7.6) makes brief comments about carrying out univariate inferential tests using computers.

7.1 PRINCIPLES OF INTERVAL ESTIMATION

If we have information about the distribution of a population from which a sample has been selected, we have a reasonable idea of the sampling distributions of the various parameters that describe that population. Using this information, for any point estimate of a parameter, we can construct an upper and/or a lower confidence limit on either side of the estimate. These determine the *confidence interval* about the parameter estimate. We designate the upper and lower limits by U and L respectively and express the probability associated with the parameter falling within this interval by Pr (the *confidence coefficient*). The *confidence level* will be $100 \cdot Pr$.

Thus, for example, in a particular problem concerned with a sample of road frontage measurements, we obtain a sample mean of 5.45 meters. Because the sample mean is usually the best estimate of the population mean, we infer that 5.45 is our best estimate of the true population mean. However, based on our knowledge of the distribution of the sample means, we can make an additional statement that we are 95 percent confident ($Pr = .95$) that the population mean lies between 5.03 m and 5.87 m. We can write this as a probability statement:

$$P(5.03 < \mu < 5.87) = .95$$

or, in general form,

$$P(L < \theta < U) = Pr \tag{7.1}$$

where θ is any estimated parameter.

Although the unknown parameter (θ) must, strictly speaking, be either inside or outside the interval, we can use phrases like "we are 95 percent confident" to represent the probability concept that over a large number of experiments (drawing of samples and estimating the unknown parameter) the interval calculated using the sample data would cover the true value of the parameter. Five times out of 100, we would expect our point estimate, by chance, to be greater or smaller than the true parameter value so that the interval calculated around the estimate would not contain this true value. The concept is illustrated in Figure 7.1.

Most of the examples in this chapter will have a confidence coefficient of .95, but any level can be used, depending on the degree of confidence we wish to associate with the interval. By increasing its value to $Pr = .99$, we could be much more confident that the true value of the parameter lies within

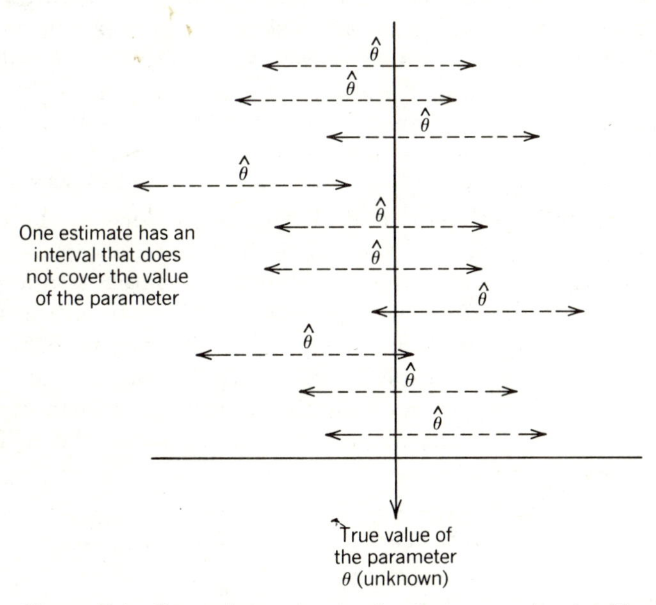

Figure 7.1 Ten point-estimates for the parameter θ with their confidence intervals.

the interval, but, on the other hand, the interval itself would have to be much larger. Confidence coefficients that are frequently used are .90, .95, and .99.

Interval Estimation Concepts

The position of upper and lower limits around the true value of a parameter depends on the (unknown) shape of the population distribution from which the sample was obtained. However, the key to interval estimation lies not in some way of estimating what the population distribution of scores looks like, but in using the sampling distribution of the parameter. This is best illustrated by looking at the estimation of the population mean. The central limit theorem tells us that the sampling distribution of sample means from any population with a mean μ and a variance σ^2 gets nearer to a normal distribution with mean μ and variance σ^2/n as the sample size gets larger (regardless of the distribution of the parent population). Thus, in a sample from any population with a sufficiently large sample size, our single sample mean can be considered to be one of the means making up the theoretical sampling distribution of sample means. We can then compare the position of this sample mean with a hypothesized theoretical mean. Alternatively, we can use the sampling distribution to look at the intervals around parameters.

To illustrate these concepts, we use data from the Cumberland County Dairy Farm Survey. For the variable Herd Size, we assume our sample comes from a population of scores that is normally distributed with an unknown mean, but with a known variance. Our estimate of the population mean is 90.2 cows, and our known variance is (we assume) 1050. It is unlikely (but not impossible) that we would know the variance and it is useful at this point to assume a known variance, but we will relax this assumption later. Our aim is to construct confidence intervals about the point estimate of mean Herd Size of 90.2 cows. To do this, we use the standard normal probability function in its tabled cumulative form (Table C at the end of the book) as we did in fitting an expected normal distribution in Chapter 6. Recall that we can determine the probability associated with any interval of scores by calculating the area under the curve between the upper and lower limits of that interval. For example, to find the probability that a single value of X comes from the interval -1.5 Z score to $+.5$ Z score, we need only to determine from the table the cumulative probability ($F(z)$) up to the points -1.5 and $+.5$ and subtract one from the other (Figure 7.2). From the tables:

Z Score	*F*(*z*)
+ .5	.6915
−1.5	.0668
	.6247

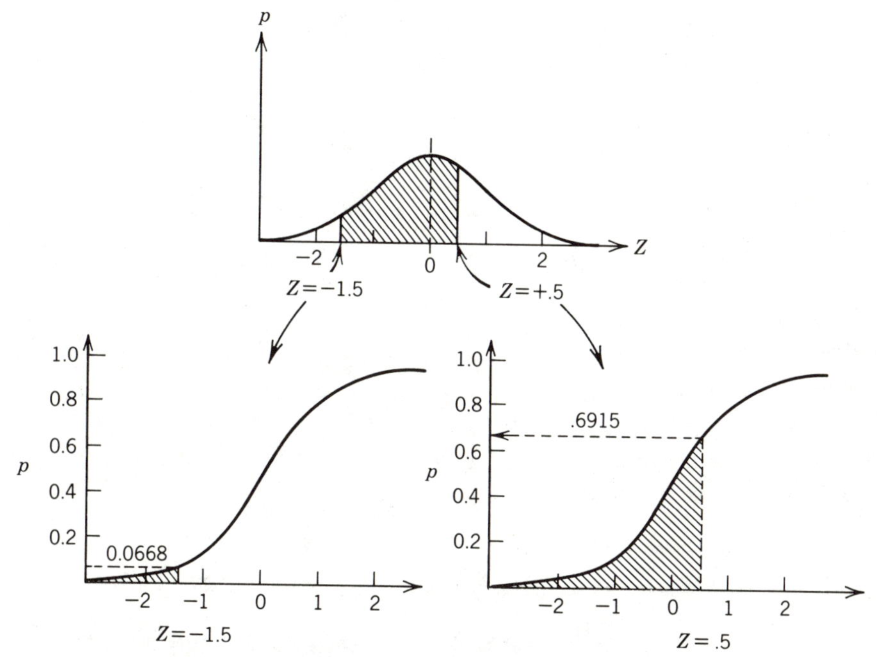

Figure 7.2 Setting up an interval.

That is, the probability that any score of X has a value between a Z score of -1.5 and $+.5$ is .6247. The same sort of calculation can be used for any interval.

Now consider this situation from the opposite point of view—we know the probability and wish to calculate the Z scores. We establish an interval about the mean ($Z = 0$), in which 95 percent of all Z_i scores lie; the probability is .95 that any particular score lies within that interval (Figure 7.3). Using our symbols L and U to denote the lower and upper limits of this interval, and Pr for the probability in the general case, we can say that

$$P(L < Z_i < U) = Pr = .95$$

and, therefore,

$$P(Z < L) + P(Z > U) = 1.0 - Pr = .05$$

But as the standard normal probability function is symmetric about the mean (and assuming we want to make our interval symmetric about the mean),

$$P(Z < 0) = P(Z > 0) = .5$$

and, therefore,

$$P(Z < L) = P(Z > U) = \frac{1.0 - Pr}{2} = .025$$

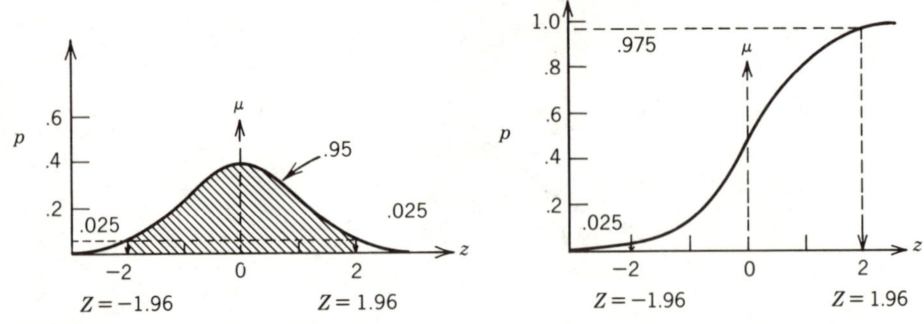

Figure 7.3 Interval about the mean.

To use the cumulative distribution form of the standard normal function we can express this last statement as

$$P(Z < L) = .025$$

and

$$P(Z < U) = 1.0 - .025 = .975$$

Using the tabled values (Table C), the Z scores associated with these probability values are

$$\text{lower limit: } Z_L = [Z, p = .025] = -1.96$$

$$\text{upper limit: } Z_U = [Z, p = .975] = +1.96$$

Thus, we are 95 percent confident (the probability is .95) that a particular Z score lies in the interval -1.96 to $+1.96$.

Recalling that

$$Z_i = \frac{X_i - \mu}{\sigma}$$

we have

$$Z_L = \frac{L - \mu}{\sigma}$$

and

$$Z_U = \frac{U - \mu}{\sigma}$$

Thus

$$L = \mu + \sigma Z_L$$

and

$$U = \mu + \sigma Z_U$$

But as Z_L is the same as Z_U except for the sign difference, we can summarize this in one statement: the upper and lower limits are given by

$$\mu \pm \sigma Z \tag{7.2}$$

Using the foregoing concepts, we can establish confidence limits for the example cited at the beginning of this section. We do not know the population mean, but we have an estimate of it, that is, $\overline{X}$. Assuming that the population of Herd Size scores is distributed normally, or invoking the central limit theorem, we know that this sample mean forms one element of a sampling distribution of sample means that is normally distributed with mean μ and variance σ^2/n. Chances are that the value of $\overline{X}$ is close to μ, but it may not be. We can convert this normally distributed sampling distribution into standard normal form. In the current situation, however, we are concerned with $\overline{X}$ scores rather than X_i scores, and the standard deviation of this distribution is not σ but $\sigma/\sqrt{n}$. This standard deviation of the sampling distribution is called the *standard error of the mean*, and is usually denoted by $\sigma_{\overline{X}}$. The Z score for any sample mean is given by

$$Z = \frac{\overline{X} - \mu}{\sigma/\sqrt{n}} = \frac{\overline{X} - \mu}{\sigma_{\overline{X}}} \tag{7.3}$$

and these Z scores will be distributed following the standard normal distribution. Using $\overline{X}$ as the estimate for μ, we can determine lower and upper limits around this sample value by utilizing the standard normal distribution in the form of

$$\overline{X} \pm \frac{\sigma}{\sqrt{n}} Z = \overline{X} \pm \sigma_{\overline{X}} Z$$

That is, the lower and upper limits for the estimated population mean are:

(estimate of population mean) ± (standard error of the mean)

· [(Z value at probability of limits)]

In general form, for the construction of intervals about any parameter, the lower and upper limits are given by

(parameter estimate) ± (standard error of the sample statistic)

· [(value of the sampling distribution at limits)]

Note that this abbreviated form of describing both upper and lower limits with one equation assumes the interval is symmetric about the parameter estimate. This would be true only if the sampling distribution of the sample statistic were symmetric, as in the standard normal and the t distributions.

To avoid confusion, it is useful to introduce a simple piece of nomenclature for specifying the value of the sampling distribution at the limits to use in the calculation. The portion of the equation that involves the obtaining of a single value from a tabled sampling distribution is enclosed in square brackets. Within the square brackets in sequence is the coded name of the probability function describing the sampling distribution (for example Z for the standard normal, t for the t distribution), the probability value to be used, and finally, the degrees of freedom associated with the parameters specifying the function. For example,

$$[t, p = .975, df = 15]$$

specifies a single value obtained for the t distribution at the $p = .975$ position with 15 degrees of freedom.

Structuring a Confidence Interval

We now outline the steps in the calculation of the confidence limits for the example we have been using. A formal outline involving five basic steps will be given, although in practice, only the calculated limits would normally be stated.

Two-sided Confidence Interval for the Mean of a Normally Distributed Population (σ^2 known)

1. ***Probability Statement***

$$P(L < \mu < U) = Pr = .95$$

2. ***Assumptions***

   (i) $X_i \sim N(\mu, \sigma^2)$, or n is large.

   This means that $\overline{X} \sim N(\mu, \sigma^2/n)$, which would occur when either $X_i \sim N(\mu, \sigma^2)$ or n is sufficiently large enough to invoke the central limit theorem.

   (ii) $\sigma^2 = 1050$; μ unknown ($\overline{X} = 90.2$)

   (iii) random sample, $n = 37$

3. ***Interval Structure***

$$\overline{X} + \frac{\sigma}{\sqrt{n}} \left[Z, p = \frac{1 + Pr}{2}\right]$$

$$\overline{X} \pm \frac{\sigma}{\sqrt{n}}(1.96)$$

4. ***Calculation***

$$90.2 \pm \frac{\sqrt{1050}}{\sqrt{37}}(1.96)$$

$$\begin{aligned} L &= 90.2 - \frac{\sqrt{1050}}{\sqrt{37}}(1.96) \\ &= 90.2 - (5.33)(1.96) \\ &= 79.8 \\ U &= 90.2 + 10.4 \\ &= 100.6 \end{aligned}$$

5. ***Conclusion***

$$P(79.8 < \mu < 100.6) = .95$$

That is, we are 95 percent confident that the population mean lies between 79.8 cows and 100.6 cows. A more correct statement is that 95 out of 100 such intervals will include the population mean.

There is a variety of ways to express this result in summary form, but the two most common methods are these:

1. As an interval, mean Herd Size: 90.2 cows (±10.4) with 95 percent confidence.

2. By stating the standard error, mean Herd Size: 90.2 cows (standard error 5.3).

7.2 INTERVAL ESTIMATION FOR POPULATION PARAMETERS

Interval Estimation around the Arithmetic Mean

It is common to estimate a sample mean and its confidence interval, but we seldom know the value of the population variance. The Z distribution (the standard normal distribution) can only be applied when the population variance is known or when n is very large. Although $(\overline{X} - \mu)/(\sigma/\sqrt{n})$ is distributed according to the standard normal distribution, the equivalent statistic $(\overline{X} - \mu)/(s/\sqrt{n})$ is not (using the sample standard deviation). However, $(\overline{X} - \mu)/(s/\sqrt{n-1})$ or $(\overline{X} - \mu)/(\hat{\sigma}/\sqrt{n})$ is distributed according to the t *distribution* with a single parameter equal to the degrees of freedom, $n - 1$ (Table D). As n, the sample size, increases the t distribution approaches the normal. The *estimated standard error of the mean* is usually designated by $s_{\overline{X}}$ and equals $\hat{\sigma}/\sqrt{n}$ or $s/\sqrt{n-1}$. The assumption that the X_i are distributed normally or that the sample size is sufficiently large to invoke the central limit theorem still applies, although smaller sample sizes are better handled by the t distribution.

We will illustrate this with the same example, but the population variance is unknown. Again, the interval construction is set out formally. From the sample of 37 farms, the Herd Size averaged 90.2 cows and the sample variance was 992. Calculating the 95 percent confidence interval:

Two-sided Confidence Interval for the Mean of a Normally Distributed Population (σ^2 unknown)

1. ***Probability Statement***

$$P(L < \mu < U) = Pr = .95$$

2. ***Assumptions***

   (i) $X_i \sim N(\mu, \sigma)$ or n is large

   (ii) μ unknown ($\overline{X} = 90.2$); σ^2 unknown ($\hat{\sigma}^2 = 992$)

   (iii) random sample, $n = 37$

3. ***Interval Structure***

$$\overline{X} \pm \frac{\hat{\sigma}}{\sqrt{n}} \qquad \left[t, p = \frac{1 + Pr}{2}, df = n - 1\right]$$

$$\overline{X} \pm \frac{\hat{\sigma}}{\sqrt{n}} \qquad \left[t, p = .975, df = 36\right]$$

$$\overline{X} \pm \frac{\hat{\sigma}}{\sqrt{n}} (2.042)$$

As frequently occurs, an exact tabled value of t (or any other sampling distribution) cannot be found because of the limited size of the tables. Two possibilities exist: (1) interpolate between values, or (2) select a value that makes the inferential process more rigorous, by selecting a lower value for the degrees of freedom, in this case 30 rather than 36.

4. ***Calculation***

$$90.2 \pm \frac{\sqrt{992}}{\sqrt{37}} (2.042)$$

$$L = 90.2 - (5.178)(2.042)$$

$$= 90.2 - 10.6$$

$$= 79.6$$

$$U = 90.2 + 10.6$$

$$= 100.8$$

5. ***Conclusion***

$$P(79.6 < \mu < 100.8) = .95$$

That is, we are 95 percent confident that the population mean Herd Size is between 79.6 and 100.8 cows.

In some situations, we may be interested in providing a one-sided interval, rather than the two-sided interval just described. We can construct either a lower limit (i.e., we are 95 percent confident that the value of the parameter is greater than L), or an upper limit (i.e., we are 95 percent confident that the value of the parameter is less than U). The three types of confidence intervals for the mean are illustrated in Figure 7.4 and for the greater-than, one-sided interval, an abbreviated outline using the same example is presented.

One-sided (Greater-than) Confidence Interval for the Mean of a Normally Distributed Population (σ^2 unknown)

1. ***Probability Statement***

$$P(\mu > L) = Pr = .95$$

2. ***Assumptions***

(i) $X_i \sim N(\mu, \sigma^2)$ or n is large

(ii) μ unknown ($\overline{X} = 90.2$); σ^2 unknown ($\hat{\sigma}^2 = 992$)

(iii) random sample, $n = 37$

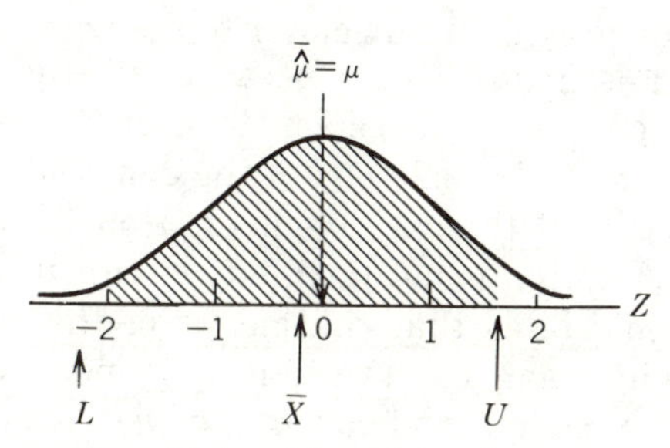

Two-sided confidence interval

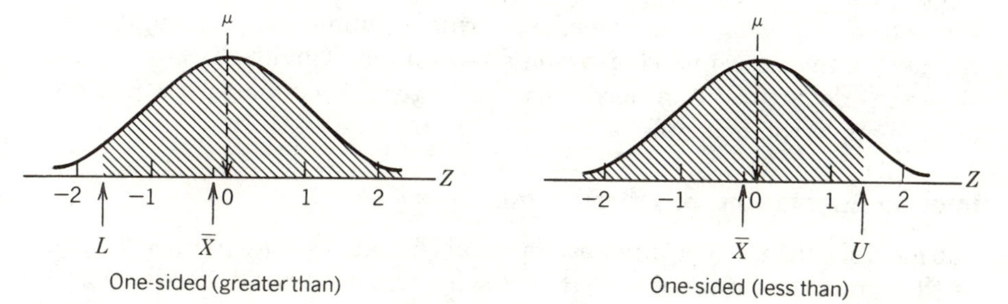

One-sided (greater than)

One-sided (less than)

Figure 7.4 The three types of interval estimation.

3. ***Interval Structure***

$$\overline{X} - \frac{\hat{\sigma}}{\sqrt{n}} \qquad [t, p = Pr, df = n - 1]$$

$$\overline{X} - \frac{\hat{\sigma}}{\sqrt{n}} \qquad [t, p = .95, df = 36]$$

$$\overline{X} - \frac{\hat{\sigma}}{\sqrt{n}} (1.697)$$

4. ***Calculation***

$$L = 90.2 \frac{\sqrt{992}}{\sqrt{37}} (1.697)$$

$$= 90.2 - 8.8$$

$$= 81.4$$

5. ***Conclusion***

$$P(\mu > 81.4) = .95$$

That is, we are 95 percent confident that the population mean Herd Size is greater than 81.4 cows.

The one-sided intervals provide limits closer to the point estimate of the mean than the two-sided estimate. This is because we are focusing on just one end of the distribution. All the 5 percent chance of getting a sample mean outside the limit is placed at one end of the distribution.

The above tests assume that the variable is distributed normally. However, even when this assumption is not justified, we can still use the tests. The central limit theorem states that the means of the samples from a population of any form (with mean μ and variance σ^2) tends towards a normal sampling distribution with mean μ and variance σ^2/n, as n, the sample size increases. Thus, if the sample size is *not extremely small*, the above methods can be used with little error. There is no exact minimum sample size—this varies from situation to situation. As a rule of thumb with a sample size of at least 30, the error introduced would be minimal. Obviously, however, the larger the departure from normality the larger the sample size would have to be to apply the principle of the central limit theorem.

Interval Estimation for Other Parameters

The logic of calculating confidence intervals for other parameters is the same as that outlined for the mean. In the same way that the t distribution describes the distribution of a statistic (i.e., $(\overline{X} - \mu)/(\hat{\sigma}/\sqrt{n})$ relating the sample mean to the population mean, it can be shown that the chi-square distribution relates the sample variance to the population variance. In this situation, it is the statistic $(n - 1)\hat{\sigma}^2/\sigma^2$ that has a chi-square distribution with $n - 1$ degrees of freedom. This statistic can be used to construct confidence intervals about the sample variance (or standard deviation). Again, it is assumed that the variable being considered is distributed normally with mean μ and variance σ^2. Unfortunately, any departure from normality can introduce considerable error into the estimate.

Confidence intervals for other measures are also available. For large samples of size n from a normal distribution, it can be shown that the sampling distribution of the sample median becomes normally distributed with a mean equal to the population median and a standard error of $1.253\sigma/\sqrt{n}$. However, the main value of the median is an estimate of the central point of distributions that depart from normality. Approximate confidence intervals for the median that can be constructed for *any* continuous distribution (as long as the sample size is not too small) use the sample size to get estimates of the position of the upper and lower limits in a ranked distribution.

Most of the intervals we have discussed have been concerned with fitting intervals around various statistics obtained from an assumed normal population. Two of these, the mean and the variance, are the parameters of the normal probability function. Thus, confidence intervals for a and b, the parameters of a normal probability function, are simply the confidence intervals for the mean and standard deviation (square root of the variance) of the preceding section. Confidence intervals for the lognormal are available in Aitchison and Brown (1957).

In the discussion of the construction of intervals for the summary parameters, all of the tests assumed to varying degrees that the population being sampled was normally distributed. Violation of this assumption leads to a certain amount of error, usually unknown. Frequently, these constructions are applied without any consideration of the underlying assumptions with the resulting possibility of considerable error and misinterpretation.

In recent years, there has been an interest in determining tests that are distribution free; that is, that make no assumptions about how the variable being examined is distributed. From one point of view, such tests are superior, but they are usually less efficient and less straightforward. Although the emphasis has been on the development of tests that are concerned with inferential consideration of differences between or among samples (where the violation of the normality assumption may be more serious), some have attempted to provide distribution-free tests for the summary parameters with which we have been concerned. One example was the test for the median already mentioned. No others are examined here. An interesting argument for the use of the distribution-free approach is provided by Edgington (1969). He outlines simple methods for calculating confidence intervals for population parameters, regardless of the type of distribution, which are especially useful when the population size is known and is small, and the sample size is relatively large. Some distribution-free tests are outlined in later sections of this book.

Estimating the Minimum Sample Size

In Section 5.1, we suggested that the availability of time and money is often the determining factor in deciding sample size. We noted, however, that if we are willing to make a number of assumptions about both our sample design and the population from which we are sampling, then we can obtain an estimate of the minimum sample size required.

The key to doing so is in making use of confidence interval structures. Let us examine the first structure that was outlined—that of a two-sided interval around the arithmetic mean when the population variance is known and we can use the normal probability function. This was given as

$$\overline{X} \pm \frac{\sigma}{\sqrt{n}} [Z \text{ value}]$$

If we take this stated interval and solve the equation for n, we get

$$n = \left(\frac{\sigma[Z \text{ value}]}{\text{Interval}} \right)^2$$

That is, if we "know" the interval, the Z value, and σ, we can determine the minimum sample size n.

Let us illustrate this by using an example. If, in the Cumberland County Farm Survey, we regard the variable Herd Size as being the key variable, and an estimate of mean Herd Size as being the fundamental summary parameter, we can base our required sample size on a confidence interval about this mean. From previous farm surveys (or any other source of information), we make the decision to estimate the mean herd size with an error margin (confidence interval) of ± 10 cows, and we want to be 95 percent confident (the confidence level) that the true mean lies within the interval of the estimated mean ± 10 cows. One further important assumption or decision needs to be made. We must estimate the population standard deviation σ. This is the most difficult decision because the other assumptions and decisions are more a part of the sample design and fairly standard methods can be usually applied. From a pilot survey (ideally), previous surveys (very important), or just an educated guess (quite common), we decide on a population standard deviation, 32.4 cows in our example.

With a confidence level of 95 percent (allowing for a 5 percent error) the appropriate Z value is

$$[Z, p = 0.975] = 1.96$$

and our estimate of the minimum sample size to satisfy the limits is

$$n = \left(\frac{(32.4)(1.96)}{10} \right)^2$$

$$\cong 40$$

The illustration using the normal distribution to fit a confidence interval around the arithmetic mean is not an unusual one. However, the same construction could be applied to confidence intervals around any summary parameter, although those most commonly used are the ones outlined, and an equivalent structure for intervals around a proportion. One advantage of using normal distribution based constructions, rather than the t distribution, for example, is that the normal distribution does not require an estimate of sample size itself to obtain the appropriate probability statistic.

In a practical situation, as well as the problem of estimating any unknown parameters (σ in our example), the main problem is in deciding which variable to use as the key indicator in the survey. While in some circumstances, this may be easily determined, in most surveys, it is not. A common structure then, is to carry out the process we have illustrated for a selected set of variables and then to make a subjective decision based on the different estimated minimum sample sizes that would be obtained. In any event, whatever the method used in deciding on sample size, it is important that *after* the survey, the results are reported with an indication of the error margins—by constructing confidence intervals or tabling standard errors.

7.3 PRINCIPLES OF HYPOTHESIS TESTING

Hypothesis testing involves concepts similar to the construction of confidence intervals, although the aims of the two techniques are different. To reiterate, the construction of a confidence interval about a sample mean implies an infinite number of hypotheses tests in which the value of the population mean takes on every possible value between the upper and lower limits. Hypothesis testing, on the other hand, tests how close the sample mean is to an assumed population mean by using the sampling distribution of a test statistic, and fails to reject or rejects the hypothesis of no difference with a certain level of probability. As noted earlier, confidence limits are more useful in analyzing the reliability of a parameter from a single sample, and hypothesis testing more applicable in testing the differences between the values of a parameter from two or more samples.

At first, the setup of a hypothesis test may seem strange to someone unfamiliar with inductive statistical methods. First, the term "hypothesis" is used in a quite strict manner. It is defined as a simple statement about a future event and is stated in such a way that it can be rejected by the application of a simple probabilistic test. Second, in many circumstances, it may seem that the hypothesis being tested is the "wrong" one, that is, we are hoping that it will not be "proved." Setting up statistical hypotheses in this way is based on the fact that hypotheses cannot be proved or disproved, but only rejected or not rejected. With this approach, we may be able to reject all the other possible alternatives to the one result in which we are interested and thus make a satisfactory conclusion.

The Basis of Hypothesis Testing

Hypothesis testing is concerned with the outcome of an experiment. In geography, the experiment usually takes the form of the selection of a sample, and the calculation of a sample statistic as an estimate of a parameter of the population. There are four general steps.

1. All possible outcomes of the experiment are anticipated *in advance of the test.* To guarantee this, we structure the hypothesis being tested and its alternatives such that together they cover all possible outcomes of the test, even if some are not very likely.
2. *Prior to the test* the method of procedure that would determine which of the possible results occurred is fixed.
3. *Prior to the test* the possible results are divided into two categories: those that would result in the rejection of the hypothesis and those that would not.
4. The test is carried out, the outcomes noted, and a decision made whether or not to reject the hypothesis. This step is the actual hypothesis test.

We note from the foregoing steps that all decisions are made prior to the test, and the test is carried out automatically once it is set up. This may seem a little unrealistic, but, for the moment, we will present hypothesis testing as a strictly formal operation where the steps must be followed implicitly. Later in this section, we will relax the strict *a priori* nature of the procedure.

As mentioned in step 1, all outcomes of the experiment are anticipated before the test is carried out. The hypothesis being tested forms one of several possible outcomes of the experiment. The primary hypothesis is called the *null hypothesis*. All other possible outcomes are included in an *alternate* (or *research*) *hypothesis*. In the specific tests to be discussed in the next section, the alternative hypothesis will be kept general to include all possible values not covered by the null hypothesis. For the moment, however, we will consider alternate hypotheses that include only those values that realistically can be taken by the parameter being tested (we still assume that the null and alternate hypotheses together cover all possible outcomes of the experiment).

Probability forms the basis of hypothesis testing just as it did for confidence levels. The sampling distribution of a statistic estimated from a sample is assumed to follow a certain probability distribution. The properties of the probability function can be used to construct a critical value for a rejection region for the hypothesis being tested, and probabilities can be assigned to the types of errors that might result from the hypothesis test. Two types of errors may occur in rejecting or not rejecting the null hypothesis (Table 7.1). A Type I error results when the null hypothesis is rejected (i.e., "accept" the alternate hypothesis), when it is actually true. We will denote the probability

TABLE 7.1 Possible Results from Hypothesis Test

Decision	True Situation: Null Hypothesis Is True	True Situation: Null Hypothesis Is False
Accept null hypothesis	No error	Type II error
Reject null hypothesis	Type I error	No error

of making this type of error by α (Greek lowercase alpha). This is the *significance level* of the test. There is an obvious similarity between α in hypothesis testing and $1 - Pr$ in confidence interval construction. Although the methods are similar and the definition of α and Pr are structured in a similar manner, the two separate nomenclatures will be retained. A Type II error occurs when we fail to reject the null hypothesis when it is actually false. Using β (Greek lowercase beta) to represent the probability of making a Type II error, we will call $1 - \beta$ the *power of the test.* It is not possible to minimize both types of error. Ideally, the probability of making both these types of errors would be set before the test is made. More usually, we decide only on a value for α and the power of the test is not taken into account. However, in some social sciences (psychology, for example) the power of the test plays an important role.

Consider a simple example. Suppose we are interested in testing the null hypothesis that the mean of a population from which we have sampled is equal to a certain value (designated as μ_0). That is, the null hypothesis is $\mu = \mu_0$. As a single alternative to the null hypothesis, we set up an alternate hypothesis that the mean is equal to some other value (μ_1); that is, our alternate hypothesis is $\mu = \mu_1$. If one of the two values μ_0 or μ_1 were, in fact, the true population mean, it would be positioned at the center of the sampling distribution as shown in Figure 7.5. Individual sample means would then be distributed around the μ_0 and μ_1, as shown in the figure. To make a decision on which hypothesized result to accept, we must place a boundary between the two distributions. In doing so, we define α and β. The boundary is located by calculating the value for the sampling distribution of the null hypothesis at some chosen α value.

In the second example, we take the same null hypothesis (that $\mu = \mu_0$), but we expand the alternate hypothesis to include all values greater than μ_0 (i.e., $\mu > \mu_0$). Our sample space for the null hypothesis remains the same, with α defined as above (Figure 7.6). However, the sample space for the alternate hypothesis could occur with its center anywhere to the right of μ_0. The value of β would vary as well, and the closer the value of μ to μ_0, the greater the probability of making a Type II error for a fixed value of $\sigma_{\overline{X}}$. The value of β is thus a function of the value of the true parameter (in this case μ).

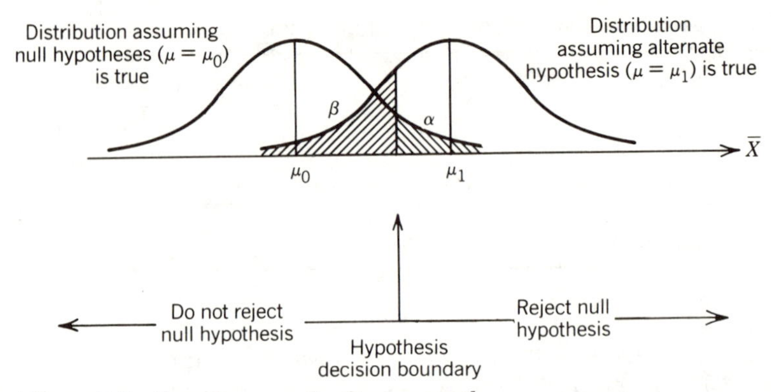

Figure 7.5 Sample space for first example.

The third situation, which we will maintain throughout this section, is to define the alternate hypothesis as including all values of the parameter not included in the null hypothesis. In the example, the alternate hypothesis will be that the mean (μ) does not equal the hypothesized value (μ_0). In this situation, the probability of making a Type I error, as illustrated by the sample space (Figure 7.7), could occur on either tail of the sampling distribution (a two-tailed test), where the probability associated with making an error in either tail would be $\alpha/2$.

Once the sampling distribution has been decided (depending on which test statistic is used) the construction of the rejection region based on the value of α is a simple matter. But, delimiting the β value is a more complicated procedure. Methods are available, but because there is more concern about making a Type I error than a Type II error, we will emphasize type I errors. To reiterate, the two types are related, and by delimiting one (the Type I error) and deciding on a sample size n, the probability of making a Type II error is indirectly determined.

Setting Up Hypothesis Tests

For pedagogical purposes, we elaborate all the stages of the test using the variable Herd Size from the Cumberland County Dairy Farm Survey. In practice, we would present only the results of the hypothesis test. With a

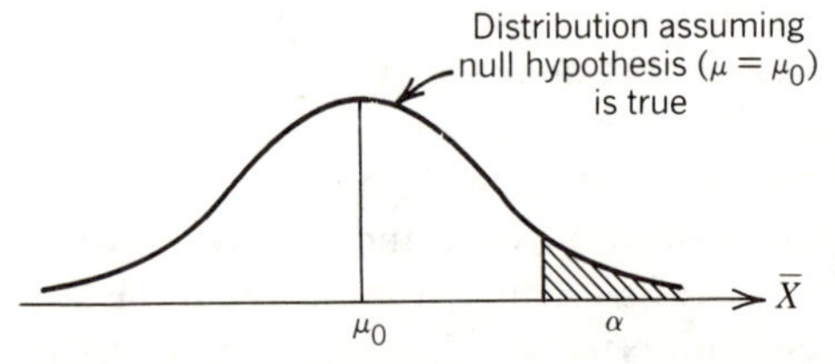

Figure 7.6 Sample space for second example.

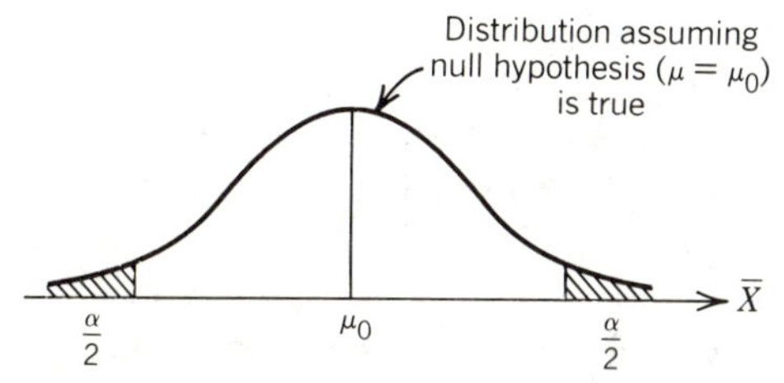

Figure 7.7 Sample space for third example.

sample mean of 90.2 cows, we assume, as before, that the variance is known and equal to 1050. We want to test for the likelihood that this sample comes from a population whose mean is equal to 95, a value that was derived from additional information outside the sample survey. The null hypothesis is, therefore, that the true mean of the population from which the sample was taken is 95 cows.

1. ***Hypotheses.*** The first step is to set up the hypotheses. We use *Ho* to designate the null hypothesis, and *Ha* for the alternate hypothesis. We maintain the alternate hypothesis in general form so that together, the two hypotheses cover all possible values. Thus,

$$Ho\text{: } \mu = \mu_0 \qquad \text{versus} \qquad Ha\text{: } \mu \neq \mu_0$$

or, in our example,

$$Ho\text{: } \mu = 95 \qquad \text{versus} \qquad Ha\text{: } \mu \neq 95$$

2. ***Assumptions.*** The assumptions involved in the test need to be carefully stated, as the results of the test (rejection or acceptance) apply only to the test as it is structured. It is also appropriate to state the α-value. The basic assumptions are:
   (i) $X_i \sim N(\mu, \sigma^2)$ or n is large
   (ii) $\sigma^2 = 1050$; μ unknown ($\bar{X} = 90.2$)
   (iii) random sample, $n = 37$
   (iv) $\alpha = .05$ (other significance levels can be used)

3. ***Test Statistic.*** The test statistic involves calculating a value from the sample whose sampling distribution is known, and which can be used to "test" the value of the parameter. In the present example, the statistic $(\bar{X} - \mu)/(\sigma/\sqrt{n})$ has a sampling distribution given by Z, the standard normal distribution with mean 0 and variance 1. The test statistic is, therefore,

$$Z = \frac{\bar{X} - \mu}{\sigma/\sqrt{n}}$$

4. ***Rejection Region.*** The size of the rejection region is defined by the α value which we set in (2). The α value, or level of significance, states that there are 5 chances in 100 that in a hypothesis test of this type we would reject a true null hypothesis—the hypothesis of no difference between the sample mean and the population mean—if they are within a specified distance apart. That is, we must choose some critical value beyond which we would not accept the hypothesis of no difference. Using the α value, this critical value is given by

$$Z \leq [Z, p = \alpha/2] \text{ in the left-hand tail}$$

and

$$Z \geq [Z, p = 1 - \alpha/2] \text{ in the right-hand tail}$$

But as the Z distribution is symmetrical about the mean, we can simplify the rejection region to

$$\begin{aligned} |Z| &\geq [Z, p = 1 - \alpha/2] \\ &\geq [Z, p = .975] \\ &\geq 1.96 \end{aligned}$$

5. ***Calculation.*** All the foregoing steps should be carried out before the sample is drawn. It now remains to calculate the value of Z and draw some conclusions. In the example with $\overline{X} = 90.2$, $\mu_0 = 95$, $\sigma^2 = 1050$, and $n = 37$,

$$\begin{aligned} Z &= \frac{\overline{X} - \mu_0}{\sigma/\sqrt{n}} \\ &= \frac{90.2 - 95}{\sqrt{1050}/\sqrt{37}} \\ &= \frac{-4.8}{5.33} \\ &= -.90 \end{aligned}$$

6. ***Conclusion.*** It is convenient in a formal outline of hypothesis tests to state the conclusion in two forms: first, as a statistical statement of the results, and second, as a general verbal conclusion. Thus,

$$|-.90| \ngeq 1.96.$$

Therefore, we fail to reject the null hypothesis and conclude that the sample could have come from a population with a mean Herd Size of 95 cows.

This conclusion applies only to the particular hypothesis under consideration, given the stated assumptions and significance level cited. If the absolute value of our calculated value of Z were greater than the tabled value of 1.96, our conclusion could have been expanded to, "we reject the null hypothesis and accept the alternate that the mean of the population from which the sample was obtained was not equal to 95 cows." Of course, in this example, we already know that the 95 percent confidence interval was 79.8 − 100.6 cows and contained the sample mean; thus, it is not surprising that we failed to reject the null hypothesis.

The hypothesis test just outlined is for a *two-tailed test*—where Type I (α) error is apportioned to the two tails of the distribution and our alternate hypothesis is simply that $\mu \neq \mu_0$. If we wanted to give a direction to the hypothesis structure so that, for example, we could test that $\mu \geq \mu_0$, we would have used a *one-tailed test*. The testing structure is the same except that the rejection region is placed entirely on one side of the sampling distribution sample space and the rejection region becomes $Z \geq [Z, p = 1 - \alpha]$. A two-tailed test, for example, at the $\alpha = .05$ level would have the same test statistic, rejection region, and conclusion as a one-tailed test at the $\alpha = .025$ level.

7.4 HYPOTHESIS TESTING FOR POPULATION PARAMETERS

Hypothesis Tests for the Arithmetic Mean

The situation outlined in the last section, in which the population variance is known and can be used in hypothesis tests of the value of the population mean, is usually unrealistic. However, as discussed in Section 7.2, the statistic $(\overline{X} - \mu_0)/(\hat{\sigma}/\sqrt{n})$ was shown to be distributed as a t distribution with $df = n - 1$ degrees of freedom. This statistic can be used in constructing a hypothesis test for the mean when the population variance is not known. To illustrate the method, the example from the Cumberland County Dairy Farm Survey for the variable Herd Size is again used, where $\overline{X} = 90.2$, $\mu_0 = 95$, $\hat{\sigma}^2 = 992$, and $n = 37$.

Hypothesis Test for the Mean of a Normal Population (σ^2 unknown)

1. ***Hypotheses***

$$Ho: \mu = \mu_0 \quad \text{versus} \quad Ha: \mu \neq \mu_0$$

$$Ho: \mu = 95 \quad \text{versus} \quad Ha: \mu \neq 95$$

2. ***Assumptions***

(i) $X_i \sim N(\mu, \sigma^2)$ or n is large

(ii) σ^2 unknown ($\hat{\sigma}^2 = 992$); μ unknown ($\overline{X} = 90.2$)

(iii) random sample, $n = 37$

(iv) $\alpha = .05$

3. ***Test Statistic***

$$t = \frac{\overline{X} - \mu_0}{\hat{\sigma}/\sqrt{n}}$$

4. ***Rejection Region***

$$|t| \geq [t, p = 1 - \alpha/2, df = n - 1]$$
$$\geq [t, p = .975, df = 36]$$
$$\geq 2.042$$

5. ***Calculation***

$$t = \frac{90.2 - 95}{\sqrt{992}/\sqrt{37}}$$
$$= \frac{-4.8}{5.178}$$
$$= -.927$$

6. ***Conclusion***

$$|-.927| \ngeq 2.042$$

Therefore, we fail to reject the null hypothesis and conclude that the sample could have come from a population with a mean Herd Size of 95 cows.

The similarity with confidence interval structure and the ease of moving from one to the other is clear. In the above example, any value for μ_0 between 95 percent confidence limits of 79.6 and 100.8 would have yielded the same result.

The one-sided test for the mean would be identical, but with the rejection region doubled in size and concentrated at one end of the distribution.

Hypothesis Tests for Other Parameters

Hypothesis tests can be structured for the parameters of any probability function, but apart from those for the mean and variance of the normal probability function, these are rarely used. Moreover, in single sample situations, we tend to use the confidence interval structure (or merely the statement of the standard error) unless we have a specifically known value to test against. The only additional test to be mentioned is for the parameter p of the binomial distribution. In some circumstances, it is useful to test whether a sample proportion can be considered as deriving from a population with a specified proportion. The calculation using the full binomial expansion follows a method similar to that outlined in Section 6.5, but it is obviously tedious. With the normal approximation, the calculation becomes a simple manipulation that can be performed on the data either in frequency form or as proportions. The test statistic is:

$$Z = \frac{P_s - P_\mu}{\sigma_{P_s}}$$

Where P_μ is the equivalent of μ and P_s the equivalent of $\overline{X}$, and where $\sigma_{P_s} = \sqrt{P_u(1 - P_u)/n}$ = the standard error of a proportion. The more useful two-sample case will be developed in greater detail.

The discussion of Z and t tests and the effects of sample size can be confusing to a beginning student and the following point-by-point list is an attempt to reiterate and codify what has been developed to this point.

1. We are concerned with a variable X_i that has a certain distribution and a certain mean μ and variance σ^2. This distribution could be normally distributed ($X_i \sim N(\mu, \sigma^2)$) or it might follow some nonnormal distribution.

2. If $X_i \sim N(\mu, \sigma^2)$ then any $\overline{X}$, the mean of a sample of size n taken from X_i, would be a member of a sampling distribution of sample means that is also $\sim N(\mu, \sigma^2/n)$.

3. Even if X_i is *not* $\sim N(\mu, \sigma^2)$, but follows some nonnormal distribution then $\overline{X} \sim N(\mu, \sigma^2/n)$ as long as n is large enough (the central limit theorem)

4. Following from either (2) or (3), the sampling distribution of sample means is $\sim N(\mu, \sigma^2/n)$. The standard deviation of this sampling distribution, $\sigma/\sqrt{n} = \sqrt{\sigma^2/n}$ is called the *standard error of the mean*. We can *standardize* (convert to Z scores) this distribution in the usual way by subtracting the mean and dividing by the standard deviation. Thus individual sample means ($\overline{X}$) then would have Z scores from

$$Z_{\overline{X}} = \frac{\overline{X} - \mu}{\sigma_{\overline{X}}}$$

The distribution of resulting standardized scores would follow the standard normal distribution with mean = 0 and variance = 1, that is,

$$Z_{\overline{X}} \sim SN(0, 1) \tag{7.4}$$

5. To carry out the standardization of (4) and therefore be able to use the standard normal distribution, requires us to know the value of $\sigma_{\overline{X}}$ and therefore to know σ^2, the variance of the original distribution of X_i. However, in most circumstances, we will *not* know the value of σ^2, therefore we would have to estimate it using $\hat{\sigma}^2$, the sample variance. This is only justifiable if the sample size were large—say, greater than 25 or 30, for example.

6. With small samples, we cannot use the standard normal distribution, as we would have little faith in using $\hat{\sigma}^2$ as our estimate of σ^2. However, there is another sampling distribution—the *t distribution*—that does not require a knowledge of σ^2, but instead incorporates the sample variance $\hat{\sigma}^2$. We can standardize an individual sample mean ($\overline{X}$) by

$$t_{\overline{X}} = \frac{\overline{X} - \mu}{s_{\overline{X}}} \tag{7.5}$$

where $s_{\overline{X}} = \sqrt{\hat{\sigma}^2/n} = \hat{\sigma}/\sqrt{n}$ = *the estimated standard error of the mean.* The $t_{\overline{X}}$ scores follow the t distribution, whose only parameter is the degrees of freedom that will equal $n - 1$. The only major requirement for its use is that the original distribution of scores approximates a normal distribution.

7. If we have both a small sample *and* a nonnormal distribution, we cannot use the Z distribution or the t distribution.

8. For convenience, because the t distribution gets closer and closer to the standard normal distribution as n gets larger and therefore any differences between the two become small, it is common practice to use t distribution methods for both small samples (if the distribution is approximately normal) and for large samples (whether or not the distribution is normal).

7.5 TESTS FOR GOODNESS OF FIT

In Section 6.3 some simple, mainly graphical, methods were introduced to assess the degree of correspondence between an observed distribution of frequencies and an expected distribution of frequencies based on a particular probability function. We can now introduce the principal methods for a more formal assessment of the level of the relationship.

Chi-Square Test for Goodness of Fit

A common method is the chi-square test for goodness of fit. The test rests on the assumption that the sum of the differences between the observed and expected frequencies (squared and standardized by dividing by the expected value) for the classes is distributed approximately as a chi-square variable. Thus, the tabled values of the chi-square distribution (Table E) can be used to assess the difference between the observed and expected distributions.

The chi-square statistic is

$$\chi^2 = \sum_{j=1}^{k} \frac{(O_j - E_j)^2}{E_j} \tag{7.6}$$

which is distributed as χ^2 with degrees of freedom:

$$df = k - p - 1$$

where

k = the number of classes

p = the number of parameters estimated in the calculation of the expected distribution

and

O_j = the observed frequency for class j

E_j = the expected frequency for class j

Calculation time can be saved by manipulating the summation term in Equation 7.6 to give

$$\chi^2 = \sum_{j=1}^{k} \frac{(O_j)^2}{E_j} - n \tag{7.7}$$

where

n = the size of the sample

The chi-square statistic is susceptible to large, somewhat artificial increases in value when the expected values are very small, and thus it is usual to require an expected value of at least 3 (some authors recommend 5) for each class. For the classes at the two tails of the distribution, it may be convenient (and not too problematic) to allow values as small as 1.0. For the Herd Size variable of the Cumberland County Dairy Farm Survey, the observed and expected frequencies have already been tabulated (Table 6.4). A number of classes at each end of the distribution have expected values less than 1.0 and so must be grouped together to satisfy the foregoing suggestion. In the example, this resulted in combining the lowest class with the second lowest and the three highest classes being grouped together. The resulting distribution has seven classes (Table 7.2). As many classes as possible should be used as the degrees of freedom, and therefore the accuracy of the test will decrease as the number of classes decreases. The calculated value for chi square can be compared with a tabled value, and if the calculated value is greater than the tabled value (for a specified level of significance), we can assume that the differences between the observed and expected frequencies cannot be attributed to chance alone.

The hypothesis test is outlined for the variable Herd Size (Table 7.2), and a summary of the results of the tests for the other three variables examined in Chapter 6 is presented in Table 7.3. We note that for the Farm Area variable fitted with a normal (using log X) curve, the value of p (the number of estimates made) would be 2 (the mean and the variance) for the Years Feed Purchased variable it would be 1 (λ was estimated); and for the Pasture Assessment Index, it would be 1 (p was estimated).

Chi-Square Test for Goodness of Fit

1. ***Hypothesis***

$Ho: O_j = E_j$ versus $Ha: O_j \neq E_j$

that is, *Ho:* {that the X_i form a random sample from a normal population with $\mu = 90.2$ and $\sigma^2 = 992$} versus *Ha:* that they do not

2. ***Assumptions***

(i) $\dfrac{(O_j - E_j)^2}{E_j}$ is distributed as a χ^2 variable

(ii) $\overline{X} = 90.2 = \mu \qquad \hat{\sigma}^2 = 992 = \sigma^2$

(iii) random sample, $n = 37$, assigned to 7 classes

(iv) $\alpha = .05$

3. ***Test Statistic***

$$\chi^2 = \sum_{j=1}^{k} \frac{(O_j - E_j)^2}{E_j} = \sum_{j=1}^{k} \frac{(O_j)^2}{E_j} - n$$

4. ***Rejection Region***

$$\begin{aligned} \chi^2 &\geq [\chi^2\, p = 1 - \alpha,\ df = k - p - 1] \\ &\geq [\chi^2\, p = .95,\ df = 4] \\ &\geq 9.488 \end{aligned}$$

5. ***Calculation***

TABLE 7.2 The Chi-Square Calculation

Class	O_j	E_j	O_j^2/E_j
<40	1	2.07	.48
40–59	4	4.17	3.84
60–79	8	7.62	8.40
80–99	10	9.15	10.93
100–119	8	7.67	8.34
120–139	3	4.22	2.13
≥140	3	2.11	4.27
TOTAL	37	37.01	38.39

$$\chi^2 = 38.39 - 37 = 1.39.$$

TABLE 7.3 χ^2 Goodness-of-Fit Tests for Three Variables

Variable	Function Fitted	Parameters Estimated	Number of Classes	χ^2	*df*, (χ^2 .95)	Conclusion
Farm Area	Normal (using log X)	$\mu\sigma^2$	7	1.62	4, 9.49	Fail to reject *Ho*
Years Feed Purchased	Poisson	λ	4	1.54	2, 7.81	Fail to reject *Ho*
Pasture Assessment Index	Binomial	p	8	1.66	6, 12.6	Fail to reject *Ho*

6. *Conclusion*

$$1.39 \not> 9.488$$

Therefore, we fail to reject the null hypothesis and conclude that the observed distribution does not differ significantly from the expected, and we can assume the sample comes from a population distributed normally with mean 90.2 and variance 992.

Kolmogorov–Smirnov Test for Goodness of Fit

Although the chi-square test is a frequently encountered test for the goodness of fit of a distribution, another test, the *Kolmogorov–Smirnov*, is more powerful than the chi-square test in certain situations. It is especially useful when the number of classes is small and when we can estimate any unknown parameters independently of the distributions. Its use will be illustrated by examining the goodness of fit of the Poisson distribution to the variable Years Feed Purchased (Table 6.9). Values for the Kolmogorov–Smirnov statistic D are presented in Table F.

Kolmogorov–Smirnov Test for Goodness of Fit

1. *Hypotheses*

$$Ho\colon O_j = E_j \quad \text{versus} \quad Ha\colon O_j \neq E_j$$

that is, *Ho*: {that the X_i form a random sample from a Poisson population with $\lambda = .65$} versus *Ha*: that they do not

2. *Assumptions*

(i) D is distributed following the Kolmogorov–Smirnov statistic
(ii) $\lambda = .65$
(iii) random sample, $n = 37$, assigned to 4 classes
(iv) $\alpha = .05$

3. *Test Statistic*

$$D = \max|F(x) - \text{CRF}_O|$$

where

$F(x) = \text{CRF}_E =$ cumulative relative frequency (expected)

$\text{CRF}_O =$ cumulative relative frequency (observed)

TABLE 7.4 Kolmogorov–Smirnov Calculation

	Observed			Expected			
Class	F_O	CF_O	CRF_O	F_E	CF_E	CRF_E	$CRF_E - CRF_O$
0	21	21	.5676	19.31	19.31	.5219	.0457
1	11	32	.8649	12.55	31.86	.8611	.0038
2	3	35	.9459	4.08	35.94	.9714	.0255
3	2	37	1.0000	1.04	36.98	.9995	.0005

4. ***Rejection Region***

$$D \geq [D, 1 - \alpha, df = n]$$
$$\geq [D, p = .95, df = 37]$$
$$\geq .23$$

5. ***Calculation.*** The value to compare with the tabled D value involves calculating the difference between the observed cumulative frequency (CFR_O) and the expected cumulative frequency (CRF_E). The largest of these differences .0457 in the example in Table 7.4, is the calculated D value and is compared with the tabled D values (Table F).

6. ***Conclusion***

$$.0457 \not\geq .23$$

Therefore, we fail to reject the null hypothesis and conclude that the observed distribution does not differ significantly from the expected, and we can assume that the sample comes from a population distributed following a Poisson distribution with $\lambda = .65$.

7.6 STATISTICAL INFERENCE WITH COMPUTERS

Most statistical packages in use today operate in a batch-processing mode (the researcher submits a file of input information to be processed) rather than in an interactive mode where the process is guided through its various steps. Such a structure does not lend itself to the formalized approach adopted in this chapter. However, both SPSS and SAS have interactive versions that are increasingly available, and, more importantly, the wealth of statistical processing systems being written for microcomputer systems provides for an interactive approach to many of the inferential techniques introduced.

The topics covered in the past two chapters have been introduced for two reasons: to outline inferential methods for univariate variables and distributions and to introduce inferential methods in general, as an introduction to the remaining chapters of this book. In fact, univariate inferential situations other than as a preliminary step are quite rare—bivariate and multivariate tests dominate most statistical packages.

Point estimation of population parameters by sample statistics is nearly always assumed by a statistical program. That is, the assumption is usually that the data being processed are from a sample, and the results presented are the best estimates of the relevant population parameters. Confidence intervals are not usually reported, but standard errors are, thus intervals can be easily constructed if required. Hypothesis tests for summary parameters are also not usually provided, but again, these can be derived easily from the standard errors. However, when recalculations are made using the output from a computer package, it is important to know how the particular statistic is calculated. Most importantly, we need to know whether the reported sample variance was estimated using n or $n - 1$.

Although fitting probability functions and testing goodness of fit are not generally available in major general-purpose packages, there are exceptions. For example, SPSS through the NPAR (nonparametric) program provides a Kolmogorov–Smirnov test for fitting the uniform, normal, or Poisson functions.

There is one by-product of the greater reliance on computer methods that needs to be mentioned, since it will be used extensively throughout the remainder of the book. This is a decrease in the formality and *a priori* nature of hypothesis testing. Rather than consider all the decision structure of a hypothesis test being fixed before the analysis is performed, most programs output a probability value representing the critical value for α (a Type I error), on either side of which the null hypothesis (usually only implied) would be rejected or failed to be rejected. Thus, to a large degree, the decision making is left until after the test, although the researcher can still have a fixed α value in mind, and only the yes/no question is answered after the test. More commonly, researchers are now reporting results with the critical α level stated to give an indication of the significance level of the result.

References and Readings

Inferential Methods

Aitchison, J. and J. A. C. Brown (1957) *The Lognormal Distribution,* Cambridge University Press: London.

Blalock, H. M. (1979) *Social Statistics,* Revised Second Edition, McGraw-Hill: New York.

Dixon, W. J. and F. J. Massey (1957) *Introduction to Statistical Analysis,* McGraw-Hill: New York.

Edgington, E. S. (1969) *Statistical Inference: The Distribution Free Approach,* McGraw-Hill: New York.

Krumbein, W. and F. A. Graybill (1965) *An Introduction to Statistical Models in Geology,* McGraw-Hill: New York.

Siegel, S. (1956) *Nonparametric Statistics for the Behavioral Sciences,* McGraw-Hill: New York.

CHAPTER 8

An Introduction to Bivariate Relationships

8.1 Testing Differences of Summary Measures:
Difference of means tests; Tests for matched pairs; Difference of proportions tests.

8.2 Contingency Tests and Associated Structures:
Contingency tables and the chi-square statistic; Measures of association based on the contingency table.

8.3 Analysis of Variance:
One-way analysis of variance; Structure of the model; Extensions and assumptions; Tests for differences among several means; Intraclass correlation.

Until now, we have limited our discussions to single samples, usually using the example of a sample of farms from Cumberland County. However, much of the power of statistical methods arises not from computing single-point estimates or confidence intervals, but rather from the ability to distinguish between samples and to make inferences about the differences between the populations from which these samples were drawn. This is especially true for those sciences concerned with experimentation in the stricter sense of the word. For example, we might wish to analyze the effect of a particular fertilizer on crop yield using a sample of farms. There is a variety of ways we could design this experiment, but all are concerned with the comparison of the results from two subsamples—one where fertilizer has been applied and one where it has not. Although this type of situation is not unknown in geography, it is rare for a geographic problem to be structured in this manner. More commonly the geographer has obtained a sample from a population in which the sampled elements might fall into two (or more) particular groups, and interest is in the comparison of the scores of one group with the other. Examples include geomorphological slope measurements on two rock types, the level of newspaper readership in two cities, and the level of unemployment in two counties. As we will see, these situations are not too different from the experimental one just mentioned, and they all can be considered as a two-variable (bivariate) problem. Both the fertilizer example and the three other examples mentioned are concerned with testing the relationship between the scores of one variable and the scores of another dichotomous or binary variable—the latter's scores indicating two treatments, two groups, two locations, etc.

There is a large number of testing structures that can be used to assess these bivariate situations. The choice of which structure to use depends on a variety of factors. We can, however, make a broad distinction between *testing for a relationship* between two variables, and providing statistical *measures of association* of two variables. Our emphasis in this chapter and the next will be on assessing relationships—either testing to see whether the distributions of scores of two variables are similar, or, more simply, to test whether a particular summary measure has similar values from two distributions. Alternatively, we might be interested in obtaining a value for a measure of association which describes the covariation of the two variables. Broadly speaking, we call these latter measures *correlation coefficients*, although this term is usually restricted to measures for variables at the higher levels of measurement. The two different approaches come together when in some circumstances, we test the association measure against some hypothetical value—thereby indirectly assessing a relationship.

To a large degree, the choice of structure also depends on the level of measurement of each of the two variables. With a variety of combinations of levels possible, and with a number of suggested measures or structures at each combination, there has been a considerable proliferation of techniques. Only the major ones are discussed in this chapter and the next.

In this chapter, we examine three of the most frequently used testing structures: testing for differences of two summary measures (two means or two proportions), assessing joint frequency distributions of two variables (contingency structures), and testing for the differences of three or more means (one-way analysis of variance). The difference of two-means test is merely a special case of the one-way analysis of variance structure, and it is this last structure that proves to be of greatest practical application for many situations. Along with simple linear regression and correlation analysis—to be examined in detail in the next chapter—it forms the basis for the analysis of multivariate situations that constitute the remainder of this text.

8.1 TESTING DIFFERENCES OF SUMMARY MEASURES

Our concern in this section is testing the difference of arithmetic means obtained from two samples or groups. A variety of different forms of this test are available and we will discuss the major ones. We will also provide a test for difference of proportions—a situation in which the summary measure of interest is simply the proportion of a sampled set of elements that has a certain attribute. Are the proportions for the two samples or groups significantly different? The testing structure is basically identical to the test for two means.

Difference of Means Tests

The *two-sample means test* (or simply the *difference of means test*) is concerned with the situation where we have samples from two populations and wish to test to see if we can conclude that the two populations have different means. This situation may arise in a number of ways.

First, with two populations defined, we take independent random samples of each population, and for each we calculate a mean for a single metric variable. Using the sample means, we wish to know if the two population means are different.

Second, with one population defined we have two variables—one a metric variable, and the other a binary variable that divides the population into two subpopulations. We want to know if the two subpopulation means are significantly different. In most circumstances, we can apply the testing structure for the situation described above to this subpopulation structure.

Third, with one population defined, we have two variables—each variable attempts to measure the same attribute of the population elements. We wish to test whether the population means estimated by the two different variables are the same or different. For each element, then, we have a *matched pair* of scores. This situation requires a different testing structure to that outlined above, and we present this as a special case toward the end of this section.

Before we outline tests for the difference between two means, we need to examine a theorem which, together with an extension of the central limit theorem, provides the necessary basis for the testing structure. The theorem states

> **if independent random samples of sizes n_1 and n_2 are drawn from populations that are normal, $N(\mu_1,\sigma_1^2)$ and $N(\mu_2,\sigma_2^2)$, then the sampling distribution of the difference between the sample means $(\overline{X}_1 - \overline{X}_2)$ will be normal, $N(\mu_1 - \mu_2, \sigma_1^2/n_1 + \sigma_2^2/n_2)$.**

Again, as in the single-sample case, the theorem can be generalized to populations of any distribution form if n_1 and n_2 are large—$(\overline{X}_1 - \overline{X}_2)$ approaches normality as n_1 and n_2 increase.

The requirement of independent random samples would be fulfilled in the first of the situations just outlined—two populations, two separate random samples. More importantly, the requirement would also be fulfilled in the second situation. Not only is each element and its score independent of other elements through random selection of elements in a sample, but also any combination of elements are independent of any other combination. Thus, when a population of scores has been subdivided into two subpopulations by a binary variable, the two subpopulations can be regarded as independent selections. The requirement of independence would not, however, hold for the matched pairs situation.

The form of the sampling distribution may be a little harder to conceptualize. Our statistic is the difference of the two sample means, and it seems reasonable to think that the mean of this sampling distribution will be the difference of the two population means ($\mu_1 - \mu_2$). However, we might also expect the variance of the sampling distribution to be the difference of the two population variances—instead we add them. A formal development of why this is so is not possible here, but it makes some intuitive sense. We are looking at the standard error of the difference of two means and with both means varying independently, this would produce a larger standard error than the standard error of each of the population means alone.

To illustrate the use of the difference of means tests, consider the problem of two farm types in Cumberland County. We are interested in whether there is a difference in the price (value) of dairy farms and mixed farms. The summary statistics for the analysis are presented in Table 8.1.

Our testing structure requires that we know something about the variances of the populations or subpopulations from which the samples were taken. This is because we need to provide an estimate of the variance of the sampling distribution of differences, namely ($\sigma_1^2/n_1 + \sigma_2^2/n_2$). Two situations arise: If we can reasonably assume that $\sigma_1^2 = \sigma_2^2$ then we can provide a *pooled estimate* of the standard error; if we cannot assume that $\sigma_1^2 = \sigma_2^2$, then we have to make a much cruder estimate based on *separate estimates* for each variance and then combine them. Tests for difference of means under both assumptions will be presented—the choice of which one to use requires a decision on the homogeneity (equality) of the two subpopulation variances.

Although a number of tests are available to assist in the decision on homogeneity (or otherwise) of variance, none is fully satisfactory. The most commonly used test for variances takes the ratio of the two sample variances and tests it against an F statistic. By placing the larger of the two sample variances in the numerator (to get a ratio ≥ 1.0) the test simplifies to *Ho:* $\sigma_1^2 = \sigma_2^2$ reject if

$$F = \frac{\hat{\sigma}_1^2}{\hat{\sigma}_2^2} \geq [\mathrm{F}, \mathrm{p} = 1 - \alpha/2, df = (n_1 - 1),(n_2 - 1)] \tag{8.1}$$

Table G at the end of the book contains F values for probabilities of .95, .975, and .99. If we fail to reject the null hypothesis then we can safely use the pooled variance estimate; if we reject it, then we must rely on the less satisfactory separate variance estimate. This test of the variances is adequate for our purposes when the sample sizes n_1 and n_2 are not too dissimilar,

TABLE 8.1 Farm Value for Two Farm Types

	Dairy Farms	Mixed Farms
Average farm value ($) $\overline{X}$	122,000	93,000
Standard deviation ($) $\hat{\sigma}$	33,000	45,000
Sample size n	37	31

and the two subpopulation distributions are not too nonnormal. We also note that the foregoing test provides a reasonably satisfactory method for the testing for difference between two population variances.

Using the data from Table 8.1, we calculate the $\hat{\sigma}_1^2/\hat{\sigma}_2^2$ statistic with an F structure at $\alpha = .05$, as

$$\frac{\hat{\sigma}_1^2}{\hat{\sigma}_2^2} = \frac{(45000)^2}{(33000)^2} = 1.86$$

1.86 is not greater than $[F, p = .975, df = 30, 36] = 1.97$, (interpolated), therefore we fail to reject the null hypothesis and assume that the two population variances are not dissimilar. Note that as we required a two-tail test because there was no prior knowledge of the relative sizes of the population variances, we had to halve the probability value we used to enter the table of F by using this form of the test.

We will outline the difference of means test using the pooled variance estimate but mention the alternate structure (separate variances) in the appropriate place.

Hypothesis Test for Difference of Means ($\sigma_1^2 = \sigma_2^2$ *assumed*)

1. ***Hypotheses.*** We state that the population means are equal, *Ho:* $\mu_1 = \mu_2$. Alternatively, we could say that $\mu_1 - \mu_2 = 0$. *Ha:* $\mu_1 \neq \mu_2$. Presumably, we are interested in whether dairy farms that, in general, involve a more intensive use of land have higher valuations than mixed farms. In such cases, we state the hypothesis that we wish to disprove, thus our null hypothesis is that there is no difference. The null hypothesis is, in effect, stating that there is no relationship between type of farm and farm value.

2. ***Assumptions***

   (i) Independent random samples of size $n_1 = 37$ and $n_2 = 31$

   (ii) Normal populations $X_{1i} \sim N(\mu_1, \sigma_1^2)$ and $X_{2i} \sim N(\mu_2, \sigma_2^2)$

   (iii) $\sigma_1^2 = \sigma_2^2$ (pooled variance test)

   (iv) Level of significance $\alpha = .05$, two-tail test

3. ***Test Statistic.*** Because σ is unknown, we will use a modification of the t test already used for single samples.

$$t = \frac{\overline{X}_1 - \overline{X}_2 - (\mu_1 - \mu_2)}{\hat{\sigma}_{\overline{X}_1 - \overline{X}_2}}$$

since we assume that $\mu_1 = \mu_2$ we write

$$t = \frac{\overline{X}_1 - \overline{X}_2}{\hat{\sigma}_{\overline{X}_1 - \overline{X}_2}} \tag{8.2}$$

where $\hat{\sigma}_{\overline{X}_1 - \overline{X}_2}$ is the estimate of the standard error of the difference between the sample means. From the theorem stated at the beginning of this section

$$\sigma_{\overline{X}_1 - \overline{X}_2} = \sqrt{\frac{\sigma_1^2}{n_1} + \frac{\sigma_2^2}{n_2}}$$

By assuming that $\sigma_1^2 = \sigma_2^2 = \sigma^2$, this equation can be rewritten as

$$\sigma_{\overline{X}_1 - \overline{X}_2} = \sigma\sqrt{\frac{1}{n_1} + \frac{1}{n_2}}$$

$$= \sigma\sqrt{\frac{n_1 + n_2}{n_1 n_2}} \tag{8.3}$$

To derive an estimate for σ, called a pooled estimate, we take an average of the sample variances.

Thus:

$$\hat{\sigma} = \sqrt{\frac{(n_1 - 1)\hat{\sigma}_1^2 + (n_2 - 1)\hat{\sigma}_2^2}{n_1 + n_2 - 2}} \tag{8.4}$$

Because we have two parameters, we lose two degrees of freedom, $n_1 + n_2 - 2$. By combining the structure given in Equation 8.3 with our pooled estimate of σ in Equation 8.4, we obtain an estimate of the standard error of the difference between sample means,

$$\hat{\sigma}_{\overline{X}_1 - \overline{X}_2} = \sqrt{\frac{(n_1 - 1)\hat{\sigma}_1^2 + (n_2 - 1)\hat{\sigma}_2^2}{n_1 + n_2 - 2}}\sqrt{\frac{n_1 + n_2}{n_1 n_2}} \tag{8.5}$$

For the alternate structure where we cannot assume $\sigma_1^2 = \sigma_2^2$, we get a more approximate estimate, using the separate sample variances

$$\hat{\sigma}_{\overline{X}_1 - \overline{X}_2} = \sqrt{\frac{\hat{\sigma}_1^2}{n_1} + \frac{\hat{\sigma}_2^2}{n_2}} \tag{8.6}$$

4. ***Rejection Region***

$$|t| \geq [t, p = 1 - \alpha/2, df = n_1 + n_2 - 2] \qquad (8.7)$$

$$\geq [t, p = .975, df = 66]$$

$$\geq 2.00 \text{ (for } df = 60\text{, the closest lower tabled value to 66)}$$

The rejection region cited is the one used for a pooled estimate of the variance. It can also be used in the separate variance structure as long as the two sample sizes are not small and not too dissimilar. If they are, then alternative testing structures which modify the degrees of freedom would have to be used.

5. ***Calculation.*** For ease of hand calculation, we use the data in thousands (000s).

$$\hat{\sigma}_{\bar{X}_1-\bar{X}_2} = \sqrt{\frac{36(33)^2 + 30(45)^2}{37 + 31 - 2}}\sqrt{\frac{37 + 31}{(37)(31)}}$$

$$= \sqrt{\frac{39204 + 60750}{66}}\sqrt{\frac{68}{1147}}$$

$$= (38.92)(0.243)$$

$$= 9.46$$

$$t = \frac{122 - 93}{9.46}$$

$$= 3.07$$

6. ***Conclusion.*** Because the calculated t value (3.07) is greater than the tabled value (2.00), we reject the null hypothesis of no difference. We can reasonably conclude that there is a difference in value of dairy farms and mixed farms.

Because we postulated that the valuation of dairy farms was higher than mixed farms, it would have been possible to use a one-tail test rather than a two-tail test. Remember from Chapter 7 that the one-tail test shifts all the critical region to one or other tail of the distribution. At the same level of significance (t .95, 66) the tabled value of t is 1.671. There is no difference in our conclusion in this case, but it is obvious that situations might arise in which you will reach different conclusions.

As we noted in the last chapter, most computer packages do not follow the formal approach to hypothesis testing that we have adopted, but instead simply report the critical α level that would separate rejecting and failing to reject an implied hypothesis. Table 8.2 illustrates the results we would

TABLE 8.2 SPSS Results for *T* Test

VARIABLE	NUMBER OF CASES	MEAN	STANDARD DEVIATION	STANDARD ERROR	F VALUE	2-TAIL PROB.
FARMVALUE						
GROUP 1	31	93.0000	45.000	8.080	1.86	0.076
GROUP 2	37	122.0000	33.000	5.425		

POOLED VARIANCE ESTIMATE			SEPARATE VARIANCE ESTIMATE		
T VALUE	DEGREES OF FREEDOM	2-TAIL PROB.	T VALUE	DEGREES OF FREEDOM	2-TAIL PROB.
3.07	66	0.002	2.98	53.99	0.002

have obtained for the farm income example we have been examining, if we had run it through the SPSS program T-TEST. Three hypotheses are indirectly tested: an F test for the hypothesis $\sigma_1^2 = \sigma_2^2$ and t tests for the hypothesis $\mu_1 = \mu_2$ under both the pooled variance and separate variance structures. Assuming an α level of .05, our interpretation of these results would proceed as follows. From the F test for equality of variances, α critical (=.076) is greater than .05; therefore, we fail to reject the hypothesis that $\sigma_1^2 = \sigma_2^2$ and conclude that the two variances are not dissimilar and that we can use the pooled variance t test. From this t test, the α-critical value of .002 is less than .05; therefore, we reject the hypothesis that $\mu_1 = \mu_2$ and conclude that the means are significantly different.

Tests for Matched Pairs

In some instances, a particular type of test for the difference of means is required. Consider the following problem. The Cumberland County administration is concerned with the effect of farm practices on soil erosion in the county. To measure the degree of soil erosion on a particular farm requires either an intensive and time-consuming field survey or a rapid study of aerial photographs. As an administrator, are you justified in using the less costly and more time-efficient aerial photograph analysis to obtain estimates of mean erosion levels in the county? What you need to do is test if the means obtained by the two measurement methods are significantly different. However, it would not be correct to use a difference of means test because we have calculated two measurements of the variable erosion level for one farm at a time. In this instance, the samples are not independent of one another, and we refer to this analysis as a test of matched pairs. As another example, consider a set of carefully "matched" cities on a variety of characteristics with the aim of analyzing the effect on voter registration of mass mailings to registered voters versus door-to-door canvassing. Which technique is most effective? To reiterate, a difference of means test is not appropriate because we do not have two sets of cases that have been selected independently.

We examine the soil erosion problem first outlined, where X_1 is a measure of soil erosion derived from field inspection and X_2 from aerial photographic analysis.

1. ***Hypotheses***

$$Ho: \mu_1 = \mu_2 \qquad \text{versus} \qquad Ha: \mu_1 \neq \mu_2$$

or

$$\mu_D = \mu_1 - \mu_2 = 0$$

2. ***Assumptions***
   (i) Random sample of size $n = 15$ (pairs) of farms
   (ii) Normal populations $X_{1i} \sim N(\mu_1, \sigma_1^2)$ and $X_{2i} \sim N(\mu_2, \sigma_2^2)$
   Therefore, $D_i \sim N(\mu_1 - \mu_2, \sigma_D^2/n)$
   where $D_i = X_{1i} - X_{2i}$ and σ_D^2/n is the standard error of the difference
   (iii) Level of significance $\alpha = .05$, two-tail test

3. ***Test Statistic***

$$t = \frac{\overline{D}}{S_D/\sqrt{n}} \tag{8.8}$$

where

$$S_D/\sqrt{n} = \text{the estimate of } \sqrt{\sigma_D^2/n}$$

and

$$S_D = \sqrt{\frac{\sum_{i=1}^{n} (D_i - \overline{D})^2}{n - 1}}$$

or, in calculation form,

$$S_D = \sqrt{\frac{\sum_{i=1}^{n} D_i^2 - \frac{\left(\sum_{i=1}^{n} D_i\right)^2}{n}}{n - 1}} \tag{8.9}$$

4. ***Rejection Region***

$$|t| \geq [t, p = 1 - \alpha/2, df = n - 1]$$
$$|t| \geq [t, p = .975, df = 14]$$
$$|t| \geq 2.145$$

5. ***Calculation.*** Using the summations from Table 8.3, we get

$$\overline{X}_1 = \frac{760}{15} = 50.67 \qquad \overline{X}_2 = \frac{795}{15} = 53.00$$

and

$$\overline{D} = \frac{-35}{15} = -2.33$$

(and we note that $\overline{D} = \overline{X}_1 - \overline{X}_2$)

From Equation 8.9,

$$S_D = \sqrt{\frac{533 - \dfrac{(-35)^2}{15}}{14}}$$
$$= 5.68$$

TABLE 8.3 Calculation of Matched Pair Differences

	X_{1i}	X_{2i}	D_i	D_i^2
	25	22	3	9
	35	41	−6	36
	20	24	−4	16
	40	37	3	9
	0	0	0	0
	55	58	−3	9
	60	70	−10	100
	75	85	−10	100
	80	87	−7	49
	90	84	6	36
	100	95	5	25
	80	88	−8	64
	25	21	4	16
	40	40	0	0
	35	43	−8	64
TOTAL	760	795	−35	533

Therefore,

$$t = \frac{-2.33}{5.68} \cdot \sqrt{15}$$

$$= -1.59$$

6. *Conclusion*

$$|-1.59| \not> 2.145$$

We fail to reject the null hypothesis and conclude that the means are not significantly different. Therefore there is little to differentiate the two methods for estimating mean erosion levels.

Difference of Proportions Tests

We are often faced with percentaged or proportionate responses rather than mean values. This requires a slightly different test, although the overall structure remains the same. In the following problem, we are concerned with the results of a survey of newspaper readership in isolated rural areas of a county. A survey showed that 65 households out of a sample of 180 in the northern part of the county subscribed to a major city newspaper, while 80 out of a sample of 150 took the same paper in the southern part of the county (closer to the major city). Is there a real difference in the proportions?

1. ***Hypotheses.*** No difference in readership *Ho*: $P_1 = P_2$; that is, the proportions in readership in the southern and northern parts of the county are equal, apart from random sample fluctuations; versus *Ha*: $P_1 \neq P_2$.

2. ***Assumptions***
   (i) Level of measurement is dichotomous (binary)
   (ii) Independent random samples of size $n_1 = 180$, $n_2 = 150$
   (iii) $\sigma_1^2 = \sigma_2^2$
   (iv) Normal sampling distribution (remember that when we are dealing with a dichotomy, we are dealing with a binomial distribution. When n is sufficiently large, we know that the sampling distribution of means will be approximately normal. If n is small, other tests based on the binomial would have to be used).
   (v) Level of significance $\alpha = .05$.

3. ***Test Statistic.*** The testing structure is similar to the t test for the difference of means with the assumption of equal variances, as outlined in the previous section.

The test statistic is

$$Z = \frac{p_1 - p_2}{\hat{\sigma}_{p_1 - p_2}} \tag{8.10}$$

where $p_1 = \hat{P}_1$ and $p_2 = \hat{P}_2$ are the proportions from samples 1 and 2. and $\hat{\sigma}_{p_1-p_2}$ is the estimate of $\sigma_{p_1-p_2}$, the standard error of the sampling distribution of the difference of proportions P_1 and P_2.

With the variance of a binomial distribution (Chapter 5) equal to $p(1 - p)$, and following from the t test for difference of means, it can be shown that

$$\sigma_{p_1-p_2} = \sqrt{\frac{(P_1)(1 - P_1)}{n_1} + \frac{(P_2)(1 - P_2)}{n_2}}$$

But with the assumption that the two variances are equal, that is,

$$P_1(1 - P_1) = P_2(1 - P_2) = p(1 - p)$$

$$\sigma_{p_1-p_2} = \sqrt{p(1 - p)}\sqrt{\frac{n_1 + n_2}{n_1 n_2}} \tag{8.11}$$

Our estimate for $\sigma_{p_1-p_2}$ simply involves estimating p, a pooled proportion. Therefore,

$$\hat{\sigma}_{p_1-p_2} = \sqrt{\hat{p}(1 - \hat{p})}\sqrt{\frac{n_1 + n_2}{n_1 n_2}} \tag{8.12}$$

and an estimate for p can be found by weighting the sample estimates of P_1 and P_2 by sample size

$$\hat{p} = \frac{n_1 p_1 + n_2 p_2}{n_1 + n_2} \tag{8.13}$$

4. ***Rejection Region***

$$\begin{aligned} |z| &\geq [z, p = 1 - \alpha/2] \\ &\geq [z, p = .975] \\ &\geq 1.96 \end{aligned}$$

5. ***Calculation***

$$\hat{P}_1 = p_1 = \frac{65}{180} = .36 \qquad \hat{P}_2 = p_2 = \frac{80}{150} = .53$$

Using Equation 8.13,

$$\hat{p} = \frac{180(.36) + 150(.53)}{180 + 150}$$

$$= .44$$

Using $\hat{p}$ in Equation 8.12,

$$\hat{\sigma}_{p_1 - p_2} = \sqrt{(.44)(.56)}\sqrt{\frac{180 + 150}{(180)(150)}}$$

$$= (.496)(.111)$$

$$= .055$$

Therefore, the test statistic from Equation 8.10 is

$$Z = \frac{.36 - .53}{.055}$$

$$= \frac{-.17}{.055}$$

$$= -3.09$$

6. ***Conclusion***

$$|-3.09| > 1.96$$

Therefore, we reject the null hypothesis and conclude that there is a difference in readership in the two parts of the county.

8.2 CONTINGENCY TESTS AND ASSOCIATED STRUCTURES

In the first section of this chapter, we examined the assessment of the relationship between a metric or enumerated variable and a binary variable (differences of means test) and a difference of proportions test to assess the relationships between two binary variables. We now want to examine a testing structure designed for the analysis of two variables derived from a classification. That is, the variables can be binary, nominally, or ordinally classified. However, we are concerned with assessing the *frequency* of scores, thus the structure can be applied to any situation where scores or groups of scores have been assigned frequencies. It is thus used at all levels of measurement.

The technique we are referring to is usually described as the construction of *contingency tables* and the testing structure as the *chi-square contingency* test. It has an extremely wide variety of applications and we have, in fact, already made use of it in examining the goodness-of-fit of observed frequency distributions to probability functions (Section 7.5). We now wish to examine the structure from a more general point of view.

Contingency Tables and the Chi-Square Statistic

We are concerned with *cross-tabulation* of scores. That is, we have two variables (note that the ideas easily extend to more than two variables) where each variable consists of a number of classes. These classes may be derived from a classification measurement process or they may have been created from variables measured by metrication or enumeration for the construction of frequency distributions. Each class will have a certain frequency, and the pattern of frequencies across the classes is an important characteristic of the variable. However, our interest is more in the joint frequency distribution of the two variables—the pattern of frequencies of combinations of classes. This is best illustrated by using a simple example—in a survey of land capability in an area, we have information for 350 parcels of land as to their Severity of Erosion (in four ordinal classes) and their Land Cover (in three nominal classes). A two-way cross-tabulation of the two variables provides a table of observed frequencies (O_j) (Table 8.4).

One thing we are interested in is whether there is a relationship or interaction between the two variables, or alternatively, whether we can assume that they are *statistically independent*. If they are the latter, we expect the same proportions of, for example, grassland cover to be severely eroded as scrub covered or forest covered. That is, as the overall proportion of parcels that are severely eroded is 50/350 = .143, we expect .143 of the 140 grass covered (= 20.0) to be severely eroded; .143 of the 120 scrub covered (= 17.1), etc. The same argument can be applied to the columns (140 of the parcels are grass covered; we expect 140/350 of the 50 severely eroded parcels to be grass covered = 140/350 × 50 = 20.0, etc.). In this manner, a table of expected frequencies (E_j) can be calculated based on the uniform distribution

TABLE 8.4 Land Cover versus Severity of Erosion

	Land Cover			
Erosion	Grass	Scrub	Forest	Total
Severe	30	10	10	50
Moderate	50	30	20	100
Slight	50	60	40	150
Negligible or none	10	20	20	50
TOTAL	140	120	90	350

TABLE 8.5 Expected Frequencies for Land Cover versus Severity of Erosion

Erosion	Grass	Scrub	Forest	Total
Severe	20.0	17.1	12.9	50
Moderate	40.0	34.3	25.7	100
Slight	60.0	51.4	38.6	150
Negligible or none	20.0	17.1	12.9	50
TOTAL	140	120	90	350

of marginal totals across the cells (Table 8.5). The calculation for each cell is

$$E_j = \frac{\text{row total} \times \text{column total}}{n} \tag{8.14}$$

An exact test for independence of the two variables is difficult to apply, although in the case of a contingency table with only two columns and two rows, these are readily available. However, if n is sufficiently large, the chi-square statistic provides a reasonably good approximation. As before (Chapter 7), we give the χ^2 statistic as:

$$\chi^2 = \sum_{j=1}^{k} \frac{(O_j - E_j)^2}{E_j} \tag{8.15}$$

$$= \sum_{j=1}^{k} \frac{O_j^2}{E_j} - n \tag{8.16}$$

We have deliberately kept the subscripting in the form $\Sigma_{j=1}^{k}$ where k is the number of cells in the contingency table. A more specific form of our r (number of rows) by c (number of columns) contingency table would be to express the summation as

$$\sum_{i=1}^{r} \sum_{j=1}^{c} \frac{(O_{ij} - E_{ij})^2}{E_{ij}} \tag{8.17}$$

However, by keeping it general, we emphasize the point that the statistic can be applied to any situation where there are observed and expected frequencies—in one dimension, two dimensions, or ever-higher dimensions.

The degrees of freedom associated with the calculated χ^2 statistic can be found from examination of the contingency table, where it is obvious that for each row and each column, one cell will be dependent on the other cells' scores and the marginal total. Therefore, the $r \times c$ cells have $(r - 1) \times (c - 1)$ degrees of freedom.

The Chi-Square Contingency Test
We will illustrate the testing structure by using the example we have outlined:

1. ***Hypotheses.*** We can express this in a number of ways. For example,

$$Ho\colon O_j = E_j \qquad \text{versus} \qquad Ha\colon O_j \neq E_j$$

or,

Ho: Severity of Erosion is statistically independent of Land Cover, versus
Ha: the two variables are statistically dependent

2. ***Assumptions***
(i) Level of measurement: (at least) two nominal scales
(ii) $\Sigma(O_j - E_j)^2/E_j$ is distributed as a χ^2 variable
(iii) Independent random sample of size $n = 350$ (the independent refers to the independence of selection of the two variables' scores)
(iv) $\alpha = .05$

3. ***Test Statistic***

$$\chi^2 = \sum_{j=1}^{k} \frac{(O_j - E_j)^2}{E_j} = \sum_{j=1}^{k} \frac{O_j^2}{E_j} - n$$

4. ***Rejection Region***

$$\begin{aligned} \chi^2 &\geq [\chi^2, p = 1 - \alpha, df = (r - 1)(c - 1)] \\ &\geq [\chi^2, p = .95, df = 6] \\ &\geq 12.592 \end{aligned}$$

5. ***Calculation.*** The calculations in Table 8.6 give a χ^2 value of 375.46 − 350 = 25.46.

Table 8.6 Table of O_j^2/E_j for Chi-Square Calculation

	45.00	5.85	7.75	58.60
	62.50	26.24	15.56	104.30
	41.67	70.04	41.45	153.16
	5.00	23.39	31.01	59.40
TOTAL	154.17	125.52	95.77	375.46

6. ***Conclusion***

$$25.46 > 12.592$$

We reject the null hypothesis and conclude that the $O_j \neq E_j$. There is, therefore, a statistical relationship between the Severity of Erosion and

Land Cover. We note also that the relative contributions of individual class combinations to the chi-square statistic can be gauged by their cell entries in Table 8.6.

Measures of Association Based on the Contingency Table

The test we have outlined establishes that there is, or is not, a relationship between the two variables. It does not provide us with a measure of the strength of the relationship. In many circumstances, a simple indication of a relationship is not sufficient—and some assessment of the degree of association is required.

A large variety of different *measures of association* have been derived to fit different levels of measurement and different situations. Ideally, a measure of association should have a large value when the degree of association is high, and a small value when the association is low. In addition, it should be independent of the units by which the variables are measured (and the size of the sample) to give a standardized dimensionless measure. Most (but not all) measures have been designed to give a range of values from 0 (no association) to +1.0 (complete association); or alternatively, if the direction of the relationship can be ascertained, from −1.0 (perfect negative association) through 0 (no association) to +1.0 (perfect positive association).

In the construction of contingency tables, the χ^2 statistic itself provides a measure of association, but the value it attains is a product not only of the strength of the association but also of the size of the sample. To provide a realistic measure of association it needs to be standardized in some way. Table 8.7 lists four measures based on the χ^2 statistic that attempt to provide a standardized structure. For the example we have been using, with a χ^2 value of 25.46 and a sample size of $n = 350$.

$$\phi^2 = \frac{25.46}{350} = .07$$

$$T^2 = \frac{25.46}{350} \sqrt{(4-1)(3-1)} = .03$$

$$V^2 = \frac{25.46}{350(3-1)} = .04$$

$$C = \frac{25.46}{25.46 + 350} = .26$$

All these measures based on the χ^2 statistic are somewhat arbitrary, but they can be useful in comparing the strength of association between sets of variables in similar experimental situations. Often ϕ^2, T^2 and V^2 are expressed in their square-root form, which gives a result more comparable with the contingency coefficient.

TABLE 8.7 Chi-Square Based Measures of Association

Measure	Symbol	Calculation	Range of Values
Phi coefficient	ϕ^2	$\dfrac{\chi^2}{n}$	0 to 1 for $r = c = 2$ 0 to $\min((r-1), (c-1))$ for larger tables
Tschuprow's T	T^2	$\dfrac{\chi^2}{n\sqrt{(r-1)(c-1)}}$	0 to 1 if $r = c$ 0 to <1 if $r \neq c$
Cramer's V	V^2	$\dfrac{\chi^2}{n \min((r-1)(c-1))}$	0 to 1
Contingency coefficient	C	$\sqrt{\dfrac{\chi^2}{\chi^2 + n}}$	0 to <1

8.3 ANALYSIS OF VARIANCE

In this book, we are going to examine analysis of variance, sometimes shortened to ANOVA, as an extension of two sample testing. Analysis of variance is a sophisticated statistical technique with many subtesting structures. Only the simplest model is presented here, analyzing the variation for only one variable. More complex models for several variables are regularly used in the fields of agricultural science, medicine, and public health but they are infrequently used in geography. Part of the explanation for this is related to the rather restrictive assumptions of the more complex models—assumptions of equal sample sizes for example—which are more easily fulfilled in fields that can control sample size with experimental designs.

Differences of means tests were used in Section 8.1 to test for differences in two samples. Analysis of variance is useful in situations where we have more than two samples. If we think of the difference of means test as a test of a relationship between a dichotomous nominal scale (farm type) and a metric scale (farm value), then we can think of analysis of variance as the relationship between a nominal scale of three or more classes and a metric scale. Although the analysis of variance model can be viewed as an extension of the differences of means test, it is also possible to view it as a model in which the variance in the metric variable is related to, or explained by, the *categories* of the nominal scaled variable. Thus, $y = f(x)$ where x is a set of groups or categories. This description of analysis of variance is like that of regression analysis where the variance in one variable—the dependent variable—is explained by one or more independent variables. We will draw out the similarities between analysis of variance and regression as we proceed.

One-Way Analysis of Variance

One-way analysis of variance, or simple analysis of variance, is concerned with differences between samples or classes, and uses the means for each of these samples or classes to summarize their characteristics. However, the procedure emphasizes sums of squares and estimates of variance rather than focusing on the difference of means. In earlier chapters, we introduced the notion of the sums of squares of a variable. We defined the corrected sums of squares of a variable as:

$$\text{TSS} = \sum_{i=1}^{n} (Y_i - \overline{Y})^2$$

or the sum of the squared differences of each observation from the mean. For ease of presentation, we will drop the corrected adjective and speak only of the sums of squares. (Note that we use Y for the variable whose variation we are interested in analyzing. This follows from our approach that views Y as a function of X. In the analysis of variance, X will be the categorization or groups into which Y is divided.)

The underlying principle of analysis of variance involves partitioning or splitting the total variance of a variable into two parts: a part that occurs within the divisions of the nominal variable (the groups, categories, or classes) that we will call the *within-class variance*, and a part that is related to differences between the groups, categories, or classes (the *between-class variance*). The within-class variance is calculated by considering each category or class separately and computing the sum of the differences between the individual scores and each category or class mean in turn. The sum over all classes is the total within-class variance. The between-class variance is calculated by computing the difference between each class mean and the total or grand mean, weighted by the sample size of each category. Alternative names for the between-class variation and the within-class variation are *explained* and *unexplained variation*, respectively—terms that we will use later. The adjective "explained" refers to the fact that the amount of variation that we can attribute to differences between samples (or classes) is a measure of how much variation is explained by, or statistically dependent on, the different samples or classes. Internal variation with the classes is left "unexplained" by the class structure.

The test of whether the variance is due mostly to within-class variation (thus not explained by the category structure) or between-class variation (or differences between classes and thus explained by the categories) is determined by the ratio of the between-class variance to the within-class variance. An important assumption in this test is that there is no difference in the populations that underlie the samples being analyzed. The sample means of the categories should only differ due to sampling error and the distribution of these differences should be normally distributed. If the population means are actually different, the sample means will differ by more than expected by the normal distribution. If the hypothesis of no difference (Ho) is incorrect, the between-class variance will be greater than the average within-class variance. This discussion is illustrated in Figure 8.1. Again, we require independent random samples, normal population distributions (if samples are small), and equal population standard deviations. After developing the test, these assumptions will be discussed further.

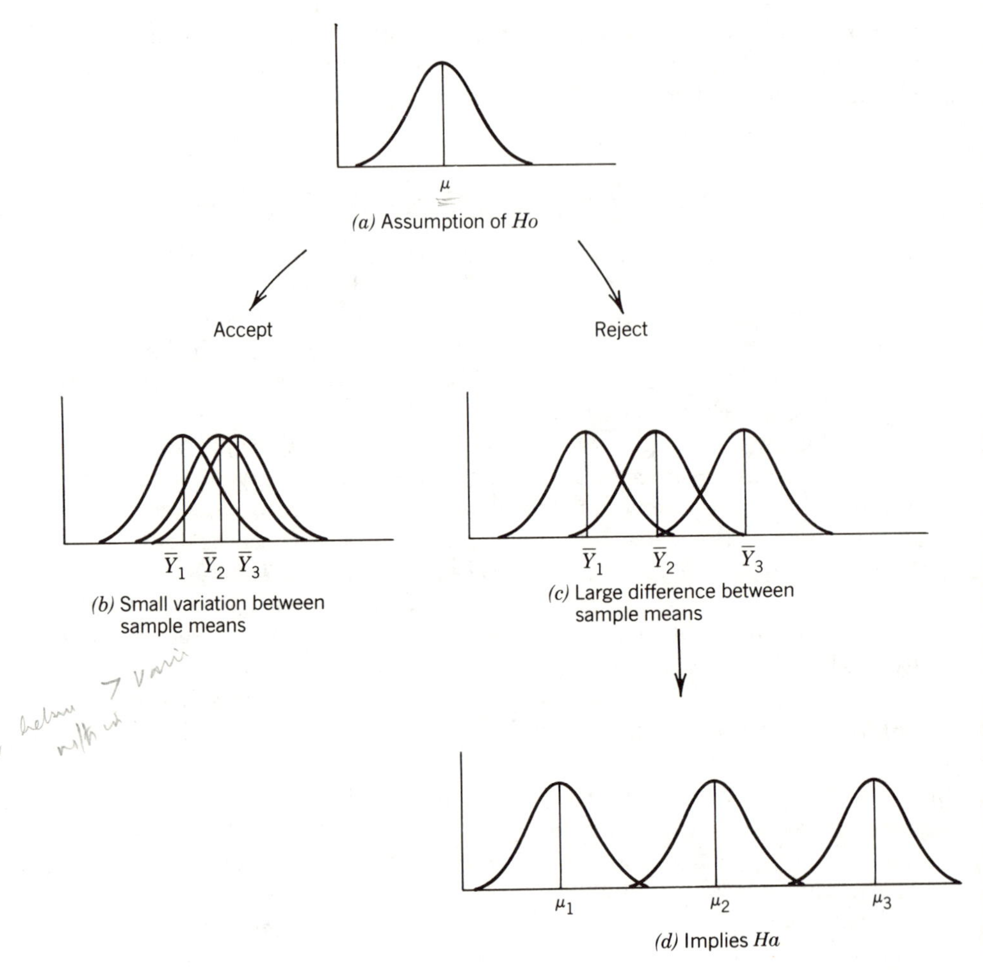

Figure 8.1 Diagram of *Ho* and *Ha* in analysis of variance.

Structure of the Model

A structure for the test of the ratio of between to within variance for three categories is set up in Table 8.8. At this point, we introduce the double summation sign ΣΣ and the identification of scores by their row and column position. Thus, Y_{ij} identifies (by tradition) the scores for the ith row and jth column. Y_{34} is the third score (row) for the fourth class. The operator

$$\sum_{i=1}^{m} \sum_{j=1}^{n} Y_{ij}$$

yields the total sum for all Y_{ij} scores. The operation involves setting the outer operator $\Sigma_{i=1}^{m}$ (sometimes abbreviated to Σ_i—sum over all i) to 1 and

TABLE 8.8 The Structure for the Analysis of Variance

	Categories or Classes			
	Class 1	Class 2	Class 3	Total
S	Y_{11}	Y_{12}	Y_{13}	
C	Y_{21}	Y_{22}	Y_{23}	
O	.	.	.	
R	.	.	.	
E	.	.	.	
S	Y_{n1}	Y_{n2}	Y_{n3}	
Sums	$\sum_{i=1}^{n_1} Y_{i1}$	$\sum_{i=1}^{n_2} Y_{i2}$	$\sum_{i=1}^{n_3} Y_{i3}$	$\sum_{j=1}^{3} \sum_{i=1}^{n_j} Y_{ij}$
Means	$\overline{Y}_1$	$\overline{Y}_2$	$\overline{Y}_3$	$\overline{Y}_g$
Sample size	n_1	n_2	n_3	n_t

$\overline{Y}_g$ = grand mean or mean of the total.
n_t = total sample size.

then summing j from 1 to n. The next step is to advance i to 2 and repeat the summation over all j values, $j = 1$ to n. Thus, to sum the following values,

$$\begin{matrix} 1 & 4 & 5 \\ 3 & 5 & 3 \\ 6 & 6 & 1 \\ 7 & 2 & 1 \end{matrix}$$

requires

$$\sum_{i=1} \sum_{j=1}^{3} = (1 + 4 + 5) + \sum_{i=2} \sum_{j=1}^{3} = (3 + 5 + 3)$$
$$= 10 \qquad\qquad + 11$$
$$+ \sum_{i=3} \sum_{j=1}^{3} = (6 + 6 + 1) + \sum_{i=4} \sum_{j=1}^{3} = (7 + 2 + 1)$$
$$+ 13 \qquad\qquad + 10 \qquad\qquad = 44$$

In our example, allowing for different numbers of elements in each of the samples or classes, the overall summation (see Table 8.8) would be given by

$$\sum_{j=1}^{3} \sum_{i=1}^{n_j} Y_{ij}$$

meaning that for each of the three classes add up the scores, then add the three class totals together.

We require sums of squares $(Y_i - \overline{Y})^2$ for the total and for the groups or classes and between the classes. Thus we need:

SST. $$\sum_i \sum_j (Y_{ij} - \overline{Y}_g)^2 = \text{total sums of squares}$$ — grand mean

SSW $$\Sigma \sum_{i=1}^{n_j} (Y_{ij} - \overline{Y}_j)^2 = \text{within class sums of squares for each class } j$$

SSB $$\Sigma \sum_j n_j(\overline{Y}_j - \overline{Y}_g)^2 = \text{the between class sums of squares}$$

where n_j is the size of the sample (or the weight) of each class. Without proof, we can write:

$$\sum_i \sum_j (Y_{ij} - \overline{Y}_g)^2 = \sum_j \sum_i (Y_{ij} - \overline{Y}_j)^2 + \sum_j n_j(\overline{Y}_j - \overline{Y}_g)^2 \tag{8.18}$$

Total sums of squares = within sum of squares + between sum of squares

SST = SSW + SSB

Note in the above expression that we have ordered the within-sum of squares as $\Sigma_j\Sigma_i$, which is the reverse of the usual order. This emphasizes that we are summing down the column in each case, which is somewhat easier to visualize. The end result of using $\Sigma_i\Sigma_j$ or $\Sigma_j\Sigma_i$ will, in this case, be identical.

To obtain estimates of the total within- and between-class variance, we divide by the appropriate degrees of freedom. For SST, we used one estimate (the mean), therefore the total degrees of freedom are $n_t - 1$. For SSW we used three estimates (the three means), therefore the within-class degrees of freedom are $n_t - 3$ or generally when k is the number of classes, $n_t - k$. For the SSB, the degrees of freedom are $k - 1$, the number of classes minus 1. We refer to these last two values as the mean square between and the mean square within.

Thus,

$$\text{Mean square between (MSB)} = \sum_j n_j \frac{(\overline{Y}_j - \overline{Y}_g)^2}{k - 1} \tag{8.19}$$

$F = \frac{MSB}{MSW}$

$$\text{Mean square within (MSW)} = \frac{\sum_j \sum_i (Y_{ij} - \overline{Y}_j)^2}{n_t - k} \tag{8.20}$$

Example of Testing Structure

With the data in Table 8.9, we can illustrate the structure of a one-way analysis of variance. Our example is a simple one with just three classes and an equal number of observations (9) in each class.

1. ***Hypotheses.*** We usually state our null hypothesis as $Ho{:}\ \mu_1 = \mu_2 = \mu_3$ and the alternative $Ha{:}$ not all μ_j are equal, but this requires a little further explanation. As we have implied, we assume that the three populations are normally distributed, and we must also assume that

TABLE 8.9 Wheat Yields for Three Soil Types

Sandy Soil	Clay Soil	Loam Soil	Total
18	21	21	60
20	23	17	60
16	14	24	54
21	24	23	68
22	21	25	68
18	20	22	60
19	19	23	61
17	21	26	64
21	23	23	67
Σ 172	186	204	562
$\bar{Y}$ 19.1	20.7	22.7	20.8
n 9	9	9	27

their variances are equal ($\sigma_1^2 = \sigma_2^2 = \sigma_3^2$). Therefore, by hypothesizing that $\mu_1 = \mu_2 = \mu_3$, we are in reality testing whether the three populations are *identical* (same means, same variances, same distributions). Note also that the alternate hypothesis states that not all means are equal—that is, they could all be different, or alternatively, two could be the same but different from a third. This point will be taken up later.

2. ***Assumptions***

   (i) Level of measurement is metric for yield, nominal for soil type
   (ii) Independent random samples, $n_1 = n_2 = n_3 = 9$
   (iii) $Y_{ij} \sim N(\mu_j, \sigma_j^2)$ for $j = 1, 2, 3$
   (iv) Population variances equal ($\sigma_1^2 = \sigma_2^2 = \sigma_3^2$)
   (v) Level of significance $\alpha = .05$

3. ***Test Statistic.*** The test for a significant difference among the means is given as the ratio of the MSB/MSW. The test statistic follows the F distribution.

$$F = \frac{\text{MSB}}{\text{MSW}} \tag{8.21}$$

We note that the F distribution cannot have values less than 0, and values less than unity indicate greater unexplained than explained variation.

4. ***Rejection Region***

$$F \geq [F, p = 1 - \alpha, df = (k - 1)(n_t - k)]$$
$$\geq (F, p = .95, df = 2{,}24]$$
$$\geq 3.40$$

5. ***Calculation.*** For the data in Table 8.9 we have

$$\text{SSW} = \sum_j \sum_i (Y_{ij} - \overline{Y}_j)^2 = \sum_i (Y_{i1} - 19.1)^2 + \sum_i (Y_{i2} - 20.7)^2 + \sum_i (Y_{i3} - 22.7)^2$$

$$= 32.89 + 70.01 + 54.01$$

$$= 156.91$$

$$\text{SSB} = \sum_j n_j(\overline{Y}_j - \overline{Y}_g)^2 = 9(19.1 - 20.8)^2 + 9(20.7 - 20.8)^2 + 9(22.7 - 20.8)^2$$

$$= 26.01 + .09 + 32.49$$

$$= 58.59$$

$$\text{SST} = \sum_i \sum_j (Y_{ij} - \overline{Y}_g)^2 = 214.08$$

Of course, you require only two of the foregoing values and the other can be found by subtraction. Although the calculations above show very clearly what is happening in the development of SSW, SSB, and SST, the calculation of SSW using group means can lead to inaccuracies, as any rounding of the group means will be multiplied through. Calculating formulas for SSB and SST are:

$$\text{SST} = \sum_{j=1}^{k} \sum_{i=1}^{n_j} (Y_{ij} - \overline{Y}_g)^2 = \sum_{j=1}^{k} \sum_{i=1}^{n_j} Y_{ij}^2 - \frac{\left(\sum_{j=1}^{k} \sum_{i=1}^{n_j} Y_{ij}\right)^2}{n_t} \quad (8.22)$$

$$\text{SSB} = \sum_{j=1}^{k} \frac{\left(\sum_{i=1}^{n_j} Y_{ij}\right)^2}{n_j} - \frac{\left(\sum_{j=1}^{k} \sum_{i=1}^{n_j} Y_{ij}\right)^2}{n_t} \quad (8.23)$$

and SSW = SST − SSB.

Table 8.10 presents the results in summary form.

6. ***Conclusion.*** The rejection region from the F table with 2 and 24 degrees of freedom is 3.40. We conclude that there is a significant difference

TABLE 8.10 Table of Analysis of Variance Results

	Sums of Squares	*df*	Mean Squares	*F*
Between	58.59	2	29.30	4.48
Within	156.91	24	6.54	
Total	214.08[a]	26		

[a] Note the small difference due to the use of a rounded grand mean of 20.8.

TABLE 8.11 Setup for SPSS ONEWAY ANOVA

```
RUN NAME          WHEAT YIELD VERSUS SOIL TYPE
VARIABLE LIST     YIELD SOIL
INPUT MEDIUM      CARD
N OF CASES        27
INPUT FORMAT      FIXED (F2.0,1X,F1.0)
ONEWAY            YIELD BY SOIL (1,3)
STATISTICS        ALL
READ INPUT DATA
18 1
20 1
....
21 2
23 2
....
21 3
17 3
....
FINISH
```

Note: Dependent variable (YIELD) is the variable you want to analyze, the independent variable is the coded or category variable (SOIL); that is, the independent variable is the classification or grouping you have imposed on your data. The (min, max) values are the range of the coded variable (i.e., (1, 3) indicates three groups).

between the wheat yields on the three different soil types. The tabled value of $F[2,24] = 3.40$ at the .05 level of significance is less than the calculated F value of 4.48.

A number of package routines provide ANOVA programs of varying sophistication. The program ONEWAY in SPSS is used to show the setup (Table 8.11) and results (Table 8.12) for the wheat yield example already presented manually.

Extensions and Assumptions

Although analysis of variance can be extended to more than one dimension—that is, for example, we could consider the wheat yield data for soil types and presence or absence of irrigation—each increase in complexity makes it more difficult to satisfy the assumptions. Although we will not examine the extension of one-way analysis of variance to higher-order models, some brief comments on the general ANOVA model and its assumptions are in order.

The most complete discussion of the methods and extensions of the analysis of variance is contained in Scheffé (1959). Several of his observations on the assumptions of normality and the equality of variances are worth reiterating. With respect to normality, Scheffé suggests that if the n's are large and there are equal-sized groups, the assumption can be relaxed. Moderate departures are tolerable even if there are unequal groups, but it is best in these cases if the variances are approximately equal. And stating this in reverse—this also means that the assumption of equality of variances

TABLE 8.12 Computer-Derived Results

– O N E W A Y –

VARIABLE YIELD
BY VARIABLE SOIL

ANALYSIS OF VARIANCE

SOURCE	D.F.	SUM OF SQUARES	MEAN SQUARES	F RATIO	F PROB.
BETWEEN GRC UPS	2	57.1852	28.5926	4.374	0.0240
WITHIN GROUPS	24	156.8888	6.5370		
TOTAL	26	214.0740			

GROUP	COUNT	MEAN	STANDARD DEVIATION	STANDARD ERROR	MINIMUM	MAXIMUM	95 PCT CONF INT FOR MEA		
GRP01	9	19.1111	2.0276	0.6759	16.0000	22.0000	17.5526	TO	20.669
GRP02	9	20.6667	2.9580	0.9860	14.0000	24.0000	18.3929	TO	22.940
GRP03	9	22.6667	2.5981	0.8660	17.0000	26.0000	20.6696	TO	24.663
0TOTAL	27	20.8148	2.8694	0.5522	14.0000	26.0000	19.6797	TO	21.949
0	FIXED EFFECTS MODEL		2.5568	0.4920			19.7993	TO	21.830
0	RANDOM EFFECTS MODEL			1.0291			16.3870	TO	25.242

RANDOM EFFECTS MCDEL – ESTIMATE OF BETWEEN COMPONENT VARIANCE 2.4506

TESTS FCR HOMCGENEITY CF VAFIANCES

COCHRANS C = MAX. VARIANCE/SUM(VARIANCES) = 0.4462, P = 0.595 (APPROX.)
BARTLETT-BOX F = 0.531, P = 0.588
MAXIMUM VARIANCE / MINIMUM VARIANCE = 2.128

The ONEWAY routine and the reported F values in SPSS assume a fixed effects model. That is, levels of categories of the independent variable (soil types in our case) can be considered either fixed or random. If the categories are considered to be a random selection of a larger set of categories, the model is a random effects model. Clearly, the soil types are not a random selection and we have a fixed effects model.

can be relaxed if the groups are equal in size and for unequal groups moderate departures are tolerable. Testing the equality of variance assumption is problematic, and a number of tests to ascertain if we can assume that this is true has been suggested. The most frequently used test—Bartlett's—has been challenged by Box (1954), and none of the alternatives are generally accepted. The discussion by Scheffé (1959, pp. 352–356) should be consulted for an elaboration on this and other assumptions.

Earlier in this chapter, we outlined a t test for the difference of two means, centering our discussion on the test where we assume $\sigma_1^2 = \sigma_2^2$. This test is, of course, just the limiting case for the one-way analysis of variance with $k = 2$. The results of the difference of means test are identical to the results we would obtain if we carried out a one-way analysis of variance, although the way of getting to the results may seem very different. The calculated F statistic from the analysis of variance would be equal to the calculated t statistic *squared*, from the difference of means test. Similarly, the tabled F value at a significance level of $\alpha(p = 1 - \alpha)$ and $(k - 1)(=1)$ and $(n_t - k)(=n_t - 2)$ degrees of freedom would have the same value as the tabled t value *squared* at a significance level of $\alpha(p = 1 - \alpha/2)$ and $(n_t - 2)$ degrees of freedom. The difference in our entry point to the tables—$p = 1 - \alpha$ in the case of F and $p = 1 - \alpha/2$ in the case of t—results from the different way we have to enter the F-statistic table.

Tests for Differences Among Several Means

If the F value is significant after completing an analysis of variance, we know that the means are significantly different from one another, but we do not know which means are different from which other means. It might be tempting simply to apply a succession of t tests to all possible pairs of means. Apart from the tedium of such an activity, a moment's reflection will suggest why this would not be satisfactory. If we had 10 means (from 10 classes) and compared each mean with every other, there would be 45 possible pairs ($10 \times 9/2$), and our knowledge of significance testing tells us that just from chance alone, we might expect 5 percent of the pairs of means to differ significantly—from 1 to 2 in our case—even if there were no significant differences. To overcome this problem, a variety of tests have been designed. The tests included in SPSS, in order of increasing conservativeness (i.e., the likelihood of rejecting *Ho*) are as follows:

1. Least-significant difference
2. Duncan's multiple range test
3. Student–Newman—Keuls
4. Tukey's honestly significant difference
5. Modified least-significant difference
6. Scheffé's test

We will outline Scheffé's test and a modification of the Tukey procedure, which is easily calculated and readily interpreted.

In the Scheffé test, two means are declared significantly different (for unequal cell sizes) if

$$|\overline{Y}_i - \overline{Y}_j| \geq \sqrt{(k-1)[F, p = 1 - \alpha, df = k - 1, n_t - k]}\sqrt{\text{MSW}\left(\frac{1}{n_i} + \frac{1}{n_j}\right)} \quad (8.24)$$

or (for equal cell sizes) if:

$$|\overline{Y}_i - \overline{Y}_j| \geq \sqrt{(k-1)[F, p = 1 - \alpha, df = k - 1, n_t - k]}\sqrt{2\,\text{MSW}/n} \quad (8.25)$$

where $[F, p = 1 - \alpha, df = k - 1, n_t - k]$ is the critical value of an F distribution with $k - 1$ (numerator) and $n_t - k$ (denominator) degrees of freedom, MSW = mean square within (also called mean square error) and k = the number of means.

One of the real advantages of the Scheffé's test is that it finds significant differences between pairs of means *only* if the F test is statistically significant.

As an illustration, we apply Scheffé's test to the results of the analysis of variance. Thus, for equal cell sizes:

$$|\overline{Y}_i - \overline{Y}_j| \geq \sqrt{(k - 1)[F\alpha, k - 1, n_t - k]}\sqrt{2 \text{ MSW}/n}$$

$$|22.7 - 19.1| \geq \sqrt{(3 - 1)(3.40)}\ \sqrt{2\ (6.54)/9}$$

$$3.6 \geq (2.61)(1.21)$$

$$3.6 \geq 3.16$$

Because 3.6 is greater than 3.16, we conclude that the mean wheat yields on sandy and loam soils are significantly different. The same analysis for sandy and clay and clay and loam soils show that those means are not significantly different from each other.

A modified and easily calculated Tukey test involves calculating gaps in a set of ordered means and evaluating deviant means. Tukey defines a significant gap as

$$(t_{.05})(\sqrt{2})(S_{\bar{y}}) \tag{8.26}$$

where $S_{\bar{y}}$ is the standard error of the mean and can be calculated from $\sqrt{\text{MSW}}/\sqrt{n}$, and n is the group sample size.

In the wheat yield study, the gap is calculated as

$$(1.99)(1.41)\left(\frac{\sqrt{6.54}}{\sqrt{9}}\right) = 2.38$$

Because there is no gap this size or larger between the sample means arranged in order (19.1, 20.7, 22.7), we cannot conclude that there are significant gaps.

Tukey's test for a deviant mean can be calculated for any group of three or more means and uses the Z distribution.

For three means

$$Z = \frac{\dfrac{(\overline{Y}_j - \overline{Y}_g)}{S_{\bar{y}}} - \dfrac{1}{2}}{3\left(\dfrac{1}{4} + \dfrac{1}{df}\right)} \tag{8.27}$$

For more than three means

$$Z = \frac{\dfrac{(\overline{Y}_j - \overline{Y}_g)}{S_{\bar{y}}} - \dfrac{6}{5}\log k}{3\left(\dfrac{1}{4} + \dfrac{1}{df}\right)} \tag{8.28}$$

where $\overline{Y}_j$ is any mean in the group and $\overline{Y}_g$ is the overall mean, k is the number of means and df is the degrees of freedom of the MSW estimate. Again, using the wheat yield data:

$$Z = \frac{\dfrac{(22.7 - 20.8)}{.85} - .5}{3\left(\dfrac{1}{4} + \dfrac{1}{24}\right)}$$

$$= \frac{2.24 - .5}{3(.2917)}$$

$$= \frac{1.74}{.875}$$

$$= 1.99$$

Because the calculated Z is greater than the tabled Z at the .05 level of significance, the mean for loam soil is significantly different from the others in the group. These tests must be used with caution but they yield further insights into the nature of differences amongst means. That Scheffé's test and the Tukey tests show only loam soil to be different emphasizes this soil type as an outlier. In general, however, the Scheffé tests are more conservative, the Ho is rejected less often. In these situations, the statistical conclusions must be set within a wider research context, and the design of the experiment, to produce substantive conclusions.

Intraclass Correlation

In our discussion of contingency tables (Section 8.2), we noted that the chi-square test only gives a yes/no answer to significant differences. The same can be said of the analysis of variance, and especially so as the structure enables small differences to be significant when the sample size is large. A number of measures have been devised to enable assessment of the degree of relationship, primarily to enable comparison with other results. The general name *intraclass correlation* has been applied to these measures. Most, but not all, have a range from 0 to 1 (perfect association). Three commonly found measures are described as follows.

The simplest measure is the *correlation ratio* (E^2), where

$$E^2 = \frac{\text{Between SS}}{\text{Total SS}} = \frac{\text{explained variation}}{\text{total variation}} \tag{8.29}$$

In our example (from Table 8.10).

$$E^2 = \frac{58.59}{214.08} = 0.274$$

Although widely used, this measure, when based on sample data, underestimates the population correlation ratio if the sample size is small. An alternative measure is the *unbiased correlation ratio* (η^2, Greek lowercase *eta*) where

$$\eta^2 = 1 - \frac{\text{Within MS}}{\text{Total SS}/n_t - 1} \tag{8.30}$$

For our data

$$\eta^2 = 1 - \frac{6.54}{214.08/26} = .206$$

One final measure, perhaps the most commonly used, is the *intraclass correlation coefficient* (r_i).

$$r_i = \frac{\text{Between MS} - \text{Within MS}}{\text{Between MS} + (\bar{n} - 1)\text{ Within MS}} \tag{8.31}$$

where $\bar{n} = n_t/k$, the mean number of observations per class. Using the same example, this gives us

$$r_i = \frac{29.30 - 6.54}{29.30 + (9 - 1)(6.54)} = .279$$

References and Readings

1. Bivariate Tests and Analysis of Variance

Box, G. E. P. (1954) "Some theorems on quadratic forms applied in the study of analysis of variance, Problems, I. Effect of inequality of variance in the one way classification," *Annals of Mathematical Statistics* 25:290–302.

Bryan, R. B. (1974) "A simulated rainfall test for the prediction of soil erodibility," *Zeitschrift fur Geomorphologie* (Supplement) 21:138–150.

Cohen, Y. S. and B. J. L. Berry (1975) *Spatial Components of Manufacturing Change*, University of Chicago, Department of Geography Research Paper number 125.

Scheffe, H. (1959) *The Analysis of Variance*, Wiley: New York.

Strahler, A. N. (1950) "Equilibrium theory of erosional slopes approached by frequency distribution analysis." *American Journal of Science* 248:673–696, 800–814.

2. Nonparametric Alternatives

Conover, W. J. (1971) *Practical Nonparametric Statistics*, Wiley: New York.

Lewis, P. (1977) *Maps and Statistics*, Methuen: London.

Siegel, S. (1956) *Nonparametric Statistics for the Behavioral Sciences*, McGraw-Hill: New York.

CHAPTER 9
The Simple Linear Regression Model

Although the emphasis in Chapter 8 was on testing for differences, we were implicitly analyzing simple relationships. That is, we were examining questions of the type: Is there a relationship between farm type and farm value? Or, is newspaper readership related to regional characteristics? Although we mentioned a wide variety of techniques in that chapter, we mainly examined only two: testing for the difference of means and its extension to the simple one-way analysis of variance. Both of these techniques are very important, but their use is much more limited than the use of the correlation and regression techniques discussed in this chapter. Simple linear correlation and regression (as for the analysis of variance the "simple" refers to the analysis of covariation between only two variables) are important techniques in their own right, but they are also the basis for virtually all the techniques presented in the remainder of this text. Moreover, it can be shown that difference-of-means tests, analysis of variance, and correlation and regression are all different parts of the same approach—the application of the general linear model.

The general linear model is concerned with situations where we are examining how a single variable is functionally dependent on one or more independent variables and when the form of the functional relationship is linear. Its operational structure varies for different situations based on the level of measurement of the independent variables. For the simple regression situation with a single metric measured independent variable, the linear model results in a graphed straight-line relationship between two variables. We will spend some time examining the simple linear regression model and its use of the linear function, because we need to use these principles in later chapters. It would have been possible to delay the discussion of analysis of variance until after this more general discussion on the linear model, but its earlier introduction allows us to talk about analysis of variation in the regression situation with greater ease.

We begin our discussion of the simple linear model (regression) by expanding our set of concepts on functional relationships, concentrating our attention on the linear function (Section 9.1). We then consider the steps in analyzing a functional relationship (Section 9.2). Detailed discussions on regression analysis (Section 9.3) and correlation analysis (Section 9.4) are followed by an outline of the various inferential tests available for assessing the simple linear regression model (Section 9.5).

To illustrate functional relationships, regression, and correlation methods, we introduce a further detailed example—the Northland Drainage Basin Survey set up as an SPSS data set (Tables 9.1 and 9.2). From this survey, we pick out one pair of variables (Runoff and Rainfall) to use as a basic example throughout this chapter.

TABLE 9.1 File Setup for Northland Drainage Basin Survey

```
RUN NAME          NORTH(1) - BASIC DATA SET UP
FILE NAME         NORTH NORTHLAND DRAINAGE BASIN STUDY
VARIABLE LIST     AREA LENGTH ELEVATN FOREST RUNOFF RAINFALL SLOPE
                  ROCKTYPE
INPUT MEDIUM      CARD
N OF CASES        26
INPUT FORMAT      FIXED (F8.2,F8.1,6F8.0)
PRINT FORMAT      AREA (2) LENGTH (1)
VAR LABELS        AREA       AREA OF BASIN - KM**2/
                  LENGTH     MAXIMUM LENGTH OF BASIN - KM/
                  ELEVATN    ELEVATION OF MOUTH OF BASIN - M/
                  FOREST     % BASIN FORESTED ESTIMATED FROM TOPO MAP -
                             %/
                  RUNOFF     MEAN ANNUAL SPECIFIC DISCHARGE - MM/
                  RAINFALL   MEAN ANNUAL RAINFALL NEAR TO BASIN CENTER -
                             MM/
                  SLOPE      ANGLE HIGHEST POINT AND BASIN MOUTH -
                             DEGREES/
                  ROCKTYPE   PREDOMINANT ROCKTYPE IN BASIN - CODED
LIST CASES        CASES=26/VARIABLES=AREA TO ROCKTYPE
FREQUENCIES       INTEGER=ROCKTYPE(1,3)
READ INPUT DATA
```

TABLE 9.2 Northland Drainage Basin Survey Data

1NORTH(1) - BASIC DATA SET UP

FILE NORTH (CREATION DATE = 07/20/85) NORTHLAND DRAINAGE BASIN STUDY

0 CASE-N	AREA	LENGTH	ELEVATN	FOREST	RUNOFF	RAINFALL	SLOPE	ROCKTYPE
1	11.71	4.1	1080.	100.	1156.	1605.	16.	3.
2	31.82	5.2	1756.	100.	1378.	2202.	17.	3.
3	18.57	3.8	2006.	73.	1560.	2036.	12.	3.
4	7.74	2.1	1257.	68.	947.	1868.	18.	3.
5	15.20	3.0	486.	4.	356.	1520.	8.	2.
6	2.31	1.5	1070.	88.	454.	1276.	11.	1.
7	4.95	3.3	1600.	75.	654.	1710.	10.	1.
8	11.06	4.0	1380.	48.	473.	1510.	10.	1.
9	12.23	2.8	335.	2.	127.	1285.	5.	2.
10	8.16	2.3	1195.	50.	651.	1522.	14.	2.
11	17.07	4.0	955.	10.	485.	1585.	10.	2.
12	24.35	3.6	946.	100.	750.	1682.	16.	1.
13	24.10	4.8	1806.	93.	978.	1984.	18.	3.
14	9.68	3.3	780.	50.	641.	1594.	11.	1.
15	3.42	1.2	600.	50.	757.	1693.	10.	1.
16	3.88	2.6	350.	46.	734.	1612.	13.	1.
17	36.00	4.9	2166.	65.	1149.	1853.	14.	3.
18	5.74	2.8	1050.	85.	753.	1477.	11.	3.
19	7.72	2.8	1175.	70.	647.	1416.	11.	1.
20	4.12	1.8	330.	1.	360.	1520.	9.	2.
21	3.13	1.4	949.	98.	610.	1500.	10.	3.
22	0.74	0.9	635.	50.	467.	1262.	10.	1.
23	5.88	2.4	1250.	50.	531.	1558.	12.	1.
24	8.21	2.8	790.	67.	366.	1352.	8.	2.
25	2.67	1.8	980.	10.	476.	1525.	14.	1.
26	12.63	3.4	610.	70.	485.	1599.	9.	2.

9.1 BIVARIATE FUNCTIONAL RELATIONSHIPS

In Chapter 1 we defined a functional relationship as an equivalence between two sets of parameters, one set dependent on the other set. When talking about frequency distributions in Chapter 2, we used the functional relationships model to describe how frequencies of scores can be considered as functionally dependent on the scores themselves (or classes of scores) for the independent variable x, and may be described by various probability functions (Chapter 4). We now continue with the use of functional relationships by considering the bivariate relationship where we have two measured (metric or enumerated) variables. In most examples that we consider, this would be two variables that have been measured as part of a survey or as a result of an experimental process.

Functions

We use an example from the Northland Drainage Basin Survey to illustrate some of the important characteristics of functions. From general principles, we consider Runoff (annual specific runoff) to be dependent (in part) on the Rainfall (annual rainfall) of the drainage basin. Runoff is thus our dependent variable (usually designated by y), and Rainfall is our single independent variable (usually designated by x). We can write this relationship in general functional form as:

$$y = f(x): \text{Runoff} = f(\text{Rainfall}) \tag{9.1}$$

We know that this relationship is not exact—other variables contribute to the physical explanation of runoff, and errors would be involved in the measurement of both variables. However, other things being equal, we ex-

TABLE 9.3 Domain and Range for the Runoff/Rainfall Functional Relationship

	Domain	Range
Theoretical	0 to $+\infty$	0 to $+\infty$
Sampled	1262 to 2202	127 to 1560

pect Runoff to have a set of scores that vary in some systematic manner according to the variability of the scores for the variable Rainfall.

A function may be defined in terms of the possible values it may take on. The maximum and minimum values for the x variable define the *domain* of the function, and those for the y variable, the *range* of the function.[1] A distinction may be made between the theoretical range and domain of a function and the sampled functional values (Table 9.3).

Types of Functions

A large number of bivariate functional types have evolved in mathematics, and a simple review is useful. Only a few, however, are used in general statistical applications. The following notation is used:

y = the dependent variable

x = the independent variable (in this chapter we are limiting our discussion to a single independent variable. Most of these expressions can, however, be expanded to handle more than one x variable

a = a parameter of the function (i.e., a measure of the functional relationship). a will be subscripted if there are several parameters

1. *Constant Functions*

$$y = f(x) = a \tag{9.2}$$

for example,

$$y = 2$$

2. *Polynomial Functions* A polynomial is an expression in the general form

$$a_n x^n + a_{n-1} x^{n-1} + \cdots + a_1 x + a_0 = \sum_{j=0}^{n} a_j x^j$$

[1]You will notice throughout this section (9.1) that we will use lower case y and x to designate dependent and independent variables because we are discussing variables in general. In section 9.2 and following we will use upper case X and Y for specific variables.

where $n > 0$ and $a_n \neq 0$ and the expression is called a *polynomial of the nth degree.*

For example, a fourth-degree polynomial

$$y = f(x) = 2x^4 + 3x^2 - x + 7$$

that is,

$$y = (2)x^4 + (0)x^3 + (3)x^2 + (-1)x + 7$$

Some important polynomial functions are:

Linear polynomial function (first degree)

$$y = f(x) = a_1x + a_0 \tag{9.3}$$

for example,

$$y = 2x + 3$$

Quadratic polynomial function (second degree)

$$y = f(x) = a_2x^2 + a_1x + a_0 \tag{9.4}$$

for example,

$$y = x^2 + 2x + 3$$

Cubic polynomial function (third degree)

$$y = f(x) = a_3x^3 + a_2x^2 + a_1x + a_0 \tag{9.5}$$

for example,

$$y = x^3$$

3. *Rational Functions* Defined as a quotient of two polynomials, for example,

$$y = f(x) = \frac{x^2 - 3x + 1}{x + 2}$$

4. *Algebraic Functions* Defined in terms of polynomials and roots of polynomials, for example,

$$y = f(x) = \frac{x - 1}{x\sqrt{x^2 + 1}}$$

5. *Nonalgebraic (Transcendental) Functions* The most important are:

Trigonometric Functions in which the functional relationship is described by sine, cosine, tangent, or other trigonometric ratios, for example,

$$y = f(x) = \sin x$$

Logarithmic Functions Described by a logarithmic transformation of x

$$y = f(x) = a \log_e x$$

for example,

$$y = \log_{10} x$$

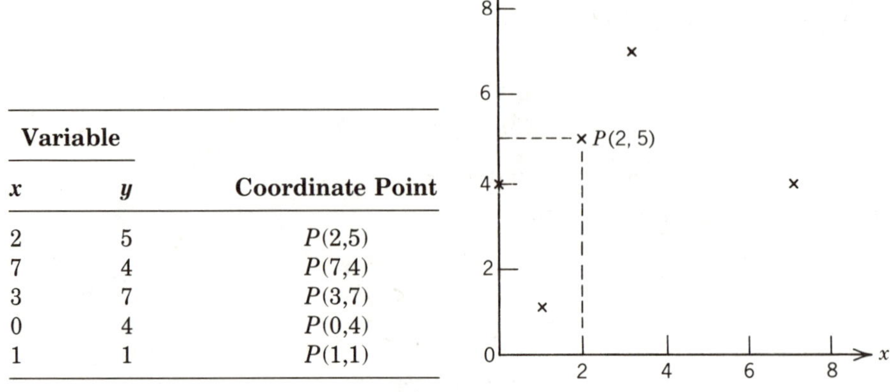

Variable		
x	y	Coordinate Point
2	5	$P(2,5)$
7	4	$P(7,4)$
3	7	$P(3,7)$
0	4	$P(0,4)$
1	1	$P(1,1)$

Figure 9.1 Construction of a scatter diagram.

Exponential Functions Described by an exponential transformation on x

$$y = f(x) = a_0 a_1^x$$

for example,

$y = 4e^x$, where e = the base of natural logarithms

or

$$y = 2^x$$

Graphing Functional Relationships

The basic properties of the rectangular (Cartesian) coordinate system were outlined in Chapter 2 where frequencies (y) were plotted against scores or classes of scores for a variable (x). We use the same structure to represent scores for two variables: the dependent (y) and independent (x) variables. Two situations can arise.

First, if we have a set of scores for two variables x and y, we can plot the location of these scores on a graph. Such a graph is called a *scatter diagram* or simply a *scattergram*. The process is illustrated in Figure 9.1. The first point, for example, with scores of $x = 2$ and $y = 5$ and therefore coordinate point $P(2,5)$, is positioned 2 units along the x axis and 5 units up the y axis.

Second, if the relationship is already expressed by a functional equation, this can be illustrated by a similarly constructed graph. A set of values of x covering a range of scores are selected and the corresponding values for y are calculated. The resulting coordinate points are plotted on the graph and joined by a smooth line to represent the function. Specific examples for six commonly encountered functions are portrayed in Figure 9.2.

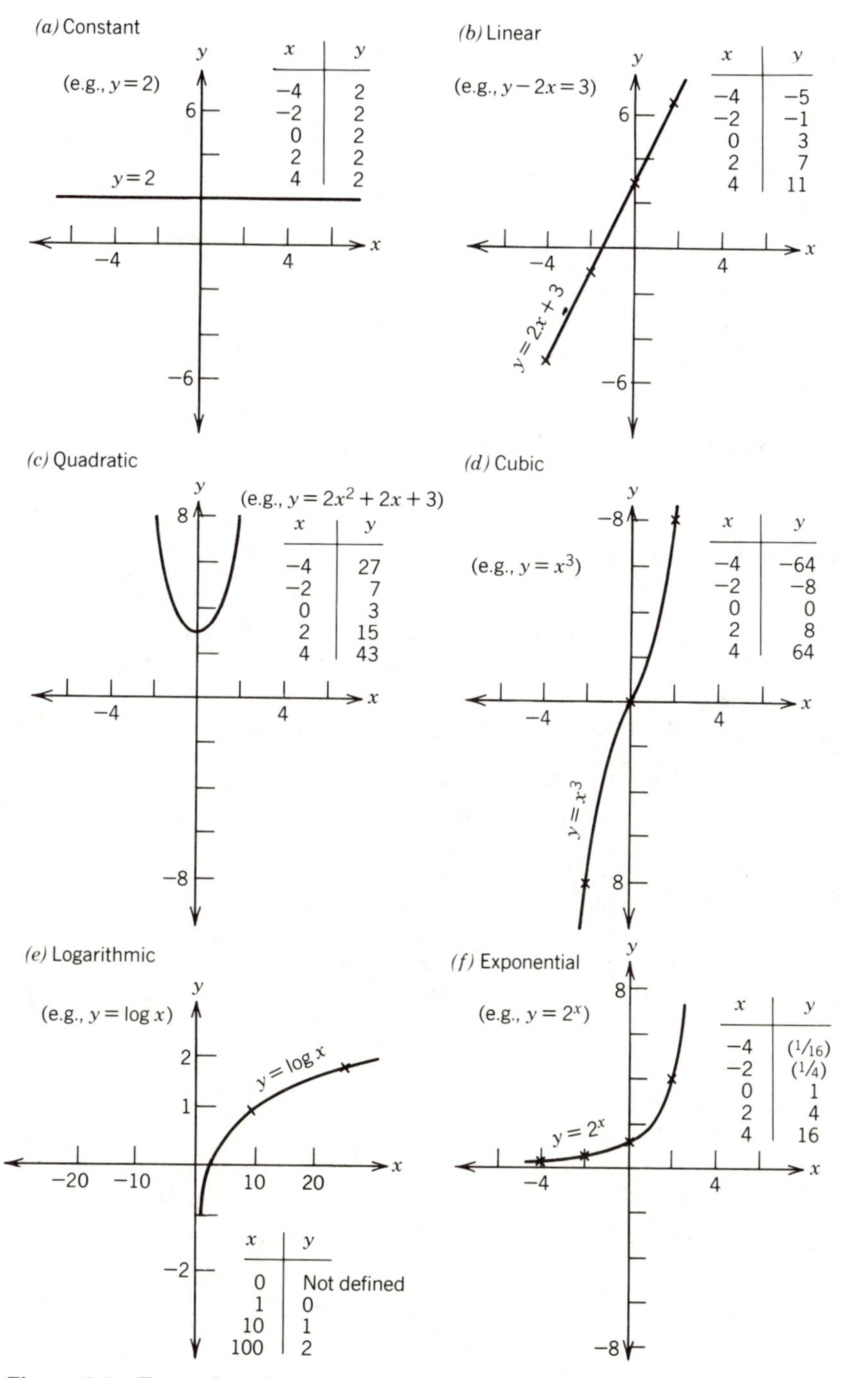

Figure 9.2 Examples of graphs of some important functions.

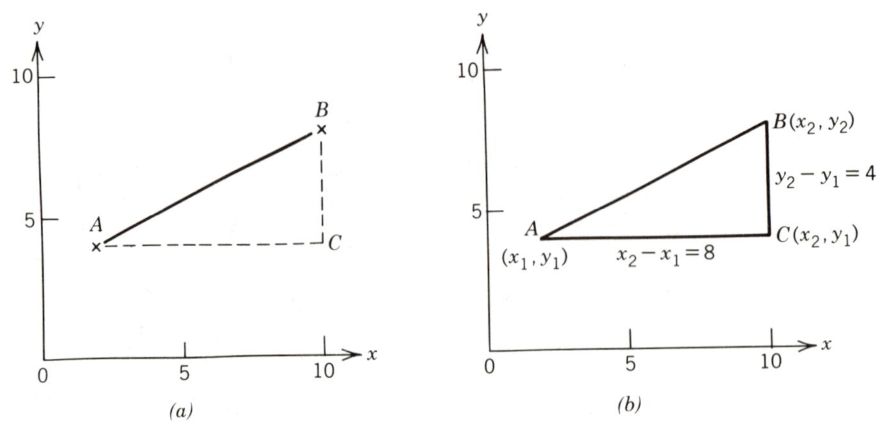

Figure 9.3 Illustration of the distance formula.

Properties of the Linear Function

The linear function, represented by a *straight-line* relationship (Figure 9.2*b*) is so basic to bivariate and multivariate statistics that it is worthwhile digressing for a moment to look at some of the simple trigonometric and algebraic properties of straight lines on rectangular graphs.

The *distance* between any two points $A(x_1,y_1)$ and $B(x_2,y_2)$ is given by

$$\text{Distance } A \text{ to } B = \text{Distance } B \text{ to } A = |AB| = \sqrt{(x_2 - x_1)^2 + (y_2 - y_1)^2} \quad (9.6)$$

This is illustrated by using an example (Figure 9.3). The two points selected are $A(2,4)$ and $B(10,8)$. If we create a point C that completes a right-angled triangle ABC, C must take on the x coordinate of B ($x_2 = 10$) and the y coordinate of $A(y_1 = 4)$. Its coordinates are, therefore, $C(x_2,y_1)$. Looking at the vertical line BC, its length must be

$$|BC| = y_2 - y_1 = 8 - 4 = 4$$

and looking at the horizontal line AC whose length must be

$$|AC| = x_2 - x_1 = 10 - 2 = 8$$

From the Pythagorean rule for right-angled triangles, the length of the hypotenuse AB must therefore be

$$\begin{aligned} |AB| &= \sqrt{|BC|^2 + |AC|^2} \\ &= \sqrt{(y_2 - y_1)^2 + (x_2 - x_1)^2} \\ &= \sqrt{4^2 + 8^2} = \sqrt{80} = 8.9 \end{aligned}$$

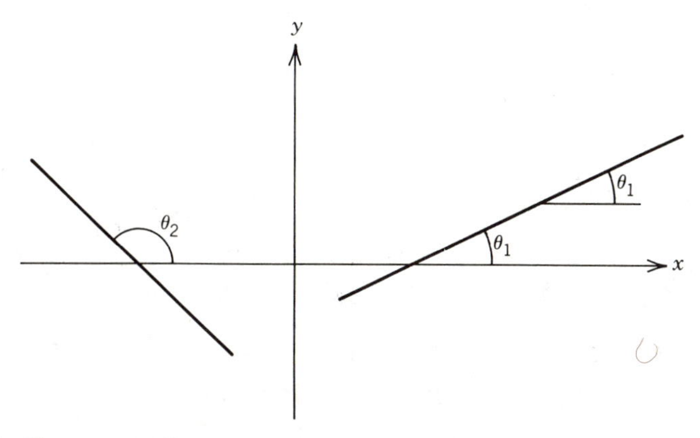

Figure 9.4 Slope of the line on a graph.

Note that we are dealing with the sum of squared values, and thus it does not make any difference which point we call A and which we call B.

The slope of a line on a graph (relative to the x axis) is given by (Figure 9.4):

$$\text{Slope} = m = \tan\theta \tag{9.7}$$

The angle θ may, therefore, range from 0° to 180°. Thus, if the graph of a straight line were drawn, the angle could be measured with a protractor, and the tangent of the angle (m) found by looking up a table of tangents. The possible range of values of m is from 0 to $+\infty$ (0° to 90°) and 0 to $-\infty$ (180° to 90°) (Figure 9.5). Note that as the tangents of the angles 0° to 90°

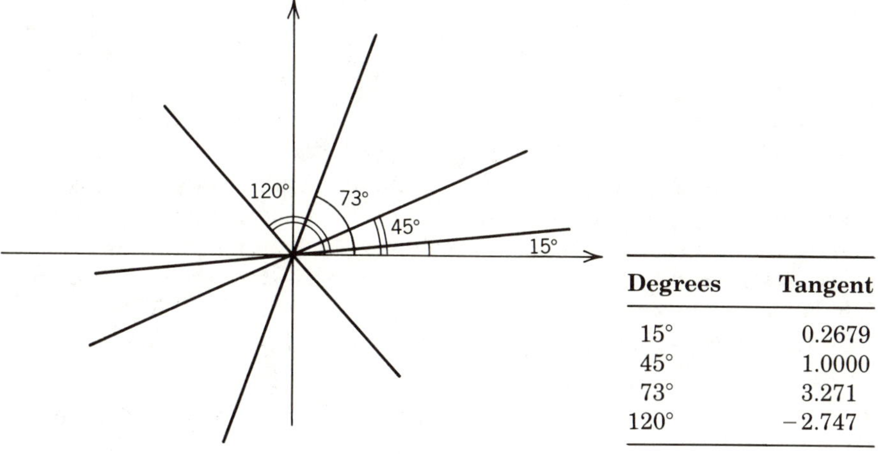

Degrees	Tangent
15°	0.2679
45°	1.0000
73°	3.271
120°	−2.747

Figure 9.5 Examples of angles and their tangents.

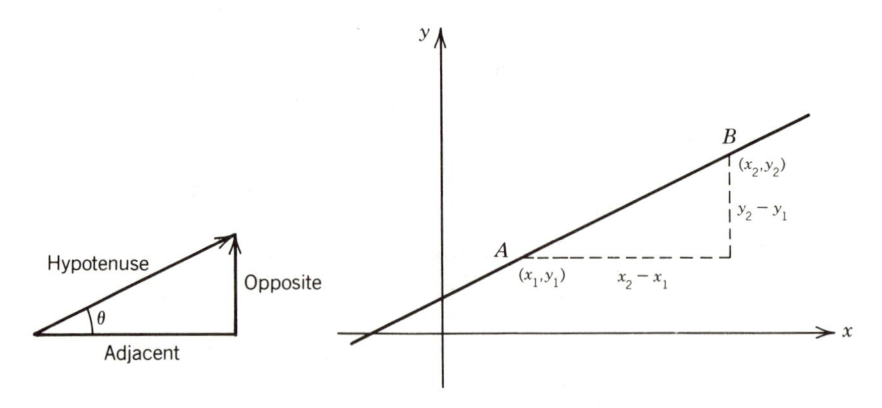

Figure 9.6 Definition of the tangent in a right-angled triangle.

are identical to those from 180° to 90° except that the sign is the opposite, the tangent of an obtuse angle (i.e., one greater than 90°) can be found by taking the tangent of (180° − the angle) and inserting a negative sign.

$$\text{For } (90° < \theta < 180°) \qquad \tan\theta = -\tan(180° - \theta) \tag{9.8}$$

However, the method of finding the tangent by measurement is tedious and inaccurate and can be avoided by using a simple property of a right-angled triangle (Figure 9.6)

$$\tan\theta = \frac{\text{opposite}}{\text{adjacent}} \tag{9.9}$$

Using the nomenclature developed in the discussion of the Distance Formula, it can be shown that:

$$\tan\theta = \frac{y_2 - y_1}{x_2 - x_1} \tag{9.10}$$

Points A and B can be selected at any convenient position along the line (Figure 9.7).

Again, the order of selection of points makes no difference to the result. (Check this in the second example by taking as the points $C(-1,3)$ and $D(-4,8)$.) An important summary conclusion can be noted: lines sloping upwards to the right have *positive* slopes and those sloping upwards to the left have *negative* slopes.

The functional equation for the straight line (the linear function) was given as:

$$y = a_1 x + a_0 \tag{9.11}$$

This equation has two important graphical and interpretative properties:

1. $a_1 = m =$ the slope of the line

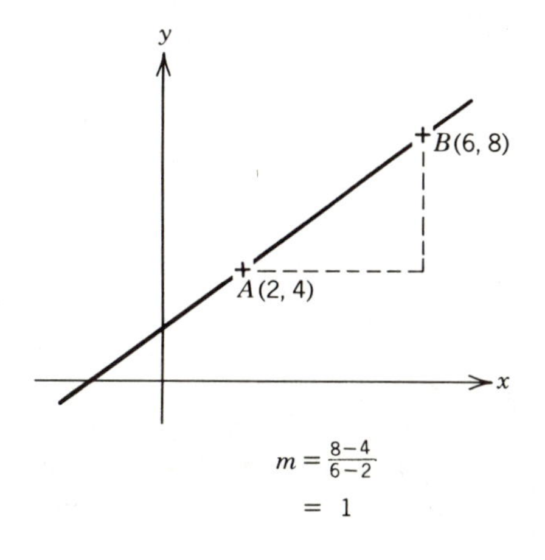

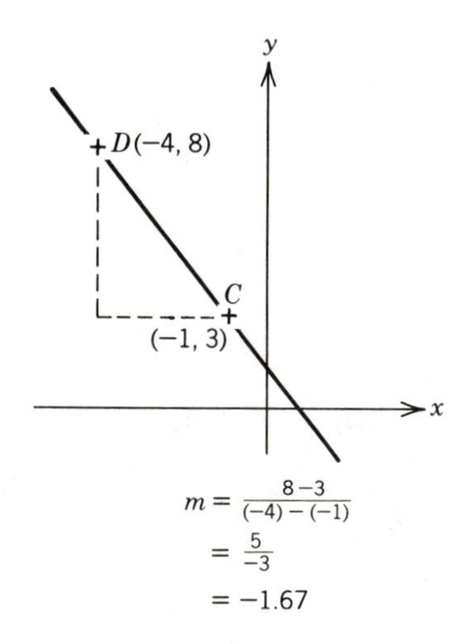

Figure 9.7 Examples of the calculation of the tangent.

2. a_0 = the y intercept, the value of y where the line crosses the y axis or, in other words the value of y when $x = 0$. This can be shown by inserting a value of 0 for x in Equation 9.11.

$$y = a_1 (0) + a_0$$

$$= a_0$$

The equation (9.11) is often written using m (the slope) for a_1 and c (the y intercept) for a_0 to give:

$$y = mx + c \tag{9.12}$$

This expression is called the *slope-intercept* form of the linear equation because the parameters m and c directly describe the slope and intercept values of the straight line. Equation 9.11 is identical to 9.12 with only a change of nomenclature for the parameters.

In certain circumstances, it is convenient to have the same equation expressed in a different form. In the *general* form of the linear function, all the terms are moved to the left-hand side, and the right-hand side is set to zero. If we have fractional values for m and/or c, these are usually multiplied out to give whole numbers. For example, the slope-intercept equation

$$y = 2.5x + 3$$

becomes

$$5x - 2y + 6 = 0$$

A third method of expressing the same linear equation is the *point-slope* form. Using $P(x,y)$ for any point on the line (i.e., not a fixed point), and

$Q(x_1, y_1)$ a known fixed point, and if m, the slope, is known then as

$$m = \frac{y - y_1}{x - x_1}$$

by multiplying out

$$y - y_1 = m(x - x_1) \tag{9.13}$$

We can move freely from one form to another depending on the requirements of the task. This is illustrated by example in the two following sections.

Sketching the graph of a known function involves the equation in slope-intercept form. For example, sketch the graph of the equation $3x + 4y - 6 = 0$.

$$4y = -3x + 6$$

$$y = -\tfrac{3}{4}x + \tfrac{3}{2}$$

That is,

$$\text{Slope} = m = -\tfrac{3}{4}$$

$$y \text{ intercept} = \tfrac{3}{2}$$

and the graph can be easily drawn (Figure 9.8).

Finding the equation for a line involves knowing the coordinate values of two points on the line, or if one point and the slope of the line are known, we can use the point-slope form to obtain the equation. Because the former situation includes the latter, we illustrate the method by example with two known points.

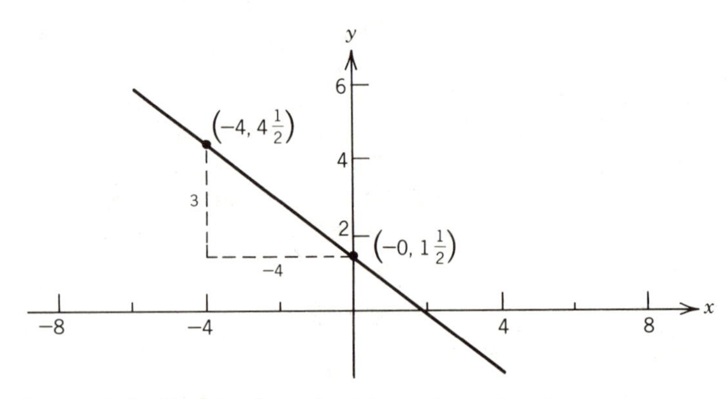

Figure 9.8 Sketch of graph of $3x + 4y - 6 = 0$.

Find the equation of the straight line that passes through the two points (−1,4) and 5,6).

$$m = \frac{y_2 - y_1}{x_2 - x_1}$$

$$= \frac{6 - 4}{5 - (-1)}$$

$$= \frac{2}{6} = \frac{1}{3}$$

$$y - y_1 = m(x - x_1)$$

Taking (−1,4) as (x_1,y_1) (the same result would be obtained using the point (5,6)).

$$y - 4 = \tfrac{1}{3}(x - (-1))$$

$$= \tfrac{1}{3}(x + 1)$$

$$y = \tfrac{1}{3}x + \tfrac{1}{3} + 4 \qquad (9.14)$$

$$= \tfrac{1}{3}x + 4\tfrac{1}{3}$$

(in slope-intercept form)

or multiplying Equation 9.14 by 3

$$3y - 12 = x + 1$$

$$x - 3y + 13 = 0 \text{ (in general form)}$$

9.2 STEPS IN THE ANALYSIS OF A FUNCTIONAL RELATIONSHIP

Having sketched some of the background for examining bivariate relationships, we are in a position to present the more formal aspects of the analysis. We do so in a rather structured manner, and for the moment keep the approach general rather than to narrow it down to an examination of only linear relationships. We differentiate four basic steps in the analysis of functional relationships. These four steps are sequential, but not all steps are necessary in particular analyses. In the analysis of the relationship between two variables, we first select the most appropriate type of function and second, use the observed data to produce a functional equation for the type of function selected—an equation that summarizes the observed relationship. This is the process of estimating the unknown parameters of the function; or, looking at it graphically, of fitting a curve of known functional form to the data; or, in general terms—*regression* (considered in Section 9.3). Third, examine the degree to which the hypothesized relationship (the regression equation) fits the data—basically the process of *correlation* (Section 9.4), and fourth, with sampled data, test various results from regression and correlation for significance (Section 9.5).

Selection of the Type of Function

Perhaps the most difficult problem in the analysis of relationships is determining which function is most appropriate. If the function selected is not suitable for the data, then further analysis (regression and correlation) will be of little value. The method is by trial and error, but the following points may be helpful in making a decision.

An analysis of the variables making up the relationship may provide some clue to the type of function; for example, in an analysis of population growth over time, it has been generally shown that population grows in some sort of exponential manner rather than in a linear one.

With a simple functional relationship—that is, between two variables only—the data can be plotted in the form of a scatter diagram. By comparing the general trend of the scatter of points with graphs of known functions (such as in Figure 9.2) the most appropriate function can be selected. This simple technique is recommended as a basic step in the analysis of bivariate relationships and as a preliminary step in the analysis of more complex multivariate relationships. With the growing use of computers (including graphic portrayal), the tedious nature of this technique has been removed and all statistical packages have one or more ways to obtain two-variable scatter diagrams. In the example of the Northland Drainage Basin Survey, the variable Runoff (a potential dependent variable) was considered as a function of a set of possible independent variables. Each potential pair of variables was initially examined by use of the scatter diagram. The file setup (Table 9.4) and an example of one of the plots—Runoff versus Rainfall

TABLE 9.4 SPSS Setup for Scattergram

```
RUN NAME          NORTH(4) - SCATTERGRAM RUNOFF V RAINFALL
FILE NAME         NORTH NORTHLAND DRAINAGE BASIN STUDY
VARIABLE LIST     AREA LENGTH ELEVATN FOREST RUNOFF RAINFALL SLOPE
                  ROCKTYPE
INPUT MEDIUM      CARD
N OF CASES        26
INPUT FORMAT      FIXED (F8.2,F8.1,6F8.0)
PRINT FORMAT      AREA (2) LENGTH (1)
VAR LABELS        AREA        AREA OF BASIN - KM**2/
                  LENGTH      MAXIMUM LENGTH OF BASIN - KM/
                  ELEVATN     ELEVATION OF MOUTH OF BASIN - M/
                  FOREST      % BASIN FORESTED ESTIMATED FROM TOPO MAP -
                              %/
                  RUNOFF      MEAN ANNUAL SPECIFIC DISCHARGE - MM/
                  RAINFALL    MEAN ANNUAL RAINFALL NEAR TO BASIN CENTER -
                              MM/
                  SLOPE       ANGLE HIGHEST POINT AND BASIN MOUTH -
                              DEGREES/
                  ROCKTYPE    PREDOMINANT ROCKTYPE IN BASIN - CODED
SCATTERGRAM       RUNOFF(100,1600) WITH RAINFALL(1250,2250)
OPTIONS           6
STATISTICS        ALL
READ INPUT DATA
```

(Figure 9.9) illustrates the process. In the example, the plot suggested a linear relationship.

It is the linear function, however, that is assumed in most geographic work. Although its applicability is questionable in some instances, its basic use stems from a number of reasons. First, because of the very nature of their distributions, many pairs of variables have basically linear relationships. Second, the linear function is the simplest—the mathematics involved in its calculation are the least complex, and it has greater statistical application than any other function. In fact, most statistical techniques of association assume a linear relationship. Third, certain functions (especially the logarithmic and exponential) can be transformed very simply into a linear equivalent form, which can then be used in any technique requiring a linear relationship. For example, using the following set of five data points (Table 9.5) (with their logarithms to the base 10), a scatter diagram could be constructed (Figure 9.10).

Examination of the scatter of points in the diagram shows a curvilinear relationship between the variables. In this instance, the X variable takes on a logarithmic range of values. By taking the logarithm of the X values, the relationship can be transformed to a linear relationship. This can be checked either by changing the scale of the X axis to a logarithmic one (see Figure 9.11*a*) or by converting the X values to their logarithms and plotting the log value of X on the X axis (Figure 9.11*b*). Any statistical analysis using the above data would use the raw Y values and the log X values.

Many studies in geography assume a linear relationship without attempting to test other possible relationships between the variables. There is little justification for such an approach.

9.3 LINEAR REGRESSION ANALYSIS

We are concerned with the analysis of the relationship between two variables where one variable is (statistically) dependent on the other. Both variables are measured on a metric scale, and the relationship between them is assumed to be linear. A method to fit a first-degree polynomial function (the linear function) to a set of observed paired scores is required. This technique is *simple linear regression analysis*. It is the basis of much of what follows and we examine its construction and operation in detail.

Derivation of the Regression Model

In Section 9.1, the mathematical model describing the linear function was variously described as:

$$y = a_1x + a_0$$

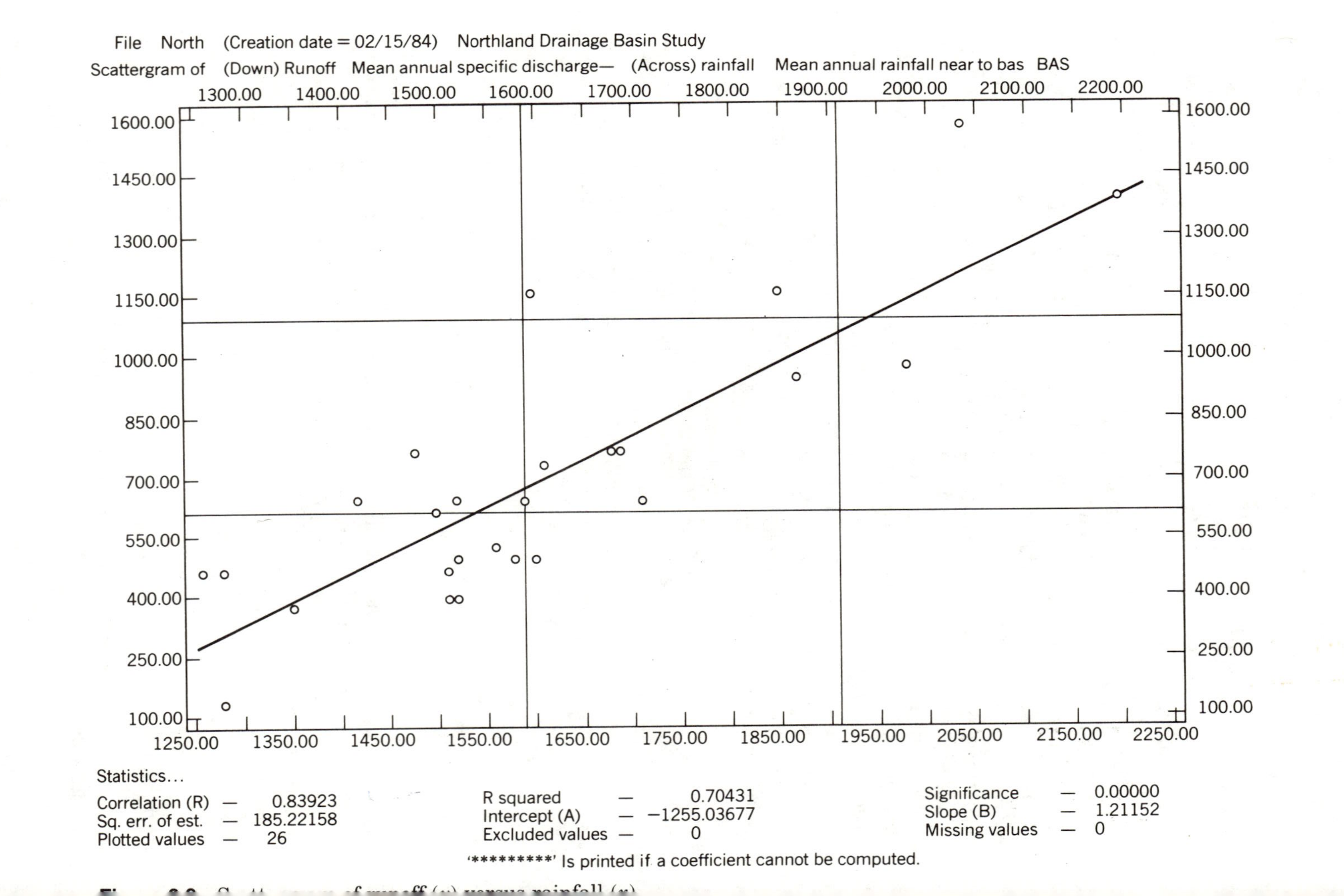
File North (Creation date = 02/15/84) Northland Drainage Basin Study
Scattergram of (Down) Runoff Mean annual specific discharge— (Across) rainfall Mean annual rainfall near to bas BAS
1300.00 1400.00 1500.00 1600.00 1700.00 1800.00 1900.00 2000.00 2100.00 2200.00
1600.00 1450.00 1300.00 1150.00 1000.00 850.00 700.00 550.00 400.00 250.00 100.00
1250.00 1350.00 1450.00 1550.00 1650.00 1750.00 1850.00 1950.00 2050.00 2150.00 2250.00
Statistics...
Correlation (R) — 0.83923
Sq. err. of est. — 185.22158
Plotted values — 26
R squared — 0.70431
Intercept (A) — −1255.03677
Excluded values — 0
Significance — 0.00000
Slope (B) — 1.21152
Missing values — 0
'*********' Is printed if a coefficient cannot be computed.

TABLE 9.5 Example of Linear Transformation by the Use of Logarithms

X	Y	log X
140	2.5	2.1461
540	3.0	2.7324
1070	4.0	3.0294
17	1.3	1.2304
25	1.8	1.3979

and

$$y = mx + c$$

where the parameters $a_1 = m =$ the slope of the line and $a_0 = c =$ the y intercept of the line (the value of y when $x = 0$).

Following common statistical nomenclature, we rewrite this mathematical model by reversing the two terms on the right-hand side (which makes no difference to the relationship) and introducing two new symbols for the two parameters. We use β_0 and β_1 (Greek lowercase *beta*) for the y intercept and slope parameters, respectively. Thus

$$Y = \beta_0 + \beta_1 X$$

This change in terminology is necessary in order to keep our terminology consistent with other texts. For an individual observation of the variable X and its associated y value, the relationship is expressed

$$Y_i = \beta_0 + \beta_1 X_i \tag{9.15}$$

However, this mathematical expression implies an exact relationship, that is, the function of x predicts a value of y exactly equal to the observed value

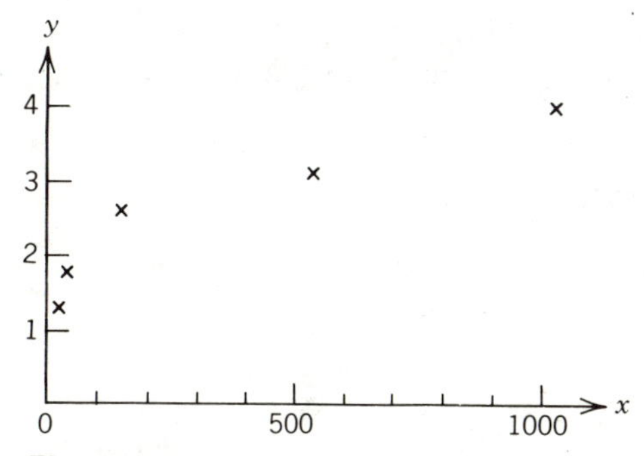

Figure 9.10 Scatter diagram of the data from Table 9.5.

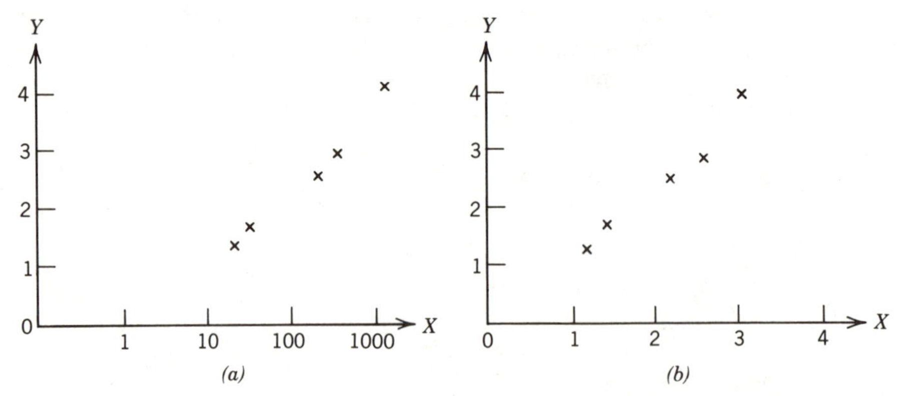

Figure 9.11 Logarithmic scatter diagrams for the example from Table 9.5.

of the dependent variable. In statistical analysis, such an exact relationship would not be achieved, thus the model must be expanded to allow for some error of prediction. This is done by adding an error term ϵ (greek lowercase *epsilon*) to the equation to form the *statistical model* of the simple linear relationship.

$$Y_i = \beta_0 + \beta_1 X_i + \epsilon_i \tag{9.16}$$

where

Y_i = the value of the dependent variable for the ith observation

X_i = the value of the independent variable for the ith observation, assumed to be measured without error

β_0, β_1 = the unknown constant parameters of the system (i.e., a_0, a_1)

ϵ_i = the error term for the ith observation, which is assumed to be a random variable with a mean of zero, and a variance of σ_E^2 (this assumption will be taken up later).

Using observed data and a least-squares technique, an approximation is made to this statistical model. This is the *regression* (or estimating or predicting) model

$$\begin{aligned} \hat{Y}_i &= \hat{\beta}_0 + \hat{\beta}_1 X_i \\ &= b_0 + b_1 X_i \end{aligned} \tag{9.17}$$

where

X_i = as before

b_0, b_1 = the estimates of the parameters of the statistical model (β_0, β_1)

$\hat{Y}_i$ = the estimate of the value of the dependent variable for the ith observation of X

The difference between the mathematical, statistical, and regression equations of the linear relationship can be illustrated by an example from the Northland Drainage Basin Survey. Twenty-six drainage basins were examined to see if Runoff (Y) was dependent on Rainfall (X) (Table 9.6). We noted that it would be impossible to obtain an exact mathematical relationship between these two variables, but some sort of linear relationship is suspected. The 26 pairs of data (X and Y values) were used in a regression analysis (to be examined shortly) and the following estimates of the parameters were obtained:

$$b_0 = -1256 \text{ (to 0 decimal places)}$$

$$b_1 = 1.212 \text{ (to 3 decimal places)}$$

Thus, the regression equation becomes

$$\hat{Y}_i = -1256 + 1.212X_i \tag{9.18}$$

For any one particular basin, it is clear that the prediction is not exact. Take, for example, sample basin 10 (see Table 9.6)

$$X_{10} = 1522 \text{ (rainfall in millimeters)}$$

$$Y_{10} = 651 \text{ (specific runoff in millimeters)}$$

From Equation 9.18,

$$\begin{aligned} \hat{Y}_{10} &= -1256 + 1.212(1522) \\ &= -1256 + 1845 \\ &= 589 \text{ mm} \end{aligned}$$

Thus, the regression equation has underpredicted the Runoff for this basin. Using the statistical model relationship and assuming that the b_0 and b_1 statistics are good and accurate estimators of β_0 and β_1, the statistical equation becomes:

$$Y_i = -1256 + 1.212(X_i) + e_i$$

where

$$e_i = \hat{\epsilon}_i = \text{the estimate of the error term for the } i\text{th observation}$$

For Y_{10},

$$Y_{10} = -1256 + 1.212(X_{10}) + e_{10}$$

therefore,

$$651 = -1256 + 1.212(1522) + e_{10}$$

and

$$\begin{aligned} e_{10} &= 651 - 589 \\ &= 62 \text{ mm} \end{aligned}$$

TABLE 9.6 Data for Calculation of Regression Coefficients

Observation Number (i)	Rainfall (X_i)	Runoff (Y_i)	Rainfall² (X_i^2)	Runoff² (Y_i^2)	Rainfall × Runoff (X_iY_i)
1	1605	1156	2,576,025	1,336,336	1,855,380
2	2202	1378	4,848,804	1,898,884	3,034,356
3	2036	1560	4,145,296	2,433,600	3,176,160
4	1868	947	3,489,424	896,809	1,768,996
5	1520	356	2,310,400	126,736	541,120
6	1276	454	1,628,176	206,116	579,304
7	1710	654	2,924,100	427,716	1,118,340
8	1510	473	2,280,100	223,729	714,230
9	1285	127	1,651,225	16,129	163,195
10	1522	651	2,316,484	423,801	990,822
11	1585	485	2,512,225	235,225	768,725
12	1682	750	2,829,124	562,500	1,261,500
13	1984	978	3,936,256	956,484	1,940,352
14	1594	641	2,540,836	410,881	1,021,754
15	1693	757	2,866,249	573,049	1,281,601
16	1612	734	2,598,544	538,756	1,183,208
17	1853	1149	3,433,609	1,320,201	2,129,097
18	1477	753	2,181,529	567,009	1,112,181
19	1416	647	2,005,056	418,609	916,152
20	1520	360	2,310,400	129,600	547,200
21	1500	610	2,250,000	372,100	915,000
22	1262	467	1,592,644	218,089	589,354
23	1558	531	2,427,364	281,961	827,298
24	1352	366	1,827,904	133,956	494,832
25	1525	476	2,325,625	226,576	725,900
26	1599	485	2,556,801	235,225	775,515
TOTAL	41,746 ΣX_i	17,945 ΣY_i	68,364,200 ΣX_i^2	15,170,077 ΣY_i^2	30,431,572 ΣX_iY_i

Note: the Y_i^2 column will be used later.

The error term is thus the difference between the observed and estimated value for Y (i.e., $e_i = \hat{\epsilon}_i = Y_i - \hat{Y}_i$) and is commonly referred to as the *residual* of the regression equation.

The Least-Squares Technique and the Estimation of the Coefficients

The method by which b_0 and b_1 are estimated in the regression equation is usually by the *least-squares technique*. This technique provides estimates of β_0 and β_1 (i.e., b_0 and b_1) such that an expression S, which is equal to the sum of all the residuals squared, is minimized

$$S = \sum_{i=1}^{n} \epsilon_i^2 = \sum_{i=1}^{n} (Y_i - \hat{Y}_i)^2 \tag{9.19}$$

where

$$n = \text{the sample size}$$

Substituting for $\hat{Y}_i$,

$$S = \sum_{i=1}^{n} [Y_i - (b_0 + b_1 X_i)]^2$$

$$= \sum_{i=1}^{n} (Y_i - b_0 - b_1 X_i)^2 \tag{9.20}$$

Minimizing the deviations can be illustrated graphically (Figure 9.12). The vertical distance from the value of Y_i to the regression line is equal to the error term ϵ_i or $(Y_i - \hat{Y}_i)$. The aim of the least-squares regression analysis is to estimate values for b_0 (the Y intercept) and b_1 (the slope of the line) such that these deviations from the regression line are at a minimum. They are squared to remove the negative values.

Minimizing a value is a common problem in calculus, and can be achieved by finding the partial derivatives of S with respect to b_0 and b_1. Because X_i and Y_i are known results, S is a function of the unknowns b_0 and b_1. The minimum value of S occurs when the partial derivatives of S with respect to b_0 and b_1 equal zero. (Note that $\Sigma = \Sigma_{i=1}^{n}$.)

$$\text{Thus for } S = \Sigma(Y_i - b_0 - b_1 X_i)^2$$

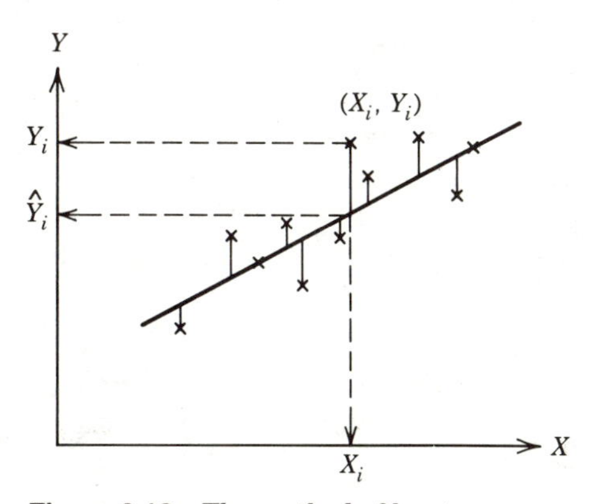

Figure 9.12 The method of least squares.

Expanding the square and separating the summations

$$S = \Sigma Y_i^2 - \Sigma Y_i b_0 - \Sigma Y_i b_1 X_i - \Sigma b_0 Y_i$$
$$+ \Sigma b_0^2 + \Sigma b_0 b_1 X_i - \Sigma Y_i b_1 X_i + \Sigma b_0 b_1 X_i + \Sigma (b_1 X_i)^2$$

This can be simplified to

$$S = \Sigma Y_i^2 - 2b_0 \Sigma Y_i - 2b_1 \Sigma X_i Y_i + nb_0^2 + 2b_0 b_1 \Sigma X_i + b_1^2 \Sigma X_i^2$$

Finding the partial derivatives of S with respect to the two unknowns and setting these equal to zero

$$\frac{\delta S}{\delta b_0} = -2\Sigma Y_i + 2nb_0 + 2b_1 \Sigma X_i = 0 \tag{9.21}$$

$$\frac{\delta S}{\delta b_1} = -2\Sigma X_i Y_i + 2b_0 \Sigma X_i + 2b_1 \Sigma X_i^2 = 0 \tag{9.22}$$

Dividing by 2 and setting the negative terms on the right-hand side of the equation yields

$$nb_0 + b_1 \Sigma X_i = \Sigma Y_i \tag{9.23}$$

$$b_0 \Sigma X_i + b_1 \Sigma X_i^2 = \Sigma X_i Y_i \tag{9.24}$$

Equations 9.23 and 9.24 are called *normal equations* (which have nothing to do with a normal distribution). Solving for b_0:

$$nb_0 = \Sigma Y_i - b_1 \Sigma X_i$$

$$b_0 = \frac{\Sigma Y_i - b_1 \Sigma X_i}{n}$$

$$= \bar{Y} - b_1 \bar{X} \tag{9.25}$$

Solving for b_1, substitute $b_0 = (\Sigma Y_i - b_1 \Sigma X_i)/n$ in Equation 9.24.

$$\left(\frac{\Sigma Y_i - b_1 \Sigma X_i}{n}\right) \Sigma X_i + b_1 \Sigma X_i^2 = \Sigma X_i Y_i$$

Multiplying out:

$$\frac{\Sigma X_i \Sigma Y_i}{n} - \frac{b_1 (\Sigma X_i)^2}{n} + b_1 \Sigma X_i^2 = \Sigma X_i Y_i$$

Isolating the b_1 values

$$b_1 \left(\Sigma X_i^2 - \frac{(\Sigma X_i)^2}{n}\right) = \Sigma X_i Y_i - \frac{\Sigma X_i \Sigma Y_i}{n}$$

We get

$$b_1 = \frac{\Sigma X_i Y_i - \dfrac{\Sigma X_i \Sigma Y_i}{n}}{\Sigma X_i^2 - \dfrac{(\Sigma X_i)^2}{n}} \tag{9.26}$$

Equations 9.25 and 9.26 provide our basic solutions for the estimates b_1 and b_0, although we note that as an alternative, Equation 9.26 is often written (multiplying top and bottom by n)

$$b_1 = \frac{n\Sigma X_i Y_i - \Sigma X_i \Sigma Y_i}{n\Sigma X_i^2 - (\Sigma X_i)^2} \tag{9.27}$$

Thus, to calculate the regression equation for a set of data, the following are required: ΣX_i, ΣY_i, $\Sigma X_i Y_i$, and ΣX_i^2. The process is illustrated with the Runoff/Rainfall data from the Northland Drainage Basin Survey (Table 9.6) that were plotted as a scattergram in Figure 9.9. The calculations yield:

$$\bar{X} = \frac{41746}{26} = 1606 \text{ mm} \qquad \bar{Y} = \frac{17{,}945}{26} = 690 \text{ mm}$$

$$b_1 = \frac{30{,}431{,}572 - \dfrac{(41{,}746)(17{,}945)}{26}}{68{,}364{,}200 - \dfrac{(41{,}746)^2}{26}}$$

$$= \frac{30{,}431{,}572 - 28{,}812{,}768}{68{,}364{,}200 - 67{,}028{,}020}$$

$$= \frac{1{,}618{,}804}{1{,}336{,}180} = 1.212$$

Therefore,

$$b_0 = 690 - 1.212\,(1606)$$
$$= 690 - 1946 = -1256$$

Therefore, the regression equation is:[2]

$$\hat{Y}_i = -1256 + 1.212X_i$$

This line was plotted on Figure 9.9. Before examining some of the properties of this regression line, we make two small but important diversions to examine the concepts of covariability and partitioning the sum of squares.

[2]Note that we will often get slightly varying results depending on what calculator or computer we are using. This is especially true when the data consist of large values—or at least large numbers of significant digits. In our example with four significant digits for each variable and the example worked on a calculator, we had to watch for overflow of the registers and inadvertent rounding.

Covariability, Covariation, and Covariance

In Chapter 3, we introduced the concepts of variability, variation and variance. We wish to expand these concepts to include the *covariability* of two variables. Reviewing the original concepts using the examples of the variables Rainfall (X) and Runoff (Y):

	Rainfall (X_i)	Runoff (Y_i)
Total variability = the sum of squares	$\Sigma X_i^2 = 68{,}364{,}200$	$\Sigma Y_i^2 = 15{,}170{,}077$
Variation = the corrected total sum of squares	$\Sigma(X_i - \bar{X})^2 = \Sigma X_i^2 - \frac{(\Sigma X_i)^2}{n}$	$\Sigma(Y_i - \bar{Y})^2 = \Sigma Y_i^2 - \frac{(\Sigma Y_i)^2}{n}$
	$= 68{,}364{,}200 - \frac{(41{,}746)^2}{26}$	$= 15{,}170{,}077 - \frac{(17{,}945)^2}{26}$
	$= 1{,}336{,}180$	$= 2{,}784{,}576$
Variance ($\hat{\sigma}^2$)	$\hat{\sigma}_x^2 = \frac{\Sigma(X_i - \bar{X})^2}{n - 1}$	$\hat{\sigma}_y^2 = \frac{\Sigma(Y_i - \bar{Y})^2}{n - 1}$
	$= \frac{1{,}336{,}180}{25}$	$= \frac{2{,}784{,}576}{25}$
	$= 53{,}447.20$	$= 111{,}383.04$

In the same way that variability (and the various measures derived from it) is concerned with analyzing the variation of scores for a single variable, covariability is concerned with the joint variability of two sets of scores. We define *total covariability* as (uncorrected) sum of products $= \Sigma X_i Y_i$

$$\begin{aligned} \textit{Covariation} &= \text{the corrected sum of products of } X \text{ and } Y \\ &= \Sigma(X_i - \bar{X})(Y_i - \bar{Y}) \\ &= \Sigma X_i Y_i - \frac{\Sigma X_i \Sigma Y_i}{n} \qquad (9.28) \\ &= \text{(uncorrected) sum of products} - \text{correction for the means} \end{aligned}$$

With covariation defined as the sum of the products of the deviations from the means (Equation 9.28), if X_i and Y_i both have positive departures from their means, they will contribute positively to the covariation (Figure 9.13). If X_i and Y_i both have negative departures from their means, then again they will contribute positively to the covariation ($-ve$ times $-ve$ $=$ $+ve$). However, if X_i has a positive departure and Y_i a negative departure (or vice versa), then the contribution to covariation will be negative. The covariation thus measures the "joint size" of X and Y (relative to their means); unlike the variation, it may total to a positive or negative number.

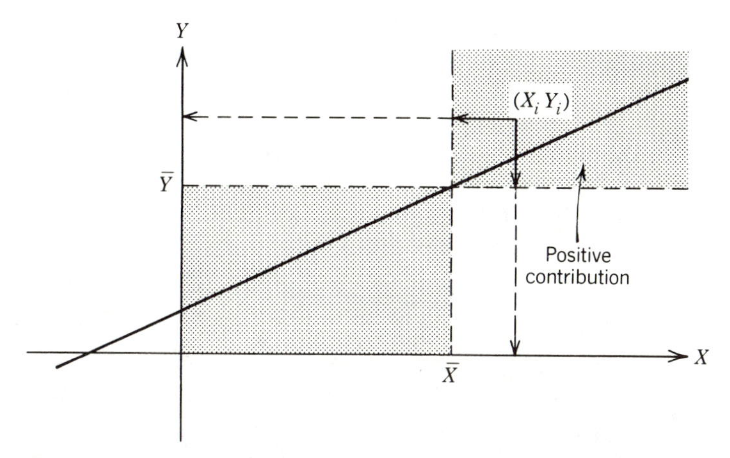

Figure 9.13 Derivation of covariability.

Because the magnitude of the covariability and covariation measures not only the magnitude of the joint variability of the scores but also the number of observations n (the larger the sample size, the larger the covariation), an "average" measure is required and this is the *covariance* (the equivalent of the variance).

$$\textit{Covariance} = \hat{\sigma}_{xy} = \frac{\text{covariation}}{n-1} = \frac{\Sigma(X_i - \bar{X})(Y_i - \bar{Y})}{n-1} \tag{9.29}$$

Alternatively, we could divide by n to obtain the covariance (s_{xy}) of a sample. In our example:

$$\text{Total covariability} = \Sigma X_i Y_i = 30{,}431{,}572$$

$$\text{Covariation} = \Sigma X_i Y_i - \frac{\Sigma X_i \Sigma Y_i}{n}$$

$$= 30{,}431{,}572 - \frac{(41{,}746)(17{,}945)}{26}$$

$$= 1{,}618{,}804$$

$$\text{Covariance} = \hat{\sigma}_{xy} = \frac{1{,}618{,}804}{25} = 64{,}752.16$$

Partitioning the Sum of Squares of the Dependent Variable

Regression computations may be considered as the process of partitioning the corrected total sum of squares of the dependent variable Y into separate parts—each part of which has important properties. Such a partitioning is simply an *analysis of variance* (or ANOVA) as outlined in Chapter 8. For

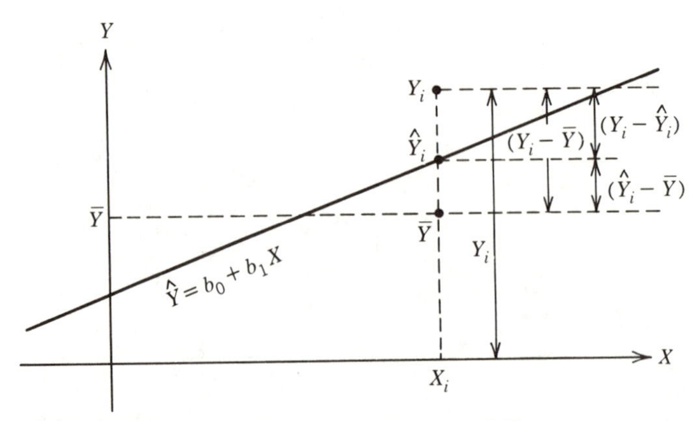

Figure 9.14 Partitioning the deviations of Y.

regression, the deviations from the mean can be partitioned into two parts (Figure 9.14).

$$(Y_i - \bar{Y}) = (\hat{Y}_i - \bar{Y}) + (Y_i - \hat{Y}_i) \tag{9.30}$$

In Figure 9.14, we have placed Y_i in a convenient location above the regression line to make this relationship clearer. Summing and squaring (the other terms drop out) yields

$$\Sigma(Y_i - \bar{Y})^2 = \Sigma(\hat{Y}_i - \bar{Y})^2 + \Sigma(Y_i - \hat{Y}_i)^2$$

SST = SS due to fitting the regression line (= the explained variation = SSR)

+ the residual (or error) sum of squares (= the unexplained variation = SSE)

$$\text{SST} = \text{SSR} + \text{SSE} \tag{9.31}$$

The calculation of $\Sigma(Y_i - \bar{Y})^2$ and $\Sigma(Y_i - \hat{Y}_i)^2$ is tedious and can be simplified to

$$\text{SSR} = \frac{\left(\Sigma X_i Y_i - \dfrac{\Sigma X_i \Sigma Y_i}{n}\right)^2}{\left(\Sigma X_i^2 - \dfrac{(\Sigma X_i)^2}{n}\right)} \tag{9.32}$$

and

$$\text{SSE} = \text{SST} - \text{SSR}$$

The complete set of calculations for the Rainfall/Runoff example is given in Table 9.7.

TABLE 9.7 Partitioning the Sum of Squares

Observation Number (i)	Observed Y (Y_i)	Predicted Y ($\hat{Y}_i$)	Deviation from $\bar{Y}$ ($Y_i - \bar{Y}$)	Variation of Y $(Y_i - \bar{Y})^2$	Explained Y ($\hat{Y}_i - \bar{Y}$)	Residual of Y ($Y_i - \hat{Y}_i$)	Explained Variation ($\hat{Y}_i - \bar{Y}_i$)	Unexplained Variation $(Y_i - \hat{Y}_i)^2$
1	1156	689	466	216,979	−1	467	1	218,089
2	1378	1413	688	473,083	723	−35	522,454	1,225
3	1560	1212	870	756,569	522	348	272,286	121,104
4	947	1008	257	65,951	318	−61	101,003	3,721
5	356	586	−334	111,683	−104	−230	10,856	52,900
6	454	291	−236	55,786	−399	163	159,353	26,569
7	654	817	−36	1,310	127	−163	16,081	26,569
8	473	574	−217	47,171	−116	−101	13,500	10,201
9	127	302	−563	317,183	−388	−175	150,691	30,625
10	651	589	−39	1,536	−101	62	10,239	3,844
11	485	665	−205	42,103	−25	−180	635	32,400
12	750	783	60	3,577	93	−33	8,614	1,089
13	978	1149	288	82,835	459	−171	210,507	29,241
14	641	676	−49	2,420	−14	−35	201	1,225
15	757	796	67	4,464	106	−39	11,196	1,521
16	734	698	44	1,919	8	36	61	1,296
17	1149	990	459	210,507	300	159	89,886	25,281
18	753	534	63	3,945	−156	219	24,395	47,961
19	647	460	−43	1,865	−230	187	52,987	34,969
20	360	586	−330	109,025	−104	−226	10,856	51,076
21	610	562	−80	6,430	−128	48	16,433	2,304
22	467	274	−223	49,814	−416	193	173,214	37,249
23	531	633	−159	25,341	−57	−102	3,271	10,404
24	366	383	−324	105,099	−307	−17	94,366	289
25	476	593	−214	45,877	−97	−117	9,446	13,689
26	485	682	−205	42,103	−8	−197	67	38,809
TOTAL	17,945 ΣY_i	17,945 $\Sigma \hat{Y}_i$	0 $\Sigma(Y_i - \bar{Y})$	2,784,576 $\Sigma(Y_i - \bar{Y})^2$	0 $\Sigma(\hat{Y}_i - \bar{Y})$	0 $\Sigma(Y_i - \hat{Y}_i)$	1,962,598 $\Sigma(\hat{Y}_i - \bar{Y})^2$	823,650 $\Sigma(Y_i - \hat{Y}_i)^2$

TABLE 9.8 Least-Squares Table

Source of Variation	SS	*df*	MS
Regression	$SSR = \Sigma(\hat{Y}_i - \bar{Y})^2$	1	$MSR = SSR$
Error (residual)	$SSE = SST - SSR$	$n - 2$	$MSE = \dfrac{SSE}{n - 2}$
TOTAL	$SST = \Sigma(Y_i - \bar{Y})^2$	$n - 1$	

The partitioned sum of squares can be presented in an ANOVA table—commonly called a *Least-Squares Regression Table*—in the same manner used in the ANOVA structure in the last chapter. The general format is given in Table 9.8 and our worked example in Table 9.9.

The MSE term, the residual mean square, has important properties. It provides a measure of average departure of the predicted points from the observed points. We give it the symbol S_E^2 and note that it is the estimate of the variance of the error term (σ_E^2) in the statistical model (Equation 9.16). The square root of S_E^2 (i.e., S_E) reduces the error term to the same units as the observed Y variable. To this term we give the name the *standard error of estimate* for the regression of Y on X. We take up its use, and the importance of the ANOVA table, later in the chapter.

Properties of the Regression Line

We are now in a position to state some of the more important properties of the regression equation $Y = b_0 + b_1X$.

First, the equation can be used to predict a value for Y for any value of X. Using our example to predict Runoff for a Rainfall of 2000 mm, we get

$$\hat{Y} = -1256 + 1.212\ (2000)$$
$$= 1168 \text{ mm}$$

Predictions using X values within the domain of X are called interpolation, and outside the domain they are called extrapolation. Because the model is

TABLE 9.9 Least-Squares Table for Runoff/Rainfall Example (Computer-Generated Results)

Source of Variation	SS	*df*	MS
Regression	1,961,207	1	1,961,207
Error (residual)	823,369	24	34,307
TOTAL	2,784,576	25	

fitted to data from within the domain of X, there is usually no statistical justification for extrapolation methods. We are also more confident about interpolations near the mean of X than near the extremes of the domain (a point taken up in Section 9.5).

Second, the regression line passes through the point defined by $\bar{X}$ and $\bar{Y}$ (a handy fact to remember when locating a regression line on a scattergram). This can be easily verified (e.g., using Equation 9.25).

Third, the Y-intercept term, b_0, is in the units of Y and gives the value of Y when $X = 0$. In our example, the value of $b_0 = -1256$ indicates that when Rainfall (X) is 0, the predicted Runoff ($\hat{Y}$) is -1256 mm (which illustrates the lack of justification for extrapolation of the model beyond the domain of X!). One small point to note—when using the Y intercept to plot the regression line on a graph, it refers to the interception of the line with the Y axis only if the Y axis is drawn at the point of $X = 0$. When the domain of the variable X is far removed from an X of 0, it is often convenient to draw the Y axis through the X axis at values other than 0. Such, is the situation illustrated in Figure 9.9.

Fourth, the slope of the regression line, b_1, is in gradient form and measures the rate of change of the variable Y (in Y units) per unit change of X. In our example, both X and Y happen to be measured in millimeters and the b_1 value of 1.212 means 1.212 mm increase in Y per 1 mm increase in X (i.e., as Rainfall increases 1 mm the predicted Runoff increases by 1.212 mm).

Fifth, the slope coefficient b_1 has the range of values that were discussed in Section 9.1 of this chapter, when looking at properties of straight lines on graphs. Values can be either positive or negative. Positive values of b_1 have lines sloping upwards to the right—as X increases so does Y; negative values of b_1 have lines sloping down to the right—as X increases, Y decreases. The angle of the line on the graph will depend on the magnitude of the b_1 value and the relative scales of the two axes. Some examples are given in Figure 9.15. When $b_1 = 0$, the line is horizontal and the simple linear model is no longer applicable (Y does not change linearly as X increases). Instead, we have the constant model $\hat{Y} = b_0$.

Sixth, from our observations on covariation and partitioning the sum of squares of Y, we note some interesting and important relationships with the regression coefficient b_1.

From Equation 9.26

$$b_1 = \frac{\Sigma X_i Y_i - \dfrac{\Sigma X_i \Sigma Y_i}{n}}{\Sigma X_i^2 - \dfrac{(\Sigma X_i)^2}{n}}$$

$$b_1 = \frac{\text{covariation}}{\text{variation}_x} \tag{9.33}$$

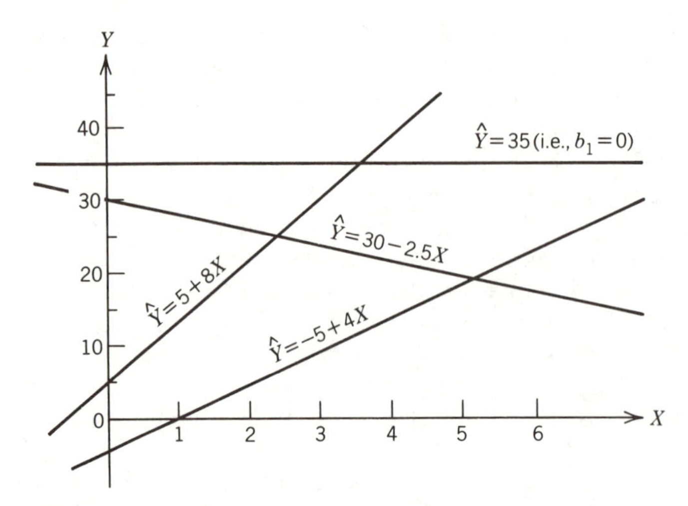

Figure 9.15 Regression line examples.

and dividing top and bottom by $n - 1$ or n

$$b_1 = \frac{\text{covariation}}{\text{variation}_x} = \frac{\hat{\sigma}_{xy}}{\hat{\sigma}_x^2} = \frac{s_{xy}}{s_x^2} \tag{9.34}$$

With the variation always positive, b_1 will be positive or negative following the covariation. And from Equation 9.32

$$\text{SSR} = \frac{\left(\Sigma X_i Y_i - \dfrac{\Sigma X_i \Sigma Y_i}{n}\right)^2}{\Sigma X_i^2 - \dfrac{(\Sigma X_i)^2}{n}}$$

$$= \frac{(\text{covariation})^2}{\text{variation}_x} \tag{9.35}$$

And by substituting Equation 9.33 into 9.35,

$$\text{SSR} = b_1\,(\text{covariation}) = b_1(n - 1)\hat{\sigma}_{xy} = b_1(n)s_{xy} \tag{9.36}$$

The SSR will always be positive, as if b_1 is negative, so is the covariation.

Finally, the regression line described above is derived from regressing Y on X. It is a directional relationship, where Y is a linear function of X. The key component of that function is b_1, the slope of the regression line defined (above) as the covariation of X and Y ratioed against the variation of X (Equation 9.33). In some circumstances, it is conceivable to consider the dependence of the two variables in the opposite direction. That is, retaining the same symbols for Y and X, we are saying that $X = f(Y)$. If we label the estimators for the regression of Y on $X, b_0(y.x)$ and $b_1(y.x)$ we can use $b_0(x.y)$

and $b_1(x.y)$ for the $X = f(Y)$ regression and these estimators can be calculated from

$$b_1(x.y) = \frac{\Sigma X_i Y_i - \dfrac{\Sigma X_i \Sigma Y_i}{n}}{\Sigma Y_i^2 - \dfrac{(\Sigma Y_i)^2}{n}} \tag{9.37}$$

and

$$b_0(x.y) = \bar{X} - b_1(x.y)\bar{Y} \tag{9.38}$$

The two regressions are closely related and as

$$b_1(y.x) = \frac{\hat{\sigma}_{xy}}{\hat{\sigma}_x^2} = \frac{s_{xy}}{s_x^2}$$

and

$$b_1(x.y) = \frac{\hat{\sigma}_{xy}}{\hat{\sigma}_y^2} = \frac{s_{xy}}{s_y^2}$$

Therefore,

$$b_1(y.x) = b_1(x.y)\frac{\hat{\sigma}_y^2}{\hat{\sigma}_x^2} = b_1(x.y)\frac{s_y^2}{s_x^2} \tag{9.39}$$

and

$$b_1(x.y) = b_1(y.x)\frac{\hat{\sigma}_x^2}{\hat{\sigma}_y^2} = b_1(y.x)\frac{s_x^2}{s_y^2} \tag{9.40}$$

Using the Runoff/Rainfall example and Equations 9.37 and 9.38

$$\begin{aligned} b_1(x.y) &= \frac{1{,}618{,}806}{2{,}784{,}576} \\ &= .581 \\ b_0(x.y) &= 1605.62 - .581\ (690.19) \\ &= 1205 \text{ mm} \end{aligned}$$

The regression equation becomes $\hat{X}_i = 1205 + .581Y_i$. This equation can be compared to the equation for Y on X either on separate graphs with the axes reversed (Figure 9.16), or on the same graph (Figure 9.17). Note that both lines pass through the point of intersection of $\bar{X}$ and $\bar{Y}$.

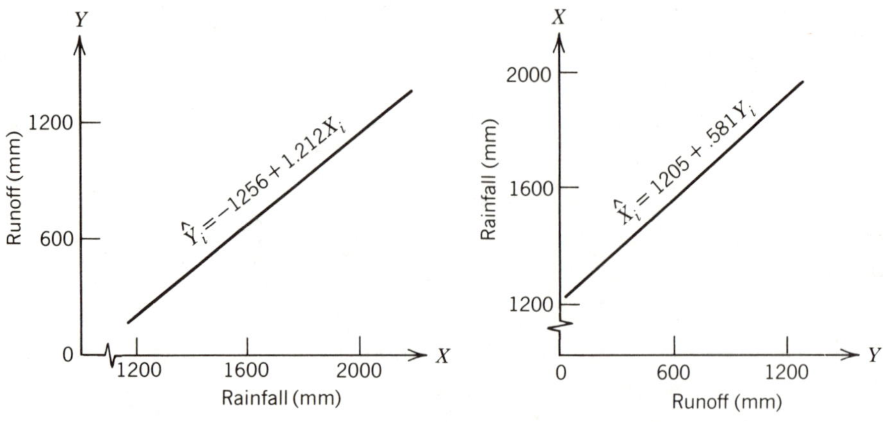

Figure 9.16 Graphs of Y on X and X on Y for runoff/rainfall example.

9.4 LINEAR CORRELATION ANALYSIS

After choosing the type of function and obtaining estimates for its parameters, the third sequential step in the analysis of a functional relationship is to examine how well the regression equation fits the observed data. This is the aim of correlation analysis: to provide a measure of how adequately the model fits the data. It is also possible to bypass the regression estimates. Our interest might simply be in assessing the relationship of two variables and only to obtain a measure of association.

A desirable measure of association between two variables should have the following properties:

1. A large value when the degree of association is high; and a small value when the association is low, and

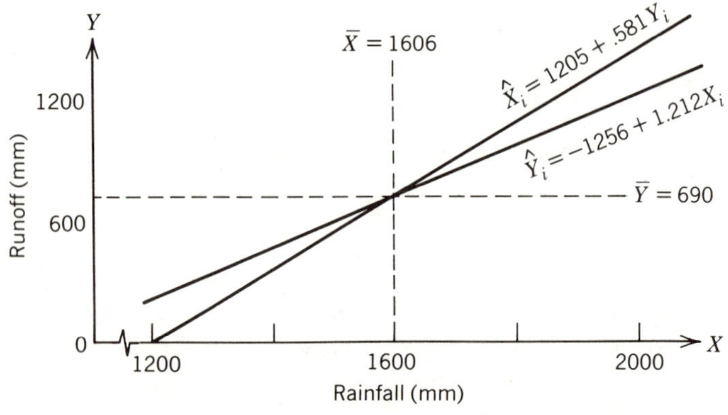

Figure 9.17 Comparison of Y on X and X on Y graphs.

2. The measure should be independent of the units by which the variables are measured to give a standardized dimensionless measure enabling comparison of correlations.

The Simple Linear Correlation Coefficient

In the linear regression model, the starting point for a measure of correlation is from the analysis of variance table. In this table, we have apportioned the total variation of the dependent variable (SST) into that part "explained" by the regression model (SSR), and the remainder left unexplained (SSE). We define a correlation measure, which we call r^2, as

$$r^2 = \frac{\text{explained variation}}{\text{total variation}}$$

$$= \frac{\text{SSR}}{\text{SST}} \tag{9.41}$$

reducing the SSR and SST terms back to their variability structures. Equation 9.41 is equal to

$$r^2 = \frac{(\text{covariation})^2}{(\text{variation}_x)(\text{variation}_y)} \tag{9.42}$$

and dividing both numerator and denominator by $n - 1$ or n

$$r^2 = \frac{(\text{covariance})^2}{(\text{variance}_x)(\text{variance}_y)}$$

This provides a measure that has a property additional to many measures of association—it shows the direction of the relationship; the covariation and covariance are positive or negative depending on the slope of the regression line (b_1). Thus, if we take the square root of r^2 and attach the sign of the covariability term to it, we have a measure that is large ($+1$ or -1) when the association is high, 0 when it is nonexistent, is dimensionless, and indicates the direction of the relationship by its sign.

We call this coefficient the *simple linear correlation coefficient* (r), or *Pearson's product-moment correlation coefficient.* Assuming that we are calculating it from basic data (and not after we have calculated the regression equation), we provide the following computing equations:

$$r^2 = \frac{\left(\Sigma X_i Y_i - \dfrac{\Sigma X_i \Sigma Y_i}{n}\right)^2}{\left(\Sigma X_i^2 - \dfrac{(\Sigma X_i)^2}{n}\right)\left(\Sigma Y_i^2 - \dfrac{(\Sigma Y_i)^2}{n}\right)} \tag{9.43}$$

and

$$r = \pm\sqrt{r^2} \text{ the sign depending on the sign of } \left(\Sigma X_i Y_i - \frac{\Sigma X_i \Sigma Y_i}{n}\right) \tag{9.44}$$

or, alternatively, taking the square root of top and bottom

$$r = \frac{\Sigma X_i Y_i - \dfrac{\Sigma X_i \Sigma Y_i}{n}}{\sqrt{\left(\Sigma X_i^2 - \dfrac{(\Sigma X_i)^2}{n}\right)} \sqrt{\left(\Sigma Y_i^2 - \dfrac{(\Sigma Y_i)^2}{n}\right)}} \tag{9.45}$$

or using the terminology for covariance and standard deviations

$$r = \frac{\hat{\sigma}_{xy}}{\hat{\sigma}_x \cdot \hat{\sigma}_y} = \frac{S_{xy}}{S_x \cdot S_y} \tag{9.46}$$

The calculation of the correlation coefficient r from basic data, therefore, requires ΣX_i, $\Sigma Y_i \Sigma X_i^2$, ΣY_i^2, $\Sigma X_i Y_i$—the same set of summations as required for the calculation of b_1, with the addition of the summation of Y_i^2. Using the data from the Rainfall/Runoff analysis presented in Table 9.6, the calculation in full is:

$$r = \frac{\Sigma X_i Y_i - \dfrac{\Sigma X_i \Sigma Y_i}{n}}{\sqrt{\left(\Sigma X_i^2 - \dfrac{(\Sigma X_i)^2}{n}\right)} \sqrt{\left(\Sigma Y_i^2 - \dfrac{(\Sigma Y_i)^2}{n}\right)}}$$

$$r = \frac{30{,}431{,}572 - \dfrac{(41{,}746)(17{,}945)}{26}}{\sqrt{(68{,}364{,}200 - \dfrac{(41{,}746)^2}{26}} \sqrt{(15{,}170{,}077 - \dfrac{(17{,}945)^2}{26}}}$$

$$= \frac{1{,}618{,}804}{\sqrt{1{,}336{,}180}\sqrt{2{,}784{,}576}}$$

$$= .8392$$

and

$$r^2 = (.8392)^2 = .7043$$

Properties of the Simple Linear Correlation Coefficient

The first property of the simple linear correlation coefficient is that the values of r range from -1.0 (perfect negative correlation) with all the coordinate points of X and Y falling exactly on the regression line and the slope of the line is negative, through 0 where the coordinate points are scattered randomly over the graph or where the points lie on a regression line which is horizontal, to +1.0 (perfect positive correlation) where all the coordinate points fall exactly on the regression line and the slope of the line is positive.

Second, the square of the correlation coefficient, r^2, is often referred to as the *coefficient of determination* and indicates the proportion of the total variation "explained" by the regression line. It is commonly expressed as a percentage. $1 - r^2$ gives the proportion of variation left unexplained.

Third, in the derivation of r and r^2 there is no distinction between dependent and independent variables; the correlation is only a measure of association of the two variables. Thus, strictly speaking, the term *explained variation* is a bit of a misnomer and perhaps it should more correctly be referred to as "associated" variation.

Fourth, correlation coefficients of ±1 indicate perfect linear correlations (positive or negative), but most coefficients will lie somewhere between these values. With population data, the r^2 value provides a measure of the degree of linear correlation, and $1 - r^2$ is a measure of the lack of fit of the linear model. However, with sampled data, the situation is less clear. We need some sort of testing structure—to test, for example, that r is significantly different from a value of zero—to conclude that the two variables are linearly correlated. We examine these tests in the next section, but we note now that the significance of a correlation coefficient (i.e., whether or not it is significantly different from zero) will increase as r gets closer to +1 or −1, but it will also change with sample size. For example, an r value of .40 may be classed as not significantly different from zero for a sample size of 12, but would be highly significant for a sample size of 200.

Fifth, remember that the correlation measure r is a linear correlation, and a value of r near zero indicates that there is no linear correlation between the two variables. That does not mean that they are unrelated, for there may be a high nonlinear correlation between them. We illustrate this with a simple example (Table 9.10). If we calculate a linear correlation with all five data points, we get an r value of 0 despite the obvious squared rela-

TABLE 9.10 Simple Nonlinear Relationship

X	Y	XY	X^2	Y^2
−2	4	−8	4	16
−1	1	−1	1	1
0	0	0	0	0
1	1	1	1	1
2	4	8	4	16
0	10	0	10	34

tionship between X and Y. More realistically, if we leave off the first pair of points, we get a linear relationship with a regression line sloping up to the right. The r value is still, however, only .55.

Sixth, the correlation measures are closely related to the regression and analysis of variation measures that we have already discussed, since all are concerned with the covariance of X and Y. There are many ways to describe this correspondence but, for the moment, we note the following relationships.

From

$$r^2 = \frac{\text{explained variation}}{\text{total variation}} \tag{9.47}$$

and

$$r = \frac{\hat{\sigma}_{xy}}{\hat{\sigma}_x \hat{\sigma}_y} = \frac{s_{xy}}{s_x s_y}$$

We can show

$$r = b_1(y.x) \frac{\hat{\sigma}_x}{\hat{\sigma}_y} = b_1(y.x) \frac{s_x}{s_y}$$

$$b_1(y.x) = r \frac{\hat{\sigma}_y}{\hat{\sigma}_x} = r \frac{s_y}{s_x} \tag{9.48}$$

$$r^2 = b_1(y.x) \cdot b_1(x.y) \tag{9.49}$$

$$\text{SSR} = r^2 \text{ (SST)} \tag{9.50}$$

$$\text{SSE} = (1 - r^2)\text{(SST)} \tag{9.51}$$

The Correlation Matrix

We have been examining the relationship of two variables X and Y and the correlation coefficient that measures it. With a research problem centered around a suite of variables, we would repeat the calculation process many times to obtain a set of correlation coefficients among all pairs of variables. As by-products of multivariate analyses and from separate programs, sets of intercorrelations can be output in the form of a *correlation matrix*. Using the Northland Drainage Basin Survey as an example, the deck setup for input to the SPSS correlation program is given in Table 9.11 and the resulting output in Table 9.12.

The diagonal of the table, from the top left corner, is the correlation of each variable with itself ($r = +1.0$). The triangle of the data above that line contains the same information as the triangle below the line (row for column). The significance values (p values) will be discussed in the next section. Options to output the variance-covariance matrix (variances in the diagonal) in addition to the correlations are available in many programs.

TABLE 9.11 SPSS Setup for Pearson Correlation

```
RUN NAME          NORTH(3) - CORRELATION MATRIX
FILE NAME         NORTH NORTHLAND DRAINAGE BASIN STUDY
VARIABLE LIST     AREA LENGTH ELEVATN FOREST RUNOFF RAINFALL SLOPE
                  ROCKTYPE
INPUT MEDIUM      CARD
N OF CASES        26
INPUT FORMAT      FIXED (F8.2,F8.1,6F8.0)
PRINT FORMAT      AREA (2) LENGTH (1)
VAR LABELS        AREA       AREA OF BASIN - KM**2/
                  LENGTH     MAXIMUM LENGTH OF BASIN - KM/
                  ELEVATN    ELEVATION OF MOUTH OF BASIN - M/
                  FOREST     % BASIN FORESTED ESTIMATED FROM TOPO MAP -
                             %/
                  RUNOFF     MEAN ANNUAL SPECIFIC DISCHARGE - MM/
                  RAINFALL   MEAN ANNUAL RAINFALL NEAR TO BASIN CENTER -
                             MM/
                  SLOPE      ANGLE HIGHEST POINT AND BASIN MOUTH -
                             DEGREES/
                  ROCKTYPE   PREDOMINANT ROCKTYPE IN BASIN - CODED
PEARSON CORR      AREA TO SLOPE
OPTIONS           3
STATISTICS        ALL
READ INPUT DATA
```

Alternative Correlation Measures

Throughout this chapter we have emphasized the simple linear regression model, and in this section the simple linear, or Pearson's correlation coefficient. There are, a wide variety of other methods and approaches to the assessment of bivariate relationships, some of which were discussed in Chapter 8. One discussed below, can be considered as an alternative to the simple linear correlation measure *r*. *Spearman's rank-order correlation coefficient* (r_s), as the name indicates, is designed for measuring the association of variables whose scores are in rank order. These may have been derived from an original ordination measurement process or from an ordinal classification

TABLE 9.12 Correlation Matrix for Northland Drainage Basin Survey

- - - - - - - - - - - - - P E A R S O N   C O R R E L A T I O N   C O E F F I C I E N T S -

	AREA	LENGTH	ELEVATN	FOREST	RUNOFF	RAINFALL	SLOPE
AREA	1.0000 (26) P=*****	0.8581 (26) P=0.000	0.5850 (26) P=0.002	0.2484 (26) P=0.221	0.5582 (26) P=0.003	0.6713 (26) P=0.000	0.4147 (26) P=0.035
LENGTH	0.8581 (26) P=0.000	1.0000 (26) P=*****	0.5990 (26) P=0.001	0.2811 (26) P=0.164	0.5333 (26) P=0.005	0.6302 (26) P=0.001	0.3708 (26) P=0.062
ELEVATN	0.5850 (26) P=0.002	0.5990 (26) P=0.001	1.0000 (26) P=*****	0.5193 (26) P=0.007	0.7281 (26) P=0.000	0.6676 (26) P=0.000	0.5581 (26) P=0.003
FOREST	0.2484 (26) P=0.221	0.2811 (26) P=0.164	0.5193 (26) P=0.007	1.0000 (26) P=*****	0.5966 (26) P=0.001	0.3715 (26) P=0.062	0.5340 (26) P=0.005
RUNOFF	0.5582 (26) P=0.003	0.5333 (26) P=0.005	0.7281 (26) P=0.000	0.5966 (26) P=0.001	1.0000 (26) P=*****	0.8392 (26) P=0.000	0.6992 (26) P=0.000
RAINFALL	0.6713 (26) P=0.000	0.6302 (26) P=0.001	0.6676 (26) P=0.000	0.3715 (26) P=0.062	0.8392 (26) P=0.000	1.0000 (26) P=*****	0.6651 (26) P=0.000
SLOPE	0.4147 (26) P=0.035	0.3708 (26) P=0.062	0.5581 (26) P=0.003	0.5340 (26) P=0.005	0.6992 (26) P=0.000	0.6651 (26) P=0.000	1.0000 (26) P=*****

(COEFFICIENT / (CASES) / SIGNIFICANCE)

with a large number of classes, or may even be derived from variables measured by enumeration or metrication. For the higher-level measures, we may wish to decrease the emphasis on the absolute values and differences of the scores and substitute instead their simple rank order.

Spearman's rank-order correlation provides a measure very similar to Pearson's correlation coefficient, with the same range of values and an identical interpretation. But it is based on the differences of ranks between the two variables rather than the values of the scores themselves and their covariance. The rank-order correlation coefficient is defined as:

$$r_s = 1 - \frac{6 \sum_{i=1}^{n} d_i^2}{n(n^2 - 1)} \tag{9.52}$$

where

d_i = the difference in rank between pairs of scores

We can illustrate the process of calculating r_s by taking the same example of Rainfall and Runoff, but substituting rank-order scores for the original scores. The scores for each of the two variables (see Table 9.6) were ranked from 1 to 26 (the sample size), from largest (rank 1) to smallest (rank 26). Tied ranks were assigned the arithmetic mean of the ranks they would have received. The data are arrayed in Table 9.13 in rank order for the Rainfall variable. Therefore,

$$\begin{aligned} r_s &= 1 - \frac{6(602)}{26(26^2 - 1)} \\ &= 1 - .2058 \\ &= .7924 \end{aligned}$$

This compares with the Pearson linear correlation coefficient of $r = .7043$.

9.5 ASSESSMENT OF THE SIMPLE LINEAR REGRESSION MODEL

There are two aspects to assessing the value of a simple linear model for a particular data set. First, it is important to assess whether the assumptions of the model have been violated. Second, with sample data it is critical to assess the significance of the estimated coefficients.

Assumptions of the Simple Linear Model

In Section 9.3 we outlined the derivation of the simple linear regression equation. A number of assumptions were mentioned and more discussion of these assumptions is now needed. However, a full discussion of all these, and some other, assumptions will not take place until Chapter 11 after the

TABLE 9.13 Rank-Order Correlation Coefficient for Rainfall and Runoff

Observation Number (i)	Rank-Order Rainfall	Rank-Order Runoff	d_i	d_i^2
2	1	2	−1.0	1.00
3	2	1	1.0	1.00
13	3	5	−2.0	4.00
4	4	6	−2.0	4.00
17	5	4	1.0	1.00
7	6	11	−5.0	25.00
15	7	7	.0	.00
12	8	9	−1.0	1.00
16	9	10	−1.0	1.00
1	10	3	7.0	49.00
26	11	17.5	−6.5	42.45
14	12	14	−2.0	4.00
11	13	17.5	−4.5	20.25
23	14	16	−2.0	4.00
25	15	19	−4.0	16.00
10	16	12	4.0	16.00
5	17.5	25	−7.5	56.25
20	17.5	24	−6.5	42.25
8	19	20	−1.0	1.00
21	20	15	5.0	25.00
18	21	8	13.0	169.00
19	22	13	9.0	81.00
24	23	23	.0	.00
9	24	26	−2.0	4.00
6	25	22	3.0	9.00
22	26	21	5.0	25.00
TOTAL			.0	602.00

simple linear regression model is expanded into the multiple linear regression model.

First, we assumed that the values of the independent variable X were measured *without error*. This comes from the basic properties of the model—the simple linear model is structured to analyze the *response* of one variable (Y) under fixed values for another variable (X). These fixed values of X are either controlled, or measured without error. Under experimental conditions, we assign values to a variable X and then measure the response to these assigned values of X by another variable Y. For the same value of X we would presumably obtain slightly varying values for Y—the differences of these values for Y are incorporated in the error term. Thus, the only "error" involved in the situation outlined above is the error in measurement of the Y values. This experimental situation rarely occurs in the types of analyses that concern geographers. However, before examining the effects of modifications to the fixed-X model, we must outline further assumptions of this simpler situation.

Second, for a given value of X the associated Y values are *normally* and *independently* distributed about a mean μ_{yx} with a variance $\sigma^2_{y|x}$. This is equivalent to assuming that the ϵ_i of Equation 9.16 are normally and independently distributed with a mean of 0. In addition, the variances $\sigma^2_{y|x}$ of the Y values for each X are identical for all X and can be denoted by σ^2_E. We can summarize these two parts of the assumption by restating the statistical equation in the form:

$$Y_{ij} = \beta_0 + \beta_1 X_i + \epsilon_{ij} \qquad (9.53)$$

where

$i = 1, \ldots, k$ (the number of different X values)
$j = 1, \ldots, n_i$ (the number of values of Y associated with the ith value of X)
and the ϵ_{ij} are normally and independently distributed with mean zero and variance σ^2_E. These two parts of the assumption have now been reduced to assumptions about the error term regardless of the values of the X values, thus Equation 9.53 can be written as it was originally presented:

$$Y_i = \beta_0 + \beta_1 X_i + \epsilon_i$$

with the same assumptions as above.

Given only the two assumptions outlined, the failure of the simple linear model to fit the observations is a function solely of the measurement error (assuming the model is properly specified). The mean square due to error (the unexplained variance) can then be used as the estimate of σ^2_E. But this estimation might be in error if the model is inadequate. Thus, there is a third assumption: the relationship between the two variables is a *linear* relation. If we are not sure of the relationship, then our error is made up of two parts—the error of the measurement of the Y values and the error due to the lack of fit of the linear model. Tests are available to assess the lack of fit of the model and separate this error from the experimental error.

The first assumption was concerned with measurement error and the simple linear model based on fixed-X values. Many geographic problems are not concerned with the assessment of the response of a Y variable to a set of selected X values, but simply with the varying value of Y as X varies. Most commonly, the X and Y values have been obtained from a random sample. The assumption of fixed and accurate values of X, therefore, needs to be relaxed. What effect will this relaxation have on the assumptions of the simple linear model? This question is very difficult to answer, but we can use the linear model successfully if we make one further assumption and restate the first assumption. This *fourth* assumption is that the random pairs of X and Y values come from a *bivariate normal distribution.* This assumption is difficult to test, but if both the X and Y variables have approximate normal distributions, the linear model can be used. Our restated first assumption becomes "that there are no errors of measurement in both X and Y." If there are errors of measurement in Y then the error term could be overestimated; if there are errors of measurement in X then the slope of the regression line (b_1) could be underestimated; if there are errors of measurement in both X and Y the consequences are not easily determined. We are considering not only measurement or calculation errors, but errors in the structure of the variable. Is the actual measure used for either the X or the Y variable really measuring what has been hypothesized in the underlying functional relationship?

If care is taken in setting up the analysis, the linear regression model is a very powerful and versatile technique. However, the assumptions involved must be isolated and tested before any tests of the significance of the results can be performed. We discuss the testing of these various assumptions when the linear model is expanded to its multivariate form. The reason for doing this is that the simple linear model is the simplest form of the multiple linear regression model. The assumptions for the latter include all the assumptions of the simple linear model plus some additional ones. Instead, we outline inferential tests for sampled data and comment on a test for the linearity assumption.

Confidence Intervals and Hypothesis Tests

In the regression analysis, we have estimated values for the parameters β_0 and β_1 (i.e., b_0 and b_1). For a given X_i, we have predicted a value of Y_i (i.e., $\hat{Y}_i$). In a sample from a population, we have an estimate (r) of the true population linear correlation that we call ρ (Greek lowercase *rho*). There are many circumstances in which we wish to place confidence intervals around these point estimates, or alternatively, to test their values against some hypothesized values. We will outline the general structure of a number of tests (without illustrating their derivation) and provide a few comments on each.

The key to fitting confidence intervals to point estimates and testing their values is the standard error of the point estimate. Point estimates and their estimated standard errors for six population parameters are listed in Table 9.14. Note that all the standard errors mentioned are functions of the standard error of estimate S_E (remember that r is also a function of S_E). Using these standard errors, the major confidence interval constructions are outlined in Table 9.15 and hypothesis tests in Table 9.16.

Of all these testing structures, only the confidence interval for β_1 (and to a lesser extent β_0) are frequently reported. The hypothesis test of $\beta_1 = 0$ (or alternatively, and identically, of $\rho = 0$) is in most circumstances, the only hypothesis test carried out. These will be illustrated in the next section containing a worked example.

There is one point that we need to make clear, even if it introduces the risk of overemphasizing the obvious. Different texts often present the basic

TABLE 9.14 Estimated Standard Errors

Population Parameter	Point Estimate	Estimated Standard Error
Error term, ϵ_i	e_i	$S_E = \sqrt{\frac{\Sigma(Y_i - \hat{Y}_i)^2}{n - 2}} = \sqrt{\text{MSE}}$
Slope, β_1	b_1	$S_{b_1} = \sqrt{\frac{S_E^2}{\Sigma(X_i - \overline{X})^2}} = \sqrt{\frac{S_E^2}{(n - 1)s_X^2}}$
Y intercept, β_0	b_0	$S_{b_0} = \sqrt{S_E^2 \left[\frac{\Sigma X_i^2}{n\Sigma(X_i - \overline{X})^2}\right]}$
Y, when $\hat{Y}_k$ is the predicted value of Y for a given X_k	$\hat{Y}_k$	$S_{\hat{y}_k} = \sqrt{S_E^2 \left[\frac{1}{n} + \frac{(X_k - \overline{X})^2}{\Sigma(X_i - \overline{X})^2}\right]}$
Correlation coefficient, ρ	r	$S_r = \sqrt{\frac{1 - r^2}{n - 2}}$
Transformed correlation coefficient, r $Z_r = \frac{\log_e(1 + r) - \log_e(1 - r)}{2}$	Z_r	$S_{z_r} = \frac{1}{\sqrt{n - 3}}$
Rank-order correlation coefficient, ρ	r_s	$S_{r_s} = \frac{1}{\sqrt{n - 1}}$

TABLE 9.15 Confidence Interval Structures

Slope	$b_1 \pm S_{b_1}\left[t, p = \dfrac{1 + Pr}{2}, df = n - 2\right]$
Y Intercept	$b_0 \pm S_{b_0}\left[t, p = \dfrac{1 + Pr}{2}, df = n - 2\right]$
Predicted Y_k for given X_k	$\hat{Y}_k \pm S_{\hat{Y}_k}\left[t, p = \dfrac{1 + Pr}{2}, df = n - 2\right]$
Correlation coefficient	$Z_r \pm S_{z_r}\left[Z, p = \dfrac{1 + Pr}{2}\right]$

TABLE 9.16 Some Hypothesis Test Structures

Hypothesis	Test Statistic	Rejection Region
Slope Ho: $\beta_1 = 0$	$t = \dfrac{b_1}{S_{b_1}}$	$[t, p = 1 - \alpha/2, df = n - 2]$
Slope Ho: $\beta_1 = 0$	$F = \dfrac{\text{MSR}}{\text{MSE}}$	$[F, p = 1 - \alpha, df = 1, n - 2]$
Y Intercept versus hypothesized β_0 Ho: $\beta_0 = \beta_k$	$t = \dfrac{b_0 - \beta_k}{S_{b_0}}$	$[t, p = 1 - \alpha/2, df = n - 2]$
Slope versus hypothesized β_1 Ho: $\beta_1 = \beta_k$	$t = \dfrac{b_1 - \beta_k}{S_{b_1}}$	$[t, p = 1 - \alpha/2, df = n - 2]$
Correlation coefficient Ho: $\rho = 0$	$t = \dfrac{r}{S_r} = \dfrac{r\sqrt{n - 2}}{\sqrt{1 - r^2}}$	$[t, p = 1 - \alpha/2, df = n - 2]$
Correlation coefficient Ho: $\rho = 0$	$F = \dfrac{r^2}{S_r^2} = \dfrac{r^2(n - 2)}{1 - r^2}$	$[F, p = 1 - \alpha, df = 1, n - 2]$
Correlation coefficient versus hypothesized value Ho: $\rho = \rho_k$ (i.e., $Z_r = Z_{p_k}$)	$Z = \dfrac{Z_r - Z_{p_k}}{S_{z_r}}$	$[Z, p = 1 - \alpha/2]$
Rank-order correlation coefficient Ho: $\rho = 0$	$Z = \dfrac{r}{S_{r_s}}$	$[Z, p = 1 - \alpha/2]$

test for the linear relationship in various ways—unfortunately, these often appear to be completely different. In fact, they are all identical. The basic test for linearity (versus the 'constant' model) can be constructed as:

1. *Ho:* $\beta_1 = 0$ using a t-distribution structure.
2. *Ho:* $\beta_1 = 0$ using an F-distribution structure.
3. *Ho:* $\rho = 0$ using a t-distribution structure.
4. *Ho:* $\rho = 0$ using an F-distribution structure.

The following statistics are used in these tests, expressed in a variety of ways (obtained from Table 9.16)

$$t = \frac{b_1}{S_{b_1}} = \frac{r}{S_r} = \frac{r\sqrt{n - 2}}{\sqrt{1 - r^2}} \tag{9.54}$$

$$F = \frac{\text{MSR}}{\text{MSE}} = \frac{r^2}{S_r^2} = \frac{r^2\,(n - 2)}{1 - r^2} \tag{9.55}$$

and, of course, F with 1 and $n - 2$ degrees of freedom (at $p = 1 - \alpha$ level) $= (t)^2$ with $n = 2$ degrees of freedom (at $p = 1 - \alpha/2$ level). In hand calculations, the last form of t or F is the easiest to derive, and with t tables being the most versatile, the form

$$t = \frac{r\sqrt{n - 2}}{\sqrt{1 - r^2}}$$

is the preferred use. However, when we move on to multiple regression the F statistic will be required because there will be more than one parameter estimated in the denominator. The ANOVA table format proves to be much more versatile for testing a variety of structures, as will be illustrated shortly.

A Worked Example

The example we have been using throughout this chapter has been the assessment of the relationship between Runoff (Y) and Rainfall (X). Preliminary calculations (Table 9.6) gave us the following summations:

$$\Sigma X_i = 41{,}746$$

$$\Sigma Y_i = 17{,}945$$

$$\Sigma X_i^2 = 68{,}364{,}200 = \text{variability}_x$$

$$\Sigma Y_i^2 = 15{,}170{,}077 = \text{variability}_y$$

$$\Sigma X_i Y_i = 30{,}431{,}572 = \text{covariability}_{xy}$$

From this we derived

$$\Sigma X_i^2 - \frac{(\Sigma X_i)^2}{n} = 1{,}336{,}180 = \text{variation}_x$$

$$\Sigma Y_i^2 - \frac{(\Sigma Y_i)^2}{n} = 2{,}784{,}576 = \text{variation}_y$$

$$\Sigma X_i Y_i - \frac{\Sigma X_i \Sigma Y_i}{n} = 1{,}618{,}804 = \text{covariation}_{xy}$$

Calculating the regression equation

$$b_1 = \frac{\Sigma X_i Y_i - \dfrac{\Sigma X_i \Sigma Y_i}{n}}{\Sigma X_i^2 - \dfrac{(\Sigma X_i)^2}{n}} = \frac{\text{covariation}_{xy}}{\text{variation}_x}$$

n = 26

$$= \frac{1{,}618{,}804}{1{,}336{,}180} = 1.212$$

and

$$b_0 = \overline{Y} - b_1\overline{X} = \frac{17{,}945}{26} - 1.212\,\frac{(41{,}746)}{26}$$

$$= 690 - 1.212(1606)$$

$$= -1256$$

yields

$$\hat{Y}_i = -1256 + 1.212X_i$$

Calculating the correlation coefficient:

$$r = \frac{\Sigma X_i Y_i - \dfrac{\Sigma X_i \Sigma Y_i}{n}}{\sqrt{\left(\Sigma X_i^2 - \dfrac{(\Sigma X_i)^2}{n}\right)}\sqrt{\left(\Sigma Y_i^2 - \dfrac{(\Sigma Y_i)^2}{n}\right)}} = \frac{\text{covariation}_{xy}}{\sqrt{\text{variation}_x}\sqrt{\text{variation}_y}}$$

$$= \frac{1{,}618{,}804}{\sqrt{1{,}336{,}180}\sqrt{2{,}784{,}576}}$$

$$= .8392$$

and

$$r^2 = (.8392)^2 = .7043$$

For convenience, we construct an abbreviated ANOVA table for the least-squares solution, although note that for the basic tests we need not carry

TABLE 9.17 Abbreviated ANOVA Table for Worked Example

Source of Variation	SS[a]	df	MS
Regression (SSR)	r^2(SST) = 1,961,207	1	MSR = 1,961,207
Error (SSE)	$(1 - r^2)$(SST) = 823,369	24	MSE = S_E^2 = 34,307
TOTAL (SST)	Variation = 2,784,576	25	

[a]Note that r^2(SST) when SST is calculated by hand from Table 9.6 yields 1,962,598, which is slightly greater than the computer results above.

out this step. From the above calculations, we construct Table 9.17 using the relationships noted in Section 9.4.

$$\text{The standard error of estimate} = \sqrt{S_E^2} = \sqrt{34307} = 185.22 \text{ mm}$$

To find the confidence intervals for b_1 and b_0, we find

$$S_{b_1} = \sqrt{\frac{S_E^2}{\text{variation}_x}} = \sqrt{\frac{34,307}{1,336,180}} = 0.160$$

$$S_{b_0} = \sqrt{S_E^2 \frac{\text{variability}_x}{n(\text{variation}_x)}} = \sqrt{34,307 \frac{68,364,200}{26(1,336,180)}}$$

$$= 259.828$$

At the 95 percent confidence level with $[t, p = .975, df = 24] = 2.064$, for the slope coefficient β_1, we get

$$1.212 \pm .160(2.064)$$

$$1.212 \pm .330$$

for the intercept coefficient β_0, we get

$$-1256 \pm 259.828(2.064)$$

$$-1256 \pm 536$$

To test the slope coefficient β_1 using the null hypothesis of $\beta_1 = 0$, we could use the ANOVA table (Table 9.17) to get

$$F = \frac{\text{MSR}}{\text{MSE}} = \frac{1,961,207}{34,307}$$

$$= 57.17$$

which with a tabled value of $[F, p = .95, df = 1,24] = 4.26$, we reject the null hypothesis of $\beta_1 = 0$ and conclude that the slope coefficient β_1 is significantly different from 0.

Alternatively, we could use the correlation coefficient (without bothering to construct an ANOVA table) and calculate

$$t = \frac{r\sqrt{n-2}}{\sqrt{1-r^2}} = \frac{.8392\sqrt{24}}{\sqrt{1-.7043}}$$

$$= 7.56$$

or

$$F = \frac{r^2(n-2)}{1-r^2}$$

$$= 57.17$$

and we note that $(t = 7.56)^2 \cong (F = 57.17)$. For the t statistic of 7.56 compared to the tabled value of $[t, .975, df = 24] = 2.064$, we again reject the null hypothesis (of $\rho = 0$) and conclude that at the 95 percent significance level, the correlation coefficient is significantly different from zero.

These calculations are produced as part of the output of a packaged statistical program. Simple linear regression results from SPSS are included in SCATTERGRAM (see Figure 9.9) and the REGRESSION program, designed to produce structures for several independent variables, can also be used. The deck setup (Table 9.18) and the output (Table 9.19) show the values for the Runoff/Rainfall problem we have been examining.

Confidence Intervals Around Predicted *Y* Values

In some circumstances, it is useful to determine confidence intervals around certain predicted Y values. This is especially true when we use the linear model to establish a relationship between Y and X and we then want to use

TABLE 9.18 SPSS Setup for Simple Regression

```
RUN NAME        NORTH(5) - SIMPLE LINEAR REGRESSION RUNOFF V RAINFALL
FILE NAME       NORTH NORTHLAND DRAINAGE BASIN STUDY
VARIABLE LIST   AREA LENGTH ELEVATN FOREST RUNOFF RAINFALL SLOPE
                ROCKTYPE
INPUT MEDIUM    CARD
N OF CASES      26
INPUT FORMAT    FIXED (F8.2,F8.1,6F8.0)
PRINT FORMAT    AREA (2) LENGTH (1)
VAR LABELS      AREA       AREA OF BASIN - KM**2/
                LENGTH     MAXIMUM LENGTH OF BASIN - KM/
                ELEVATN    ELEVATION OF MOUTH OF BASIN - M/
                FOREST     % BASIN FORESTED ESTIMATED FROM TOPO MAP -
                           %/
                RUNOFF     MEAN ANNUAL SPECIFIC DISCHARGE - MM/
                RAINFALL   MEAN ANNUAL RAINFALL NEAR TO BASIN CENTER -
                           MM/
                SLOPE      ANGLE HIGHEST POINT AND BASIN MOUTH -
                           DEGREES/
                ROCKTYPE   PREDOMINANT ROCKTYPE IN BASIN - CODED
REGRESSION      VARIABLES=AREA TO SLOPE/
                REGRESSION=RUNOFF WITH RAINFALL(2) RESID=0
STATISTICS      ALL
READ INPUT DATA
```

TABLE 9.19 Simple Regression Output from Regression Program

```
                                                                                                  REGRESSION LIST   1
DEPENDENT VARIABLE..     RUNOFF      MEAN ANNUAL SPECIFIC DISCHARGE - MM

VARIABLE(S) ENTERED ON STEP NUMBER  1..     RAINFALL    MEAN ANNUAL RAINFALL NEAR TO BASIN CENTE

MULTIPLE R              0.83923                   ANALYSIS OF VARIANCE     DF       SUM OF SQUARES        MEAN SQUARE          F
R SQUARE                0.70431                   REGRESSION                1.      1961207.20236      1961207.20236     57.16633
ADJUSTED R SQUARE       0.69199                   RESIDUAL                 24.       823368.83610        34307.03484
STANDARD ERROR        185.22158

----------------- VARIABLES IN THE EQUATION -----------------        ------------- VARIABLES NOT IN THE EQUATION -------------

VARIABLE           B           BETA      STD ERROR B       F         VARIABLE      BETA IN      PARTIAL    TOLERANCE          F

RAINFALL       1.211516      0.83923      0.16024       57.166
(CONSTANT)    -1255.037

                                                     SUMMARY TABLE

VARIABLE                                          MULTIPLE R  R SQUARE  RSQ CHANGE   SIMPLE R          B                BETA

RAINFALL   MEAN ANNUAL RAINFALL NEAR TO BASIN CENTE  0.83923   0.70431    0.70431     0.83923       1.211516           0.83923
(CONSTANT)                                                                                      -1255.037
```

that modeled relationship to predict Y values for particular interpolated X values. The standard error for Y and its confidence interval structure was outlined in Tables 9.14 and 9.15. We illustrate the procedure as follows using our worked example on Runoff/Rainfall relationships.

To construct a 95 percent confidence interval about the predicted Runoff value when Rainfall $(X_k) = 1400$ mm:

$$\hat{Y} = -1256 + 1.212(1400)$$

$$= 440.8 \text{ mm}$$

From Table 9.14, our estimate of the standard error for that point estimate is

$$S_{\hat{y}_k} = \sqrt{S_E^2\left[\frac{1}{n} + \frac{(X_k - \overline{X})^2}{\Sigma(X_i - \overline{X})^2}\right]}$$

$$= \sqrt{34{,}307\left[\frac{1}{26} + \frac{(1400 - 1606)^2}{1{,}336{,}180}\right]}$$

$$= 49.082$$

Therefore, our confidence interval becomes

$$\hat{Y}_k = 440.8 \pm 49.082(2.064)$$

$$= 441 \pm 101$$

Of considerable visual, and often practical, use is to extend the idea of the confidence interval for one value of X, calculate it for a number of values, and establish *confidence bands* about the prediction equation. These could be graphed to provide a confidence chart for the model, enabling confidence estimates to be read for any value of X. The process is illustrated using the same example. Confidence intervals were calculated for five selected values of X: 1262 (the minimum recorded X), 1400 (a convenient point between the

TABLE 9.20 Selected Prediction Confidence Intervals

Rainfall (X_k)	Predicted Runoff ($\hat{Y}_k$)	Predicted Standard Error ($S_{\hat{Y}_k}$)	95% Confidence Interval	95% Confidence Lower	95% Confidence Upper
1262	273.5	66.014	±136.3	137.2	409.8
1400	441.8	49.082	±101.3	339.5	542.1
1606	690.5	36.325	± 75.0	615.5	765.5
1900	1046.8	59.488	±122.8	924.0	1169.6
2202	1412.8	102.175	±210.9	1201.9	1623.7

minimum value and the mean), 1606 (mean X), 1900 (a convenient point between the mean and the maximum X), and 2202 (maximum X). The calculations are given in Table 9.20 and plotted on a graph (Figure 9.18) containing the fitted regression line, and then joined with smooth lines to define the confidence band. Notice that the confidence interval is smallest at the point of mean X and becomes increasingly wider as we move farther from the mean. The bands are symmetrical in either direction.

Tests for Lack of Fit of the Linear Model

Before completing our discussion of the simple linear regression model, there is one topic that is convenient to introduce at this point. While the tests we have examined were concerned with detecting a significant relationship in the population, they were also tests of whether we could conclude that there was a linear relationship between X and Y compared to no relationship (i.e., $\beta_1 = 0$, or $\rho = 0$) in the population. However, these tests do not tell us if

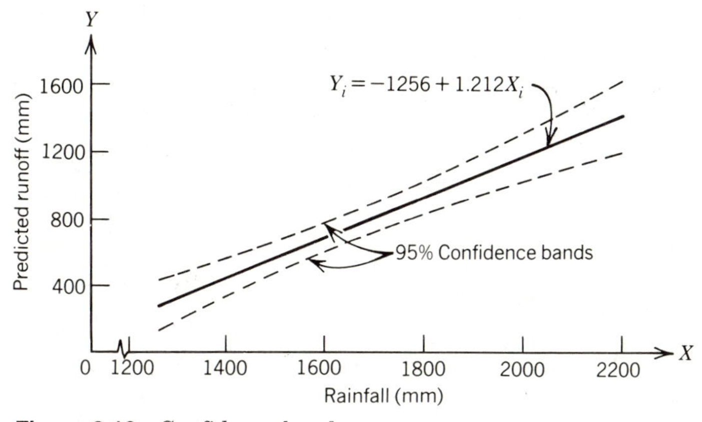

Figure 9.18 Confidence bands.

the linear relationship fits better than other nonlinear models. Even if we have found that the linear model fits significantly, there may be other models that have higher levels of explanation and smaller amounts of residual error. The residual error term in the linear model might therefore partly consist of lack of fit of the linear model. Tests are available to assess the lack of fit of the linear model and to separate this error from the pure error. Two approaches can be used. If it is possible in some way to manipulate the values for the variable X (as might be possible in the fixed-X model), then we can assess the relative contributions of pure error and lack of fit by replicating (repeating) the analysis with the same set of X values. By repetition with the same X values but with different measured Y values for some of the X values, the pure error contribution can be assessed, and the error due to lack of fit is found by subtraction.

An alternative method is to use the one-way analysis of variance technique for testing the difference of means. This is best illustrated with a simple hypothetical example (Figure 9.19) where we observe an obviously nonlinear relationship between Y and X, and where the X variable is fixed and, for convenience, uniformly spaced along the X axis with a similar number of Y observations for each X. The analysis of variance technique was concerned with the sum of the squared deviations from each of the class means. For each class (represented in our example by a single X score), the sum of squares about the class mean will be smaller than the sum of squares around any other number—including, obviously, the sum of squares around the least squares estimated Y points. The closer the class means are to the least-squares line, the closer will be the two sums of squares and the more linear will be the relationship. We can use this as the basis for a test for linearity.

The correlation ratio E^2 (Section 8.3) can be used as a simple measure of association of two variables in the analysis of variance situation, and is

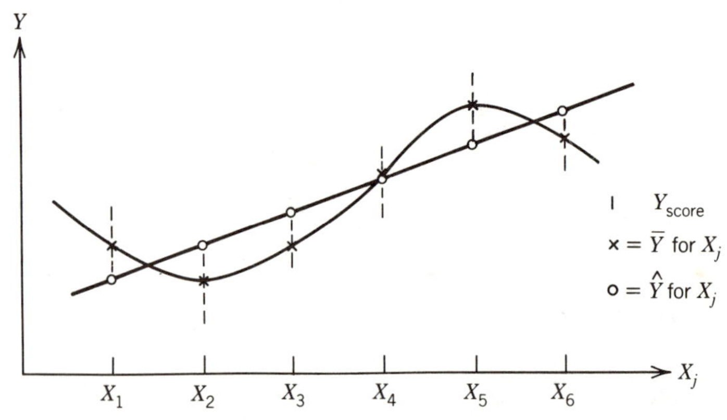

Figure 9.19 Deviations from class means and least-squares estimates.

TABLE 9.21 ANOVA Structure for Testing Nonlinearity

Source of Variability	Sum of Squares	*df*	Mean Square
Explained variation	SSB $= E^2(\text{SST})$	$k - 1$	
By linear model	SSR $= r^2(\text{SST})$	1	
Nonlinear residual	$(E^2 - r^2)(\text{SST})$	$k - 2$	$\dfrac{(E^2 - r^2)(\text{SST})}{k - 2}$
Unexplained variation	SSW $= (1 - E^2)(\text{SST})$	$n - k$	$\dfrac{(1 - E^2)(\text{SST})}{n - k}$
Total variation	SST	$n - 1$	

n = number of observations.
k = number of classes.

defined as

$$E^2 = \frac{\text{between SS}}{\text{total SS}} = \frac{\text{SSB}}{\text{SST}}$$

That is, the "explained" sum of squares is a proportion of the total sum of squares. It has a theoretical range of values between 0 (no between sum of squares—the class means are identical) to 1.0 (no within sum of squares—the scores within a class are identical, but the classes differ). This measure is directly analogous to the coefficient of determination (r^2) in the linear model, which was defined as

$$r^2 = \frac{\text{regression SS}}{\text{total SS}} = \frac{\text{SSR}}{\text{SST}}$$

We can use these two measures to assess linearity, or more specifically, the lack of it. The E^2 gives the proportion of variability of Y explained by class differences (without recourse to any direction of relationship); r^2 gives the proportion explained by the linear model; and $E^2 - r^2$ gives the proportion of variation not explained by the linear relationship. The testing structure can best be illustrated by an expanded ANOVA table (Table 9.21), and by providing the SS expressions in a practical calculation form. Of particular interest is the "nonlinear residual"—if this is insignificant, then the relationship between X and Y is assumed to be approximately linear. The ratio of the two relevant mean squares provides an F statistic that can be used to test the significance of the nonlinear contribution. The SST cancels out in the F ratio, and the F statistic simply becomes

$$F = \frac{(E^2 - r^2)(n - k)}{(1 - E^2)(k - 2)} \tag{9.56}$$

and this can be tested against a tabled F of $[F, p = 1 - \alpha, df = (k - 2), (n - k)]$.

TABLE 9.22 ANOVA for Runoff by Five Classes

Source of Variability	Sum of Squares	*df*	Mean Square
Between classes	SSB = 2,058,594	4	514,648.5
Within classes	SSW = 725,983	21	34,570.6
TOTAL	SST = 2,784,577	25	

With a fixed-X model, the derivation of classes for testing the linear fit is straightforward. However, for the random-X model, classes would need to be created. We illustrate the process by testing for linearity in the Runoff/Rainfall example. Five classes were created for the 26 observations by ranking the Rainfall (X) data and grouping the associated Y scores with five scores per class except for six for the central class. An analysis of variance for difference of means was carried out on these five classes of Y scores with the results presented in Table 9.22. Using these results,

$$E^2 = \frac{2{,}058{,}594}{2{,}784{,}576} = .7393$$

These can then be combined with the ANOVA from the regression and the r^2 data. For illustrative purposes, the results are presented in the form of the combined ANOVA (Table 9.23), although we could simply derive this from Equation 9.56.

$$F = \frac{(.7393 - .7043)(21)}{(1 - .7393)(3)} = .94$$

With a tabled value of $[F, p = .95, df = 3, 21] = 3.07$ we fail to reject the null hypothesis of no contribution of the nonlinear term and can assume that there is no significant nonlinear contribution to the model.

TABLE 9.23 Results of Nonlinearity Test

Source of Variability	Sum of Squares	*df*	Mean Square
Explained variation	2,058,594	4	
By linear model	1,961,207	1	
Nonlinear residual	97,387	3	32,462.3
Unexplained variation	725,983	21	34,570.6
Total variation	2,784,577	25	

References and Readings

1. The Simple Linear Regression Model

Ezekiel, M. and K. A. Fox (1959) *Methods of Correlation and Regression Analysis,* Wiley: New York.

Johnston, J. (1972) *Econometric Methods,* Second Edition, McGraw-Hill: New York.

Poole, M. A. and P. N. O'Farrell (1971) "The assumptions of the linear regression model," *Institute of British Geographers, Transactions* 52:145–158.

2. Applications

Clements, D. W. (1978) "Utility of linear models in retail geography," *Economic Geography* 54:17–25.

Mark, D. M. and T. K. Peucker (1978) "Regression analysis and geographic models," *Canadian Geographer* 22:51–64.

Pinch, S. P. (1978) "Patterns of local housing authority allocation in greater London between 1966 and 1973," *Transactions, Institute of British Geographers* (N.S.) 3:35–54.

Snedecor, G. W. and W. G. Cochran (1966) *Statistical Methods,* Iowa State University Press: Ames.

Taylor, P. J. (1975) *Distance Decay Models in Spatial Interaction,* Concepts and Techniques in Modern Geography (CATMOG) 2: Geo Abstracts: Norwich, England.

CHAPTER 10
The General Linear Model—Multiple Regression

The simple linear model allowed us to assess the influence of one variable on another. However, almost all variables are influenced by more than one variable. Thus, the dollar amount of sales in the Central Business District is a function not only of the size of the city, but of the number of competing shopping centers, the accessibility to downtown, and a host of other factors. Or, specific runoff is not only a product of the amount of rainfall but is related to the nature of the groundcover and the soil structure, for example. The introduction of additional variables reduces *stochastic error*—the error that arises because of the inherent irreproducibility of physical and social phenomena. Stochastic error is defined as the effect on Y of many omitted variables, each with an individually small effect.

The simple linear model needs to be expanded to allow for the consideration of additional explanatory variables. Such an extension is not difficult, but will require us to structure our analyses with greater care, because the set of assumptions in using the linear model must also be extended. In the next two chapters, we focus on the *multiple regression model*—the application of the general linear model to the situation in which the dependent variable is a function of two or more independent variables measured at metric or enumerated scales.

10.1 THE MULTIPLE REGRESSION MODEL

In multiple regression, the dependent variable Y is regressed against two or more independent variables. To do this, we extend the mathematical model used for simple regression by adding a term for each of the additional variables. However, we also make a change in symbolism from that used in the last chapter by using the idea of *expected values* introduced in Chapter 4. With the expected value format, we can combine most of the characteristics of the mathematical and statistical model used in Chapter 9 into one structure, just adding in the error term (ϵ_i) when dealing with individual sets of observations. This change will also bring us in line with most texts on multiple regression. Our mathematical model for the multiple regression situation thus becomes

$$\begin{aligned} E(Y_i) &= \beta_0 + \beta_1 X_{i1} + \beta_2 X_{i2} + \ldots + \beta_k X_{ik} \\ &= \beta_0 + \sum_{j=1}^{k} \beta_j X_{ij} \end{aligned} \tag{10.1}$$

where

$$k = \text{the number of independent variables}$$

and

$$i = \text{the element number, as before}$$

When $k = 1$, the equation reverts back to its simple regression form, although with the extra subscript (j) to allow for the multiple regression expansion.

We illustrate the development of the multiple regression model by using examples where $k = 2$. The extension to situations with greater numbers of independent variables is operationally simple but difficult to illustrate or to provide worked examples. Assume we are interested in variations in Wheat Yield (Y) in response to Fertilizer application (X_1) and annual Rain-

TABLE 10.1 Data for Wheat Yield Example

Y_i (Wheat Yield)	X_{i1} (Fertilizer)	X_{i2} (Rainfall)
45	50	30
50	100	33
55	150	34
70	200	34
75	250	31
75	300	29
85	350	33

fall (X_2) for a fictitious sample of $n = 7$ elements (Table 10.1). The model describing expected yield is

$$E(Y_i) = \beta_0 + \beta_1 X_{i1} + \beta_2 X_{i2}$$

Diagrammatic Representation

One useful way to understand the relationships among the three variables is to graph the relationship space for Wheat Yield, Rainfall, and Fertilizer application. Whereas the simple regression relationship can be expressed as a two-dimensional graph (Chapter 9), with one dimension required for each variable, the multiple regression model can be illustrated only for two independent variables, that is, for three variables in total. Although we fit a line to a scatter of points for two variables, we fit a plane (a plane when $k = 2$—more generally, an n-dimensional hypersurface) to a scatter of points in multiple regression (Figure 10.1). The expected yield ($E(Y_i)$) is the point on the plane directly above the X_{i1} and X_{i2} position for Rainfall and Fertilizer. In the example we presented, the actual Y_i is "above" the plane. That is, the actual value is greater than the predicted value. The difference, or deviation, between the actual and predicted values is the error term or residual. Thus, the full equation is written

$$Y_i = \beta_0 + \beta_1 X_{i1} + \beta_2 X_{i2} + \epsilon_i \tag{10.2}$$

The slope of the plane is fully described by two parameters: β_1, which indicates the slope in the X_1 direction, and β_2, which indicates the slope in the X_2 direction. To be more precise, we say that the β's represent the slopes of a set of intersecting lines that are perpendicular to the X_2 and X_1 axes (Figure 10.2) and that jointly define the regression plane. The plane is a least-squares plane and minimizes the sum of the squared deviations in the vertical or Y dimension. Each slope coefficient represents the relationship between the dependent and respective independent variables, but the re-

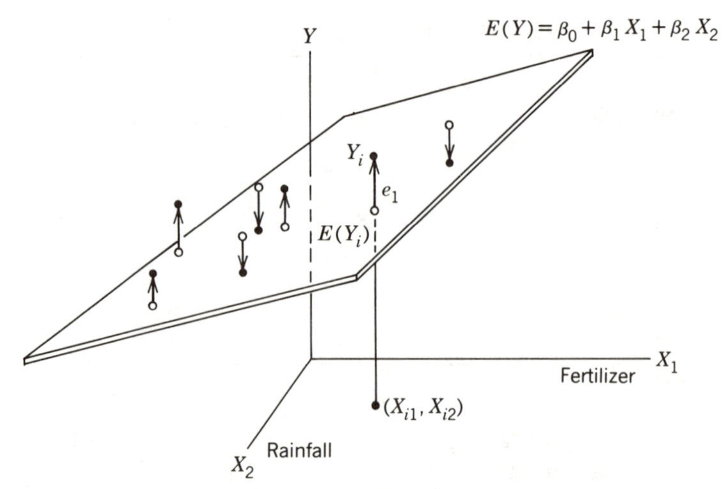

Figure 10.1 Relationship space for two independent variable example. Source: Wonnacott and Wonnacott, 1970, p. 256, by permission of John Wiley & Sons, Inc., Publishers.

lationship is not as straightforward as the interpretation of the single-slope coefficient in simple regression. It requires the introduction of the notion of partial regression coefficients.

Least-Squares Estimation

Fitting a plane to a scatter of points using the least-squares criterion (the three variable equivalent to fitting a least-squares line for two variables), involves selecting estimates of β_0, β_1, and β_2 that minimize the sum of the squared deviations between the observed Y's and the expected Y's. The procedure then is to minimize the expression

$$\begin{aligned} S &= \sum_{i=1}^{n} (Y_i - \hat{Y}_i)^2 \\ &= \sum_{i=1}^{n} (Y_i - \hat{\beta}_0 - \hat{\beta}_1 X_{i1} - \hat{\beta}_2 X_{i2})^2 = \sum_{i=1}^{n} \epsilon_i^2 \end{aligned} \tag{10.3}$$

where $\hat{\beta}_0$, $\hat{\beta}_1$, and $\hat{\beta}_2$ are our estimates of β_0, β_1, and β_2. The minimum can be found by the method outlined in Section 9.3 for the two parameter case (β_0 and β_1), which involved setting the partial derivatives of the function with respect to the unknowns equal to zero. In the present situation, it will

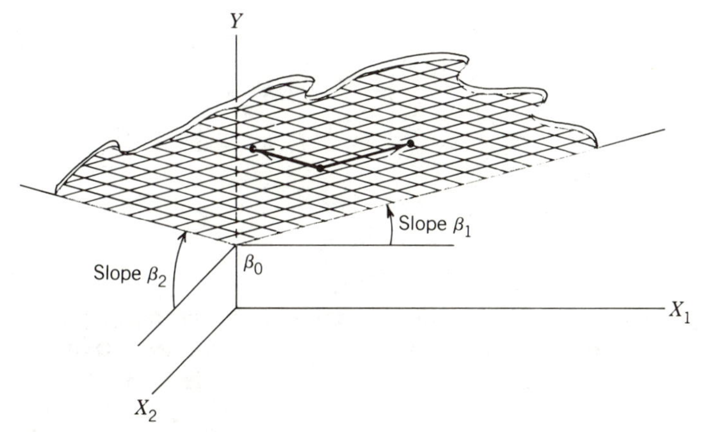

Figure 10.2 Slopes of regression plane. Source: Wonnacott and Wonnacott, 1972, p. 322, by permission of John Wiley and Sons, Inc., Publishers.

yield a set of three normal equations. We will not go through this derivation again (the method is identical) but will take up the topic later in this chapter where we use some matrix concepts that will simplify the whole task. Instead, we present the three normal equations (for the three unknowns β_0, β_1, β_2) and illustrate the solution of these equations to give estimates by a worked example.

Following from the terminology used in Chapter 9 we call our three estimates b_0, b_1, and b_2 for $\hat{\beta}_0$, $\hat{\beta}_1$, and $\hat{\beta}_2$. These are the symbols used when our estimates are derived from samples. In addition, for ease of presentation, we will ignore the i subscript for the variables and assume that the summations are all over the n cases. The normal equations derived from the minimization process will be

$$\begin{aligned} b_0 n + b_1 \sum X_1 + b_2 \sum X_2 &= \sum Y \\ b_0 \sum X_1 + b_1 \sum X_1^2 + b_2 \sum X_1 X_2 &= \sum X_1 Y \\ b_0 \sum X_2 + b_1 \sum X_2 X_1 + b_2 \sum X_2^2 &= \sum X_2 Y \end{aligned} \qquad (10.4)$$

There is one further possible simplification to the structure, a simplification that until now we have made little use of. This is because in some circumstances it tends to confuse rather than to simplify. The change involves the use of *deviations* from the mean for a variable rather than the raw scores themselves. As we mentioned in Chapter 3, most texts use the symbol lowercase x to indicate scores for the variable X when in deviation-from-the-mean form. If we use x_{i1} for $(X_{i1} - \overline{X}_1)$ and x_{i2} for $(X_{i2} - \overline{X}_2)$, our three normal equations reduce to two, and a solution to the unknown $b_0 = \hat{\beta}_0$ is found immediately.

$$b_0 = \frac{\Sigma Y}{n} = \overline{Y}$$

$$b_1 \sum x_1^2 + b_2 \sum x_1 x_2 = \sum x_1 Y \tag{10.5}$$

$$b_1 \sum x_1 x_2 + b_2 \sum x_2^2 = \sum x_2 Y$$

A simple worked example for the solution of these normal equations for the Wheat Yield example is given in Table 10.2. For comparative purposes, an SPSS analysis of the same data is reported in Table 10.3.

10.2 THE COEFFICIENTS OF THE MODEL

In contrast to the simple linear regression model that allowed us to describe, display, and interpret the various coefficients making up the model with relative ease, our task in multiple regression and correlation is a much larger and more difficult one. In this section, we examine each of the various coefficients used for regression and correlation descriptions and attempt to show how each can be interpreted. We also show how one is related to the others. We limit our examples to the two-independent-variable case.

Partial Regression Coefficients

The relationships between the b coefficients and the slope of the regression surface has already been illustrated. An understanding of this relationship is important, as ultimately this leads to an interpretation of the b coefficients. Perhaps we can best portray the role of the b coefficients by rewriting the multiple linear regression equation using an expanded system of subscripting.

$$Y = b_0 + b_{Y1.2}X_1 + b_{Y2.1}X_2 + e \tag{10.6}$$

The "dot" subscripting system proves useful whenever we want to specify more clearly the nature of the relationship. To the left of the dot are the subscripts of the variables related by the coefficient; to the right are any variables that will influence the value of the coefficient but that are not being directly assessed. Thus replacing b_1 by $b_{Y1.2}$ indicates that the b coefficient expresses a relationship between the Y variable and variable (X_1), but its value is influenced by variable (X_2).

The b values (the estimates of the population values $-$ β) in a multiple regression model are called *partial regression coefficients* or *net regression coefficients.* They express the relationship between the dependent and the independent variable while holding the other variables in the regression

TABLE 10.2 Least-Squares Estimates for Multiple Regression of Wheat Yield on Fertilizer and Rainfall

Wheat Yield (Bushels per Acre) Y_i	Fertilizer (lb) X_{i1}	Rainfall (in.) X_{i2}	$x_{i1} = (X_{i1} - \overline{X}_1)$	$x_{i2} = (X_{i2} - \overline{X}_2)$	$x_{i1}Y_i$	$x_{i2}Y_i$	x_{i1}^2	x_{i2}^2	$x_{i1}x_{i2}$
45	50	33	−150	1	−6,750	45	22,500	1	−150
50	100	30	−100	−2	−5,000	−100	10,000	4	200
55	150	34	−50	2	−2,750	110	2,500	4	−100
70	200	34	0	2	0	140	0	4	0
75	250	31	50	−1	3,750	−75	2,500	1	−50
75	300	29	100	−3	7,500	−225	10,000	9	−300
85	350	33	150	1	12,750	85	22,500	1	150
$\Sigma Y_i = 455$ $\overline{Y} = 65$	$\Sigma X_1 = 1{,}400$ $\overline{X}_1 = 200$	$\Sigma X_2 = 224$ $\overline{X}_2 = 32$	0	0	$\Sigma x_{i1}Y_i = 9{,}500$	$\Sigma x_{i2}Y_i = -20$	$\Sigma x_{i1}^2 = 70{,}000$	$\Sigma x_{i2}^2 = 24$	$\Sigma x_{i1}x_{i2} = -250$

Estimating equations 10.5:

$$9500 = 70{,}000b_1 - 250b_2$$
$$-20 = -250b_1 + 24b_2$$

Solution:

$$b_0 = 65$$
$$b_1 = \frac{9292}{67{,}396} = .1379$$
$$b_2 = \frac{3900}{6470} = .6028$$

In terms of the original X_1 and X_2:

$$\hat{Y} = 65 + .1379(X_{i1} - \overline{X}_1) + .6028(X_{i2} - \overline{X}_2)$$
$$= 65 + .1379(X_{i1} - 200) + .6028(X_{i2} - 32)$$
$$= 18.14 + .1379X_1 + .6028X_2$$

TABLE 10.3 SPSS Setup and Run for Wheat Yield, Rainfall, and Fertilizer

```
                   1 FUN NAME        YIELD ON FERTILIZER AND RAINFALL
                   2 VARIABLE LIST   V1.V2.V3
                   3 VAR LABELS      V1 WHEAT YIELD/
                   4                 V2 FERTILIZER/
                   5                 V3 RAINFALL
                   6 INPUT FORMAT    FIXED (F2.0,1X,F3.0,1X,F2.0)

                    ACCORDING TO YOUR INPUT FORMAT, VARIABLES ARE TO BE READ AS FOLLOWS

                    VARIABLE  FORMAT  RECORD     COLUMNS

                    V1        F 2. 0     1       1-    2
                    V2        F 3. 0     1       4-    6
                    V3        F 2. 0     1       8-    9
THE INPUT FORMAT PROVIDES FCR     3 VARIABLES.   3 WILL BE READ
IT PROVIDES FOR     1 RECCRDS ('CARDS') PER CASE.  A MAXIMUM OF     9 'COLUMNS' ARE USED ON A RECORD.

                   7 N OF CASES      7
                   8 REGRESSION      VARIABLES=V1,V2,V3/
                   9                 REGRESSION=V1 WITH V2 TO V3(2)
                  10 OPTIONS         20

* * * * * * * * * * * * * * * * * * * * * *  M U L T I P L E   R E G R E S S I O N  * * * * * * * * * * * * * *   VARIABLE LIST   1
                                                                                                                 REGRESSION LIST 1
DEPENDENT VARIABLE..    V1           WHEAT YIELD
VARIABLE(S) ENTERED ON STEP NUMBER  1..    V3           RAINFALL
                                           V2           FERTILIZER

MULTIPLE R          0.98043                    ANALYSIS OF VARIANCE    DF     SUM OF SQUARES     MEAN SQUARE          F
R SQUARE            0.96125                    REGRESSION              2.       1297.68161        648.84080       49.6070
ADJUSTED R SQUARE   0.94187                    RESIDUAL                4.         52.31839         13.07960
STANDARD ERROR      3.61657

---------------- VARIABLES IN THE EQUATION -----------------        ------------ VARIABLES NOT IN THE EQUATION ------------

VARIABLE          B            BETA      STD ERROR B        F         VARIABLE      BETA IN      PARTIAL   TOLERANCE        F

V3            0.6027821       0.08037      0.75236        0.642
V2            0.1378671       0.99276      0.01393       97.940
(CONSTANT)   18.13756

ALL VARIABLES ARE IN THE EQUATION
```

TABLE 10.4 Data for Family Income Example

Observation Number	Y Family Income ($)	X_1 Years of Education	X_2 Weeks of Unemployment
1	20,900	10.0	2.5
2	24,500	12.0	2.0
3	16,000	7.0	4.0
4	18,000	10.5	4.0
5	16,700	8.0	5.0
6	12,400	6.0	5.5
7	17,000	8.5	5.5
8	15,200	8.5	5.0
9	18,000	10.0	4.0
10	22,500	10.0	3.5
11	24,200	11.0	2.0
12	23,000	12.0	2.0
13	18,500	9.5	2.5
14	21,000	11.5	3.0
15	16,000	8.0	5.0
16	13,000	6.0	5.0
17	17,000	8.5	5.0
18	15,200	7.0	5.0
19	21,000	10.0	2.5
20	22,200	11.0	2.0

TABLE 10.5 SPSS Setup and Run of Simple Regression for Data in Table 10.4

```
                   1 RUN NAME         INCOME ON EDUCATION AND UNEMPLOYMENT
                   2 VARIABLE LIST    V1,V2,V3
                   3 VAR LABELS       V1 INCOME/
                   4                  V2 EDUCATION/
                   5                  V3 UNEMPLOYMENT
                   6 INPUT FORMAT     FIXED(F5.0,F3.1,F2.1)

                    ACCORDING TO YOUR INPUT FORMAT, VARIABLES ARE TO BE READ AS FOLLOWS

                    VARIABLE  FORMAT  RECORD     COLUMNS

                    V1        F 5. 0     1       1-    5
                    V2        F 3. 1     1       6-    8
                    V3        F 2. 1     1       9-   10

THE INPUT FORMAT PROVIDES FOR    3 VARIABLES.    3 WILL BE READ
IT PROVIDES FOR     1 RECORDS ('CARDS') PER CASE.  A MAXIMUM OF    10 'COLUMNS' ARE USED ON A RECORD.

                   7 N CF CASES       20
                   8 REGRESSION       VARIABLES=V1,V2,V3/
                   9                  REGRESSION=V1 WITH V2,RESID=0/
                  10 OPTICNS          11,20
                  11 STATISTICS       4

DEPENDENT VARIABLE..      V1          INCOME

VARIABLE(S) ENTERED ON STEP NUMBER  1..    V2         EDUCATION

MULTIPLE R               0.91945                ANALYSIS OF VARIANCE    DF     SUM OF SQUARES       MEAN SQUARE           F
R SQUARE                 0.84539                REGRESSION              1.   208307556.60377   208307556.60377     98.41833
ADJUSTED R SQUARE        0.83680                RESIDUAL               18.    38097943.39623     2116552.41090
STANDARD ERROR        1454.83759

---------------- VARIABLES IN THE EQUATION ----------------     ------------ VARIABLES NOT IN THE EQUATION ------------

VARIABLE             B          BETA     STD ERROR B      F       VARIABLE     BETA IN    PARTIAL   TOLERANCE        F

V2              1773.208      0.91945    178.73993     98.418
(CONSTANT)      2212.830
```

equation constant. Thus, from this point of view, the b values in the multiple regression equation are different from those in the simple bivariate regression equation. Another way of expressing the meaning of the partial b values is to indicate that each b value represents the amount of change in Y for each unit of change in X_1 while holding the contribution of the other independent variables constant.

It is clear from the foregoing comments that as variables are added to a multiple regression model, the b values will *not usually* remain the same. This result is an important part of regression analysis, perhaps the most important part of the step from bivariate to multivariate analysis.

An example is the best way to clarify the contribution of the partial b values. Consider the relationship

$$\hat{Y} = b_0 + b_1 X_1$$

where Y is Family Income in dollars and X_1 is Years of Education. Using the data in Table 10.4, an SPSS run (Table 10.5) estimates the coefficients of the bivariate relationship as

$$\hat{Y} = 2212.83 + 1773.21 X_1 \qquad r^2 = .85 \tag{10.7}$$

The relationship can be plotted as in Figure 10.3.

Now we wish to increase our explanation by adding an additional variable—Unemployment—the number of weeks unemployed during the last three months. However, we want to add the new variable only after as much explanation as possible has already been established for the first variable. In essence, this requires establishing a relationship between the new variable X_2 and the residuals $(Y - \hat{Y})$ where $\hat{Y}$ has been predicted from the first variable entered in the equation. However, if the new variable (X_2) is in part related to the original variable (X_1) already in the equation, it is nec-

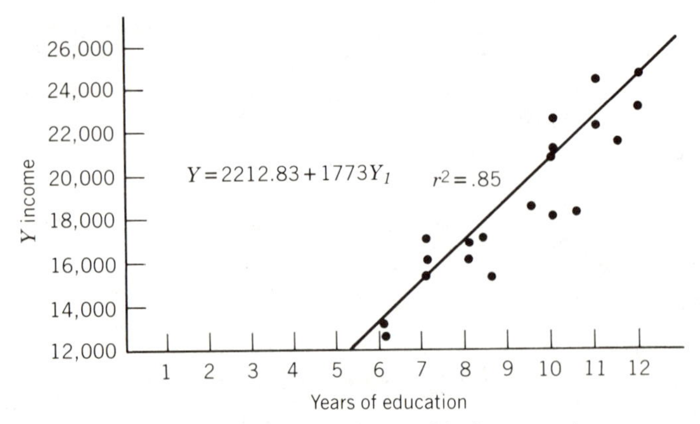

Figure 10.3 The bivariate relationship between income and education.

essary to correct for this effect. To do this, we regress the new variable on the variable already in the equation. Thus, regress X_2 on X_1 and calculate residuals $(X_2 - \hat{X}_2)$. The SPSS analysis is reproduced as Table 10.6 and the results graphed in Figure 10.4. The equation obtained was

$$\hat{X}_2 = 9.1604 - .5849X_1 \qquad r^2 = .68 \tag{10.8}$$

The equation indicates that there is a relationship between X_2 and X_1. We must allow for this in calculating the multiple regression equation. Now regress the residuals $Y - \hat{Y}$ of the dependent variable against the residuals of the new independent variable X_2.

$$\widehat{Y - \hat{Y}} = b_0 + b_1\,(X_2 - \hat{X}_2)$$

TABLE 10.6 SPSS Run for Unemployment and Education

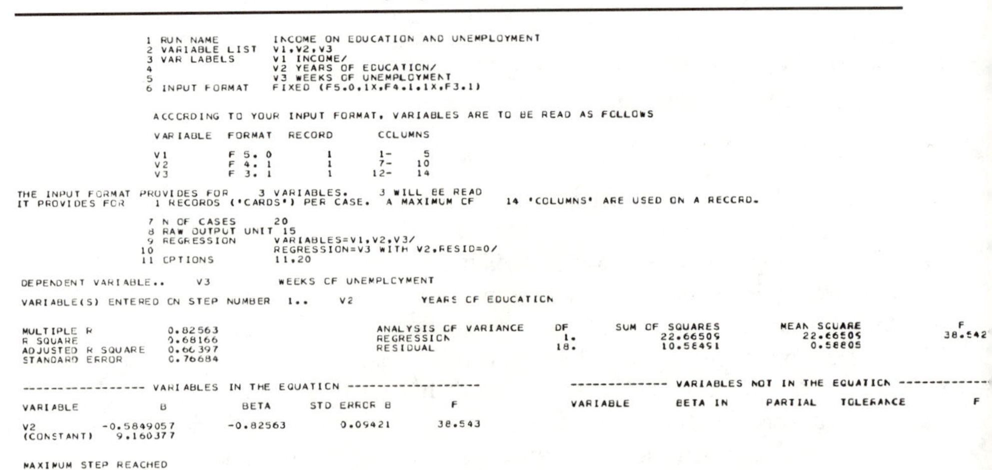

```
              1 RUN NAME         INCOME ON EDUCATION AND UNEMPLOYMENT
              2 VARIABLE LIST    V1,V2,V3
              3 VAR LABELS       V1 INCOME/
              4                  V2 YEARS OF ECUCATICN/
              5                  V3 WEEKS OF UNEMPLOYMENT
              6 INPUT FORMAT     FIXED (F5.0,1X,F4.1,1X,F3.1)

               ACCCRDING TO YOUR INPUT FORMAT, VARIABLES ARE TO BE READ AS FCLLOWS

               VARIABLE   FORMAT   RECORD      CCLUMNS

               V1         F 5. 0      1        1-    5
               V2         F 4. 1      1        7-   10
               V3         F 3. 1      1       12-   14

THE INPUT FORMAT PROVIDES FOR    3 VARIABLES.    3 WILL BE READ
IT PROVIDES FOR    1 RECORDS ('CARDS') PER CASE.  A MAXIMUM CF    14 'COLUMNS' ARE USED ON A RECCRD.

              7 N OF CASES       20
              8 RAW OUTPUT UNIT  15
              9 REGRESSION       VARIABLES=V1,V2,V3/
             10                  REGRESSION=V3 WITH V2,RESID=0/
             11 CPTIONS          11,20

DEPENDENT VARIABLE..    V3          WEEKS CF UNEMPLCYMENT

VARIABLE(S) ENTERED CN STEP NUMBER  1..    V2          YEARS CF EDUCATICN

MULTIPLE R            0.82563          ANALYSIS CF VARIANCE    DF     SUM OF SQUARES     MEAN SCUARE        F
R SQUARE              0.68166          REGRESSICN               1.       22.66509          22.66505      38.542
ADJUSTED R SQUARE     0.66397          RESIDUAL                18.       10.58491           0.58805
STANDARD ERROR        0.76684

---------------- VARIABLES IN THE EQUATICN -----------------     ------------ VARIABLES NOT IN THE EQUATICN ------------

VARIABLE          B          BETA      STD ERRCR B       F         VARIABLE     BETA IN     PARTIAL   TOLERANCE      F

V2           -0.5849057    -0.82563     0.09421      38.543
(CONSTANT)    9.160377

MAXIMUM STEP REACHED
```

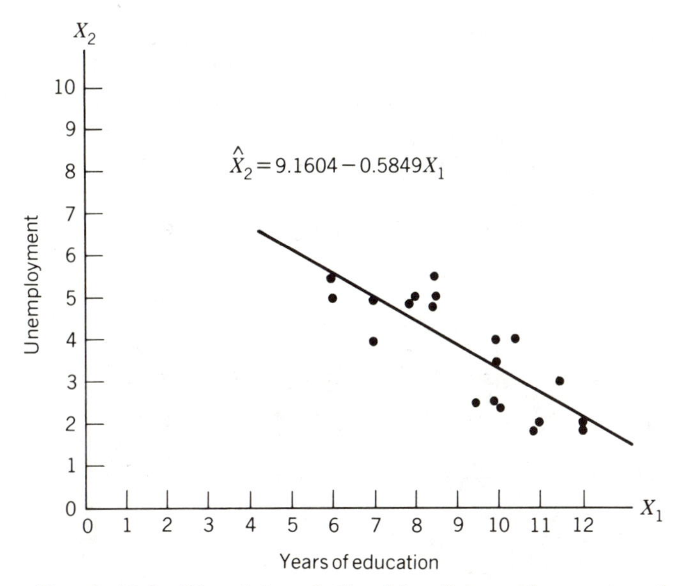

Figure 10.4 Bivariate relationship of unemployment and years of education.

But, because we regress two sets of residuals with means of zero, the regression line will therefore pass through the origin of the graph, and the y intercept is 0 and the b_0 term drops out of the equation. Using our example, we get

$$\widehat{Y - \hat{Y}} = -1101.87\,(X_2 - \hat{X}_2) \tag{10.9}$$

The results are plotted in Figure 10.5.

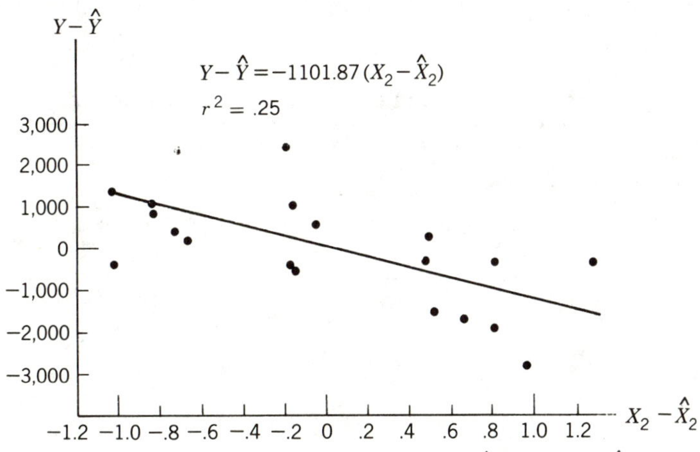

Figure 10.5 Bivariate relationship of $Y - \hat{Y}$ and $X_2 - \hat{X}_2$.

The plot in the foregoing figure is generated from the raw residuals from each of the earlier SPSS runs. This is a useful experience in setting up the SPSS runs, producing the residual set, and plotting the results. Note that the actual results may include a small positive or negative constant that can be ignored, because it derives from small rounding errors in deriving the residual output.

The final step is to substitute the results obtained earlier into Equation 10.9. Thus:

$$\text{for } \hat{Y} \text{ substitute } 2212.83 + 1773.21X_1$$

$$\text{for } \hat{X}_2 \text{ substitute } 9.1604 - .5849X_1$$

Therefore, from $\widehat{(Y - \hat{Y})} = -1101.87(X_2 - \hat{X}_2)$

$$\hat{Y} - (2212.83 + 1773.21X_1) = -1101.87[X_2 - (9.1604 - .5849X_1)]$$

Simplification yields:

$$\hat{Y} = 12306.40 + 1128.73X_1 - 1101.87X_2$$

The actual computer equation using both variables is:

$$\hat{Y} = 12306.39 + 1128.71X_1 - 1101.87X_2 \qquad \text{(Table 10.7)}$$

The coefficients would be the same even if we reversed the operation and solved first for the X_1 coefficient. Each coefficient represents the relationship of the dependent and respective independent variable after accounting for any influence of the other independent variables in the equation.

Although the b values—the partial regression coefficients—indicate the relationship of the respective independent variables with the dependent variable, they do not indicate the relative importance of each independent variable in explaining the variation in the dependent variable. Rather, it is

TABLE 10.7 Regression of Income on Education and Unemployment

```
* * * * * * * * * * * * * * * * * * * * * * * *  M U L T I P L E   R E G R E S S I O N  * * * * * * * * * * * * *   VARIABLE LIST 1
                                                                                                                     REGRESSION LIST 1
DEPENDENT VARIABLE..   V1        INCOME

VARIABLE(S) ENTERED ON STEP NUMBER  1..   V3        WEEKS OF UNEMPLOYMENT
                                          V2        YEARS OF EDUCATION

MULTIPLE R            0.94739                  ANALYSIS OF VARIANCE    DF     SUM OF SQUARES      MEAN SQUARE          F
R SQUARE              0.89754                  REGRESSION              2.     221158914.43850     110579457.21925      74.45960
ADJUSTED R SQUARE     0.88549                  RESIDUAL               17.      25246585.56150       1485093.26832
STANDARD ERROR     1218.64403

---------------- VARIABLES IN THE EQUATION -----------------        ------------ VARIABLES NOT IN THE EQUATION ------------

VARIABLE           B          BETA       STD ERROR B       F         VARIABLE       BETA IN     PARTIAL   TOLERANCE         F

V3           -1101.872     -0.40476     374.57033       8.654
V2            1128.717      0.58527     265.36049      18.092
(CONSTANT)   12306.39

ALL VARIABLES ARE IN THE EQUATION
* * * * * * * * * * * * * * * * * * * * * * * *  M U L T I P L E   R E G R E S S I O N  * * * * * * * * * * * * *   VARIABLE LIST 1
                                                                                                                     REGRESSION LIST 1
DEPENDENT VARIABLE..   V1        INCOME

                                                 SUMMARY TABLE

VARIABLE                                     MULTIPLE R  R SQUARE  RSQ CHANGE  SIMPLE R         B               BETA

V3         WEEKS OF UNEMPLOYMENT              0.88797    0.78850   0.78850    -0.88797     -1101.872        -0.40476
V2         YEARS OF EDUCATION                 0.94739    0.89754   0.10904     0.91945      1128.717         0.58527
(CONSTANT)                                                                                 12306.39
```

the *standardized partial regression coefficients* or *beta coefficients* that enable us to assess relative importance.

$$B_{Yj.k} = \frac{s_j}{s_y} b_{Yj.k} \tag{10.10}$$

where the standardized partial regression coefficient is found by multiplying the partial regression coefficient by the ratio of the standard deviation of the chosen independent variable over the standard deviation of the dependent variable. For the Wheat Yield problem, the standardized partial regression coefficients are:

$$B_{Y1.2} = \frac{108.012}{15.0} \cdot .1379 = .99299$$

$$B_{Y2.1} = \frac{2.0}{15.0} \cdot .6028 = .08037$$

The Multiple Correlation Coefficient

The multiple linear correlation coefficient, R, fulfills the same role in multiple regression that the simple correlation coefficient, r, did in bivariate regression. The multiple correlation coefficient is a measure of the (linear) goodness of fit of the least-squares surface to the scatter of data points. The simplest way to interpret the multiple correlation coefficient is to think of it as the zero-order (gross as opposed to net), or simple correlation coefficient between the actual values observed for the dependent variable and the values predicted for Y from the least-squares surface. It is therefore a correlation between $\hat{Y}$ and Y. If all the observed points lie on the regression surface, the actual and predicted values will coincide and the multiple regression will be unity. As in simple regression, the greater the scatter of points around the surface, the lower the correlation between the actual values and the estimated values. Unlike the simple correlation coefficient r, which shows the direction of the relationship of the two variables by its sign (positive or negative), R has a range from 0 to +1.0. R^2 and r^2 are directly analogous.

The multiple correlation coefficient can be derived from the simple r values, although it is rarely computed manually. For three variables, the general formula can be written:

$$R_{Y12} = \sqrt{r^2_{Y1} + r^2_{Y2.1} \cdot (1 - r^2_{Y1})}$$

or alternatively,

$$R^2_{Y12} = r^2_{Y1} + r^2_{Y2.1} \cdot (1 - r^2_{Y1}) \tag{10.11}$$

$$\begin{pmatrix}\text{Proportion}\\ \text{explained by}\\ X_1 \text{ and } X_2\end{pmatrix} = \begin{pmatrix}\text{proportion}\\ \text{explained}\\ \text{by } X_1\end{pmatrix} + \begin{pmatrix}\text{additional potential}\\ \text{proportion}\\ \text{explained by } X_2\end{pmatrix} \cdot \begin{pmatrix}\text{proportion not}\\ \text{explained}\\ \text{by } X_1\end{pmatrix}$$

where $r_{Y2.1}$ is the partial correlation of Y with X_2, holding X_1 constant (to be discussed shortly). If the correlation between the independent variables is zero then

$$R^2_{Y12} = r^2_{Y1} + r^2_{Y2} \tag{10.12}$$

There are several important observations to be made about R. First, R can never be less than any of the individual simple correlations, since it is impossible to explain less by adding additional variables. Second, the increase in the magnitude of R will be in part dependent on the degree of interrelatedness of the independent variables. If the intercorrelations among the independent variables are quite high, then the multiple R will not increase very much with each additional variable. On the other hand, independent variables that are less related will likely increase R significantly. As a corollary of the above two statements, if we wish to explain as much variation in the dependent variable as possible, we should include variables that are relatively unrelated to one another, but highly related to the dependent variable. Finally, in most computer outputs from regression analysis, a value for adjusted R^2 is reported. Adjusted R^2 is adjusted for the number of independent variables in the equation and the number of cases (the sample size). The formula used in SPSS is:

$$\text{Adjusted } R^2 = R^2 - \frac{(k - 1)}{n - k}(1 - R^2) \tag{10.13}$$

where k is the number of independent variables in the regression equation and n is the number of cases. Clearly, with small sample sizes and a large number of independent variables, adjusted R^2 can decrease with the addition of independent variables.

Partial Correlation Coefficients

An alternative explanation of the contribution of independent variables to the explanation of the variation in the dependent variable involves the use of the *partial correlation coefficient.* The square of the partial correlation coefficient is most usefully described as a measure of the extent to which that part of the variation in the dependent variable not explained by independent variables already in the equation can be explained by the addition of a new independent variable. The verbal description is quite similar to that used to explain the partial b values, (the partial regression coefficients).

Consider the correlation coefficients associated with the following regression equations and the data in Table 10.8:

$$Y = b_0 + b_1X_1$$

$$Y = b_0 + b_{Y1.2}X_1 + b_{Y2.1}X_2$$

TABLE 10.8 Data for Farm Income Example

Farm Income, dollars (Y)	Cows (X_1)	Farm Size (X_2)
19,200	30	60
16,600	2	220
25,200	20	180
12,200	10	80
11,800	2	120
18,000	12	100
16,400	10	170
17,600	18	110
17,200	10	260
15,200	4	230
20,400	22	70
21,600	28	120
19,200	12	240
14,000	2	160
16,000	15	90
22,600	23	110
15,200	2	220
14,800	9	110
19,600	20	160
16,000	28	80

where

$$Y = \text{Farm Income in dollars}$$

$$X_1 = \text{number of Cows}$$

$$X_2 = \text{Farm Size in acres}$$

The fitted equation (Table 10.9) is $\hat{Y} = 8222.31 + 354.46X_1 + 29.57X_2$. The correlation coefficients are:

$$R_{Y12} = .804$$

$$r_{Y1} = .666$$

$$r_{Y2} = .001$$

and not given in Table 10.9 but easily found, $r_{12} = -.560$.

The correlation coefficient associated with the first (simple regression) equation is ($r = .666$). The correlation coefficient associated with the multiple regression model is ($R_{Y.12} = .804$). If we square both values we have $r^2_{Y1} = .443$ and $R^2_{Y.12} = .647$. That is, the first independent variable (X_1) through a simple regression explained 44.3 percent of the variation in the dependent variable, and the addition of a second variable (X_2) increased the

TABLE 10.9 SPSS Regression of Data in Table 10.8

```
        1 RUN NAME          REGRESSION OF FARM INCCME ON COWS AND ACRES
        2 VARIABLE LIST     V1,V2,V3
        3 VAR LABELS        V1 FARM INCCME/
        4                   V2 COWS/
        5                   V3 ACRES
        6 INPUT FORMAT      FIXED (F5.0,1X,F2.0,1X,F3.0)

         ACCCRDING TO YOUR INPUT FORMAT, VARIABLES ARE TO BE READ AS FOLLOWS

         VARIABLE   FORMAT   RECORD       COLUMNS

         V1         F 5. 0      1         1-    5
         V2         F 2. 0      1         7-    8
         V3         F 3. 0      1        10-   12

THE INPUT FORMAT PROVIDES FUR     3 VARIABLES.     3 WILL BE READ
IT PROVIDES FOR     1 RECORDS ('CARDS') PER CASE.  A MAXIMUM CF    12 'COLUMNS' ARE USED ON A RECORD.

        7 N CF CASES        20
        8 REGRESSION        VARIABLES=V1,V2,V3/
        9                   REGRESSICN=V1 WITH V2,V3(2),RESID=0/
       10 STATISTICS        4


DEPENDENT VARIABLE..     V1          FARM INCOME

VARIABLE(S) ENTERED CN STEP NUMBER  1..     V3          ACRES
                                            V2          COWS


MULTIPLE R           0.80448          ANALYSIS CF VARIANCE    DF     SUM OF SQUARES       MEAN SQUARE           F
R SQUARE             0.64718          REGRESSICN               2.   141220707.62674    70610353.81337      15.59187
ADJUSTED R SQUARE    0.60563          RESIDUAL                17.    76987292.37326     4528664.25725
STANCARD ERROR    2128.06585

---------------- VARIABLES IN THE EQUATION -----------------    ------------- VARIABLES NOT IN THE EQUATION -------------

VARIABLE          B            BETA      STD ERRCR B       F     VARIABLE      BETA IN     PARTIAL   TCLERANCE        F

V3             29.57040      0.54563      9.42781       9.838
V2            354.4634       0.97145     63.47572      31.184
(CONSTANT)    8222.313

ALL VARIABLES ARE IN THE EQUATION

DEPENDENT VARIABLE..     V1          FARM INCOME
                                                        SUMMARY TABLE

VARIABLE                                 MULTIPLE R  R SQUARE  RSQ CHANGE  SIMPLE R          B              BETA

V3          ACRES                         0.00109    0.00000    0.00000    0.00109      29.57040        0.54563
V2          COWS                          0.80448    0.64718    0.64718    0.66559     354.4634         0.97145
(CONSTANT)                                                                            8222.313
```

explanation to 64.7 percent. Adding the extra variable has increased the variance, which can be explained by 20.4 percent. Whereas 55.7 percent of the variance was still to be explained in the first instance, only 35.3 percent remained after the second independent variable was added. We examine the increase with the additional variable as a proportion of the variance that was left unexplained before it was added, and find the ratio 20.4/55.7 or 0.364. Thus 36.4 percent of the variance left unexplained by Cow numbers is explained by adding the variable Farm Size. Taking the square root gives the coefficient of partial correlation, which is $r_{Y2.1} = .603$ (approximately).

A series of calculations using the simple correlation coefficients will formalize what we have just described.

$$r_{Y2.1} = r_{Y2} - \frac{(r_{Y1})(r_{12})}{\sqrt{1 - r_{Y1}^2}\sqrt{1 - r_{12}^2}}$$

where the dot, as before, in $r_{Y2.1}$ indicates the partial correlation of variables Y and X_2, holding X_1 constant. Thus, for Y and X_2, holding X_1 constant,

$$r_{Y2.1} = \frac{.001 - (.666)(-.560)}{\sqrt{1 - (.666)^2}\sqrt{1 - (-.560)^2}}$$

$$= .605 \text{ (The difference is due to rounding in the calculations.)}$$

With partial correlations explained, we can now illustrate the explanation for R^2 given in Equation 10.11, using the above example, and also show the relationship between these two correlation measures. From

$$r_{Y1}^2 = (.666)^2 = .443$$

$$1 - r_{Y1}^2 = .557$$

and

$$r_{Y2.1}^2 = (.605)^2 = .366$$

Therefore, from Equation 10.11,

$$R_{Y12}^2 = .443 + (.366)(.557)$$
$$= .647$$

10.3 MATRICES

Although we have established the nature of the general linear model and its solution, the simplicity of solving the general linear regression model is enhanced with the use of elementary matrix operations. It is not necessary to refer to the theory of matrices in order to make use of some simple operational structures, and the next several pages outline some basic matrix operations followed by the representation of the general linear model in matrix terms.

A matrix is a table of letters or numbers enclosed by a pair of square brackets (or something similar), subject to certain rules of operation. Matrices are most often referred to by boldface symbols, usually uppercase capital letters. For example,

$$\mathbf{A} = \begin{bmatrix} a_{11} & a_{12} & a_{13} \\ a_{21} & a_{22} & a_{23} \\ a_{31} & a_{32} & a_{33} \\ a_{41} & a_{42} & a_{43} \\ a_{51} & a_{52} & a_{53} \end{bmatrix} \quad \text{or} \quad \mathbf{A} = \begin{bmatrix} 3 & 6 & 1 \\ 4 & 8 & 3 \\ 2 & 1 & 9 \\ 4 & 2 & 5 \\ 8 & 1 & 9 \end{bmatrix} \tag{10.14}$$

The advantage of matrices is that we can refer to a whole table of numbers or letters with a single symbol. The numbers or letters in a matrix are called its *elements,* and the *order* of a matrix is the number of rows and columns, usually written $\mathbf{A}_{4\times 3}$ or ${}_4\mathbf{A}_3$. By convention, rows are referred to before columns.

Kinds of Matrices

One of the simplest ways to master the different kinds of matrices that are used in statistical analysis is to consider a variety of matrices as special cases of one another. Thus, a *square matrix* is a special matrix in which the number of rows equals the number of columns. In a square matrix, the elements a_{11}, a_{22}, . . . are called the *diagonal* elements. The remainder are the *off-diagonals*. A *symmetric matrix*, usually a derived matrix rather than a data matrix, is a special square matrix in which the elements of the corresponding columns and rows are interchangeable. That is, $a_{ij} = a_{ji}$. For example,

$$\mathbf{S} = \begin{bmatrix} 11 & 2 & 3 \\ 2 & 5 & 10 \\ 3 & 10 & 6 \end{bmatrix}$$

Two common symmetric matrices used in statistical analysis are the correlation matrix and the variance/covariance matrix (Section 9.4).

A special symmetric matrix is a *diagonal matrix* in which all the off-diagonal elements equal zero.

$$\mathbf{D} = \begin{bmatrix} 11 & 0 & 0 \\ 0 & 5 & 0 \\ 0 & 0 & 6 \end{bmatrix}$$

And a special diagonal matrix is the *scalar matrix* in which the diagonal elements are equal, and a special scalar matrix is the *identity matrix*, with 1's in the diagonal.

$$\mathbf{I} = \begin{bmatrix} 1 & 0 & 0 \\ 0 & 1 & 0 \\ 0 & 0 & 1 \end{bmatrix}$$

The identity matrix has the same uses in matrix algebra that the number 1 has in ordinary algebra.

Matrices with only one column or one row are called *vectors*. Thus

$$\mathbf{V} = \begin{bmatrix} 3 \\ 4 \\ 5 \end{bmatrix} \text{ is a } \textit{column vector}$$

and

$$\mathbf{W} = [6 \quad 8 \quad 5] \text{ is a } \textit{row vector}$$

When a matrix has only one row and one column it is called a *scalar quantity* or simply a *scalar*.

Basic Matrix Operations

The *transpose* of a matrix is usually indicated with a prime next to the matrix symbol—**A**′ is the transpose of the matrix **A**. The transpose involves the interchanging of rows and columns. The first row becomes the first column, and so on. Thus,

$$\mathbf{A} = \begin{bmatrix} 4 & -2 & 1 \\ 4 & 2 & 3 \\ 6 & 5 & 4 \\ 2 & 8 & 2 \\ 9 & 1 & 7 \end{bmatrix} \quad \mathbf{A}' = \begin{bmatrix} 4 & 4 & 6 & 2 & 9 \\ -2 & 2 & 5 & 8 & 1 \\ 1 & 3 & 4 & 2 & 7 \end{bmatrix}$$

In general terms, the ith row becomes the ith column. It is useful to know that the transpose of a symmetric matrix is the same as the original matrix.

The *addition and subtraction* of matrices involves the adding or subtracting of the respective elements in the two matrices. To carry out such operations, the two matrices must be of the same order.

$$\mathbf{A} = \begin{bmatrix} 3 & 7 \\ 2 & 4 \\ 8 & 1 \end{bmatrix} \quad \mathbf{B} = \begin{bmatrix} 5 & 10 \\ 61 & 32 \\ 41 & 12 \end{bmatrix}$$

$$\mathbf{A} + \mathbf{B} = \begin{bmatrix} 8 & 17 \\ 63 & 36 \\ 49 & 13 \end{bmatrix} \quad \mathbf{A} - \mathbf{B} = \begin{bmatrix} -2 & -3 \\ -59 & -28 \\ -33 & -11 \end{bmatrix}$$

Matrices can also be *multiplied* (and therefore also divided) by *scalars*, simply by multiplying (or dividing) each element in the matrix by the scalar.

$$2 \cdot \begin{bmatrix} 4 & 2 \\ 6 & 0 \\ 3 & 5 \end{bmatrix} = \begin{bmatrix} 8 & 4 \\ 12 & 0 \\ 6 & 10 \end{bmatrix}$$

The *multiplication of two matrices*, however, is a little more complicated. In matrix multiplication, the order of multiplying is important and the two matrices must be conformable in the sense that the number of columns in the first (left-hand) matrix must equal the number of rows in the second (right-hand) matrix. It is quite possible for two matrices **A** and **B** to be conformable for the multiplication **AB** but not for **BA** or vice versa. Multiplication involves multiplying each element in the first row in matrix **A** by its corresponding element in the first column in matrix **B**, and adding them together. This would give the value for the element in row 1 column

1 of the resulting matrix. This process is repeated for each combination. For example,

$$\mathbf{A} = \begin{bmatrix} 3 & 3 \\ 2 & 1 \\ 1 & 9 \end{bmatrix} \quad \mathbf{B} = \begin{bmatrix} 4 & 6 & 2 \\ 5 & 1 & 3 \end{bmatrix} \quad \mathbf{C} = \mathbf{AB} = \begin{bmatrix} 27 & 21 & 15 \\ 13 & 13 & 7 \\ 49 & 15 & 29 \end{bmatrix}$$

The operation involves

$$c_{11} = a_{11}b_{11} + a_{12}b_{21}$$

$$c_{12} = a_{11}b_{12} + a_{12}b_{22}$$

and so on.

More generally,

$$c_{ij} = \sum_{k=1}^{n} a_{ik}b_{kj} \tag{10.15}$$

where

n = the number of columns in **A** = the number of rows in **B**

For matrix multiplication, the number of columns in the left-hand matrix must equal the number of rows in the right-hand matrix. The other two dimensions do not matter, but they decide the order of the resulting matrix. Thus in our example, **A** *postmultiplied* by **B** gives a 3 × 3 matrix, and **A** *premultiplied* by **B** gives a 2 × 2 matrix.

$${}_3\mathbf{A}_2 \, {}_2\mathbf{B}_3 = {}_3\mathbf{C}_3 \qquad \text{and} \qquad {}_2\mathbf{B}_3 \, {}_3\mathbf{A}_2 = {}_2\mathbf{C}_2$$

In general,

$${}_m\mathbf{A}_n \, {}_n\mathbf{B}_s = {}_m\mathbf{C}_s$$

The Inverse of a Matrix

It is not possible to divide by matrices in the same way as in ordinary algebraic operations. However, to carry out an operation that is similar to division, we use the concept of the *inverse* of a matrix. We will use the symbol $\mathbf{A}^{-1}$ for the inverse of matrix **A** and note that the algebraic operation of $X = B/A$ is replaced by the matrix multiplication $\mathbf{X} = \mathbf{BA}^{-1}$. The concept of matrix inverses is a very important one, but it is also one that is difficult to provide an adequate description for and even more difficult to calculate practical examples. We do so with only simple examples.

In considering inverses of matrices, we are concerned only with square matrices. A square matrix **A** is said to be *nonsingular* if there exists another matrix (which we will call its *inverse* and denote by $\mathbf{A}^{-1}$) such that $\mathbf{AA}^{-1} = \mathbf{A}^{-1}\mathbf{A} = \mathbf{I}$ the identity matrix. For example, the small matrix

$\mathbf{A} = \begin{bmatrix} 2 & 5 \\ 1 & 3 \end{bmatrix}$ is nonsingular and has an inverse $\mathbf{A}^{-1} = \begin{bmatrix} 3 & -5 \\ -1 & 2 \end{bmatrix}$ which when premultiplied or postmultiplied by $\mathbf{A}$ equals the identity matrix $\begin{bmatrix} 1 & 0 \\ 0 & 1 \end{bmatrix}$. If no such inverse exists, then the matrix $\mathbf{A}$ is *singular*.

For each square matrix $\mathbf{A}$ (whether singular or nonsingular), there exists a unique number called the *determinant* of the matrix, which we will denote by $d(\mathbf{A})$ (many other symbols are used). The determinant has a number of important properties that measure various relationships of the rows and columns of the matrix by manipulating combinations of elements. Put another way, it measures aspects of the *solution* of equations represented in matrix form—thus it "determines" the solvability of these equations. It can be shown that a matrix $\mathbf{A}$ is nonsingular (and therefore has an inverse $\mathbf{A}^{-1}$) if (and only if) $d(\mathbf{A}) \neq 0$.

Calculating the determinant and the inverse is a tedious process (by hand methods), and we will only illustrate the process for the simplest of examples—a 2×2 matrix. It can be shown that for the matrix

$$\mathbf{A} = \begin{bmatrix} a_{11} & a_{12} \\ a_{21} & a_{22} \end{bmatrix}$$

$$d(\mathbf{A}) = a_{11}a_{22} - a_{12}a_{21} \tag{10.16}$$

and

$$\mathbf{A}^{-1} = \frac{1}{d(\mathbf{A})} \begin{bmatrix} a_{22} & -a_{12} \\ -a_{21} & a_{11} \end{bmatrix} \tag{10.17}$$

For example,

$$\mathbf{A} = \begin{bmatrix} 8 & 6 \\ 6 & 7 \end{bmatrix}$$

therefore,

$$d(\mathbf{A}) = (8 \times 7) - (6 \times 6) = 20$$

and

$$\mathbf{A}^{-1} = \frac{1}{20} \begin{bmatrix} 7 & -6 \\ -6 & 8 \end{bmatrix}$$

$$= \begin{bmatrix} \frac{7}{20} & -\frac{6}{20} \\ -\frac{6}{20} & \frac{8}{20} \end{bmatrix}$$

And we can check our calculation

$$\mathbf{AA}^{-1} = \begin{bmatrix} 8 & 6 \\ 6 & 7 \end{bmatrix} \begin{bmatrix} \frac{7}{20} & -\frac{6}{20} \\ -\frac{6}{20} & \frac{8}{20} \end{bmatrix}$$

$$= \begin{bmatrix} 1 & 0 \\ 0 & 1 \end{bmatrix} = \mathbf{I} \text{ the identity matrix}$$

Let us list a few of the properties of determinants and inverses, some of which will prove useful later. If **A** and **B** are nonsingular matrices and **k** is a nonzero scalar, $\mathbf{A}^{-1}$, $\mathbf{kA}$, $\mathbf{A}'$, and $\mathbf{AB}$ are all nonsingular.

$$(\mathbf{A}^{-1})^{-1} = \mathbf{A}$$

$$(\mathbf{kA})^{-1} = \frac{1}{\mathbf{k}}(\mathbf{A}^{-1})$$

$$(\mathbf{A}')^{-1} = (\mathbf{A}^{-1})'$$

$$(\mathbf{AB})^{-1} = \mathbf{B}^{-1}\mathbf{A}^{-1}$$

$$d(\mathbf{A}') = d(\mathbf{A})$$

If **A** has a zero row or a zero column then d(**A**) = 0 and **A** is singular.

If **A** has a row or column that is a linear function of another row or column then d(**A**) = 0 and **A** is singular.

These last two properties in particular have direct relevance to the use of matrices in statistical analysis.

10.4 MATRIX SOLUTIONS FOR THE COEFFICIENTS OF THE GENERAL LINEAR MODEL

With a working knowledge of matrices, we can now reexamine the derivation of the least squares solution to the normal equations as discussed in Section 10.1. We illustrate this derivation with reference to a situation with only two independent variables and use as an example the Farm Income structure used previously (Table 10.8).

Rewriting the Regression Equation

Our regression equation, with two independent variables, is given by

$$\hat{Y}_i = b_0 + b_1X_{i1} + b_2X_{i2} \qquad \text{for } i = 1 \text{ to } n$$

First of all, let us rewrite this equation making two changes, changes that make no difference to the result, but make the equation easier to manipulate.

$$\hat{Y}_i = X_{i0}b_0 + X_{i1}b_1 + X_{i2}b_2 \qquad \text{for } i = 1 \text{ to } n \tag{10.18}$$

$$\text{and where } X_{i0} = 1 \text{ for all } i$$

We have reversed the order of the X scores and the b coefficients (which makes no difference to their multiplication), and we have added in a variable X_0, that because it always equals 1, will also have no effect on any calculation.

If we list all the individual observations making up our regression equations in the form of a sequence of rows of a matrix, we get

$$\begin{aligned}
\hat{Y}_1 &= X_{10}b_0 + X_{11}b_1 + X_{12}b_2 \\
\hat{Y}_2 &= X_{20}b_0 + X_{21}b_1 + X_{22}b_2 \\
\hat{Y}_3 &= X_{30}b_0 + X_{31}b_1 + X_{32}b_2 \\
&\vdots \\
\hat{Y}_n &= X_{n0}b_0 + X_{n1}b_1 + X_{n2}b_2
\end{aligned}$$

Separating on either side of the equals sign,

$$_n\begin{bmatrix} \hat{Y}_1 \\ \hat{Y}_2 \\ \hat{Y}_3 \\ \vdots \\ \hat{Y}_n \end{bmatrix}_1 = {}_n\begin{bmatrix} X_{10}b_0 + X_{11}b_1 + X_{12}b_2 \\ X_{20}b_0 + X_{21}b_1 + X_{22}b_2 \\ X_{30}b_0 + X_{31}b_1 + X_{32}b_2 \\ \vdots \\ X_{n0}b_0 + X_{n1}b_1 + X_{n2}b_2 \end{bmatrix}_1 \tag{10.19}$$

And separating the right-hand side into two multiplied matrices.

$$_n\begin{bmatrix} \hat{Y}_1 \\ \hat{Y}_2 \\ \hat{Y}_3 \\ \vdots \\ \hat{Y}_n \end{bmatrix}_1 = {}_n\begin{bmatrix} X_{10} & X_{11} & X_{12} \\ X_{20} & X_{21} & X_{22} \\ X_{30} & X_{31} & X_{32} \\ \vdots & \vdots & \vdots \\ X_{n0} & X_{n1} & X_{n2} \end{bmatrix}_3 \cdot {}_3\begin{bmatrix} b_0 \\ b_1 \\ b_2 \end{bmatrix}_1 \tag{10.20}$$

(Multiply this last matrix equation out to see how Equations 10.19 and 10.20 are identical.) Using the labels $\hat{\mathbf{Y}}$, $\mathbf{X}$, and $\mathbf{B}$ to represent the three matrices in order, our regression equation in matrix form is

$$\hat{\mathbf{Y}} = \mathbf{XB} \tag{10.21}$$

Similarly, the statistical model we are fitting will be

$$\mathbf{Y} = \mathbf{XB} + \mathbf{E}$$

where $\mathbf{E}$ is an $n \times 1$ vector of residuals.

The Normal Equations

Let us digress a moment and examine some relevant manipulations of the observed X and Y data. Using the Farm Income example (Table 10.8), and setting up the data in the form of Equation 10.20, we have

$$\mathbf{Y} = {}_{20}\begin{bmatrix} 19{,}200 \\ 16{,}600 \\ 25{,}200 \\ \vdots \\ 16{,}000 \end{bmatrix}_1 \qquad \mathbf{X} = {}_{20}\begin{bmatrix} 1 & 30 & 60 \\ 1 & 2 & 220 \\ 1 & 20 & 180 \\ \vdots & \vdots & \vdots \\ 1 & 28 & 80 \end{bmatrix}_3 \quad (\text{columns } X_0 \; X_1 \; X_2)$$

Therefore

$$\mathbf{X}'\mathbf{X} = {}_3\begin{bmatrix} 1 & 1 & 1 & \dots & 1 \\ 30 & 2 & 20 & \dots & 28 \\ 60 & 220 & 180 & \dots & 80 \end{bmatrix}_{20} {}_{20}\begin{bmatrix} 1 & 30 & 60 \\ 1 & 2 & 220 \\ 1 & 20 & 180 \\ \vdots & \vdots & \vdots \\ 1 & 28 & 80 \end{bmatrix}_3$$

$$= {}_3\begin{bmatrix} 20 & 279 & 2{,}890 \\ 279 & 5{,}531 & 34{,}130 \\ 2{,}890 & 34{,}130 & 491{,}900 \end{bmatrix}_3 = \begin{bmatrix} n & \Sigma X_1 & \Sigma X_2 \\ \Sigma X_1 & \Sigma X_1^2 & \Sigma X_1 X_2 \\ \Sigma X_2 & \Sigma X_1 X_2 & \Sigma X_2^2 \end{bmatrix}$$

$$\mathbf{X}'\mathbf{Y} = {}_3\begin{bmatrix} 1 & 1 & 1 & \dots & 1 \\ 30 & 2 & 20 & \dots & 28 \\ 60 & 220 & 180 & \dots & 80 \end{bmatrix}_{20} {}_{20}\begin{bmatrix} 19{,}200 \\ 16{,}600 \\ 25{,}200 \\ \vdots \\ 16{,}000 \end{bmatrix}_1$$

$$= {}_3\begin{bmatrix} 348{,}800 \\ 5{,}263{,}800 \\ 50{,}406{,}000 \end{bmatrix}_1 = \begin{bmatrix} \Sigma Y \\ \Sigma X_1 Y \\ \Sigma X_2 Y \end{bmatrix}$$

Recalling the normal equations (Equation 10.4) that were found by minimizing the expression $\Sigma(Y - \hat{Y})^2$

$$b_0 n + b_1 \Sigma X_1 + b_2 \Sigma X_2 = \Sigma Y$$

$$b_0 \Sigma X_1 + b_1 \Sigma X_1^2 + b_2 \Sigma X_1 X_2 = \Sigma X_1 Y$$

$$b_0 \Sigma X_2 + b_1 \Sigma X_1 X_2 + b_2 \Sigma X_2^2 = \Sigma X_2 Y$$

Placing these in matrix form and separating the b coefficients in the left-hand side.

$$
{}_3\begin{bmatrix} n & \Sigma X_1 & \Sigma X_2 \\ \Sigma X_1 & \Sigma X_1^2 & \Sigma X_1 X_2 \\ \Sigma X_2 & \Sigma X_1 X_2 & \Sigma X_2^2 \end{bmatrix}_3 \; {}_3\begin{bmatrix} b_0 \\ b_1 \\ b_2 \end{bmatrix}_1 = {}_3\begin{bmatrix} \Sigma Y \\ \Sigma X_1 Y \\ \Sigma X_2 Y \end{bmatrix}_1
$$

Finally, using the manipulations made with the X and Y data above, the normal equations reduce to the single matrix equation,

$$\mathbf{X'XB} = \mathbf{X'Y} \tag{10.22}$$

Extension to greater than two independent variables would simply add extra rows to each of the matrices.

The Least-Squares Solution

From the normal equation (Equation 10.22), we wish to solve for the vector $\mathbf{B}$. To do this we need to eliminate the $\mathbf{X'X}$ term on the left-hand side. To accomplish this we must premultiply both sides of the equation by $(\mathbf{X'X})^{-1}$. Thus, to solve for the b coefficients, the matrix $\mathbf{X'X}$, and therefore the data matrix $\mathbf{X}$, must be nonsingular.

$$\mathbf{X'XB} = \mathbf{X'Y}$$

Premultiplying by $(\mathbf{X'X})^{-1}$;

$$(\mathbf{X'X})^{-1}\,(\mathbf{X'X})\mathbf{B} = (\mathbf{X'X})^{-1}\,(\mathbf{X'Y})$$

But by definition $(\mathbf{X'X})^{-1}\,(\mathbf{X'X}) = \mathbf{I}$ (the identity matrix) and

$$\mathbf{IB} = \mathbf{B}$$

Therefore,

$$\mathbf{B} = (\mathbf{X'X})^{-1}\,(\mathbf{X'Y}) \tag{10.23}$$

is the required least-squares solution.

In our example, we already have $(\mathbf{X'Y})$ and $(\mathbf{X'X})$, but we need to find the inverse of the matrix $(\mathbf{X'X})$. We will not illustrate this calculation, but simply note that the determinant of the matrix $\mathbf{X'X}$ was found to be 1,670,107,600, and therefore $\mathbf{X'X}$ is nonsingular. It can be shown that (correct to four decimal places)

$$
(\mathbf{X'X})^{-1} = \begin{bmatrix} .9316 & -.0231 & -.0039 \\ -.0231 & .0009 & .0001 \\ -.0039 & .0001 & .0000 \end{bmatrix}
$$

Thus for

$$\mathbf{B} = (\mathbf{X}'\mathbf{X})^{-1}(\mathbf{X}'\mathbf{Y})$$

$$= \begin{bmatrix} .9316 & -.0231 & -.0039 \\ -.0231 & .0009 & .0001 \\ -.0039 & .0001 & .0000 \end{bmatrix} \begin{bmatrix} 348{,}800 \\ 5{,}263{,}800 \\ 50{,}406{,}000 \end{bmatrix}$$

$$= \begin{bmatrix} 8222.31 \\ 354.46 \\ 29.57 \end{bmatrix}$$

and our regression equation is

$$\hat{Y} = 8222.31 + 354.46X_1 + 29.57X_2$$

10.5 TESTS FOR THE REGRESSION MODEL

We wish to test for the significance of the overall equation and for the significance of the partial regression coefficients. We defer until the next chapter a discussion of the assumptions that influence the validity of these tests, but we should point out that the applicability and power of the tests we will discuss are considerably influenced by those assumptions, and a clear understanding of them is critical.

Testing the Overall Model

Testing the significance of the overall model that has been fitted to the data can be handled by an F test examining the ratio of explained to unexplained variance, in an identical manner to that outlined for simple regression. The F statistic can be derived from either the analysis of variance table for the regression (see, for example, Table 10.9) or by calculation from R, the multiple linear correlation coefficient. The two identical tests are

$$F = \frac{\text{MSR}}{\text{MSE}} \tag{10.24}$$

$$F = \frac{R^2}{1 - R^2} \frac{n - k - 1}{k} \tag{10.25}$$

where

$$n = \text{the sample size}$$

and

$$k = \text{the number of independent variables}$$

The F statistic would be tested against $[F, p = 1 - \alpha, df = k, n - k - 1]$

In the Farm Income example (Table 10.9) we obtain

$$F = \frac{\text{MSR}}{\text{MSE}} = \frac{70{,}610{,}353.81}{4{,}528{,}664.26} = 15.59$$

or

$$F = \frac{R^2}{1 - R^2} \frac{n - k - 1}{k} = \frac{.64718}{.35282} \cdot \frac{20 - 2 - 1}{2} = 15.59$$

and with $[F, p = .95, df = 2, 17] = 3.63$ we can conclude that our model is significant. The hypothesis we are testing can be stated in a variety of ways, but basically our null hypothesis is

Ho: ρ (the true multiple correlation coefficient) $= 0$ versus *Ha:* $\rho \neq 0$

or

Ho: $\beta_1 = \beta_2 = \ldots = \beta_k = 0$ versus *Ha:* at least one $\beta_j \neq 0$

Note that this is a somewhat weak hypothesis in that even small R values will give significant results. Thus, for example, the presence of one independent variable that is contributing substantially to the explanation of the variability of Y will provide a significant model, even in the presence of a number of other independent variables that contribute little. This emphasizes the importance of testing the contribution of each independent variable to the overall model (to be discussed next) and, in general, of carefully examining the level of explanation of the model as well as providing a test for overall significance.

Testing the Individual Regression Coefficients

In the simple linear regression and correlation structure, our test for the overall model was also a test for significance for the single slope parameter β_1. We can provide an extension to that structure to look at the contribution of individual β's in the multivariate situation. But, although the extension is simple, its calculation is more difficult and its interpretation can be confusing.

First, look at the testing structure that is standard from the SPSS and SAS packages. For each regression coefficient, its *standard error* is reported and in the case of SPSS, an F statistic where

$$F = \left(\frac{b_j}{S_{b_j}}\right)^2 \qquad (10.26)$$

where

S_{b_j} = the standard error of the regression coefficient b_j

or in the case of SAS a t statistic where

$$t = \frac{b_j}{S_{b_j}} \tag{10.27}$$

The two structures are, of course, identical with F simply being the square of t. These values would be tested against $[F, p = 1 - \alpha, df = 1, n - k - 1]$ and $[t, p = 1 - \alpha/2, df = n - k - 1]$, respectively. Both tests use S_{b_j} the standard error of the regression coefficient b_j which, although based on the same structure as for S_{b_1} for the simple regression, is a little more difficult to obtain.

$$S_{b_j} = \sqrt{C_{jj}\, S_E^2} \tag{10.28}$$

where C_{jj} is the entry in the diagonal of the $(\mathbf{X}'\mathbf{X})^{-1}$ matrix corresponding to the appropriate row and column for the jth variable, and S_E^2 is the MSE (the variance of the error term). For the two independent variable case, in calculation form, this is

$$S_{b1.2} = \frac{S_E}{\sqrt{\Sigma X_{1i}^2 (\Sigma X_{i1} X_{i2})^2 / \Sigma X_{i2}^2}} \tag{10.29}$$

where X_1 and X_2 represent the independent variables and $S_{b1.2}$ is the standard error of b for X_1 holding X_2 constant.

However, it is important to realize what hypothesis is actually being tested in the above t or F testing structures. We can state this in a number of ways.

Ho: β_j (given all other β_k) $= 0$ versus Ha: $\beta_j \neq 0$

Ho: R^2 change $= 0$ versus Ha: R^2 change $\neq 0$

What we are really saying is that the testing structure is testing whether the extra contribution of that particular variable X_j is adding significantly to the model, compared to the situation with the same model minus that particular X_j. Thus, it is testing the increase in the R^2 value resulting from a situation as if we had added that variable into the equation last.

This testing structure is perhaps best illustrated by performing a partitioning of the regression sum of squares into its component parts, using the Farm Income example. The relevant summary measures taken from Table 10.9 are

$$\text{SST} = 218{,}208{,}000$$

$$R_{Y12}^2 = .64718$$

$$r_{Y1}^2 = (.66559)^2$$

$$r_{Y2}^2 = (.00109)^2$$

Table 10.10 shows the ANOVA table from the multiple regression analysis summarized in Table 10.9. The two versions (a and b) show how the same

TABLE 10.10 Partitioning the Regression Sum of Squares

Source of Variability	Sum of Squares	*df*	Mean Square	*F*
(a)				
Due to regression	SSR = R^2_{Y12}(SST) = 141,220,708	2	70,610,354	15.592
Due to X_1	r^2_{Y1}(SST) = 96,668,337	1		
Due to $X_2.X_1$	SSR − r^2_{Y1}(SST) = 44,552,371	1	44,552,371	9.838
Due to error	SSE = (1 − R^2_{Y12})(SST) = 76,987,292	17	4,528,664	
TOTAL	SST = 218,208,000	19		
(b)				
Due to regression	SSR = R^2_{Y12}(SST) = 141,220,708	2	70,610,354	15.592
Due to X_2	r^2_{Y2}(SST) = 259	1		
Due to $X_1.X_2$	SSR − r^2_{Y2}(SST) = 141,220,448	1	141,220,448	31.184
Due to error	SSE = (1 − R^2_{Y12})(SST) = 76,987,292	17	4,528,664	
TOTAL	SST = 218,208,000	19		

regression sum of squares can be partitioned into two parts—one due to a simple regression, the other to the extra sum of squares explained by adding a second independent variable. The (a) version looks at the simple regression of $Y = f(X_1)$ and records that 96,668,337 of the 141,220,708 sum of squares due to regression can be considered as being derived from that simple regression. The remainder, the "extra" sum of squares due to adding X_2 given X_1 $(X_2.X_1)$, therefore derives from adding the second variable, and with 1 degree of freedom can be divided by the mean square error to give an F ratio. Version (b) does the same thing but starting with the simple regression of $Y = f(X_2)$ and the extra regression sum of squares, derives from adding X_1 given X_2. The two F statistics are the same F values we described as standard output and defined as $(b_j/S_{b_j})^2$.

The previous discussion on testing the individual regression coefficients describes the basis and methods for carrying out the standard testing of regression coefficients. There are many other forms of testing structures that can be utilized, but most are based on partitioning the regression sum of squares in some useful manner. No further tests will be considered here, but we indirectly examine some other structures in later chapters. However, before completing the present discussion there are two general comments that need to be made.

First, the testing of the individual regression coefficients would normally follow a decision that the overall model was significant. Often the results of the individual variable tests would show that one or more variables are not contributing significantly. But remember that the removal of one independent variable can make major changes to the values and significance of all the other independent variables in an equation. This observation will be taken up again in Chapter 12. Second, we have outlined the "standard" test for significance of individual regression coefficients. In summary, this tests each variable as if it were the last variable to enter the equation. This is the structure assumed by SPSS and SAS for basic output. It is not, however, the standard output for all statistical packages. For some, the F statistic reported tests the extra contribution that that variable makes compared to a model with *only the preceding variables listed* included. That is, the order the variables are listed in the analysis will control the F statistics being reported. Other packages report t statistics for one form and F statistics for another. The conclusion to this discussion is that the manual accompanying the statistical package being used should be carefully consulted to check what the reported test statistics mean.

References and Readings

1. General Introduction

Blalock, H. M. (1979) *Social Statistics*, Revised Second Edition, McGraw-Hill: New York.

Johnston, R. J. (1978) *Multivariate Statistical Analysis in Geography*, Longman: London.

Wonnacott, H. and R. J. Wonnacott (1972) *Introductory Statistics*, Wiley: New York.

2. Matrices

Davis, J. C. (1973) *Statistics and Data Analysis in Geology*, Wiley: New York.

Horst, P. (1963) *Matrix Algebra for Social Scientists*, Holt, Rinehart and Winston: New York.

3. Applications

Ferguson, R. I. (1973) "Channel, pattern and sediment type," *Area* 5:38–41.

Johnston, R. J. (1976) "On regression coefficients in comparative studies of the friction of distance," *Tijdschrift Econ. Soc. Geografiska* 67:15–28.

Mark, D. M. and T. K. Peucker (1978) "Regression analysis and geographic models," *Canadian Geographer* 22:51–64.

Massey, D. S. (1980) "Residential segregation and spatial distribution of non-labor forces population: The needy, elderly, and disabled," *Economic Geography* 56:190–199.

Mather, P. M. (1976) *Computational methods of multivariate analysis in physical geography*, Wiley: Chichester.

Mather, P. M. and S. Openshaw (1974) Multivariate methods in geographical data, *The Statistician* 23:283–308.

Wong, S. T. (1963) "A multivariate statistical model for predicting mean annual flood in New England," *Annals, Association of American Geographers* 53:298–311.

CHAPTER 11

Issues in the Application of the General Linear Model

The previous chapter outlined the model for multiple regression and interpreted the coefficients. The validity of this model and the tests of the coefficients depend on a series of assumptions. It is critical to examine these assumptions and to "correct" the model where possible in order not to violate an underlying assumption. Without an understanding of, and sensitivity toward, the assumptions, the results obtained from applying a multiple regression model may prove of little use. In Section 11.1, we first provide a summary review and extension of the assumptions already discussed for simple linear regression in Section 9.5. We then focus our attention on the analysis of residuals from regression (Section 11.2), since residuals provide us with our main source of information on possible assumption violations. Then we examine each of the assumptions in turn, describing methods for testing the assumptions and, where possible, ways of overcoming violations of them. Normality, linearity, and homoscedasticity (equal variances) are considered in Section 11.3, autocorrelation in Section 11.4, and multicollinearity in 11.5. Considerable attention is paid to the autocorrelation problem because this assumption is the one most easily violated, most difficult to recognize, and hardest to counteract. It is of particular concern to geographers, who frequently deal with spatial data in which autocorrelation may be present, and for which standard methods are not readily available. In the last section of the chapter (11.6), we take up another problem that is specifically relevant to spatial problems, that of aggregation of data and ecological correlation.

11.1 ASSUMPTIONS OF THE GENERAL LINEAR MODEL

In Chapter 9, we outlined four basic assumptions behind the use of the simple linear regression and correlation model. These were concerned with measurement error, with the pattern of the distribution of the Y scores (or the residuals), with the linearity of the relationship of Y and X, and with the normality of the Y and X distributions. We now wish to review some of these assumptions in more detail while extending them to the multiple regression and correlation model. In particular, we want to examine the assumptions of independent random variables and homoscedasticity of the Y scores (both deriving from the pattern of distribution of scores) and the linearity assumption before considering the somewhat difficult problem of normality. In moving to a multivariate situation, there is a further assumption that needs to be considered in applying the general linear model. This is the assumption concerning *collinearity,* or the relationship among the independent variables. However, we leave a discussion of this assumption until Section 11.5.

The Basic Assumptions

The assumption of *independent random variables* states that values of Y_i, the dependent variable, are statistically independent. Thus, a large value of Y_1 does not tend to make Y_2 large. Expressed another way, Y_2 is not affected by Y_1; or Y_2 is not related to Y_1. If we have sampled correctly, using a random sampling procedure, there is no obvious reason against making this assumption. However, with spatial data we will see that autocorrelation—the lack of independence—especially in the dependent variable, occurs.

A second assumption, *homoscedasticity,* states that the variance, σ^2, of Y is the same for all X_i (Figure 11.1). If the variance increased with increased values of X_i, then we could not be confident of the predictions made from our regression line. Although we usually have only one observation of X_i for each Y_i, we must imagine having a large number of Y_i values for each X_i or a small range of X_i and these "distributions" must have the same variances. If variances are not equal, the estimates, although unbiased, are inefficient (Figure 11.1).

A third assumption—that of *linearity*—relates to the means of the dependent variable for a variety of values of the independent variable. The

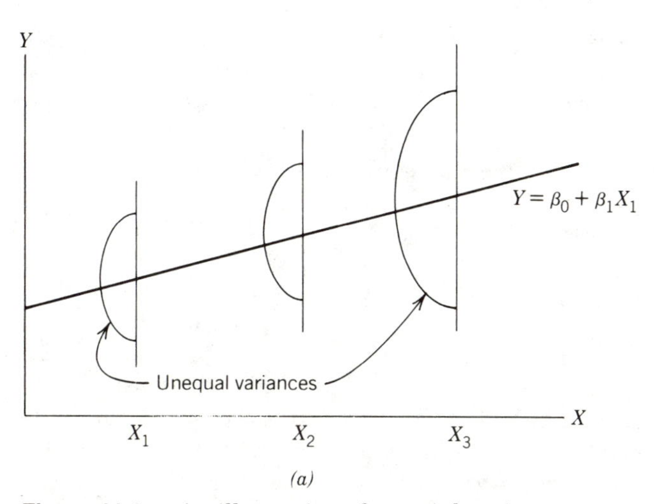

Figure 11.1*a* An illustration of unequal variances.

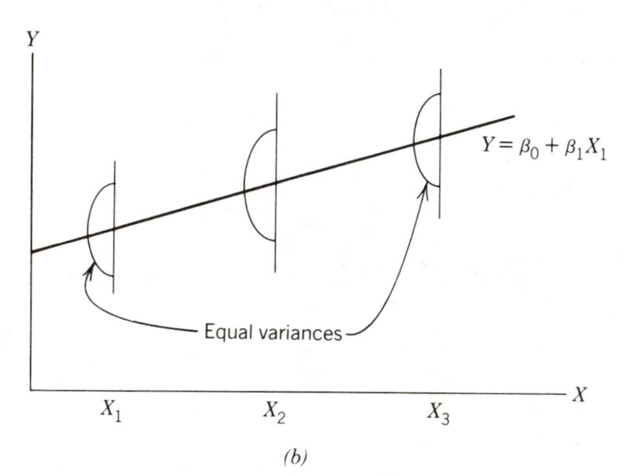

Figure 11.1*b* An illustration of equal variances.

assumption is that the means ($E(Y_i)$) lie on a straight line known as the true, or the population, regression line. If the values do not lie on a straight line, the estimate of the β values will be inefficient (Figure 11.2).

Quite often, these three assumptions are stated not in terms of Y_i, but in terms of the residual or error term e_i, where $e_i = \hat{\epsilon}_i = (Y - \hat{Y}_i)$. The residual is the quantity that is left over after the independent variables have explained as much as possible. Y_i and e_i have the same distributions, except for different means, with the mean of the e_i being 0 (summing the positive and negative deviations yields zero). More formally, we write the e_i are independent random variables with mean = 0 and variance = σ_E^2.

Normality Assumptions

A fourth assumption (*normality*) is required for testing the β's and relates to the shape of the distribution. In effect it has two parts. First, the assumption relates to the distribution of e_i, the error term; specifically, that the e_i, or equivalently the Y_i, are normally distributed. Second, it can relate to a multivariate normality assumption, that all variables are jointly normally distributed for tests of R^2.

The first three assumptions are called the weak assumptions. This distinguishes them from the strong assumptions that include, in addition to the first three assumptions, the assumption that the e_i's are normally distributed. It is possible to show with the Gauss-Markov theorem that the least-squares method yields the *b*est *l*inear *u*nbiased *e*stimates (sometimes called BLUE estimates). The Gauss-Markov theorem states that within the class of linear unbiased estimators of β_j, the least-squares estimator has minimum variance. This is the major justification for using the least-squares method in the linear regression model. In most cases, however, we want to make tests on the estimates we have derived for the regression equation, and this requires the assumption of normality of the error term, the e_i's.

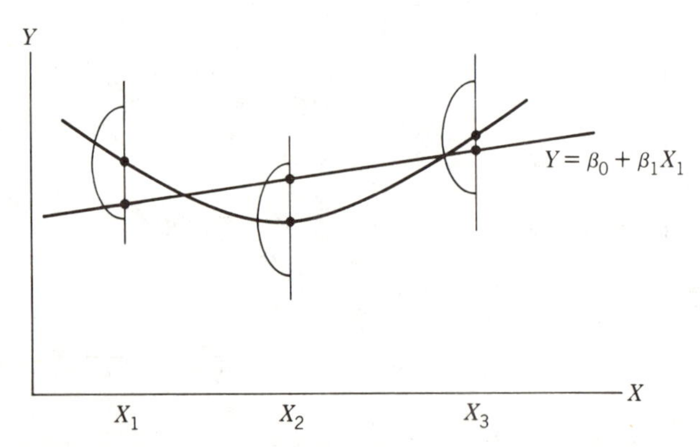

Figure 11.2 Departure from the linearity assumption.

Note again that the assumption of normally distributed e_i is the same as assuming a normally distributed Y_i. When the e_i's are normally distributed, ordinary least-squares estimates are equivalent to maximum likelihood estimates (MLE) for β_j.

An intuitive understanding of MLE is useful at this point, although we will not develop this material in any detail. Maximum likelihood estimators are used because they have the statistically attractive properties of efficiency and consistency (Section 6.2). An efficient estimator is one whose distribution has a small variance, and a consistent estimator is one which approaches the target (the population value) as the sample size increases. The MLE estimates are the hypothetical values of β_j which generate the greatest probability for the sample values we have observed. Deriving MLE's involves speculating on various possible populations and how likely each is to give rise to the sample observed. The fact that maximum likelihood estimates are identical to least-squares estimates (when Y_i is normally distributed) is further support for the use of the least-squares method. This discussion has also developed the reasoning behind the assumption of a normally distributed dependent variable.

We have made no assumptions about the independent variable, nor is it necessary for regression analysis. Our interest centers only on the estimates of β_j; β_0, and β_1 and the predicted value of Y_i. If, however, we are interested in R, the multiple correlation coefficient, then in order to test for the significance of R it is necessary to assume normally distributed independent variables. The assumption of normally distributed Y and X variables is termed *multivariate normality,* an extension of bivariate normality (Section 9.5).

Even without multivariate normality, it is possible to test for at least one aspect of regression analysis with a test of the regression coefficients. A key test in correlation analysis is a test of the null hypothesis H_0: $\rho = 0$ (H_0: $r = 0$), which requires multivariate normality. However, by testing H_0: $\beta_1 = 0$, the rejection of $\beta_1 \neq 0$ implies $\beta = 0$ and the conclusion that there is no correlation between X and Y.

11.2 THE ANALYSIS OF RESIDUALS

In most instances, the validity of the assumptions is examined through an analysis of the residuals. Graphical and statistical examination of the residual terms from least-squares regression analysis is an important tool in producing adequate regression models. Examination of the residuals can lead to model refinement in terms of transformations and new variables, the discovery of unknown influences on the model, and most critically, it can be used to test for violations of the assumptions of the linear regression model. It is in the interest of every student to undertake at least some analysis of the residuals every time a regression analysis is performed. With the advent of computers, this can be done easily and cheaply.

The Nature of Residuals

Recalling the regression model:

$$\hat{Y}_i = b_0 + b_1X_{i1} + b_2X_{i2} + \cdots + b_kX_{ik} \tag{11.1}$$

where

$\hat{Y}_i$ = the predicted dependent variable

b_0 = a constant

b_j = regression coefficients

X_{ij} = independent variable

k = the number of independent variables

and

$$e_i = Y_i - \hat{Y}_i = \text{the residual, an estimate of } \epsilon_i \text{ the unobservable error term} \tag{11.2}$$

The residual term (e_i) is, then, the difference between the actual value of the dependent variable (Y_i) and the value of the dependent variable ($\hat{Y}_i$) calculated or predicted from the regression model. It represents the amount of variation in the observed Y_i, which is "unexplained" by the regression line as in Figure 9.14 for the bivariate case.

The residual is commonly called the "error" term and includes the true error in the model as well as all the ways the fitted model fails to properly explain the observed variation. More formally, the residual equals a combination of modeling error and stochastic error.

The results can be examined in the form of,

1. the (absolute) residual $Y_i - \hat{Y}_i$
2. a relative residual $\dfrac{Y_i - \hat{Y}_i}{Y_i}$ (11.3)
3. a standardized residual $\dfrac{Y_i - \hat{Y}_i}{S_E}$ (11.4)

where

S_E = the standard error of estimate

Both the absolute and standardized residuals can be easily obtained from SPSS and SAS.

A variation on this distributional analysis of the residuals is the areal mapping of residuals (Krumbein and Graybill, 1965; Chorley and Haggett, 1968; King, 1969). Residuals, especially in the form of the standardized residual when plotted on a map, provide an excellent visual indication of idiosyncrasies in the model or failure of the model to explain all the variation in Y_i. An example of residual plots for a bivariate case is given in Figure 11.3.

While the mapped residuals can be used for modifications of the model or its regional interpretation, the current discussion of residuals is more specifically directed to examining the assumptions of the error term. To reiterate, those assumptions where e_i is our estimate of ϵ_i, are

1. $E(\epsilon_i) = 0$
2. $E(\epsilon_i^2) = \sigma_E^2$ (a constant) for all X_i
3. the ϵ_i^2s are normally distributed
4. ϵ_i is independent from ϵ_j $(i \neq j)$

Techniques of Examination

Anscombe and Tukey (1963, p. 142) suggest that one of the most useful analysis techniques is the visual examination of graphs of the residual terms, especially for time-series or spatial location data. Although most residual analyses plot the residuals against several standard factors, any plot that

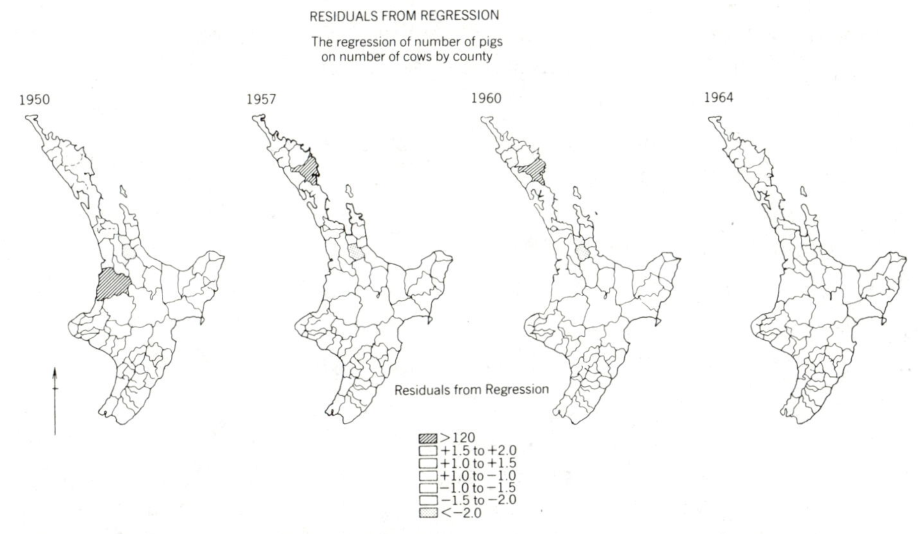

Figure 11.3 Residuals from the regression of number of pigs on number of cows by county for the North Island of New Zealand. Source: W.A.V. Clark, 1967, by permission of the New Zealand Geographical Society.

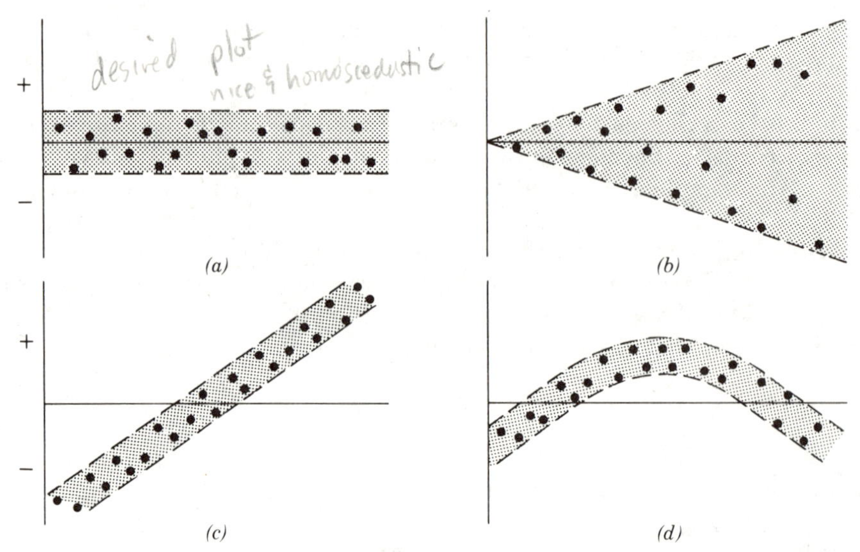

Figure 11.4 Basic forms of the residual plot.

shows potential nonhomogeneity is useful. If, by arranging any of the variables such that a plot of the residuals shows some nonrandom pattern (i.e., some systematic distribution), then the model does not contain all that is needed to completely explain the dependent variable. Wonnacott and Wonnacott (1970, p. 74) urge the researcher to "(squeeze) the residuals until they talk."

Several different types of plots are possible, but in essence, they focus on the following plots of the residuals against Y, $\hat{Y}$, *or* X.

1. Overall plots (frequency distributions) of the numerical value (absolute form) to analyze "normality."
2. Plots of the residuals in time sequence or order sequence of observation, to review for serial autocorrelation.
3. Plots of the residuals against the predicted $\hat{Y}_i$.
4. Plots of the residuals against the independent variables X_{ij}.
5. Plots of the residuals against the actual Y_i values.

The form of the residual plots (for 2 to 5 above) can take a variation of one of the forms shown in Figure 11.4. The uniform band of equal variance for all values of the plotted variable (Figure 11.4*a*) is the desired plot—any of the other forms indicates deficiency in the model in some way.

Deviant forms as shown in Figure 11.4 indicate the following problems.

(*b*) Variance changing (increasing) with values of the plotted variable, (a weighted least-squares approach is needed).

(*c*) An additional linear term is needed in the model.

(*d*) An additional quadratic term is needed in the model or a respecification (transformation) of Y_i prior to regression.

Anscombe and Tukey (1963, p. 146) consider the detection of *outliers* (major departures from the pattern of residuals) as the single most important reason for calculating and examining residuals. Most tests for examining residuals are very sensitive to the presence of outliers and they go so far as to say "...it is usually unsafe to apply a *numerical* procedure to residuals until outliers have been screened out...." Several arbitrary methods of outlier rejection are available, but Draper and Smith (1966, p. 95) urge caution. Outliers may simply be an error in measurement or data transcription, in which case it should be rejected out of hand, but it might be a signal pointing toward an unexpected combination of circumstances surrounding the experiment, or provide revealing insights about the phenomenon under study.

Although many researchers feel that visual examination of residual plots is sufficient for most general geographical research, there are mathematical statistics designed for more rigorous and exhaustive testing of residual distributions. A complete accounting of these statistics and additional references can be found in Anscombe and Tukey, (1963), Goldberger, (1962), and Malinvaux, (1970). The next section examines in greater detail and outlines the assumptions of normality, linearity, and homoscedasticity.

11.3 NORMALITY, LINEARITY, AND HOMOSCEDASTICITY

As we discussed in Section 11.1, the normality assumption is concerned with two different situations—the normality of Y, and the multivariate normality of Y with the X. But in recognizing the difference between the fixed-X model (a number of observed Y values for each value of X) and the random-X model (pairs of XY scores), we are, in fact, interested in normality from three different points of view.

1. In the fixed-X model we want to know for *each X score* if the corresponding Y scores (or the residuals) are normally distributed.

2. In the random-X model, we want to know if *all* the Y scores (or all the residuals) are normally distributed.

3. In either model, in order, to make conclusions about the multiple correlation coefficient R we need to know if all the scores (Y and X_j) are distributed according to a multivariate normal distribution. In practice, we can

TABLE 11.1 Distribution of Levels of Moisture in Beach Sand

Percent Moisture in Beach Sand	Frequency	Cumulative Frequency	Percent Cumulative Frequency
0.0–0.9	4	4	.03
1.0–1.9	14	18	.14
2.0–2.9	38	56	.44
3.0–3.9	42	98	.77
4.0–4.9	23	121	.95
5.0–5.9	6	127	1.00

Source: Krumbein and Graybill, 1965, p. 123.

approximate this by finding out if the Y variable and all the X_j variables are normally distributed.

Testing for Normality

In Chapter 6, we discussed ways of deriving an expected normal distribution. Either a visual test, accomplished by plotting the cumulative frequency distribution on normal probability paper, or more properly, an expected normal distribution using the sample mean and variance can be derived and compared with the sample distribution. For example, can we assume that the distribution of data in Table 11.1 is a normal distribution? Visually, the points are close to a straight line, although there is some deviation (Figure 11.5).

We can use the mean ($\overline{X} = 3.16$) and standard deviation ($\hat{\sigma} = 1.14$) to construct an expected normal distribution for the moisture in beach sand data. The procedure is illustrated in Table 11.2. Using the procedures in Section 6.3, we can compute the expected area under the normal curve and multiply that proportion times the total number of cases in each interval. The result is an observed and an expected distribution that can be tested with χ^2. (Note that we combine the extreme classes because the expected frequencies are less than 5.) In the example in Table 11.2, we cannot reject the hypothesis of no difference and conclude the distribution is not different from a normal distribution. The chi-square value = .2834 and the table value for χ^2 with 2 degrees of freedom at the .05 level is 5.991.

Testing for Linearity

Because the multiple regression model is additive in its parameters, the assumption of linearity is the same for multiple regression as for simple regression. Thus, *each* independent variable should have a linear relation-

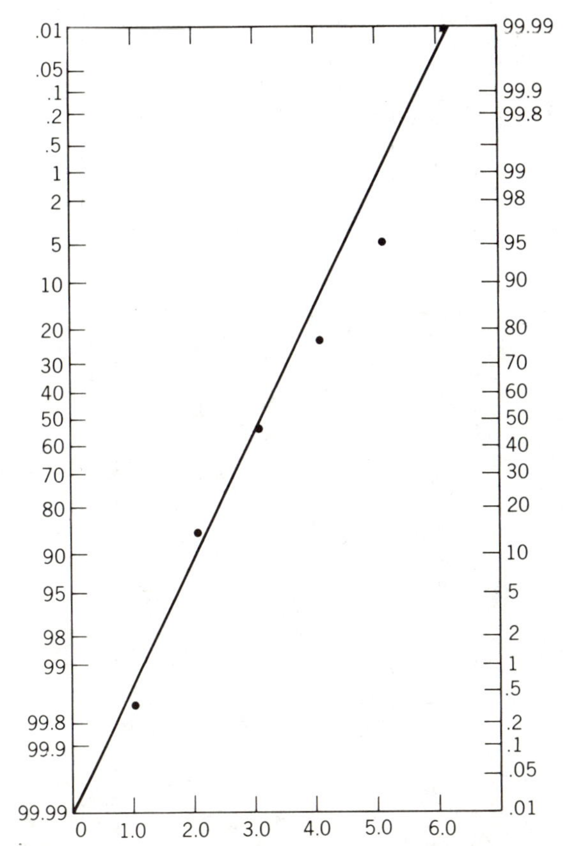

Figure 11.5 Percent moisture in beach sand.

ship with Y, and each should be tested. The test for linearity is also straightforward, although testing may be unnecessary if there is an obvious curve in the scatter diagram. We illustrated the issues for two variables in Chapter 9. If the scatter of points (imagining several values of Y for each X) is basically linear in form, then the means of some arbitrary set of groups will lie close to the regression line. If, on the other hand, the data are curvilinear in form, the means of some arbitrary groups will lie far from the regression line (again, see Figure 9.19).

In the multivariate case, we can conduct a series of tests of the dependent and each of the independent variables to establish linearity of the plane. The tests and an example were discussed in Chapter 9. To reiterate, we can calculate a total sum of squares that can be partitioned into a sum of squares around the group means correlation ratio E^2 (equivalent to within-group sum of squares in the analysis of variance routine), a sum of squares around the regression line, r^2, and an unexplained sum of squares (by the group

TABLE 11.2 Calculation of Expected Normal Distribution

Class (% Moisture)	Observed Cases	Class Boundary	$Z = \frac{X - \overline{X}}{\hat{\sigma}}$	Cumulative Area	Class Probability	Expected Frequency
				.0028	.0028	.36[a]
.0–1.9	4	.0	−2.77	.0294	.0266	3.38
1.0–1.9	14	1.0	−1.89	.1539	.1245	15.81
2.0–2.9	38	2.0	−1.02	.4443	.2904	36.88
3.0–3.9	42	3.0	−.14	.7704	.3261	41.41
4.0–4.9	23	4.0	.74	.9463	.1759	22.34
5.0–5.9	6	5.0	1.61	.9936	.0473	6.01
≥6.0		6.0	2.49	1.0000	.0064	.81
	127				1.0000	127.00

[a] 127 × .0028.

means). Thus,

$$\begin{aligned}\text{Total SS} &= \Sigma(Y_i - \overline{Y}_i)^2 \text{ with } n - 1 \text{ df}\\ \text{Explained by linear regression} &= r^2\Sigma(Y_i - \overline{Y}_i)^2 \text{ with 1 df}\\ \text{Additional explained by the group means (nonlinear)} &= (E^2 - r^2)\Sigma(Y_i - \overline{Y}_i)^2 \text{ with } k - 2 \text{ df}\\ \text{Unexplained (by group means)} &= (1 - E^2)\Sigma(Y - \overline{Y})^2 \text{ with } n - k \text{ df}\end{aligned}$$

where k = the number of classes.

The test involves the ratio of the SS explained by the nonlinear relationship to the amount unexplained by the most general model. Clearly, a large additional proportion explained by a nonlinear model of the amount unexplained by a linear general model is indicative of a nonlinear relationship. The test reduces to

$$F = \frac{(E^2 - r^2)(n - k)}{(1 - E^2)(k - 2)} \tag{11.5}$$

which can be tested against $[F, p = 1 - \alpha, df = (k - 2), (n - k)]$.

Although E^2 can be calculated for any number of groups k, a very small number of observations as a ratio of the number of groups, will make the assumption that the X scores are clustered at the midpoints an unrealistic assumption.

Homoscedasticity and Weighted Least Squares

Homoscedasticity is the assumption of equal variances. When it is not true, the estimates of the regression coefficients are unbiased but inefficient (Intriligator, 1978, p. 156). Heteroscedasticity, the lack of equal variance, often gives rise to a horn-shaped scatter of points when we plot observed Y against a selected X_j (Figure 11.6).

There is no adequate formal test for homogeneity therefore we must obtain an estimate of its presence from the examination of residuals only. A solution to this problem is given by a method called weighted least squares (WLS). WLS allows us to minimize the impact of the observations in the wide part of the horn because these add substantially to the residual sum of squares.

In ordinary least-squares simple regression, the function minimized was:

$$\Sigma(Y_i - b_0 - bX_{i1})^2$$

In weighted least squares the function minimized is:

$$\sum \frac{1}{s_i^2}(Y_i - b_0 - b_1X_{i1})^2 \tag{11.6}$$

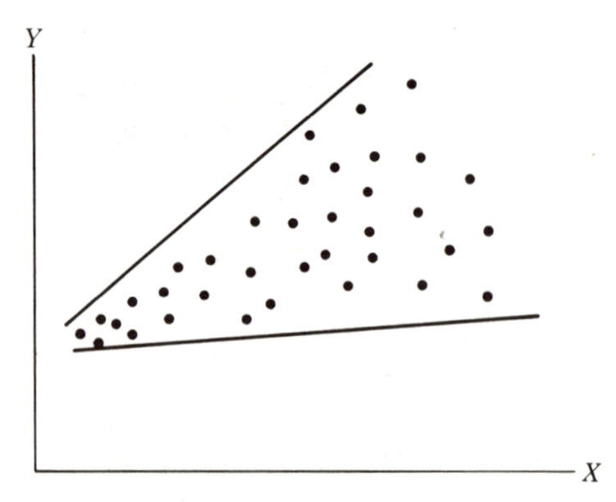

Figure 11.6 Heteroscedasticity.

where s_i^2 is the variance of the error term at a given value of X. The effect is to minimize the influence of large values of s_i^2 because in these cases the "weight" $1/s_i^2$, or the multiplying ratio, is small. However, in many instances, we do not know s_i^2 for X values, and it cannot easily be estimated without multiple values of Y for each X. There is at least one special case, however, that can be more easily solved. When the standard deviation of the error term s_E, the standard error of estimate, is thought to increase proportionately with the independent variable, then it is possible to transform the original equation by dividing through by the independent variable:

$$Y_i = b_0 + b_1 X_i \tag{11.7}$$

$$\frac{Y_i}{X_i} = b_0 \frac{1}{X_i} + b_1 + \frac{e_i}{X_i} \tag{11.8}$$

We then regress Y_i/X_i on $1/X_i$ using standard linear regression procedures. Note that the estimating equation is then $Y_i' = b_0' + b_1'X_i$, where the coefficient of $1/X_i$ is b_0' and the constant term is b_1'.

11.4 AUTOCORRELATION

It is common for many variables (economic, social, and physical) to be characterized by correlations between adjacent values in space or in time. This has been of considerable interest to economists who have developed a variety of tests for the analysis of autocorrelated time-series data. Geographers, on the other hand, have only recently directed their attention to the analysis of autocorrelation, and most of this work is still not widely incorporated into statistical analysis.

Autocorrelation, or serial correlation, as it is often called, refers to the correlation of the residuals from a regression equation. By auto or serial correlation, we are describing the situation in which any residual or error value e_i is correlated with any of the previous values (e_{i-1}, e_{i-2}, etc.). There are two situations in which this problem arises.

1. For time-series data (for example, yearly rainfall, streamflow, gross national product, and the consumer price index)
2. For spatial data (for example, house value by tract, because the house value in a tract is likely to be similar to the adjacent tract. Such observations can be made for many other socioeconomic variables).

Both conditions give rise to autocorrelation of the error term and so violate the assumption of statistically independent error terms. The effect of this violation is that although the estimates of β_0 and β_1 from ordinary least squares are unbiased, the confidence intervals around these estimates are very large. Thus, the estimators are inefficient and t and F tests are invalid. (They are wide of the true predicted value.) Although we speak of autocorrelation in the residuals, in fact this arises primarily because of autocorrelation in the dependent variable, although it is possible to have autocorrelation without the variables themselves being autocorrelated.

Testing for Serial Correlation

A major part of econometrics is concerned with the identification and solution of problems of serial correlation that arise when time series data are analyzed. An excellent discussion of the effect of serial correlation and the use of regressions of first differences (regressions of ($Y_t - Y_{t-1}$ on $X_t - X_{t-1}$)) and lagged variables is available in Wonnacott and Wonnacott (1970) and will not be pursued here. However, it is appropriate to outline the standard test for serial (time series) autocorrelation.

The Durbin–Watson test for first-order serial correlation is

$$d = \frac{\sum_{t=2}^{n} (e_t - e_{t-1})^2}{\sum_{t=1}^{n} e_t^2} \tag{11.9}$$

where e_t denotes residuals at time t.

The Durbin–Watson test is a statistic that is designed to test for first-order autocorrelation in the residuals from regressions using time-series

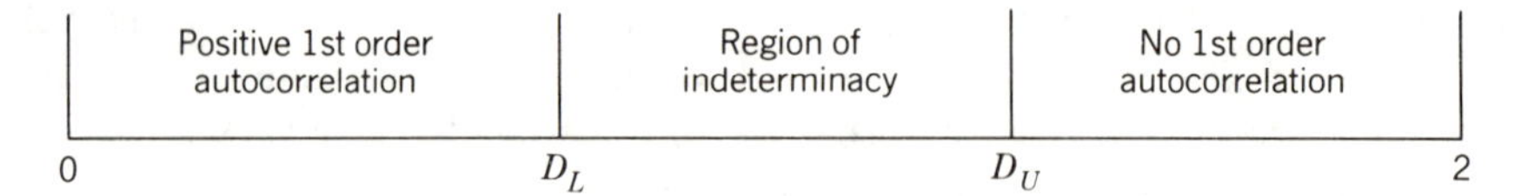

Figure 11.7 Decision intervals for the Durbin–Watson test.

data. There are upper and lower bounds for the test, D_U and D_L, for various values of n, the number of observations, and k, the number of independent variables in the equation. To use the test, evaluate d. If $d < D_L$, the lower bound, reject the hypothesis of random disturbance elements in favor of positive first-order autocorrelation. If $d > D_U$, it is likely that there is no first-order correlation (Figure 11.7). Selected values of d are given in Table H at the end of the book, larger tables may be found in Maddala (1977) and other econometrics books.

The test suffers from two problems. First, the region of ignorance, the interval between D_U and D_L presents some difficulty for the researcher. However, Intriligator says that the convention (if the calculated value falls in this range) is to reject the null hypothesis of no first-order serial correlation (Intriligator, 1978, p. 163).

The second and more fundamental problem for geographers is that the d statistic is not applicable to problems of spatial autocorrelation because it tests for association in only one direction. Thus, the Durbin–Watson test is useful for time-series data but has little practical use for geographers dealing with spatial series.

Spatial Autocorrelation

Geographers have only recently begun to examine the problems caused by spatial autocorrelation, and there are as yet only limited test statistics for the existence of spatial autocorrelation and few adequate solutions. The difficulty in geography or any study of a spatial series is that the potential autocorrelation may not be only with the earlier observations t, $t - 1$, $t - 2$, but may be with any contiguous unit. Thus, if a tract has five neighbors, all of these neighbors may be correlated with the observed tract.

As Tobler (1965) pointed out, it is not unreasonable to expect that events at one location (i,j) may be related to events at some other geographical location $(i + m, j + n)$. We are thus led to a consideration of autocorrelation of each observation with, for example, its neighbor immediately to the east, that is,

$$r(Z_{ij}, Z_{i+1,j})$$

or with the neighbor immediately north,

$$r(Z_{ij},Z_{j+1})$$

or with the neighbor two removed to the east,

$$r(Z_{ij},Z_{i+2,j})$$

and so on, where Z is some observed spatial variate.

We can readily see that the problem of spatial autocorrelation is much more difficult to handle than autocorrelation in time-series data, where there is a natural progression from past to future.

Given this problem, there is general agreement in the literature that three consequences result.

1. Standard errors of the estimates of β_0 and β_1 will not be minimized, and there may be underestimation of these variances.
2. Inferential tests (F and t) will no longer be valid.
3. Estimates of the dependent variable will be inefficient in that the sampling variance will be needlessly large.

We are dependent on the work of Cliff and Ord (1973, 1981) for much of our understanding of spatial autocorrelation—and it is only in the past five years that much progress has been made in this crucial area of spatial analysis. Even though the tests are complicated and their development is continuing, it is important to utilize even the limited tests that are presently available. In this section, we outline tests for spatial autocorrelation in the dependent (which would imply autocorrelation in the error term e_i) and independent variables, and discuss the modifications of that test for the residuals and an alternative (simpler) join count test also for the residuals.

Cliff and Ord (1981) have emphasized the I statistic as the most useful for the evaluation of spatial autocorrelation. The I statistic is defined as

$$I = \frac{n \sum_{i=1}^{n} \sum_{\substack{j=1 \\ i \neq j}}^{n} \omega_{ij} z_i z_j}{S_0 \Sigma z_i^2} \tag{11.10}$$

where ω_{ij} = the degree of contact between areas (the weights). In the simplest case, ω_{ij} = 1 if areas are contiguous and 0 otherwise. I can be simplified to

$$I = \frac{n}{2A} \frac{\Sigma(2)\ \omega_{ij} z_i z_j}{\Sigma z_i^2} \tag{11.11}$$

where the weights are 1/0 and where

$$\Sigma(2) = \sum_{i=1}^{n} \sum_{j=1}^{n} i \neq j$$

and

$$z_i \text{ and } z_j = X_i - \overline{X}, \text{ and } X_j - \overline{X}, \text{ respectively}$$

$$S_0 = \sum_{\substack{i=1 \\ i \neq j}}^{n} \sum_{j=1}^{n} \omega_{ij} = \text{total number of joins between the units of the system}$$

$$= 2A \quad \text{if } \omega_{ij} = 1$$

where n = the total number of counties or areas.

The expected value of I under the assumption of normality is

$$E(I) = -(n - 1)^{-1} \tag{11.12}$$

The second moment (under the assumption of normality) of I is

$$E(I^2) = \frac{n^2 S_1 - n S_2 + 3 S_0^2}{S_0^2 (n^2 - 1)} \tag{11.13}$$

where, in addition to previously used terminology,

$$S_1 = \tfrac{1}{2} \sum_{\substack{i=1 \\ i \neq j}}^{n} \sum_{i=1}^{n} (\omega_{ij} + \omega_{ji})^2 = 2 \text{ times } S_0 \text{ if } \omega_{ij} = 1$$

$$S_2 = \sum_{i=1}^{n} (\omega_{i.} + \omega_{.i})^2$$

where L_i is the number of areas or counties contiguous to county or area i, and

$$\omega_{i.} = \sum_{j=1}^{n} \omega_{ij}$$

$$\omega_{.i} = \sum_{j=1}^{n} \omega_{ji}$$

The second moment $E(I^2)$ is given with normality assumed. The assumption states that the variable being evaluated, X_i, is the result of n independent drawings from a normal population. (See Cliff and Ord (1981) for a discussion of the alternative randomization assumption.)

In the following simple example, the joins between units were assigned weights = 1, and only four joins (rook's case) were considered. It is possible to use weights of differing magnitudes and to use either a rook's or queen's matrix for rectangular lattices. The rook's and queen's matrices describe the matrices that are used to weight the values of the variables entering the autocorrelation function. When the weighting matrix is based on the rook's case, the value in the central cell is taken to depend on the cell values immediately to the north, south, east, and west. Each cell value is weighted as having an equal influence on the central cell. When the queen's case is used, the weighting matrix depends on the six cells that surround the central cell on a checkerboard pattern (Figure 11.8).

A simple hand-worked example can be derived from Figure 11.9 where $n = 9, A = 12$, and $\overline{X} = 2.777$. The basic calculations are provided in Table 11.3.

$$I = \frac{n}{2A} \frac{\Sigma(2)\ \omega_{ij} z_i z_j}{z_i^2} = \frac{9}{2(12)} \frac{3.4075}{13.5556}$$

$$= (.375)(.2514) = +.0943$$

$$E(I) = -(n - 1)^{-1} = -(9 - 1)^{-1} = -.125$$

$$E(I^2) = \frac{4An^2 - 8(A + D)n + 12A^2}{4(A^2)(n^2 - 1)},$$

$$D = \tfrac{1}{2}\, \Sigma L_i(L_i - 1) = \tfrac{1}{2}\,(44) = 22$$

$$= \frac{4(12)9^2 - 8(12 + 22)\ 9 + 12(12^2)}{4(12^2)(9^2 - 1)}$$

$$= \frac{5616 - 2448}{46{,}080} = \frac{3168}{46{,}080} = -.06875$$

$$\text{Standard deviation } I = \sqrt{E(I^2)}$$

$$= .262$$

$$Z = \frac{I - E(I)}{\text{S.D.}(I)} = \frac{+.0943 - (-.125)}{.262}$$

$$= \frac{+.2193}{.262}$$

$$= .837$$

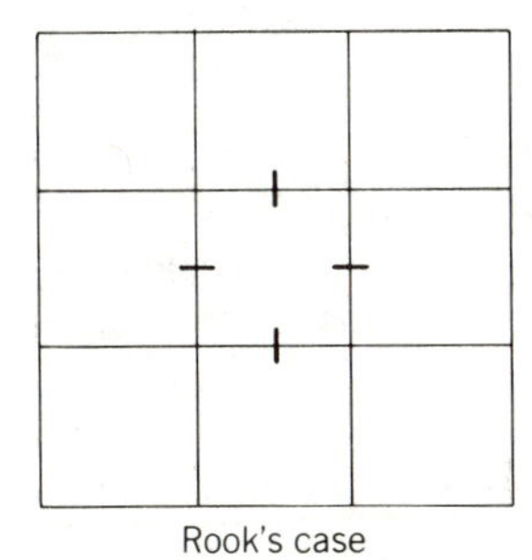

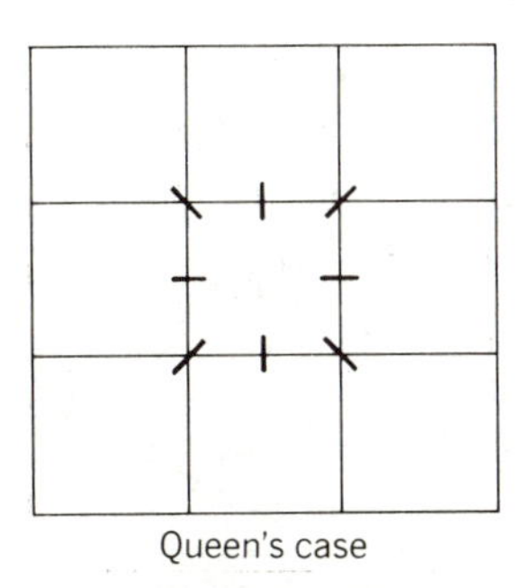

Figure 11.8 Rook's and queen's matrices indicates that cell value influences central cell.

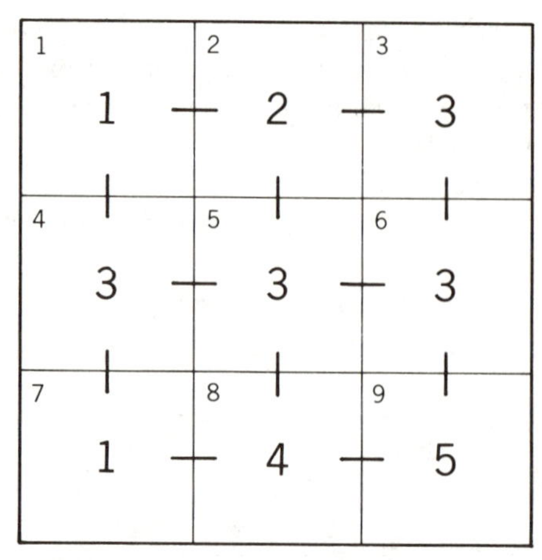

Figure 11.9 A simple data set consisting of a 3 × 3 grid joined by the rook's binary weighting matrix. Study unit grid designations are in the upper left corner of each cell, data values are the central figures. Joins (A) between study units are indicated by the dark bars.

where Z is a standard normal deviate. Spatial autocorrelation in this example is probably not statistically significant. However, the example is only illustrative, and at such a small sample size, the I standard deviate cannot be assumed to be normally distributed. With larger sample sizes (around 60 or more) z could be compared to standard normal distribution tables for hypothesis testing. A more realistic example is provided by an examination of a common urban pattern—the median value of housing in 1970 for a pattern of tracts in the City of Los Angeles. We find that the I statistic is .4469, and the standard normal deviate is 7.2 (Figure 11.10). These results clearly indicate the existence of blocks of autocorrelated cells.

Tests for Autocorrelated Residuals

The preceding section introduced tests for spatial autocorrelation of any variable, but, in most instances, our interest is in testing the residuals from a regression analysis that has utilized data for spatial units. This can arise from an ordinary bivariate or multivariate model that utilizes tract, county, or regional data, or from a trend surface analysis of point or cell data. Certainly it is possible to test for spatial autocorrelation in the dependent variable, and the existence of such autocorrelation would suggest that the e_i will be correlated, thus a test of the residuals is essential. However, it is likely in most situations that utilize data on spatial units, that there will

TABLE 11.3 Calculations for 3 × 3 Rook's Case Example

i	X_i	$z_i(=X_i - \overline{X})$	z_i^2	Joins	$\sum_{j=1}^{n} \omega_{ij} z_i z_j$	$L_i(L_i - 1)$
1	1	−1.7778	3.1605	2,4	(−1.7778)(−.7778 + .2223) = +.9876	2(2 − 1) = 2
2	2	− .7778	.6049	1,3,5	(−.7778)(−1.7778 + .2223 + .2223) = +1.037	3(3 − 1) = 6
3	3	+ .2223	.0494	2,6	(+.2223)(−.7778 + .2223) = −.1235	2(2 − 1) = 2
4	3	+ .2223	.0494	1,5,7	(+.2223)(−1.7778 + .2223 − 1.7778) = −.741	3(3 − 1) = 6
5	3	+ .2223	.0494	2,4,6,8	(+.2223)(−.7778 + .2223 + .2223 + 1.2223) = +.1977	4(4 − 1) = 12
6	3	+ .2223	.0494	3,5,9	(+.2223)(+.2223 + .2223 + 2.2223) = .5929	3(3 − 1) = 6
7	1	−1.7778	3.1605	4,8	(−1.7778)(+.2223 + 1.2223) = −2.5682	2(2 − 1) = 2
8	4	+1.2223	1.4938	5,7,9	(+1.2223)(+.2223 − 1.7778 + 2.2223) = .815	3(3 − 1) = 6
9	5	+2.2223	4.938	6,8	(+2.2223)(+1.22223 + .22223) = +3.21	2(2 − 1) = 2
TOTAL			13.5556		= 3.4075	44

Reproduced from Whitley and Clark (1978) by permission of Academic Press.

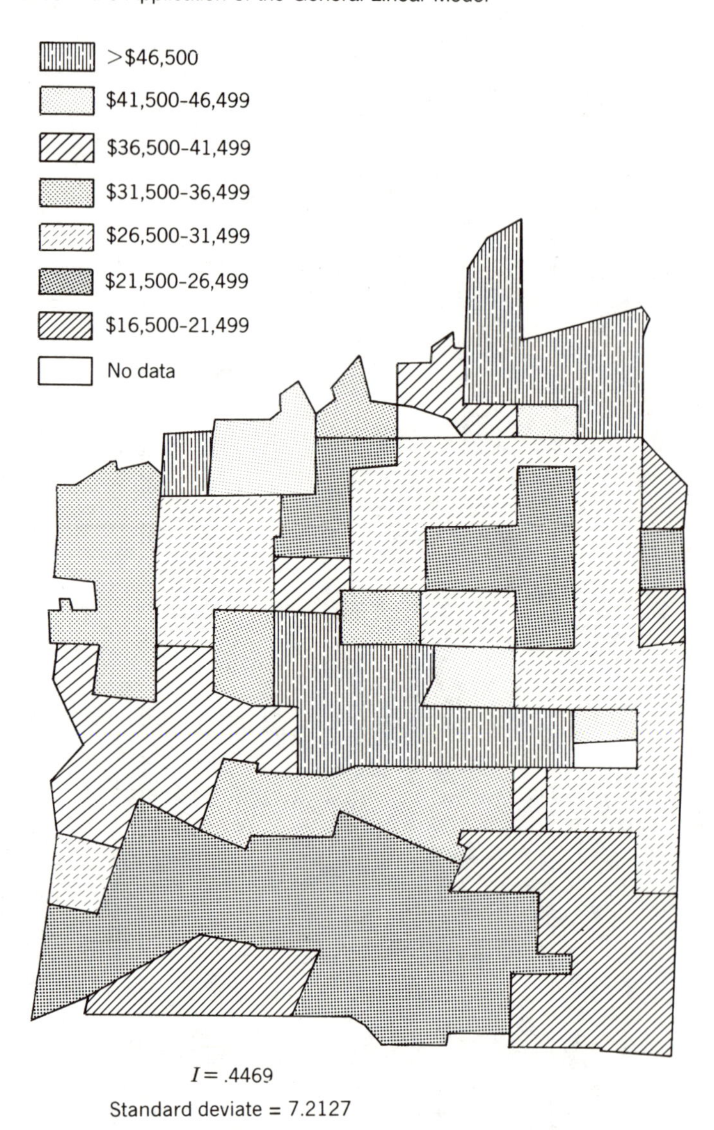

Figure 11.10 Los Angeles Hancock Park area. Median value of housing, 1970.

be spatial autocorrelation in both the dependent and independent variables (Martin, 1974) and if the independent variables are spatially autocorrelated, then the standard errors are likely to be seriously underestimated.

Spatial dependence is the rule rather than the exception and understanding it is critical to the development of models of spatial organization. If geographers are to use inferential statistics and their associated tests of significance correctly, it is crucial that they clearly understand the impact on the models of spatial dependence. There are a number of ways to do this, but prior to complicated revisions of statistical models applied to spatial data, we must be clear about measurement of existence of spatial dependence. Analyses with the I statistic will yield important information for the interpretation of the statistical results. Although we have already outlined the I statistic as a technique for examining the degree of spatial autocorrelation in a dependent or independent variable, modifications of that I statistic are required to apply the technique in multivariate analysis. In this section, we outline the I statistic for the analysis of residuals and a simpler technique, the join count statistic. We take up the latter first.

Join Count Statistics

One method for analyzing residuals from regression depends on the notions of contiguity and the simplification of the interval valued residuals to either negative or positive values. The question that is posed in a contiguity analysis is this: Are the negative and positive units (counties, tracts, regions, . . .) arranged in a definite pattern (i.e., in a nonrandom manner) or are the units distributed without pattern (i.e., randomly)?

Using N to represent negative residuals (or units) and P to represent positive residuals (or units) and defining a join as the existence of a positive nonzero link between two spatial units (or two units with a common border), we can develop a test for contiguity. The joins can be NN, PP, or PN. A comparison of the actual number of NN, PP, and PN joins with the number expected under the null hypothesis is a test of contiguity or grouping. If all the N units are clustered in one part of the map, and the P units in another, then there will be more NN and PP joins than expected, and fewer PN joins.

Dacey (1968) developed tests for NN, PP, and PN joins as part of his analysis of two- and k-color map patterns. He identified two "colors," black (B) and white (W) and the possible joins as BB, WW and BW. The normal distribution (Z values ± 1.96) can be used to test the probabilities of BB, WW, and BW joins or in our case NN, PP, and PN joins.

Dacey noted that the number of units should be large (probably greater than 50), especially if the study area is elongated, in which case p (the probability of a unit having a positive residual or a negative residual) should be large ($p > .7$). It is not necessary for all three tests to be applied. A PN test may be sufficient, although by convention all three tests are usually computed. The test statistics are

$$Z(NN) = \frac{J(NN) - \mu(NN)}{\sigma(NN)} \tag{11.14}$$

$$Z(PP) = \frac{J(PP) - \mu(PP)}{\sigma(PP)} \tag{11.15}$$

$$Z(PN) = \frac{J(PN) - \mu(PN)}{\sigma(PN)} \tag{11.16}$$

The expected number of joins for an irregular lattice is

$$\mu(NN) = p^2L$$
$$\mu(PP) = q^2L$$
$$\mu(PN) = 2pqL$$

where

$L = \frac{\Sigma L_i}{2}$, and L_i = the number of links for area or unit i

$p = \frac{N_N}{M}$, probability of an area having a positive residual

$q = \frac{N_P}{M} = 1 - p$, probability of an area having a negative

residual and N_N and N_P are the number of areas with negative and positive residuals, respectively, and M is the total number of areas. The expected variances are

$$\sigma^2(NN) = p^2L + p^3K - p^4(L + K)$$
$$\sigma^2(PP) = q^2L + q^3K - q^4(L + K)$$
$$\sigma^2(PN) = 2pqL + pqK - 4p^2q^2(L + K)$$

where

$$K = \sum_{i=1}^{n} L_i(L_i - 1)$$

Example

Data from Cliff and Ord (1981, pp. 224–225) for the residuals from a quadratic surface fitted to agricultural land values for 1978 are evaluated with join count statistics. The data for the analysis is presented in Figure 11.11 and Table 11.4.

The following calculations apply: (Number of residuals $(M) = 99$)

$$p = \frac{N_N}{M} = \frac{55}{99} = .556$$

$$q = 1 - p = 1 - .556 = .444$$

The constants are

$$L = \frac{\sum L_i}{2} = \frac{443}{2} = 221.5$$

$$K = \sum (L_i(L_i - 1)) = 1662$$

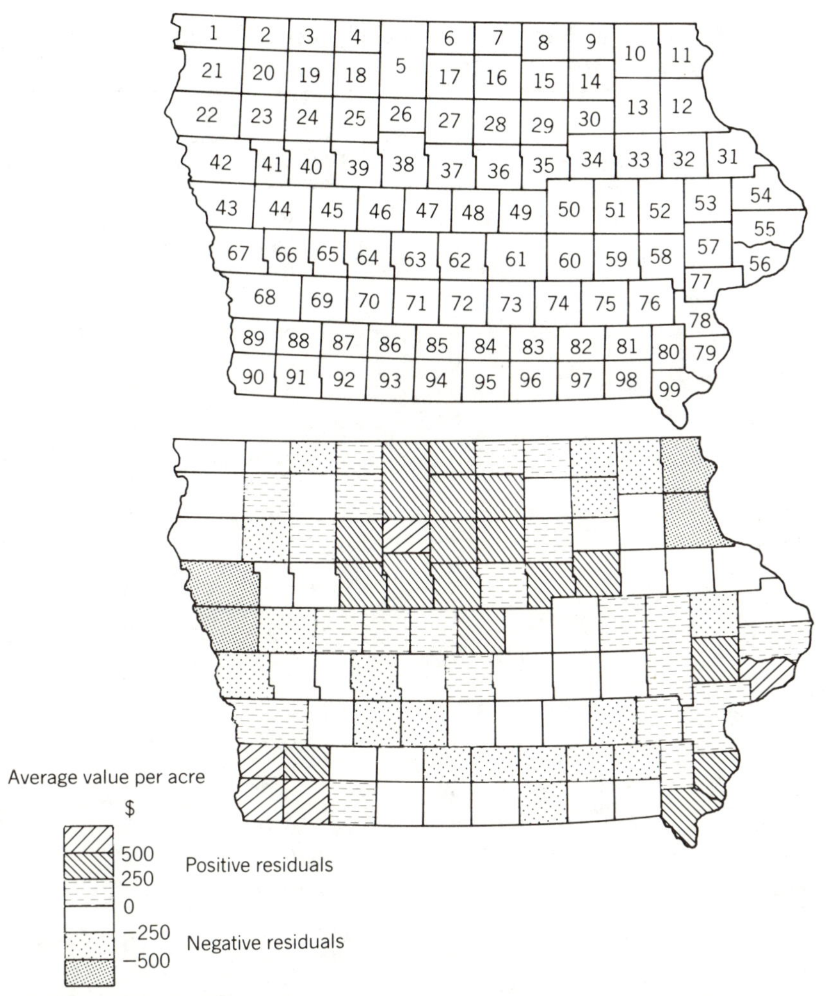

Figure 11.11 Residuals from a quadratic regression surface fitted to 1978 average land values per acre by county for the state of Iowa. Source: Cliff, A.D., Ord, J.K., *Spatial Processes,* 1981 (Pion Ltd.), by permission of the publishers.

The expected number of joins of $- -$ (NN) $+ +$ (PP) and $- +$ (PN) are

$$\mu(NN) = p^2L = (.556)^2(221.5) = 68.47$$

$$\mu(PP) = q^2L = (.444)^2(221.5) = 43.67$$

$$\mu(PN) = 2pqL = 2(.444)(.556)(221.5) = 109.36$$

TABLE 11.4 Details of Joins for a Contiguity Test of Residuals from a Trend Surface Analysis of 1978 Iowa Agricultural Data

Unit	Residual	Joins			
		NN (− −)	*PP* (+ +)	*PN* (− +)	Total
1	−	2	0	0	2
2	−	2	0	1	3
3	−	2	0	1	3
4	+	0	2	1	3
5	+	0	5	0	5
6	+	0	3	0	3
7	+	0	3	0	3
8	+	0	2	2	4
9	−	2	0	1	3
10	−	4	0	0	4
11	−	2	0	0	2
12	−	5	0	0	5
13	−	5	0	0	5
14	−	5	0	0	5
15	−	1	0	3	4
16	+	0	4	1	5
17	+	0	4	0	4
18	+	0	3	1	4
19	−	1	0	3	4
20	+	0	0	4	4
21	−	2	0	1	3
22	−	3	0	0	3
23	−	3	0	2	5
24	+	0	1	3	4
25	+	0	5	0	5
26	+	0	4	0	4
27	+	0	5	0	5
28	+	0	4	0	4
29	+	0	3	2	5
30	−	2	0	2	4
31	−	4	0	0	4
32	−	4	0	1	5
33	−	2	0	3	5
34	+	0	3	3	6
35	+	0	3	2	5
36	+	0	4	1	5
37	+	0	5	0	5
38	+	0	7	0	7
39	+	0	4	1	5
40	−	2	0	3	5
41	−	4	0	0	4
42	−	4	0	0	4

TABLE 11.4 *(Continued)*

Unit	Residual	Joins			
		NN (− −)	*PP* (+ +)	*PN* (− +)	Total
43	−	3	0	0	3
44	−	5	0	1	6
45	+	0	2	4	6
46	+	0	4	2	6
47	+	0	5	1	6
48	+	0	4	2	6
49	−	2	0	3	5
50	−	2	0	3	5
51	+	0	2	3	5
52	+	0	3	3	6
53	−	3	0	3	6
54	−	2	0	1	3
55	+	0	2	2	4
56	+	0	3	0	3
57	+	0	5	1	6
58	+	0	5	1	6
59	−	2	0	3	5
60	−	5	0	0	5
61	−	4	0	2	6
62	+	0	2	3	5
63	−	2	0	3	5
64	−	3	0	2	5
65	−	3	0	1	4
66	−	4	0	1	5
67	−	3	0	1	4
68	+	0	2	3	5
69	−	4	0	2	6
70	−	5	0	0	5
71	−	5	0	0	5
72	−	4	0	1	5
73	−	5	0	0	5
74	−	6	0	0	6
75	−	5	0	1	6
76	+	0	3	3	6
77	+	0	4	0	4
78	+	0	5	0	5
79	+	0	3	0	3
80	+	0	4	2	6
81	−	3	0	2	5
82	−	5	0	0	5
83	−	5	0	0	5
84	−	5	0	0	5
85	−	5	0	0	5
86	−	5	0	0	5

TABLE 11.4 *(Continued)*

		Joins			
Unit	**Residual**	*NN* (− −)	*PP* (+ +)	*PN* (− +)	**Total**
87	−	3	0	2	5
88	+	0	3	2	5
89	+	0	3	0	3
90	+	0	2	0	2
91	+	0	3	0	3
92	+	0	1	2	3
93	−	2	0	1	3
94	−	3	0	0	3
95	−	3	0	0	3
96	−	3	0	0	3
97	−	3	0	0	3
98	−	2	0	2	4
99	+	0	2	1	3
TOTALS		185	146	112	443

and their respective variances are

$$\sigma^2(NN) = p^2L + p^3K - p^4(L + K)$$
$$= (.556)^2(221.5) + (.556)^3 1662 - (.556)^4(221.5 + 1662)$$
$$= 174.14$$

$$\sigma^2(PP) = q^2L + q^3K - q^4(L + K) = (.444)^2(221.5) + (.444)^3(1662)$$
$$-(.444)^4(221.5 + 1662)$$
$$= 115.94$$

$$\sigma^2(PN) = 2pqL + pqK - 4p^2q^2(L + K)$$
$$= 2(.556)(.444)(221.5) + (.556)(.444)(1662)$$
$$- 4(.556)^2(.444)^2(1883.5)$$
$$= 60.51$$

With this information we now obtain the three test statistics.

$$Z(NN) = \frac{J(NN) - \mu(NN)}{\sigma(NN)} = \frac{92.5 - 68.47}{13.20} = 1.82$$

$$Z(PP) = \frac{J(PP) - \mu(PP)}{\sigma(PP)} = \frac{73 - 43.67}{10.77} = 2.72$$

$$Z(PN) = \frac{J(PN) - \mu(PN)}{\sigma(PN)} = \frac{56 - 109.36}{7.78} = -6.86$$

Two of the three values are greater than one would expect under the null hypothesis. Especially for the *PN* joins, we cannot accept *Ho*, and we conclude that contiguity is present in the pattern of residuals. The results parallel the tests with the I statistic in Cliff and Ord (1981).

The foregoing example has demonstrated the use of the technique in testing for contiguity in the residuals from a regression. The residuals are assumed to be independent random variables, a condition that is violated if contiguity is present. Although the technique just described can be used to test for patterns in residuals, the necessity of using a binary measure for the residuals means a loss of precision. A more sophisticated technique for testing for autocorrelation involves a variation of the I statistic previously developed.

Tests of Residuals With the I Statistic

The approach in this section utilizes the earlier presentation of matrices and some elementary matrix manipulation with the SAS package.

The statistic I modified for the analysis of residuals is given by

$$I = \frac{n}{S_0} \frac{\mathbf{e}'\mathbf{W}\mathbf{e}}{\mathbf{e}'\mathbf{e}} \tag{11.17}$$

where $\mathbf{W}$ is the weighting matrix, $\mathbf{e}$ the vector of residuals, and

$$S_0 = \sum_{\substack{i=1 \\ i \neq j}}^{n} \sum_{j=1}^{n} W_{ij}$$

The expected value of I is given by:

$$E(I) = -\frac{n \cdot tr\,(\mathbf{A})}{(n - k)S_0}$$

where tr = trace of the matrix, the sum of the diagonal elements of the matrix. The variance of I is given by

$$\text{Var}(I) = \frac{n^2}{S_0^2(n - k)(n - k + 1)} \times \left\{ S_1 + 2tr\,(\mathbf{A}^2) - tr(\mathbf{B})^2 - \frac{[tr\,(\mathbf{A})]^2}{n - k} \right\}$$

where

$$\mathbf{A} = (\mathbf{X}'\mathbf{X})^{-1}\mathbf{X}'\mathbf{W}\mathbf{X}$$

$$\mathbf{B} = 4(\mathbf{X}'\mathbf{X})^{-1}\mathbf{X}'\mathbf{U}^2\mathbf{X}$$

$$k = \text{number of variables}$$

$$\mathbf{U} = \frac{1}{2}(\mathbf{W} + \mathbf{W}')$$

$$S_1 = \frac{1}{2}\sum_{i=1}^{n}\sum_{\substack{j=1 \\ i \neq j}}^{n}(w_{ij} + w_{ji})^2$$

Of course, when the scaling assigns positive weights of 1 only to contiguous counties, then $S_0 = n$.

Utilizing the residuals (Table 11.5) from a cubic surface fitted to the data in Table 11.6 (see Whitley and Clark, 1985 for a discussion of the data), and a binary weighting matrix Table 11.7, it is possible to compute tests of the residuals. The following SAS matrix operations were performed:

```
 C = E'*W*E;
 D = E'*E;
 G = D** - 1*C;
 P = W + W';
 Q = P*0.5;
 Z = Q*X;
 S = X'*Z;
 T = Z'*Z;
 R = X'*X;
 N = R** - 1;
 A = N*S;
 B = N*T;
 Y = A**2;
TA = TRACE(A);
TB = TRACE(B);
TY = TRACE(Y);
```

The following calculations yield I, $E(I)$, and $\text{var}(I)$:

$$I = \frac{n}{S_0}\frac{\mathbf{e}'\mathbf{W}\mathbf{e}}{\mathbf{e}'\mathbf{e}}$$

$$= .183161$$

$$E(I) = -\frac{n \cdot tr\,(\mathbf{A})}{(n - k)S_0}$$

$$= -\frac{47 \cdot 2.91337}{44 \cdot 47}$$

$$= -.0662$$

$$\begin{aligned}
\text{Var}(I) &= \frac{47^2}{47^2(44)(46)}(78 + 5.6614 - 3.35851 - .3858) \\
&= .03948 \\
\text{S.D.}(I) &= .1987 \\
Z &= 1.25
\end{aligned}$$

Discussion[1]

In analyzing data used in an earlier trend-surface analysis, we skirted an issue of primary theoretical and methodological importance in the analysis of geographically distributed data. We emphasized that the regression model and the commonly available trend-surface packages (based on fitting a surface with ordinary least-squares regression) assume that no spatial autocorrelation is present in the data or in the residuals once the model has been fitted to the data. Because these models operate under the assumption that no interdependence is present, the ability to make inferential statements from their results rests on the satisfaction of this assumption. Yet it is exactly this interdependence or autocorrelation and the consequent lack of independence that is of interest in most spatial or locational analyses (Gould, 1970, p. 443). It is only because of such interdependence that patterning (and corresponding order and predictability) are found, and laws of culture and behavior can be derived. When an archaeologist uses the regression model to test hypotheses concerning the regular attenuation of a trade item away from its source, or the locational correlation between lithic tool types within a site, it is the notion that the variate values are spatially autocorrelated in some predictable fashion that is ultimately being tested. Thus, a fundamental conflict exists between the goals of such an analysis and the methods often employed to achieve them.

The philosophical problems inherent in the analysis of spatial data using standard inferential statistics have been addressed frequently. Inferential statistics such as regression and correlation have largely been developed to identify significant relationships in spaceless and timeless social and physical processes (i.e., under conditions where sampling can be undertaken in a population lacking both spatial and temporal parameters). Ultimately, the philosophical conflict is founded on what logicians identify as the distinction between individuation in substance and spacetime languages. In substance language, individuation—the process of defining the individual members of the population—is based on the specification of a set of properties, such as land use, rock type, or tool type. In spacetime language, which is required for any locational analysis, individuation results from the definition of the map or grid coordinates of each individual (e.g., site location or distance from a raw material source). Unfortunately, the two languages have very different properties and cannot easily be used together in the individuation

[1]This discussion is derived from Whitley and Clark (1978) by permission of Academic Press.

TABLE 11.5 Data for Analysis of Residuals from a Trend Surface

Site No.	Location X-Coord	Location Y-Coord	Date Z-Value	Predicted	Residual
1	153.000	119.000	790.000	811.700	−21.700
2	165.000	129.000	790.000	816.405	−26.405
3	144.000	124.000	769.000	811.115	−42.115
4	230.000	87.000	849.000	835.153	13.847
5	112.000	110.000	800.000	802.385	−2.385
6	187.000	27.000	810.000	833.655	−23.655
7	175.000	152.000	800.000	821.728	−21.728
8	232.000	107.000	849.000	829.514	19.486
9	51.000	140.000	844.000	870.357	−26.357
10	55.000	100.000	874.000	820.273	53.727
11	231.000	223.000	800.000	791.126	8.874
12	153.000	121.000	780.000	812.139	−32.139
13	103.000	86.000	810.000	793.094	16.906
14	228.000	24.000	884.000	846.826	37.174
15	212.000	120.000	800.000	825.585	−25.585
16	195.000	92.000	879.000	826.094	52.906
17	155.000	127.000	810.000	813.984	−3.984
18	249.000	213.000	766.000	773.307	−7.307
19	222.000	59.000	800.000	840.506	−40.506
20	96.000	89.000	805.000	794.158	10.842
21	236.000	44.000	780.000	844.552	−64.552
22	204.000	16.000	889.000	842.851	46.149
23	243.000	135.000	790.000	817.268	−27.268
24	185.000	135.000	840.000	820.581	19.419
25	105.000	39.000	756.000	790.700	−34.700
26	193.000	41.000	761.000	833.800	−72.800
27	214.000	79.000	849.000	834.526	14.474
28	221.000	84.000	820.000	834.917	−14.917
29	189.000	11.000	830.000	837.923	−7.923
30	55.000	64.000	783.000	800.686	−17.686
31	103.000	84.000	795.000	792.539	2.461
32	169.000	110.000	790.000	815.418	−25.418
33	228.000	146.000	790.000	817.877	−27.877
34	74.000	153.000	879.000	859.611	19.389
35	238.000	198.000	810.000	793.012	16.988
36	173.000	122.000	889.000	817.413	71.587
37	184.000	97.000	869.000	821.312	47.688
38	196.000	79.000	869.000	828.447	40.553
39	83.000	0.000	830.000	801.363	28.637
40	27.000	61.000	810.000	829.348	−19.348
41	234.000	105.000	835.000	829.997	5.003
42	197.000	67.000	889.000	830.887	58.113
43	216.000	100.000	849.000	830.397	18.603
44	177.000	39.000	790.000	825.942	−35.942
45	232.000	61.000	889.000	841.385	47.615
46	118.000	99.000	810.000	798.545	11.455
47	213.000	87.000	793.000	832.597	−39.597

TABLE 11.6 47 Lowland Sites and Latest Dated Monuments

Site Number	Site Name	Long Count Date	Date	Monument
1	Aguas Calientes	9.18.0.0.0	790	Stela 1
2	Aguateca	9.18.0.0.0	790	Stela 6(7)
3	Altar de Sacrificios	9.16.18.5.1	769	Stela 15
4	Benque Viejo	10.1.0.0.0	849	Stela 1
5	Bonampak	9.18.10.0.0	800	Stela 2
6	Calakmul	9.19.0.0.0(?)	810	Stela 15(16)
7	Cancuen	9.18.10.0.0	800	Stela 1
8	Caracol	10.1.0.0.0	849	Stela 17
9	Chinkultic	10.0.15.0.0	844	Stela 1
10	Comitan	10.2.5.0.0	874	Stela 1
11	Copan	9.18.10.0.0	800	Altar G
12	El Caribe	9.17.10.0.0	780	Stela 2
13	El Cayo	9.19.0.0.0	810	
14	El Palmar	10.2.15.0.0(?)	884	Stela 41
15	Ixkun	9.18.10.0.0	800	Stela 5
16	Ixlu	10.2.10.0.0	879	Stela 2
17	La Amelia	9.19.0.0.0	810	Stela 1
18	La Florida	9.16.15.0.0	766	Stela 7
19	La Honradez	9.18.10.0.0(?)	800	Stela 4
20	La Mar	9.18.15.0.0	805	Stela 3
21	La Milpa	9.17.10.0.0	780	Stela 7
22	La Muneca	10.3.0.0.0	889	Stela 1
23	Lubaantun	9.18.0.0.0	790	Altar 2
24	Machaquila	10.0.10.0.0	840	Stela 5
25	Morales	9.16.5.0.0	756	Stela 1
26	Naachtun	9.16.10.0.0(?)	761	Stela 10
27	Nakum	10.1.0.0.0(?)	849	Stela D
28	Naranjo	9.19.10.0.0	820	Stela 32
29	Oxpemul	10.0.0.0.0	830	Stela 7
30	Palenque	9.17.13.0.7	783	Tablet 96, Glyphs
31	Piedras Negra	9.18.5.0.0	795	Stela 12
32	Polol	9.18.0.0.0	790	Stela 1
33	Pusilha	9.18.0.0.0	790	Stela R
34	Quen Santo	10.2.10.0.0	879	Stela 2
35	Quirigua	9.19.0.0.0	810	Temple 1
36	Seibal	10.3.0.0.0	889	Stelae 20, 17, 18
37	Tayasal-Flores	10.2.0.0.0	869	Stela 1
38	Tikal	10.2.0.0.0	869	Stela 11, Altar 11
39	Tila	10.0.0.0.0	830	Stela A
40	Tonina	9.19.0.0.0	810	Altar 8
41	Tzmin Kax	10.0.5.0.0	835	Altar 1
42	Uaxactun	10.3.0.0.0	889	Stela 12
43	Ucanal	10.1.0.0.0	849	Stela 4
44	Uxul	9.18.0.0.0(?)	790	
45	Xultun	10.3.0.0.0	889	Stela 10
46	Yaxchilan	9.19.0.0.0	810	Lintel 10
47	Yaxha	9.18.3.0.0	793	Stela 13

Source: Reproduced from Bove 1981, by author's permission.

TABLE 11.7 Binary Weighting Matrix for Nearest Neighbor Sites

Site Number	Site Number																																														
					5					10					15					20					25					30					35					40							47
1	0	0	0	0	0	0	0	0	0	0	0	1	0	0	0	0	0	0	0	0	0	0	0	0	0	0	0	0	0	0	0	0	0	0	0	0	0	0	0	0	0	0	0	0	0	0	0
2	0	0	0	0	0	0	0	0	0	0	0	0	0	0	0	0	0	0	0	0	0	0	0	0	0	0	0	0	0	0	0	0	0	0	0	1	0	0	0	0	0	0	0	0	0	0	0
3	0	0	0	0	0	0	0	0	0	0	0	0	0	0	0	0	1	0	0	0	0	0	0	0	0	0	0	0	0	0	0	0	0	0	0	0	0	0	0	0	0	0	0	0	0	0	0
4	0	0	0	0	0	0	0	0	0	0	0	0	0	0	0	0	0	0	0	0	0	0	0	0	0	0	0	1	0	0	0	0	0	0	0	0	0	0	0	0	0	0	0	0	0	0	0
5	0	0	0	0	0	0	0	0	0	0	0	0	0	0	0	0	0	0	0	0	0	0	0	0	0	0	0	0	0	0	0	0	0	0	0	0	0	0	0	0	0	0	0	0	0	1	0
6	0	0	0	0	0	0	0	0	0	0	0	0	0	0	0	0	0	0	0	0	0	0	0	0	0	0	0	0	1	0	0	0	0	0	0	0	0	0	0	0	0	0	0	0	0	0	0
7	0	0	0	0	0	0	0	0	0	0	0	0	0	0	0	0	0	0	0	0	0	0	0	1	0	0	0	0	0	0	0	0	0	0	0	0	0	0	0	0	0	0	0	0	0	0	0
8	0	0	0	0	0	0	0	0	0	0	0	0	0	0	0	0	0	0	0	0	0	0	0	0	0	0	0	0	0	0	0	0	0	0	0	0	0	0	0	0	1	0	0	0	0	0	0
9	0	0	0	0	0	0	0	0	0	0	0	0	0	0	0	0	0	0	0	0	0	0	0	0	0	0	0	0	0	0	0	0	0	1	0	0	0	0	0	0	0	0	0	0	0	0	0
10	0	0	0	0	0	0	0	0	1	0	0	0	0	0	0	0	0	0	0	0	0	0	0	0	0	0	0	0	0	0	0	0	0	0	0	0	0	0	0	0	0	0	0	0	0	0	0
11	0	0	0	0	0	0	0	0	0	0	0	0	0	0	0	0	0	0	0	0	0	0	0	0	0	0	0	0	0	0	0	0	0	0	0	0	0	0	0	0	0	0	0	0	0	0	0
12	1	0	0	0	0	0	0	0	0	0	0	0	0	0	0	0	0	1	0	0	0	0	0	0	0	0	0	0	0	0	0	0	0	0	0	0	0	0	0	0	0	0	0	0	0	0	0
13	0	0	0	0	0	0	0	0	0	0	0	0	0	0	0	0	0	0	0	0	0	0	0	0	0	0	0	0	0	0	1	0	0	0	0	0	0	0	0	0	0	0	0	0	0	0	0
14	0	0	0	0	0	0	0	0	0	0	0	0	0	0	0	0	0	0	0	0	0	1	0	0	0	0	0	0	0	0	0	0	0	0	0	0	0	0	0	0	0	0	0	0	0	0	0
15	0	0	0	0	0	0	0	0	0	0	0	0	0	0	0	0	0	0	0	0	0	0	0	0	0	0	0	0	0	0	0	0	0	0	0	0	0	0	0	0	0	0	1	0	0	0	0
16	0	0	0	0	0	0	0	0	0	0	0	0	0	0	0	0	0	0	0	0	0	0	0	0	0	0	0	0	0	0	0	0	0	0	0	0	0	0	0	0	0	0	1	0	0	0	0
17	0	0	0	0	0	0	0	0	0	0	0	1	0	0	0	0	0	0	0	0	0	0	0	0	0	0	0	0	0	0	0	0	0	0	0	0	0	0	0	0	0	0	0	0	0	0	0
18	0	0	0	0	0	0	0	0	0	0	0	0	0	0	0	0	0	0	0	0	0	0	0	0	0	0	0	0	0	0	0	0	0	0	1	0	0	0	0	0	0	0	0	0	0	0	0
19	0	0	0	0	0	0	0	0	0	0	0	0	0	0	0	0	0	0	0	0	0	0	0	0	0	0	0	0	0	0	0	0	0	0	0	0	0	0	0	0	0	0	0	0	1	0	0
20	0	0	0	0	0	0	0	0	0	0	0	0	1	0	0	0	0	0	0	0	0	0	0	0	0	0	0	0	0	0	0	0	0	0	0	0	0	0	0	0	0	0	0	0	0	0	0
21	0	0	0	0	0	0	0	0	0	0	0	0	0	0	0	0	0	0	1	0	0	0	0	0	0	0	0	0	0	0	0	0	0	0	0	0	0	0	0	0	0	0	0	0	0	0	0
22	0	0	0	0	0	0	0	0	0	0	0	0	0	0	0	0	0	0	0	0	0	0	0	0	0	0	0	0	1	0	0	0	0	0	0	0	0	0	0	0	0	0	0	0	0	0	0

Site Number					5					10					15					20					25					30					35					40							47
23	0	0	0	0	0	0	0	0	0	0	0	0	0	0	0	0	0	0	0	0	0	0	0	0	0	0	0	0	0	0	0	0	1	0	0	0	0	0	0	0	0	0	0	0	0	0	0
24	0	0	0	0	0	0	0	0	0	0	0	0	0	0	0	0	0	0	0	0	0	0	0	0	0	0	0	0	0	0	0	0	0	0	0	1	0	0	0	0	0	0	0	0	0	0	0
25	0	0	0	0	0	0	0	0	0	0	0	0	0	0	0	0	0	0	0	0	0	0	0	0	0	0	0	0	0	1	0	0	0	0	0	0	0	0	0	0	0	0	0	0	0	0	0
26	0	0	0	0	0	0	0	0	0	0	0	0	0	0	0	0	0	0	0	0	0	0	0	0	0	0	0	0	0	0	0	0	0	0	0	0	0	0	0	0	0	0	0	1	0	0	0
27	0	0	0	0	0	0	0	0	0	0	0	0	0	0	0	0	0	0	0	0	0	0	0	0	0	0	0	0	0	0	0	0	0	0	0	0	0	0	0	0	0	0	0	0	0	0	1
28	0	0	0	1	0	0	0	0	0	0	0	0	0	0	0	0	0	0	0	0	0	0	0	0	0	0	0	0	0	0	0	0	0	0	0	0	0	0	0	0	0	0	0	0	0	0	0
29	0	0	0	0	0	1	0	0	0	0	0	0	0	0	0	0	0	0	0	0	0	0	0	0	0	0	0	0	0	0	0	0	0	0	0	0	0	0	0	0	0	0	0	0	0	0	0
30	0	0	0	0	0	0	0	0	0	0	0	0	0	0	0	0	0	0	0	0	0	0	0	0	0	0	0	0	0	0	0	0	0	0	0	0	0	0	0	1	0	0	0	0	0	0	0
31	0	0	0	0	0	0	0	0	0	0	0	0	1	0	0	0	0	0	0	0	0	0	0	0	0	0	0	0	0	0	0	0	0	0	0	0	0	0	0	0	0	0	0	0	0	0	0
32	1	0	0	0	0	0	0	0	0	0	0	0	0	0	0	0	0	0	0	0	0	0	0	0	0	0	0	0	0	0	0	0	0	0	0	0	0	0	0	0	0	0	0	0	0	0	0
33	0	0	0	0	0	0	0	0	0	0	0	0	0	0	0	0	0	0	0	0	0	0	1	0	0	0	0	0	0	0	0	0	0	0	0	0	0	0	0	0	0	0	0	0	0	0	0
34	0	0	0	0	0	0	0	0	1	0	0	0	0	0	0	0	0	0	0	0	0	0	0	0	0	0	0	0	0	0	0	0	0	0	0	0	0	0	0	0	0	0	0	0	0	0	0
35	0	0	0	0	0	0	0	0	0	0	0	0	0	0	0	0	0	1	0	0	0	0	0	0	0	0	0	0	0	0	0	0	0	0	0	0	0	0	0	0	0	0	0	0	0	0	0
36	0	1	0	0	0	0	0	0	0	0	0	0	0	0	0	0	0	0	0	0	0	0	0	0	0	0	0	0	0	0	0	0	0	0	0	0	0	0	0	0	0	0	0	0	0	0	0
37	0	0	0	0	0	0	0	0	0	0	0	0	0	0	0	1	0	0	0	0	0	0	0	0	0	0	0	0	0	0	0	0	0	0	0	0	0	0	0	0	0	0	0	0	0	0	0
38	0	0	0	0	0	0	0	0	0	0	0	0	0	0	0	0	0	0	0	0	0	0	0	0	0	0	0	0	0	0	0	0	0	0	0	0	0	0	0	0	0	1	0	0	0	0	0
39	0	0	0	0	0	0	0	0	0	0	0	0	0	0	0	0	0	0	0	0	0	0	0	0	0	0	0	0	0	0	0	0	0	0	0	0	0	0	0	1	0	0	0	0	0	0	0
40	0	0	0	0	0	0	0	0	0	0	0	0	0	0	0	0	0	0	0	0	0	0	0	0	0	0	0	0	0	0	0	0	0	0	0	0	0	0	0	0	0	0	0	0	0	0	0
41	0	0	0	0	0	0	0	1	0	0	0	0	0	0	0	0	0	0	0	0	0	0	0	0	0	0	0	0	0	0	0	0	0	0	0	0	0	0	0	0	0	0	0	0	0	0	0
42	0	0	0	0	0	0	0	0	0	0	0	0	0	0	0	0	0	0	0	0	0	0	0	0	0	0	0	0	0	0	0	0	0	0	0	0	0	1	0	0	0	0	0	0	0	0	0
43	0	0	0	0	0	0	0	0	0	0	0	0	0	0	0	1	0	0	0	0	0	0	0	0	0	0	0	0	0	0	0	0	0	0	0	0	0	0	0	0	0	0	0	0	0	0	0
44	0	0	0	0	0	1	0	0	0	0	0	0	0	0	0	0	0	0	0	0	0	0	0	0	0	0	0	0	0	0	0	0	0	0	0	0	0	0	0	0	0	0	0	0	0	0	0
45	0	0	0	0	0	0	0	0	0	0	0	0	0	0	0	0	0	0	1	0	0	0	0	0	0	0	0	0	0	0	0	0	0	0	0	0	0	0	0	0	0	0	0	0	0	0	0
46	0	0	0	0	1	0	0	0	0	0	0	0	0	0	0	0	0	0	0	0	0	0	0	0	0	0	0	0	0	0	0	0	0	0	0	0	0	0	0	0	0	0	0	0	0	0	0
47	0	0	0	0	0	0	0	0	0	0	0	0	0	0	0	0	0	0	0	0	0	0	0	0	0	0	1	0	0	0	0	0	0	0	0	0	0	0	0	0	0	0	0	0	0	0	0

process. In the formal sense, then, it is logically inappropriate to employ statistical tests based on individuation with a substance language and developed to study cultural and social process, for the analysis of locational data, where individuation requires a spacetime language and the goal is to define geographical processes.

There is, unfortunately, no simple resolution of these conflicts, perhaps underscoring Thomas's (1978) comment that the pushbutton availability of many multivariate statistics has too often resulted in their application without careful consideration of their applicability. Geographers have taken two practical approaches to this problem. The first of these involves the design of spatial models incorporating an autoregressive structure, as illustrated by Curry (1966, 1971). Hodder and Orton (1976, p. 176), in fact, have provided a mathematical description of an autoregressive model in the archaeological literature, but their very brief discussion underscores the mathematical difficulty in constructing anything other than a theoretical description of such. The practical design and specification of autoregressive models, in short, are beyond the technical capabilities of the average social scientist with a background limited to introductory calculus and an advanced course in statistics. Thus, while autoregressive models are probably the preferable approach, it is unlikely that they will be employed by any one other than the most mathematically sophisticated of the discipline.

Alternatively, pragmatic solutions to the problem of spatial autocorrelation are suggested by a reading of Matern's work (Matern, 1960). Matern drew a series of spatial samples at increasingly greater intervals. Comparison of the correlation coefficients that resulted from these series showed that the correlations decreased with the increase in width of the sampling interval. Direction also influenced the degree of reduction in the correlation coefficients. It would seem that this is a particularly profitable avenue for a practical approach. By drawing repeated samples at various distances between points, the researcher may be able to settle on a sampling interval in which the autoregressive structure would be minimized or perhaps vanish altogether. The pragmatic suggestion of a widely separated data set as a means of overcoming spatial autocorrelation in the dependent or independent variables is not always possible, but it emphasizes the associated need to focus on individual data, such as households, rather than the census tract aggregations. Thus, one solution (in the analysis of the spatial aspects of urban structure) to both the problems of aggregated units and spatial autocorrelation is to choose a sample of households within the urban region where some significant spatial separation is achieved.

11.5 MULTICOLLINEARITY

A situation arises frequently in regression analysis in which the independent variables are in some way intercorrelated. If the independent variables are highly intercorrelated, the coefficients for these variables can be misleading, and in the limiting case where the variables are perfectly correlated, the transpose of the matrix of independent variables has no inverse. That is, $\mathbf{X}'\mathbf{X}$ has a determinant of 0. Another way of stating this is that the multiple regression model assumes that the independent variables are *linearly independent* of each other so their individual contribution to the explanation of the variability of Y can be added. The greater the degree of dependence among the independent variables, as seen through their intercorrelations, the greater the violation of the assumption, and the greater the level of imprecision of the regression coefficient estimates.

The Problem Defined

The explanation of this assumption involves understanding the variances of the independent variables. Let us first examine the simple regression relationship in order to grasp the underlying principle. If the X's are bunched together, it is difficult to estimate the true regression line. In the extreme case where there is no variation in X, as Figure 11.12*a* shows, there is virtually an unlimited number of regression lines that could be fitted to the scatter of points. On the other hand, if the observations on the independent variable are spread out (Figure 11.12*b*), it is easy to estimate b_1 and the regression line.

When two independent variables have an exact linear relationship, a plot of data points for the regression will fall on a line rather than a three-dimensional plane (Figure 11.13). The researcher attempting to fit a plane to those data will find that an infinite number of planes pass through the line. The equations representing each plane will have similar R^2 values (multiple correlations) but different regression coefficients. Accurate estimation of the regression coefficient is impossible, consequently, the effects of each independent variable on the variation of the dependent variable cannot be obtained. In fact, it does not make sense to speak of the separate

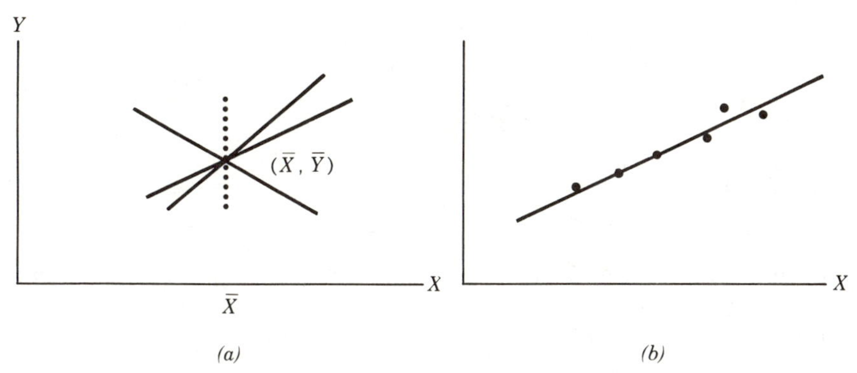

Figure 11.12 (*a*) Degenerate regression because of no spread (variation) in X, (*b*) adequate spread.

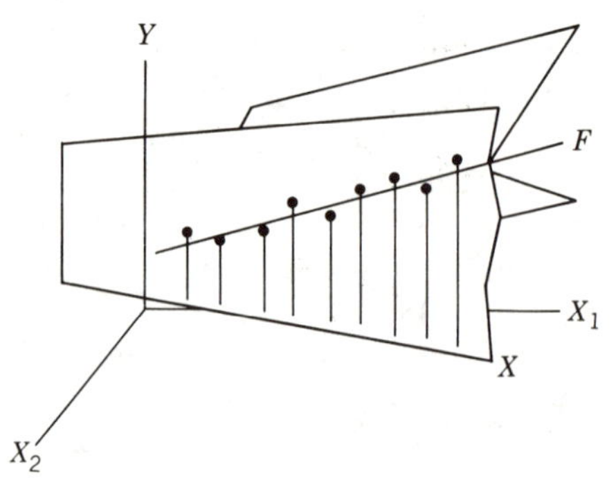

Figure 11.13 Multicollinearity. Source: Wonnacott and Wonnacott, 1970, p. 261, by permission of John Wiley & Sons, Inc., Publishers.

effects of individual variables in this situation because it is impossible to hold either of the highly collinear variables constant.

Now consider the case where the values do not all fall on a vertical plane, but come very close to it. In this situation, it is possible to fit a plane to the scatter of points, but the estimating procedure, as in the case of a cluster of points in the two-dimensional case, is very unstable. There are large variances (relative to the actual values) for the estimators b_1 and b_2 associated with X_1 and X_2. In matrix terms, if X_1 and X_2 are highly interrelated, then as we have noted, $\mathbf{X}'\mathbf{X}$ has a determinant near zero and the inverse $(\mathbf{X}'\mathbf{X})^{-1}$ is the covariance matrix of $\hat{b}$ and so will have very large variances and results in very large standard errors around $\hat{b}$.

An example will clarify the problems of multicollinearity. Assume that a geographer is interested in predicting stream flow and measures rainfall in centimeters (X_1) and inches (X_2) as independent variables. Although this is unlikely, it will illustrate the effect of collinear variables. The independent variables are exactly related:

$$X_1 = 2.54X_2$$

A regression plane fitted to the data in Table 11.8 could yield

$$\hat{Y} = b_0 + 22X_1 + 0X_2$$

An equally valid equation could be expressed as:

$$\hat{Y} = b_0 + 0X_1 + .23622X_2$$

In fact, as Wonnacott and Wonnacott (1970) demonstrate, there are a family of solutions, all proportional to one another. They can be expressed with the formula:

$$\hat{Y} = b_0 + 22\lambda X_1 + .23662(1 - \lambda)X_2$$

where λ is an arbitrary value.

TABLE 11.8 Stream Flow and Precipitation (Hypothetical Data)

Stream Flow (Cubic Feet/Second)	Rainfall (Inches)	Rainfall (Centimeters)
50	20	50.8
55	25	63.5
60	30	76.2
62	35	88.9
64	40	101.6
65	45	114.3
70	50	127.0

All the equations will give "correct" answers for the Y value, but as noted several times, it is not possible to make any statements about the relative contributions of X_1 and X_2. However, even when there is serious multicollinearity, it is still possible to have good predictions and significant overall F values. Although it is unlikely that any geographer would introduce two independent variables that are functions of one another, it is often the case that variables representing very similar measures may be introduced.

Recognizing Multicollinearity

Several methods have been proposed for the detection of multicollinearity in a set of independent variables. Often, researchers set an arbitrary r value (zero-order correlation between two variables) above which the collinearity is said to be severe. Usually this is set at $r = .8$ or $.9$.

A second method relates the simple correlation to the overall multiple correlation between the dependent variable and all the independent variables. The relationship is severe if

$$r_{ij} \geq R_y \tag{11.18}$$

where

r_{ij} = zero-order correlation between any two independent variables
R_y = multiple correlation between the dependent and all independent variables

The intercorrelation of multicollinearity is not necessarily a problem unless it is high relative to the overall degree of multiple correlation. In cases of complete multicollinearity, this method is not applicable. Another way of examining possible multicollinearity, and a method easily available through a statistic provided in both SAS and SPSS, is to measure the *tolerance* of an independent variable.

$$\text{Tolerance of } X_j = 1 - R^2(X_j \text{ with all } X_k, j \neq k) \tag{11.19}$$

This measure is simply the multiple correlation (squared) of X_j with all of the other independent variables in the regression, subtracted from 1. The tolerance value will vary from 0 (where X_j is a perfect linear combination of the other independent variables) to 1.0 (where it is uncorrelated with any of the other independent variables).

The selection of variables for a regression equation is a very important part of the model building. Every attempt should be made to select variables that will be independent additive contributors to the explanation of the variation in the dependent variable. The whole point of multiple regression is to isolate the effects of the individual regressors by controlling on other variables. However, when multicollinearity is a problem, the concept of contribution by an individual variable remains inherently ambiguous.

11.6 AGGREGATION AND ECOLOGICAL CORRELATION

It has long been known that the use of aggregate data may yield correlation coefficients exhibiting considerable bias above their values at the individual level. Blalock (1964) has shown that the regression coefficients as substitutes for the "true" microlevel estimates are most serious in terms of the causal inferences to be drawn from statistical analysis. It is now well accepted that it is incorrect to assume that relationships existing at one level of analysis will necessarily demonstrate the same strength at another level. The estimates derived from aggregate data are valid only for the particular system of observational units employed.

Much of statistical analysis in the social sciences involves the use of observational units which are aggregates of smaller units. The smallest possible unit for which an appropriate observation on some variable may be made is defined as an element or individual. But, in social research, an individual may not necessarily correspond in size and definition to a single person. Larger observational units (e.g., households, counties, or nations) may be considered as individuals in appropriate studies. An individual is assumed to be homogeneous in its appropriate characteristics. So a unit of analysis may be interpreted as an individual if the variable (or variables) being measured are constant throughout the unit, or in other words, if the measured characteristics are properties of the unit as a whole. On the other hand, when small homogeneous units are aggregated to form a larger heterogenous collection of individuals, the resultant unit is an aggregate. Observations of aggregate characteristics are generally taken as summary measurements, percentages, means or medians, and express the degree to which individual characteristics are exhibited in common by a group. Such properties are unique to the collective group but dependent on properties of each single nonmodifiable member of the group. In some instances, recorded data are available only in aggregate form, and in others, the investigator is constrained by time or costs. But, in any case, social scientists in general and geographers in particular continue to work with data that measure properties of aggregate groups.

Inference from Aggregate Data

There are three types of erroneous inferences that can occur if a researcher attempts to generalize from one level of investigation to another.

The individualistic fallacy is the attempt to impute macrolevel (aggregate) relationships from microlevel (individual) relationships. It is the classic aggregation problem first examined by economists.

Cross-level fallacies can occur when one makes inferences from one subpopulation to another at the same level of analysis.

The ecological fallacy is the opposite of the individualistic fallacy and involves making inferences from higher to lower levels of analysis. Although the ecological fallacy has been widely discussed and publicized, it is a common error in studies involving causal inference.

Robinson (1950) first attempted a rigorous explanation of the increasing magnitude of correlation coefficients with increasing size of unit, or what he termed *ecological correlation*. He differentiated between an individual correlation in which the statistical object or thing described is indivisible, and an ecological correlation in which "the statistical object is a group of persons and the variables are descriptive properties of the groups." Robinson demonstrated that there was not necessarily a correspondence between individual and ecological correlations, and that generally the latter would be larger than the former. He derived the mathematical relationship between individual and ecological correlations in terms of the analysis of covariance of a bivariate frequency distribution. He demonstrated that aggregation reduces the between unit variation in a variable, making the variable seem more homogeneous. This has the effect of increasing the magnitude of r.

Even though ecological correlations cannot be used to make valid individual level inferences, there are two situations in which aggregate variables are appropriate.

1. When the variables are functions of some common underlying causal structure inherent not in the individuals themselves but in the properties of the areas. Thus, social conflict, cultural conflict, and other sociological variables may be the underlying structures for some individual level analyses.

2. When one aggregate variable is related to another and an individual correlation would be impossible (as, for example, in the correlation between the number of physicians per capita and the infant death rate).

Blalock has attempted to specify more clearly the varying nature of the aggregation procedure, particularly in the context of linear causal analysis. He has examined four ways in which grouping procedures may affect bivariate causal relationships: (1) random grouping, (2) grouping to maximize variation in the independent variable, (3) grouping to maximize variation in the dependent variable, and (4) grouping on the basis of spatial proximity of individuals (Blalock, 1964, pp. 104–112). He concludes that in the case of random grouping, aggregation does not disproportionately affect variation in either x or y, and the use of aggregate data yields correlation and regression parameters that are unbiased estimates of the microlevel parameters.

In all other cases including the aggregation of observations, if the individual observations are not spatially autocorrelated, the aggregation procedure may be likened to random grouping, and the macrolevel parameters will be unbiased estimates of the microlevel parameters. If the individual data exhibit spatial autocorrelation, aggregation by proximity can systematically affect variation in both x and y. Of course, if the degree of autocorrelation of the original data is sufficiently strong and the aggregate unit is relatively small, a proximity grouping procedure may yield aggregates that are perfectly homogeneous internally, in which case a biased correlation between x and y is unlikely. In practical research problems, however, grouping by proximity will tend to create aggregates in which the variables of interest have been systematically affected, thus leading to biased estimates of the correlation coefficient and allowing the possibility of biased slope coefficients. Moreover, in most instances in practical research, the data are already grouped and we have little choice about the grouping procedure.

An Example of Aggregation Effects

In order to demonstrate the effect of simple proximity grouping on the correlation and regression coefficients, we present the following results drawn from a study of income and education. The analysis compares the results of fitting the model

$$\hat{Y}_i = b_0 + b_1 X_i$$

where Y = family income and X is a measure of the years of education of the head of household. Four different levels of aggregation are examined. The individual household data for the studies were obtained from a survey of the Los Angeles Metropolitan Area in 1972. A cluster sample of 1024 households yielded usable data for 952 households. Additional data were collected for 1556 of the 1597 census tracts in Los Angeles for 1970. In addition to the 952 individual units and 1556 census tract units, two governmental groupings, (the 134 Welfare Planning Council Study areas, and the 35 Regional Planning Commission Statistical Areas) were used as aggregations (Table 11.9). For the census aggregations, mean annual family

TABLE 11.9 Correlation and Slope Coefficients Derived from Household and Aggregated Data

Data Set	Method of Data Generalization	r	r^2	b
952 units: households	Not applicable	.4028	.1623	857.60
1556 units: census tracts	Tract mean	.6434	.4140	2413.64
134 units: Welfare Planning Council (groups of tracts)	Group mean	.7606	.5785	2808.21
35 units: Regional Planning Commission (groups of tracts)	Group mean	.8503	.7230	3103.62

income and median school years are used. For the tract aggregations, the group means of the aggregated tract values are used.

Inspection of the correlation coefficients reveals that in general the variations in r and r^2 caused by aggregation of the data conform quite well to the expected results predicted by Robinson and others. Aggregation of observational units on the basis of proximity leads to substantially biased correlation coefficients, with an increase in r as the level of grouping increases. Whereas r and r^2 are fairly low at the microlevel (.4028 and .1623 respectively), the macrocoefficients obtained at the tract level are substantially higher (r = .6434 and r^2 = .4140). Data derived from the proximity aggregations of tracts, i.e., the 134 Welfare Planning Council Study Areas (for Los Angeles), and the 35 Regional Planning Commission Statistical Areas (for Los Angeles), yielded macrocoefficients that are even higher than those obtained at the tract level. The system of tract aggregations having the fewest groups (35) tended to produce the highest correlation coefficients.

Because the census tracts are not true aggregates of the 952 individuals of the household data, it may be more appropriate to carry out the analysis on a specific group of the individual data. The model applied to a second proximity grouping procedure of the 952 households yielded the results in Table 11.10. Each aggregate grouping is based on proximity grouping of the

TABLE 11.10 Grouping of Individual Level Data

Number of Units	r	r^2	b
952	.4028	.1623	857.60
136	.5763	.3321	1205.58
68	.6692	.4478	1394.40
34	.7214	.5205	1520.65
17	.7014	.4919	1423.92

next lowest level. Within one level, each group consists of an equal number of individual households. The same bivariate linear model was used.

As expected, the correlation between income and education tends to increase markedly with the level of aggregation. Whereas only a weak positive relationship is exhibited at the microlevel ($r^2 = .1623$), X explains approximately 50 percent of the variation in Y at the two highest levels. Similarly, the slope coefficient, b, also tends to increase. An apparent anomaly occurs at the fifth level, where both b and r show increases below their corresponding values at the fourth level.

Discussion

Although the technical aspects of data aggregation in correlation and regression are of some inherent interest, their real importance lies in their effects on the substantive conclusions derived from an analysis of the correlation and regression coefficients in a particular substantive study. Two variables that are weakly related at the individual level may show a strong correlation as the data are aggregated. Thus, the strength and slope of the relationship are partially dependent on the size of observational unit employed and cannot be analyzed without reference to the nature of the units and the underlying method of aggregation. The implication for any statistical analyses of census tract information, including factorial ecological studies, is that the substantive conclusions should be treated with caution. Even when the census tract correlations are not used to infer individual relationships, the fact that the coefficients are inflated by an unknown magnitude, and different variables with differing degrees of homogeneity are likely to yield different magnitudes of inflation, also suggests that we recognize explicitly the possibility of bias in our analyses.

References and Readings

1. Assumptions

Norcliffe, G. B. (1969) "On the use and limitations of trend surface models," *Canadian Geographer* 13:338–348.

Poole, M. A. and P. O'Farrell (1971) "The assumptions of the linear regression model," *Transactions of the Institute of British Geographers* 52:145–158.

Thomas, D. H. (1978) "The awful truth about statistics in archeology," *American Antiquity* 43:231–244.

Wonnacott, T. H. and R. J. Wonnacott (1970) *Econometrics*, Wiley: New York.

2. Residuals

Anscombe, F. J. and J. W. Tukey (1963) "The examination and analysis of residuals," *Technometrics* 5(2):141–160.

Chorley, R. J. and P. Haggett (1968) "Trend-surface mapping in geographical research," in B. J. L. Berry and D. F. Marble (eds.), *Spatial Analysis: A Reader in Statistical Geography,* Prentice Hall: Englewood Cliffs, N.J.

Clark, W. A. V. (1967) "The use of residuals from regression in geographical research," *New Zealand Geographer,* 23:71–84.

Draper, N. R. and H. Smith (1966) *Applied Regression Analysis,* Wiley: New York.

Goldberger, A. S. (1962) "Best linear unbiased prediction in the generalized linear regression model," *Journal of the American Statistical Association* 57(298):369–375.

Intriligator, M. (1978) *Econometric Models, Techniques and Applications,* Prentice-Hall: Englewood Cliffs, N.J.

King, L. J. (1969) *Statistical Analysis in Geography,* Prentice-Hall: Englewood Cliffs, N.J.

Krumbein, W. and F. A. Graybill (1965) *Statistical Models in Geology,* McGraw-Hill: New York.

Maddala, G. S. (1977) *Econometrics,* McGraw-Hill: New York.

Malinvaux, E. (1970) *Statistical Methods of Econometrics,* North Holland: Amsterdam.

Thomas, E. N. (1968) "Maps of residuals from regression," in B. J. L. Berry and D. F. Marble (eds.), *Spatial Analysis: A Reader in Statistical Geography,* Prentice Hall: Englewood Cliffs: N.J.

3. Spatial Autocorrelation

Agteberg, F. P. (1974) *Geomathematics,* Elsevier: Amsterdam.

Bove, F. J. (1981) "Trend surface analysis and the lowland Classic Maya collapse," *American Antiquity* 46:93–112.

Cliff, A. D. and J. K. Ord (1973) *Spatial Autocorrelation,* Pion: London.

Cliff, A. D. and J. K. Ord (1981) *Spatial Processes: Models and Applications,* Pion: London.

Curry, L. (1966) "A note on spatial association," *The Professional Geographer* 18:97–99.

Curry, L. (1971) "A spatial analysis of gravity flows," *Regional Studies* 6:131–147.

Dacey, M. F. (1965) "A review of measures of contiguity for 2 and k color maps," Department of Geography, Northwestern University Technical Report #2, Spatial Diffusion Study.

Dacey, M. F. (1968) "A review of measures of contiguity for two and k-color maps," in B. J. L. Berry and D. F. Marble (eds.), *Spatial Analysis: A Reader in Statistical Geography,* Prentice Hall: Englewood Cliffs, N.J.

Gould, P. (1970) "Is 'statistix inferens' the geographical name for a wild goose," *Economic Geography* 46:439–448.

Hodder, I. and C. Orton (1976) *Spatial Analysis in Archaeology,* Cambridge University Press: Cambridge.

Martin, R. L. (1974) "On autocorrelation, bias and the use of first spatial differences in regression analysis," *Area* 6, 185–194.

Matern, B. (1960) "Spatial variation," *Meddelanden Fran Statens Skogsforsknings-institut,* 49, 1–144.

Tobler, W. R. (1965) "Computation of the correspondence of geographical patterns," *Papers, Regional Science Association* 15:131–139.

Whitley, D. S. and W. A. V. Clark (1985) "Spatial autocorrelation tests and the classic Maya collapse: Methods and inferences," *Journal of Archaeological Sciences* 12:377–395.

4. Aggregation

Blalock, H. (1964) *Casual Inferences in Non-Experimental Research,* University of North Carolina Press: Chapel Hill.

Clark, W. A. V. and K. Avery (1976) "The effects of data aggregation in statistical analysis," *Geographical Analysis* 8:428–438.

Hannan, M. T. (1971) *Aggregation and Disaggregation in Sociology,* D.C. Heath: Lexington, Mass.

Robinson, W. S. (1950) "Ecological correlation and the behavior of individuals," *American Sociological Review* 15: 351–357.

Openshaw, S. (1976) "A general method for identifying scale and aggregation effects on any statistical model of pattern and process in a spatial domain," *Advanced Applications in Probability* 8:656–657.

CHAPTER 12

Extensions of Multivariate Linear Regression Methods

In the last two chapters, we examined the multiple regression model in some detail and outlined a procedure for applying the model to a wide variety of situations. A number of restrictions in the use of the model which are dependent on the structure of the data we are analyzing have been inherent in our discussion, however. These restrictions are only indirectly reflected in the assumptions of the general linear model, and it is worthwhile stating these specifically. The multiple regression form of the general linear model we have outlined was applied to situations with the following characteristics.

1. A single dependent variable is a function of one or more independent variables.

2. Both the dependent variable and all the independent variables are measured on what we have been calling metric and enumerated (interval data) scales.

3. The contribution of the independent variables to the explanation of the dependent variable Y and the error term are all *additive* (we add the effects of each to get the total explanation).

4. The relationship of each independent variable to the dependent variable is *linear*.

These last two restrictions are contained in the linearity assumption of the multiple linear regression model, but the first two restrictions are only implied. In addition, we have used the multiple linear regression model in testing a specific structure—that is, the number and nature of the independent variables are decided before the analysis, and in reality we are merely testing a known model.

In this chapter, and also to some extent in the next, we relax some of these restrictions. In doing so, we move more towards the application of the full general linear model rather than just its regression form. However, our emphasis will remain on the use of the regression structure rather than shift to the more versatile analysis of variance testing structure. As we have mentioned before, the analysis of variance designs, although more versatile, have some additional restrictions that make them more suited to be used in experimentally controlled situations—a research structure not as commonly encountered in geography as in some other sciences.

In this chapter we examine three general adaptations of the linear regression model, plus a modification of the model for specific spatial investigations of importance to both physical and human geography. First, we examine the restriction of linearity and additivity and note that the multiple linear regression model can be used in an even wider variety of circumstances through *variable transformations* (Section 12.1). Relaxing the model testing structure when we have a large group of potential independent variables and wish to identify the "best" combination of these to explain the variability of Y, requires us to examine ways of searching for models. One of the main methods utilized is centered on *stepwise regression* (Section 12.2). The requirement of interval-scaled data can often be a problem. We can find numerous examples in the geographic literature of analyses including ordinally classified scores (stream order, for example) but we have to recognize that we are attaching metric (interval) significance to the data, which only sometimes can be justified. A more difficult problem arises when we wish to incorporate nominally classified variables (categorical data) into a regression structure, a situation that can be quite common in geography. Through the use of *dummy variables,* however, nominally classified scores can be incorporated (Section 12.3). In doing so, we move close to the analysis of variance form of the general linear model (and its extension, the analysis of covariance). In the next chapter, an example of an alternative multivariate technique (alternative to multiple linear regression) considers the regression situation with basically dichotomous (two-class nominal classification) dependent variables, rather than independent variables. In the final section of this chapter, we use the multiple linear regression model to fit "surfaces" to the variation in some observed variable (the dependent variable) as a function of spatial location (the independent variables)—a technique commonly called *trend surface analysis* (Section 12.4).

12.1 VARIABLE TRANSFORMATIONS

The multiple linear regression model states clearly that the relationship between each independent variable and the dependent variable should be *linear,* and that the contributions of the independent variables and the error term should be *additive.* Departures from these assumptions can lead to erroneous results. This has been commented on and illustrated frequently in Chapters 9 through 11. If the assumptions cannot be fulfilled then the standard regression model should not be used. Unfortunately, a lack of basic linearity between variables is common. However, a distinction needs to be made between the model of the relationship among the variables, and the statistical model used to assess the relationship. The two need not be the same. Thus, if we retain the symbols used previously, but generalize the form, we are interested in the relationship of

$$Y = f(X_1, X_2, \ldots, X_k)$$

where k = the number of observed independent variables. To analyze this relationship we can use the linear statistical model. With a new set of symbols, we can state it as

$$Y = \beta_0 + \beta_1 Z_1 + \beta_2 Z_2 + \ldots + \beta_p Z_p + \epsilon \tag{12.1}$$

where p = the number of independent variables in the equation. In the examples we have been examining, we have made a straight match of the X variable with the Z, and have used the linear model with $p = k$.

However, there are many situations in which the functional relationship we are interested in, $Y = f(X_1, X_2, X_3, \ldots, X_k)$, is either not linear in its parameters ($\beta_0, \beta_1, \ldots$), or is not additive, or both. Thus, the relationships are *nonlinear.* Through the use of *transformations,* however, many of these relationships can be restructured to make use of the linear statistical model. Such nonlinear models, capable of transformation to a linear model, are called *intrinsically linear.* While we will not concern ourselves with *intrinsically nonlinear* models (methods are available to examine these, but they are outside the scope and purpose of this book), it is worthwhile to consider some of the more common intrinsically linear models. Our interest in these models might come from two situations. First, in testing the linearity of Y with an individual X_j variable, the evidence suggests a significant departure from a linear relationship. Can we transform the X_j data in such a way that a linear relationship can be approximated and the linear regression model retained? We would therefore be concerned with bivariate models. Second, theory suggests a nonlinear relationship: Can we transform the model into a linear structure so that linear regression methods can be used to fit the nonlinear model? In this situation we could examine either bivariate or multivariate nonlinear models.

Transformations of the *X* Variable

The simplest type of transformation is performed to establish a basically linear relationship between Y and a particular X_j variable. Thus, we are concerned with bivariate situations. Two types of transformations are possible: those that change only the X variable to achieve linearity with Y and those that involve the fitting of an intrinsically nonlinear model to the data, including modification to both Y and X and the parameters. We look at the X transformation first. Although theory or previous investigation may suggest a particular transformation, the usual method is by examination of the scatter diagram and by trial-and-error fitting of commonly used transformations. Any transformation that fits can be used, but justification is usually necessary. Using the nomenclature of Equation 12.1, our transformations involve, for the relationship $Y = f(X_j)$ where the functional form appears nonlinear, the establishment of $Z_j = f(X_j)$, such that the statistical model $Y = \beta_0 + \beta_1 Z_j + \epsilon$ can be used. A large number of transformations of the X_j variable have appeared in geographical analyses at one time or another, but the following is a list of the most common:

Logarithmic	$Z_j = \log X_j \quad (\log_e \text{ or } \log_{10})$
Square root	$Z_j = \sqrt{X_j} = X_j^{1/2}$
Reciprocal	$Z_j = \dfrac{1}{X_j} = X_j^{-1}$
Square	$Z_j = X_j^2$
Arcsin square root	$Z_j = \arcsin \sqrt{X_j}$

After the examination of the relationship of Y with each independent variable X_j, our multiple regression model could contain a wide variety of transformations. Thus, for example, we could have as a model

$$Y = \beta_0 + \beta_1 Z_1 + \beta_2 Z_2 + \beta_3 Z_3 + \epsilon$$

where $Z_1 = \log_e X_1 \quad Z_2 = \sqrt{X_2} \quad Z_3 = X_3$
or in its original form,

$$Y = \beta_0 + \beta_1 \log_e X_1 + \beta_2 \sqrt{X_2} + \beta_3 X_3 + \epsilon$$

To construct the regression equation we would simply transform the X_1 and X_2 scores and run the analysis of Y on $\log_e X_1$, $\sqrt{X_2}$, and X_3.

All the transformations considered here are of the independent variables only, and the dependent side of the statistical model remains unchanged. In some circumstances, however, the transformation suggested by the scatter diagram may involve transformation of Y (with or without X). In this situation, we must be a little more careful about how we use the transformation, because we are no longer concerned with a statistical model for Y but for some transformation of Y. Also, in a multiple regression model, if we need to transform Y to satisfy linearity with a particular X_j, then all other X variables must use the same transformation of Y to satisfy their linearity. In this circumstance, we should consider the fitting of a nonlinear model to the data—we hope an intrinsically linear one.

Bivariate Nonlinear Models

Three widely used intrinsically linear bivariate models, along with their transformations, are illustrated in Figure 12.1. In each case, we have drawn them as "decay" curves with negative slopes, but if the slope coefficient in the transformed equation were positive they would illustrate "growth" curves with positive slopes. In describing these models we will use a and b as the parameters to avoid confusion. The first model, given a variety of names (but including *logarithmic*), is

$$Y = ab^X\epsilon \tag{12.2}$$

which, by taking logarithms (natural or base 10) of both sides gives us

$$\log Y = \log a + \log bX + \log \epsilon$$

which is now in the bivariate form of Equation 12.1

$$Y' = \beta_0 + \beta_1 X + \epsilon'$$

where the estimates of β_0 and β_1 can be found by regressing $\log Y$ on X and

$$\hat{a} = \text{antilog } (\hat{\beta}_0)$$

$$\hat{b} = \text{antilog } (\hat{\beta}_1)$$

$$\hat{Y}_i = \text{antilog } (\hat{Y}'_i)$$

and

$$\hat{\epsilon}_i = \text{antilog } (\hat{\epsilon}'_i)$$

Similarly for the *geometric* or *log-log* model

$$Y = aX^b\epsilon \tag{12.3}$$

$$\log Y = \log a + b \log X + \log \epsilon$$

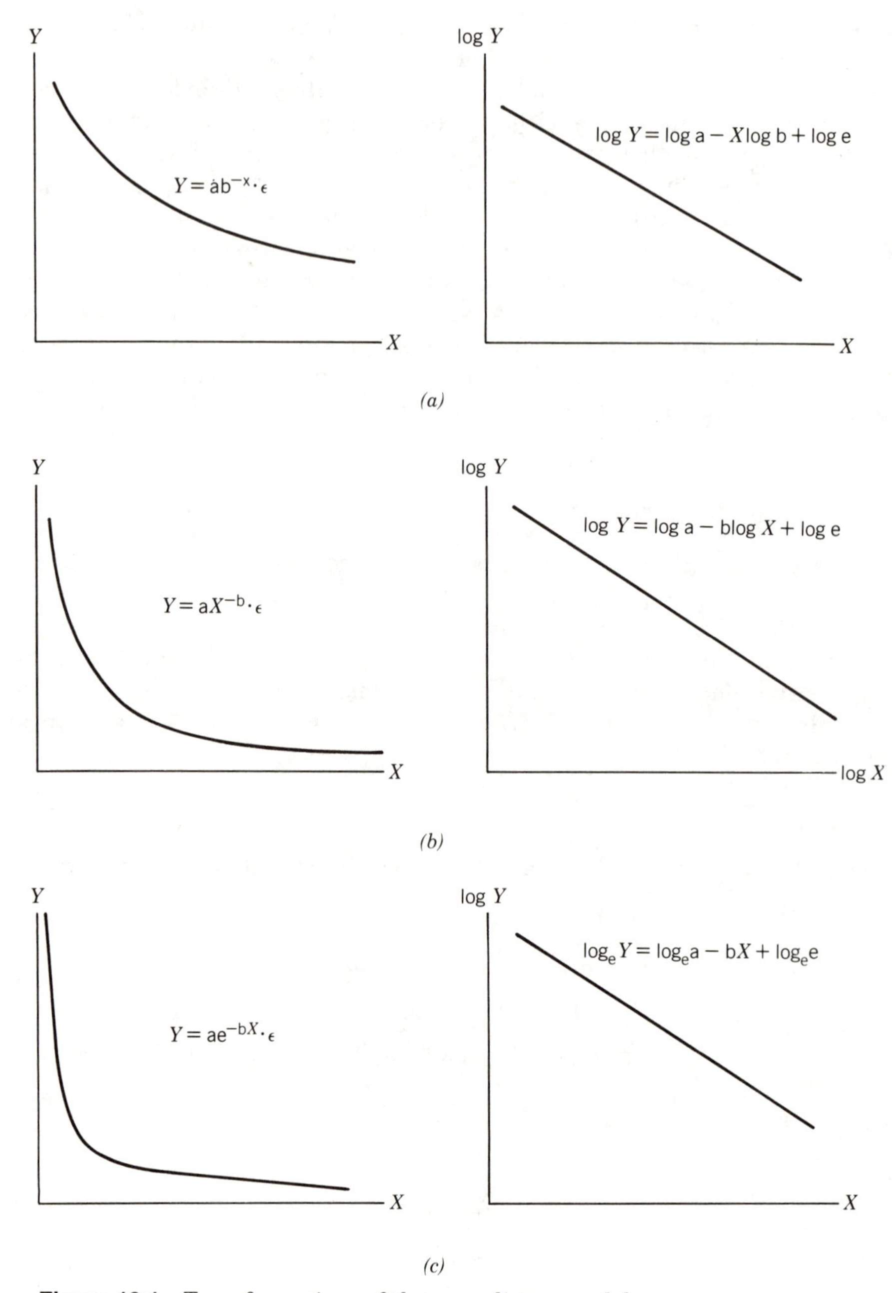

Figure 12.1 Transformations of three nonlinear models.

And using the regression of log Y on log X (natural or base 10)

$$\hat{a} = \text{antilog}\ (\hat{\beta}_0)$$

$$\hat{b} = \hat{\beta}_1$$

The third model is the *exponential* or *semilog* model

$$Y = ae^{bX}\epsilon \tag{12.4}$$

where

$$e = \text{the base of natural logarithms}$$

$$\log_e Y = \log_e a + bX + \log_e \epsilon$$

And using the regression of $\log_e Y$ on X

$$\hat{a} = \text{antilog}\ (\hat{\beta}_0)$$

$$\hat{b} = \hat{\beta}_1$$

Many other models are available, but only one illustration of a model will be given. The exponential or semilog model is frequently used to examine the decline in population density with distance from the center of a city (Figure 12.2). X is distance from the city center, and D_X is the estimated population density at point X. Population density is estimated as a negative exponential function of distance from the center D_0. The particular advantage of natural logarithms in this type of situation is that b in the transformation equation measures the percentage rate of decrease in population density for each unit change in X.

Multivariate Nonlinear Models

The same principle can be applied to multivariate situations, and many multivariate nonlinear models are intrinsically linear. The most common is the *multiplicative* model, an extension of the log-log model. For example, with three independent variables, the model

$$Y = aX_1^{b_1}X_2^{b_2}X_3^{b_3}\epsilon \tag{12.5}$$

transforms to

$$\log Y = \log a + b_1 \log X_1 + b_2 \log X_2 + b_3 \log X_3 + \log \epsilon$$

Note that the very similar (and quite realistic) model

$$Y = aX_1^{b_1}X_2^{b_2}X_3^{b_3} + \epsilon$$

with an additive error term rather than a multiplicative one is not intrinsically linear and should be solved by nonlinear methods.

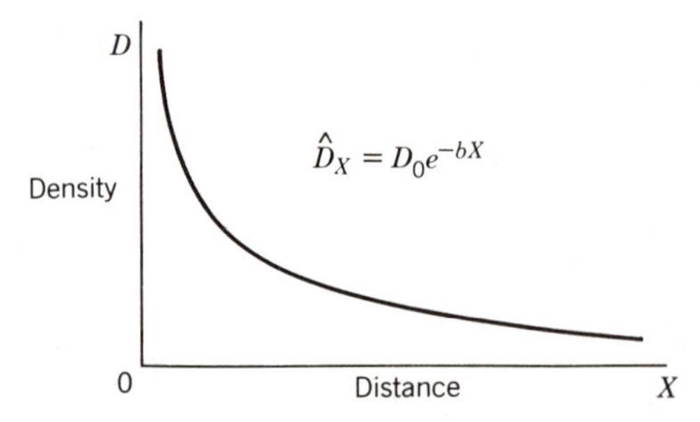

Figure 12.2 Population density with distance.

Two other widely used models are an *exponential* in two variables

$$Y = e^{a + b_1X_1 + b_2X_2} \cdot \epsilon \tag{12.6}$$

which reduces to

$$\log_e Y = a + b_1X_1 + b_2X_2 + \log_e \epsilon$$

and the *reciprocal* model

$$Y = \frac{1}{a + b_1X_1 + b_2X_2 + \epsilon} \tag{12.7}$$

transformed to

$$\frac{1}{Y} = a + b_1X_1 + b_2X_2 + \epsilon$$

In all these multivariate transformations, and in the ones described for bivariate models, the regression models use the least squares principle on the transformed scores, and thus we get different results when we apply a nonlinear least-squares structure to the data. There can be important differences.

Polynomial Regression

One of the transformations of X considered earlier was the square of X. This transformation presumably arises because the relationship of Y and X indicates that a quadratic function (a parabola) seems most appropriate. More complicated bivariate curves can be handled by taking higher-order polynomials and including these as variables in a standard regression routine. This process is generally called *polynomial regression*. Thus, for example, the bivariate relationship $Y = f(X_1)$ could be examined using a fourth-order polynomial regression model (which describes a curve with three "bends")

$$Y = \beta_0 + \beta_1 Z_1 + \beta_2 Z_2 + \beta_3 Z_3 + \beta_4 Z_4 + \epsilon$$

where

$$Z_1 = X_1 \qquad Z_2 = X_1^2 \qquad Z_3 = X_1^3 \qquad Z_4 = X_1^4$$

The methods and interpretation are the same as for the standard regression model. The model can be extended to more than one measured independent variable.

When we are more interested in fitting a model as closely as possible to the data than with the structure of the parameters, polynomial regression can be a very useful tool. For example, it is widely used in hydrology to establish relationships between water volume discharge and depth of flow. The aim is simply to predict discharge from easily obtained depth information. The statistical structure of the relationship is usually unimportant. We will look at an application of polynomial regression in the last section of this chapter.

12.2 STEPWISE REGRESSION AND THE "BEST" MODEL

Stepwise regression is a procedure for selecting one regression equation from several possible combinations of independent variables. In fact, it is both the name of a general approach and a specific procedure. Even though a set of variables has been selected because they are theoretically relevant (the critical reason for any variable selection), a smaller subset of these variables may provide a satisfactory model of the process under examination. Selecting the best subset of variables includes a trade-off between prediction and explanation—between making the best prediction possible (which suggests using a large number of independent variables to obtain reliable estimates), and keeping the model as parsimonious as possible (which allows a clearer interpretation of the interactions between the independent and dependent variables). Stepwise regression analysis is a general approach for solving this trade-off problem. However, it is important to recognize that the procedure does not give the best selection of variables in an absolute sense, only the best statistical selection of variables under certain constraints, such as maximizing the adjusted R^2. This involves maximizing R^2 while minimizing the number of independent variables.

Procedures for Selection of Independent Variables

Stepwise regression is the best known and most frequently used of several techniques for successively selecting variables for a regression equation. If there is a large number of variables, it is never possible to examine all the possible regression equations, as there would be $(2^k - 1)$ (k = the number of potential independent variables) possible equations. For even a small number of variables, 2^k is a very large number. Thus, a number of techniques that circumvent examining every regression equation have been developed. In this discussion, we will emphasize *stepwise regression* but a discussion of other techniques, including *backward elimination* and *forward selection* is found in Draper and Smith (1981) and we will illustrate forward selection. The backward elimination technique starts with all the possible independent variables included in the model, then begins to eliminate them one at a time until all that remain in the model contribute significantly. Forward selection is the opposite. It starts with one independent variable and then adds in one variable at a time, until none remain that would contribute significantly to the model. Both backward elimination and forward selection are limited. The drawback of backward elimination is related to the number of variables in the initial equation and the possibility of an ill-conditioned $\mathbf{X'X}$ matrix. For a relatively large number of variables and/or a large number of highly collinear variables, the method may not be successful. The forward selection technique (the only technique available in SPSS until 1981) does not truly evaluate the effect of a new variable on the variables already in the equation. The coefficients are recalculated, but the statistical significance of variables included earlier is not reevaluated and thus, some variables that may no longer be significant are retained in the equation.

The basic approach of stepwise regression is to add variables to the regression equation in accordance with their marginal contribution to the percent of explained variance in the dependent variable. The simplest way of adding variables is first to take the independent variable with the highest correlation with Y, then find the independent variable with the highest partial correlation and add it to the equation. The process continues until the F test of the partial is not significant. The F test on the partial is the same test we outlined in Chapter 10 to test the *extra* contribution of an independent variable. This method is the method used by both the forward selection and the stepwise approach. However, whereas the forward selection procedure adds variables based on the relative size of their partials, the stepwise procedure considers the partial of each variable in the equation as if it had just been entered and any variable whose partial is nonsignificant is removed. The advantage of this method is that a variable entered into the equation at an early stage, which is intercorrelated with other independent variables entered later, may be less useful in explaining the variation in the dependent variable. The partial F-test value can be set by the user, although most programs have default values.

The basic steps in the procedure are as follows:

1. Compute the simple correlation matrix—choose the highest simple correlation between the dependent and independent variables.

2. Choose as the second variable, from the remaining independent variables, the one with the highest partial correlation with the dependent variable.

3. Evaluate the equation $Y = b_0 + b_1X_1 + b_2X_2$ as if X_1 had been the last variable entered. Calculate its partial and test it against a preselected F value. If the partial is less than the specified F value, remove the variable from the equation.

4. Choose as the third variable, that with the remaining highest partial correlation among the remaining independent variables and repeat the evaluation step (step 3).

Although the stepwise procedure will yield a "best" selection from a set of variables entered into the program, it is important to reiterate that the procedure is no stronger than the theoretical model being used and the variables entered into the computing routine.

Examples Using the Stepwise Routine

To use the stepwise structure in its basic form, we need to specify all the potential independent variables, set any controlling values for the parameters, and then let the routine search through the variables and add (or subtract) independent variables to the model, one at each step. Usually it would be set to provide interim results step by step. This process continues until either all variables are in the model or only those that fulfill the various parameter specifications. As far as the routine is concerned, it has produced the "best" equation at this final stage.

However, the use of the procedure in this manner is rare. Instead, most researchers wish to retain more control over the process, or alternatively, wish to make the final decision on what is best. Two approaches are common. In the first, the researcher uses the default controlling parameters that in SPSS and SAS would allow almost all (if not all) independent variables to enter the equation, and then specifies an operational structure that would provide entry of the variables one at a time into the model. At the end of the computing procedure, the results are examined in terms of order of entry, the structural changes as different variables are entered, the nature of the variables left out, etc. After these considerations, a reduced model is run, and the process repeated, gradually refining the model. This approach is most suitable when we are searching for a model or want a model primarily for prediction purposes.

In many other situations, we would have more specific demands related to which variables, or groups of variables, should be included in the model. The stepwise or forward selection structure is extremely versatile as to how and when variables are entered into the series of regression equations. Maybe we wish to *force* certain variables into the structure, or we want a group of variables to be entered at the same time (because they specify a dummy variable structure, for example). We will use an example of this more structured approach to illustrate the value of the stepwise regression procedure.

Migration data are available for 10 variables for 42 Dutch cities (Table 12.1). Our interest is in explaining the mobility rate by the independent

TABLE 12.1 Migration Data for Dutch Cities for Stepwise Regression Analysis

V1	V2	V3	V4	V5	V6	V7	V8	V9	V10
0.722	29.00	84.30	41.34	1.9069	1.8771	21.49	1.	*	*
0.788	23.80	88.20	19.52	2.9998	1.8570	20.69	0.	8.	12.
0.590	29.20	89.00	55.38	1.8790	1.7190	15.38	0.	6.	11.
0.810	24.90	83.50	55.19	2.3984	1.2961	20.51	0.	9.	14.
0.801	25.00	78.90	36.05	3.0612	1.4298	20.11	0.	5.	23.
0.853	26.40	*****	33.28	******	******	*****	1.	8.	2.
0.677	25.10	82.50	27.11	1.6186	1.7236	20.01	0.	8.	10.
0.872	31.50	79.20	30.02	0.6030	5.6603	29.53	1.	*	*
0.940	25.60	82.80	48.37	2.3936	2.4957	25.53	0.	*	*
0.952	25.50	84.80	19.67	1.0180	1.9589	22.99	1.	9.	4.
0.668	23.70	88.80	23.54	3.7247	0.9906	16.10	0.	4.	18.
0.879	25.90	83.00	33.15	3.2728	2.3780	16.89	0.	10.	8.
0.187	24.30	91.40	54.33	******	2.2842	10.68	0.	6.	*
0.950	23.80	*****	47.44	2.5872	2.4641	18.37	0.	*	0.
0.795	26.80	82.50	60.74	1.9116	2.2258	21.93	0.	8.	20.
0.911	23.20	77.20	47.77	******	5.0460	50.01	1.	9.	10.
1.115	32.60	71.00	60.58	0.3804	5.0491	29.47	1.	5.	16.
*****	25.40	*****	37.89	******	******	45.22	1.	9.	18.
0.586	24.60	83.80	36.69	2.0470	******	21.46	0.	5.	10.
0.657	24.50	92.50	61.20	4.0688	2.2960	17.03	0.	4.	5.
0.632	24.50	*****	33.73	******	2.0185	20.68	0.	4.	8.
0.700	27.90	84.00	29.27	2.0184	2.8735	12.75	0.	10.	4.
0.519	22.40	82.90	*****	2.8732	2.0212	36.48	1.	10.	19.
0.780	24.60	81.00	41.30	******	1.8571	*****	1.	8.	12.
0.969	31.10	78.00	45.16	1.2645	2.7600	35.56	1.	5.	20.
0.824	24.40	*****	57.16	2.2941	2.2841	28.83	0.	6.	20.
0.879	30.30	76.90	47.03	2.0525	3.2329	24.93	0.	*	20.
0.680	24.60	93.80	34.76	******	1.2919	10.21	0.	8.	6.
0.566	24.10	85.10	24.86	0.1878	5.5579	31.10	1.	*	*
*****	28.40	85.60	45.33	2.6785	1.7451	22.36	0.	7.	14.
0.942	29.80	74.50	17.09	0.9399	4.0181	38.66	1.	7.	12.
0.651	25.20	94.80	9.51	2.7129	1.8816	15.63	0.	6.	6.
0.362	23.70	78.50	*****	1.7417	1.7331	22.15	0.	6.	18.
0.535	31.50	93.50	36.90	0.7684	4.0266	2.67	0.	9.	13.
0.851	25.10	*****	51.81	2.3062	1.7826	23.82	0.	3.	9.
0.632	26.20	87.60	47.67	******	******	*****	0.	6.	28.

TABLE 12.1 (*Continued*)

V1	V2	V3	V4	V5	V6	V7	V8	V9	V10
0.416	24.30	83.80	*****	******	******	*****	0.	7.	16.
0.705	25.40	87.60	45.09	2.0697	2.2404	24.74	0.	5.	16.
*****	21.40	88.50	3.64	1.7683	2.0495	27.41	0.	5.	22.
0.430	25.10	90.40	66.70	6.2688	0.8969	14.57	0.	7.	16.
0.495	25.18	82.38	69.05	******	******	*****	0.	4.	8.
0.808	27.98	83.54	63.82	4.0475	1.3011	17.43	0.	*	11.

VARIABLE LIST

V1 Mobility rate (total movers/total population)
V2 % Pop. 20–35 yrs.
V3 % Pop. in families
V4 % of total housing units with govt. subsidies
V5 Ratio of single to multi-family units (e.g., for the first case, for every 19,069 single family units, there are 10,000 multi-family units)
V6 Ratio of owned units to rented units
V7 % of housing stock built before 1930
V8 City received urban renewal funds (yes = 1)
V9 Number of institutions involved in allocating housing
V10 Number of criteria used by institutions in allocating housing

* = missing data

variables, but we wish to force them into the regression in the order V2, V3 (demographic variables), V5, V6, V7 (housing stock measures), V4, V8 (measures of federal policies for financial allocations to cities), and V9, V10 (measure of local government housing allocation). The hypothesis tested in the analysis is that after controlling for demographic and housing stock characteristics, federal and local policies influence the internal mobility rates across cities. An analysis of the results in Table 12.2 for the initial and end steps (the middle steps are not included) indicates that the variations in migration rates is explained largely by demographic variation, and that housing stock and policy variables add less explanation.

However, two further points are worth exploring. Consideration of the data in Table 12.1 indicates that there are numerous missing observations on particular variables, and the default in SPSS regression analysis is to remove *cases* when there are any missing data. In this example, there are only 18 available cases, and thus the results are being influenced by the very small n. An alternative (for which the summary results are given in Table 12.3) is to use "pairwise deletion" rather than case deletion. In this situation, a missing value for a particular variable means that the case will be removed from calculations involving that variable only. As the SPSS manual notes, pairwise deletion can lead to strange results. The technique

TABLE 12.2 Setup and Run for Stepwise (Forward Selection) Regression Showing the First Two and the Last "Steps"

```
 1 RUN NAME           MIGRATION DATA
 2 VARIABLE LIST      V1 TO V10
 3 VAR LABELS         V1 INTERNAL MIGRATION/
 4                    V2 POP 20-35YRS/
 5                    V3 FAMILIES/
 6                    V4 PERCENT UNITS WITH SUBSIDIES/
 7                    V5 SINGLE OR MULTI UNIT/
 8                    V6 RENT OR OWN/
 9                    V7 HOUSING BUILT PRE L930/
10                    V8 CITY RECEIVED URBAN RENEWAL MONEY/
11                    V9 NUMBER OF INSTITUTIONAL CRITERIA/
12                    V10 TOTAL NUMBER OF CRITERIA/
13 INPUT FORMAT       FIXED( 2X,F5.3,3F6.2,2F7.4,F6.2,F3.0,2F4.0)

   ACCORDING TO YOUR INPUT FORMAT, VARIABLES ARE TO BE READ AS FOLLOWS

   VARIABLE   FORMAT   RECORD      COLUMNS

   V1         F 5. 3     1          3-    7
   V2         F 6. 2     1          8-   13
   V3         F 6. 2     1         14-   19
   V4         F 6. 2     1         20-   25
   V5         F 7. 4     1         26-   32
   V6         F 7. 4     1         33-   39
   V7         F 6. 2     1         40-   45
   V8         F 3. 0     1         46-   48
   V9         F 4. 0     1         49-   52
   V10        F 4. 0     1         53-   56

THE INPUT FORMAT PROVIDES FOR    10 VARIABLES.    10 WILL BE READ
IT PROVIDES FOR     1 RECORDS ('CARDS') PER CASE.   A MAXIMUM OF     56 'COLUMNS' ARE USED ON A RECORD.

14 INPUT MEDIUM       DISK
15 N OF CASES         42
16 MISSING VALUES     V1(99.999)/V2 TO V4(999.99)/V5 TO V6(9.9999)/
17                    V7(999.99)/V8 TO V10(99)
18 REGRESSION         VARIABLES=V1 TO V10/
19                    REGRESSION=V1 WITH V2(19),V3(17),V4(9),V5(13),V6(15),
20                    V7(11),V8(7),V9(5),V10(3)
* * * * * * * * * * * * * * * * * * * * * *  M U L T I P L E   R E G R E S S I O N  * * * * * * * * * * * * *   VARIABLE LIST  1
                                                                                                                REGRESSION LIST  1
DEPENDENT VARIABLE..    V1          INTERNAL MIGRATION

VARIABLE(S) ENTERED ON STEP NUMBER  1..    V2           POP 20-35YRS

MULTIPLE R            0.34966                    ANALYSIS OF VARIANCE    DF      SUM OF SQUARES       MEAN SQUARE            F
R SQUARE              0.12227                    REGRESSION               1.           0.06027           0.06027       2.22874
ADJUSTED R SQUARE     0.06741                    RESIDUAL                16.           0.43266           0.02704
STANDARD ERROR        0.16444

---------------- VARIABLES IN THE EQUATION -----------------          ------------ VARIABLES NOT IN THE EQUATION ------------

VARIABLE          B              BETA      STD ERROR B       F          VARIABLE     BETA IN     PARTIAL    TOLERANCE        F

V2           0.2122966D-01     0.34966      0.01422        2.229        V3         -0.83559    -0.79310     0.79073      25.431
(CONSTANT)   0.1894487                                                  V4         -0.20525    -0.21432     0.95699       0.722
                                                                        V5         -0.62834    -0.49843     0.55230       4.958
                                                                        V6          0.67136     0.39623     0.30574       2.794
                                                                        V7          0.71176     0.73440     0.93447      17.563
                                                                        V8          0.83279     0.72912     0.67282      17.025
                                                                        V9         -0.00981    -0.01048     0.99997       0.002
                                                                        V10         0.05899     0.06232     0.97963       0.058

* * * * * * * * * * * * * * * * * * * * * * * * * * * * * * * * * * * * * * * * * * * * * * * * * * * * * * * * * * * * * * * *

VARIABLE(S) ENTERED ON STEP NUMBER  2..    V3           FAMILIES

MULTIPLE R            0.82120                    ANALYSIS OF VARIANCE    DF      SUM OF SQUARES       MEAN SQUARE            F
R SQUARE              0.67436                    REGRESSION               2.           0.33241           0.16620      15.53168
ADJUSTED R SQUARE     0.63094                    RESIDUAL                15.           0.16051           0.01070
STANDARD ERROR        0.10345

---------------- VARIABLES IN THE EQUATION -----------------          ------------ VARIABLES NOT IN THE EQUATION ------------

VARIABLE          B              BETA      STD ERROR B       F          VARIABLE     BETA IN     PARTIAL    TOLERANCE        F

V2          -0.1978298D-02    -0.03258      0.01006        0.039        V4         -0.18966    -0.32506     0.95659       1.654
V3          -0.2196975D-01    -0.83559      0.00436       25.431        V5         -0.38627    -0.48431     0.51193       4.290
(CONSTANT)   2.676573                                                   V6          0.42642     0.40532     0.29420       2.752
                                                                        V7          0.31396     0.35592     0.41850       2.031
                                                                        V8          0.48084     0.57823     0.47090       7.032
                                                                        V9          0.00724     0.01269     0.99944       0.002
                                                                        V10        -0.21947    -0.35830     0.86798       2.062

* * * * * * * * * * * * * * * * * * * * * *  M U L T I P L E   R E G R E S S I O N  * * * * * * * * * * * * *   VARIABLE LIST  1
                                                                                                                REGRESSION LIST  1
DEPENDENT VARIABLE..    V1          INTERNAL MIGRATION

VARIABLE(S) ENTERED ON STEP NUMBER  7..    V4           PERCENT UNITS WITH SUBSIDIES

MULTIPLE R            0.92465                    ANALYSIS OF VARIANCE    DF      SUM OF SQUARES       MEAN SQUARE            F
R SQUARE              0.85498                    REGRESSION               7.           0.42144           0.06021       8.42236
ADJUSTED R SQUARE     0.75347                    RESIDUAL                10.           0.07148           0.00715
STANDARD ERROR        0.08455

---------------- VARIABLES IN THE EQUATION -----------------          ------------ VARIABLES NOT IN THE EQUATION ------------

VARIABLE          B              BETA      STD ERROR B       F          VARIABLE     BETA IN     PARTIAL    TOLERANCE        F

V2          -0.3560232D-01    -0.58639      0.01665        4.571        V9          0.00804     0.01759     0.69339       0.003
V3          -0.1355850D-01    -0.51568      0.00546        6.156        V10        -0.00046    -0.00091     0.55513       0.000
V6           0.4485619D-01     0.29033      0.03638        1.520
V5          -0.3551247D-01    -0.29742      0.02385        2.217
V7           0.7641754D-04     0.00372      0.00501        0.000
V8           0.1702409         0.42769      0.09599        3.145
V4           0.6274246D-03     0.06585      0.00145        0.186
(CONSTANT)   2.781065

F-LEVEL OR TOLERANCE-LEVEL INSUFFICIENT FOR FURTHER COMPUTATION
                                                  SUMMARY TABLE

VARIABLE                                       MULTIPLE R  R SQUARE  RSQ CHANGE   SIMPLE R            B               BETA

V2          POP 20-35YRS                         0.34966    0.12227    0.12227     0.34966     -0.3560232D-01      -0.58639
V3          FAMILIES                             0.82120    0.67436    0.55210    -0.82068     -0.1355850D-01      -0.51568
V6          RENT OR OWN                          0.85315    0.72786    0.05350     0.49661      0.4485619D-01       0.29033
V5          SINGLE OR MULTI UNIT                 0.88106    0.77627    0.04841    -0.58099     -0.3551247D-01      -0.29742
V7          HOUSING BUILT PRE L930               0.89962    0.80932    0.03305     0.75463      0.7641754D-04       0.00372
V8          CITY RECEIVED URBAN RENEWAL MONEY    0.92319    0.85228    0.04296     0.76032      0.1702409           0.42769
V4          PERCENT UNITS WITH SUBSIDIES         0.92465    0.85498    0.00270    -0.12391      0.6274246D-03       0.06585
(CONSTANT)                                                                                      2.781065
```

TABLE 12.3 Alternative Stepwise Results with Pairwise Deletion of Missing Values

* M U L T I P L E R E G R E S S I O N * * * * * * * * * * * * * VARIABLE LIST 1
REGRESSION LIST 1

DEPENDENT VARIABLE.. V1 INTERNAL MIGRATION

SUMMARY TABLE

VARIABLE		MULTIPLE R	R SQUARE	RSQ CHANGE	SIMPLE R	B	BETA
V2	POP 20-35YRS	0.38947	0.15169	0.15169	0.38947	0.5880457D-01	0.80669
V3	FAMILIES	0.59125	0.34958	0.19789	-0.55301	-0.3823863D-02	-0.10970
V7	HOUSING BUILT PRE L930	0.62817	0.39459	0.04501	0.45709	0.2427065D-01	1.19497
V6	RENT OR OWN	0.64226	0.41250	0.01790	0.29609	-0.5128393D-01	-0.32882
V4	PERCENT UNITS WITH SUBSIDIES	0.67680	0.45806	0.04556	-0.11488	-0.3468870D-02	-0.28487
V5	SINGLE OR MULTI UNIT	0.68470	0.46881	0.01075	-0.30149	0.1004456D-01	0.06359
V8	CITY RECEIVED URBAN RENEMAL MONEY	0.69312	0.48042	0.01161	0.37866	-0.2668556	-0.63164
V9	NUMBER OF INSTITUTIONAL CRITERIA	0.69593	0.48431	0.00389	0.09907	0.8120661D-02	0.08304
V10	TOTAL NUMBER OF CRITERIA	0.79611	0.63379	0.14948	-0.05412	-0.1453207D-01	-0.50785
(CONSTANT)						-0.5872475	

must be used with care, but in situations with scattered missing data, it provides an alternative to reducing data sets to very small n's. Our emphasis here on the use of stepwise regression as a technique to order the variables already selected for the regression equation, or to constrain their order of entry, deemphasizes the differences between forward selection and true stepwise regression.

Creating Regression Models

We see from the examples that stepwise regression is a useful additional technique in our search for prediction and explanation. But, to reiterate, stepwise regression will be most appropriate when it is used to examine the order in which variables are introduced into the equation, or to assess the relative contributions of sets of variables in some preselected arrangement. However, there may be a large number of variables and little guidance as to their relative importance. It is natural to consider stepwise regression, but in cases when the computer "picks out the model," it is important to divide the sample into two parts, one for estimating the model and the other for validating the model.

The dangers of "computer models" are illustrated with the following program which generated random cases for the analysis (Table 12.4). Y is the dependent variable and X_1 to X_{15} are the independent variables. The results from this analysis show that X_3, X_7 and X_{13} are significantly different from 0, but X_3, X_7 and X_{13} are simply random variables. With real data, it is tempting to come up with explanations for the significant variables. But, utilizing a subset of data, it would be possible to validate our results and protect against incorrect emphasis on interpretations of computer selected variables. Of course, with random data, the model is nonsensical but this procedure will guard against overinterpretation with stepwise procedures.

12.3 ANALYSIS WITH DUMMY VARIABLES

In many instances, the level of measurement for one or more of the independent variables derives from classification (categorical data). Instances of the effects of different regions, different rock types, different land use types

TABLE 12.4 Hypothetical Stepwise Regression

```
// EXEC SAS
  DATA REGS ;
      ARRAY VARS Y X1-15;
      DROP CASE;
      DO CASE = 1 TO 400;
          DO OVER VARS;
              VARS = NORMAL (83383);
          END;
          OUTPUT;
      END;
  PROC STEPWISE;
      MODEL Y = X1 - X15
```

Reproduced by permission from C. Hofacker (1983), "Abuse of Statistical Packages: The Case of the General Linear Model," *American Journal of Physiology* 245:299–302.

and tenure, sex, and class differences abound. The dummy variable is one means of incorporating information into the regression equation that is not conveniently or even able to be measured on an interval scale. An equally important reason for introducing a dummy variable relates to the potential for bias in the estimated parameters when categorical effects are left out of the model. When attention is focused on a primary relationship between a dependent and independent variable, ignoring a regional effect, for example (which might appear as noise), our estimation of the proper relationship between the dependent and independent variables is clouded. Although experimental control over the regional effect may not be possible, it should be included in the regression analysis.

Specification and Use

Dummy variables are easily specified. For two effects (categories, classes, or regions), values are ordered 0 for one type and 1 for the other. By convention, the category which has a potential positive relationship with the dependent variable is coded 1. For three categories, it would require two dummy variables:

$$X_1 = 1 \qquad X_2 = 0 \text{ for category 1}$$

$$X_1 = 0 \qquad X_2 = 1 \text{ for category 2}$$

$$X_1 = 0 \qquad X_2 = 0 \text{ for category 3}$$

The role and function of dummy variables is best illustrated with an example (Table 12.5). Assume that we are interested in the rate of erosion of natural and spoil banks in river waterways. Erosion is measured in centimeters of material removed at the waterline. The dependent variable, Erosion Rate, is postulated to be a function of the median diameter (milli-

TABLE 12.5 Data Used in the Analysis for Natural and Spoil Banks

Case No.	Erosion Rate (cm)	Grain Size Median Diameter	Bank Type
1	.30	.02	0
2	.60	.04	0
3	.45	.03	0
4	.20	.02	0
5	.40	.04	0
6	.40	.04	0
7	.30	.03	0
8	.50	.04	0
9	.65	.05	0
10	.70	.06	0
11	.80	.04	1
12	.55	.01	1
13	1.00	.05	1
14	.60	.02	1
15	.60	.01	1
16	.80	.02	1
17	1.15	.06	1
18	1.20	.05	1
19	1.00	.05	1
20	.70	.02	1

meters) of the material in the banks. This Grain Size of the material, as measured by its median diameter, is an important factor in erodibility. In many circumstances, the coarser material is more easily eroded, and so rate of erosion could increase with increasing grain size.

The material composing the banks of the waterways is made up of two distinct kinds: natural banks consisting largely of clays and silts mixed with decomposed organic material, and most importantly, vegetated with marsh grass and its intricate root system. The second kind of bank—the spoil bank—is the product of dredging along the waterway bottom. The sediment, of similar size to that found in natural banks, is merely dumped on the side of the canal and is only vegetated on the upper part of the bank. The intricate root mat of the natural banks is absent, and the resistance to erosion is much lower.

Examining the scatter diagram (Figure 12.3), we note there is a clear distinction between the natural banks (○) and the spoil banks (●). To handle this double scatter of points, we will introduce a dummy variable (Bank Type), where natural banks are given the value 0 and spoil banks 1.

The interpretation and use of dummy variables is aided by working through this example. Based on the general scatter of points (Figure 12.3), we fit the simple linear regression model to all the data

$$\hat{Y} = b_0 + b_1X_1 \tag{12.8}$$

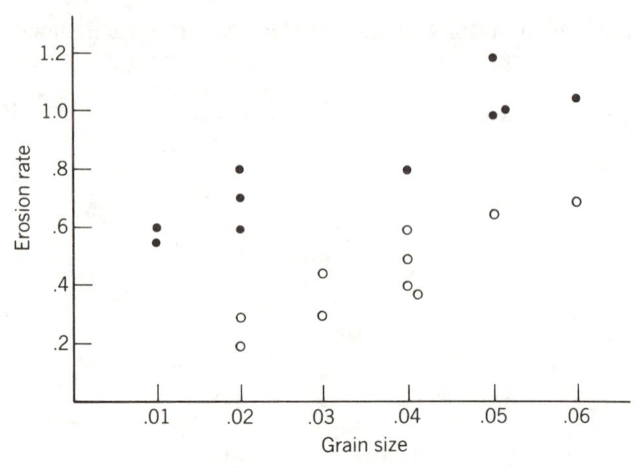

Figure 12.3 Plot of erosion rate against grain size.

where

$$Y = \text{Erosion Rate}$$

$$X_1 = \text{Grain Size}$$

The results obtained were

$$\hat{Y} = .302 + 9.787X_1$$

with

$$r^2 = .300 \ (S_E = .241)$$

but note that the regression line, despite the moderate r^2 value, is not a good fit to the data (Figure 12.4). To get better fits (but end up with two equations),

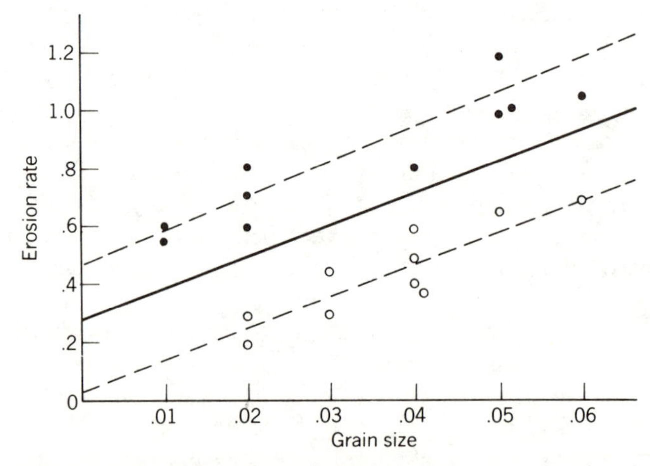

Figure 12.4 Regression of erosion rate on grain size. Dashed lines represent separate regressions for each bank type. Solid line is a simple regression for all data points.

we could run separate regression models (Equation 12.8) for each of the two types of banks. This results in the two equations.

For natural banks:

$$\hat{Y} = .017 + 11.702X_1 \tag{12.9}$$

with

$$r^2 = .805\ (S_E = .077)$$

For spoil banks:

$$\hat{Y} = .457 + 11.620X_1 \tag{12.10}$$

with

$$r^2 = .869\ (S_E = .091)$$

The two equations fit the data much better, but they do not provide a solution to the problem of how to include the obvious influence of the two bank types on our analysis of the relationship of Erosion Rate and Grain Size.

This can be done easily by including Bank Type as a dummy variable in a two independent-variable, multiple linear regression model.

$$\hat{Y} = b_0 + b_1X_1 + b_2X_2 \tag{12.11}$$

where

$$\begin{aligned} X_1 &= \text{Grain Size} \\ X_2 &= 0 \text{ if Bank Type} = \text{natural} \\ &= 1 \text{ if Bank Type} = \text{spoil} \end{aligned}$$

Before running this model let us regress Y on X_2, the dummy variable, alone. We get

$$\hat{Y} = .450 + .390X_2$$

with

$$r^2 = .507\ (S_E = .203)$$

Note that when X_2 = natural banks = 0

$$.390X_2 = 0$$

and, therefore,

$$\hat{Y} = .450 = \overline{Y}_{\text{natural}}$$

and when X_2 = spoil banks = 1

$$.390X_2 = .390$$

and, therefore,

$$\hat{Y} = .450 + .390 = .840 = \overline{Y}_{\text{spoil}}$$

Or generally, for two groups,

$$\overline{Y}_1 = b_0 \qquad \text{and} \qquad \overline{Y}_2 = b_0 + b_1$$

Regressing Y on the dummy variable alone has fit a straight line between the means of the two groups (at one unit apart on the x axis). This point will be taken up shortly.

Now look at the combined model (Equation 12.11). The results were

$$\hat{Y} = .019 + 11.645X_1 + .437X_2$$

with

$$R^2 = .925 \; (S_E = .081)$$

The interpretation and testing of the overall model and the individual regression coefficients are handled in the standard manner. It is interesting to observe the effect of the dummy variable X_2 in the equation. For natural banks, where $X_2 = 0$

$$\hat{Y} = .019 + 11.645X_1$$

For spoil banks, where $X_2 = 1$

$$\hat{Y} = .019 + 11.645X_1 + .437$$
$$= .456 + 11.645X_1$$

Comparing these results with the original separate simple linear regression estimates for the two bank types (Equations 12.9 and 12.10), shows that the three-dimensional regression plane passes closely to the individual regression lines projected in the same space.

In common with any form of multiple regression, there could be *interaction* between the two independent variables (i.e., the contribution of the two independent variables is not strictly additive). For the dummy variable structure, however, we can easily identify this by comparing the slopes of the individual regression lines for the two groups. In our example, the values were 11.702 and 11.620 for natural and spoil banks, respectively, and the two lines were nearly parallel (Figure 12.4). A simple test for interaction is to calculate the *product* of the two variables, include this composite variable in the analysis, and test for its contribution. The model would be

$$\hat{Y} = b_0 + b_1X_1 + b_2X_2 + b_3X_1X_2 \qquad (12.12)$$

In our example, this gives

$$\hat{Y} = .017 + 11.702X_1 + .440X_2 - .082X_1X_2$$

TABLE 12.6 Testing for Interaction

Source of Variability	Sum of Squares	*df*	Mean Square	*F*
Due to regression	1.38701	3	.46234	
Due to X_1 and X_2	1.38700	2	.69350	
Due to X_1X_2 given X_1 and X_2	.00001	1	.00001	.00
Due to error	.11249	16	.00703	
TOTAL	1.49950	19		

with

$$R^2 = .925\ (S_E = .084)$$

We can illustrate the effect of this structure and how it would be tested by partitioning the sums of squares regression as outlined in Section 10.5 (Table 12.6). With a tabled value of $[F, p = .95, df = 1{,}16] = 4.49$, we fail to reject the hypothesis that β_3 (=contribution of X_1X_2) = 0, and can consider that the two variables contribute independently. The interaction test used here can be applied to other pairs (or even more than two) of independent variables, and it is not only restricted to dummy variables.

Analysis of Variance and Covariance

The regression of Y on X where X is a categorical variable (derived from classification) using one or more dummy variables, is identical to the *one-way analysis of variance* as described in Section 8.3. For example, an independent variable X consisting of four classes would require three dummy variables, and the resulting regression equation is in the form

$$\hat{Y} = b_0 + b_1X_1 + b_2X_2 + b_3X_3 \tag{12.13}$$

where X_1, X_2, and X_3 are dummy variables with scores 0/1, representing in combination the four classes for the measured variable X. Apart from the equation itself, the results of this regression are identical to an analysis of variance of Y across the four classes. By using dummy variables, a four-dimensional surface is fitted through the mean-Y's of the four classes. If our dummy variables were structured in the following manner,

X_1	X_2	X_3	Class
1	0	0	1
0	1	0	2
0	0	1	3
0	0	0	4

then from Equation 12.13 by applying 0/1 scores to each of the variables

$$\bar{Y}_4 = b_0$$

$$\bar{Y}_1 = b_0 + b_1$$

$$\bar{Y}_2 = b_0 + b_2$$

$$\bar{Y}_3 = b_0 + b_3$$

Thus, the hypothesis we would be testing $Ho\colon \beta_1 = \beta_2 = \beta_3 = 0$ is identical to the analysis of variance hypothesis $Ho\colon \mu_1 = \mu_2 = \mu_3 = \mu_4$.

In the bank type example, there were only two classes. The equation we fitted was

$$\hat{Y} = .45 + .39X_2$$

and the resulting analysis of variance table for regression is given in Table 12.7a. A one-way analysis of variance to test for difference of means would have obtained the results given in Table 12.7b. Note also that with only two classes, our analysis of variance test is identical to the *t* test for the difference of two means (equal variances assumed).

The general linear model may take on many different forms, and we can often achieve the same results through a variety of methods. The analysis of variance structure we have outlined can be extended, for example, to two independent variables, both of which are categorical variables. To handle this in the regression format would require $(k_1 - 1) + (k_2 - 1)$ dummy variables where k_1 and k_2 are the number of classes in the two independent variables. The contribution of a metric variable (for example, Grain Size) and a categorical variable (for example, Bank Type) to the explanation of a dependent variable (for example, Erosion Rate) illustrates another form of

TABLE 12.7 Identical Analyses of Variance

Source of Variability	Sum of Squares	*df*	Mean Square	*F*
(a) SIMPLE REGRESSION				
$\hat{Y} = b_0 + b_1X$ where $X = 0/1$				
Due to regression	.76050	1	.76050	18.52
Due to error	.73900	18	.04106	
TOTAL	1.49950	19		
and $r^2 = .507$				
(b) ANALYSIS OF VARIANCE				
for Y with $k = 2$ classes				
Between groups	.76050	1	.76050	18.52
Within groups	.73900	18	.04106	
TOTAL	1.49950	19		
and $E^2 = .507$				

the general linear model—often called *covariance analysis*. Covariance analysis can also be extended to handle more variables. A wide variety of different structures can be tested through the general linear model—the form of testing would depend on the specific aims of the analysis and the ease with which these aims can be achieved.

12.4 TREND SURFACE ANALYSIS

Geographers have a basic interest in phenomena that can be plotted on maps. Traditional cartographic methods have emphasized the description of a variety of distributions, but the analysis of these distributions with regression methods has been more recent. The application of the general linear model to describing and interpreting spatial distributions was developed first by geologists, and its use in geography, particularly in human geography, is perhaps more limited. However, it is an important statistical method if used with care, and with consideration for the assumptions of the model.

Maps and Contouring

It is useful to step back a moment before we develop the methodology of *trend surface analysis* (the name we give to the application of the general linear model to spatial coordinate data), and consider the nature of a map and its description. A map is a two-dimensional representation of a geographic unit that portrays a set of data in terms of its coordinate locations. All maps, whether at a large or small scale, are attempts to generalize information. The data varies from being discrete (distribution of cities) to continuous (vegetation cover) and includes the portrayal of such diverse phenomena as rock types, housing types, and cloud cover. Although the map may be of discrete or continuous phenomena, it is in the process of generalization or averaging the data to be portrayed that potential problems arise. Although many maps may appear as continuous distributions, in fact, most maps are estimates of continuous functions based on discrete observations at some set of coordinate locations. Certainly, the topographic map is a good example of a map in which the contour lines are constructed from measurements made at discrete points of triangulation. Quite obviously, the accuracy and detail of the final map will be dependent on the pattern of observation points. The distribution of points is of significance both in contouring and in trend surface analysis.

Point distributions can be described as clustered, regular (sometimes uniform), or random, although the actual distribution will not be exactly equivalent to a regular clustered or uniform distribution. In many geographic analyses, we think of a continuum with clustered and regular distributions at either end and random distributions in an intermediate position. Some attempts have been made to measure these distributions but for our purposes, we will simply offer definitions of each of these distributions.

A *regular* distribution is defined as one in which the points are arranged at the intersections of a grid. In such a case, the distances between any points i and j are equal. Such a regular grid would require the points at intersections of a hexagonal lattice. The hexagonal arrangement of points is sometimes described as uniform to distinguish it from other regular (square grids, for example) arrangements of points. A *clustered* distribution is best described when a set of objects are contiguous and the degree of clusteredness can be measured with observations on the distances between any point i and its neighbors. The *random* distribution is defined as the situation when any subarea of a larger region is equally likely to receive a point. The position of a point has no influence on any other point.

Given a relatively uniform arrangement of points, it is possible to describe the spatial arrangement of these points with contour lines. The simplest and quickest method is merely to interpolate contour lines using the values associated with the points as guides. This may suffice for a casual analysis, but more elegant contouring methods are possible. Many of these come as integrated software packages, although the basic principle is the same. Essentially, the procedure is to take an arrangement of points with some associated values and draw contour lines which will translate the points into a smooth surface. The process, which is basic to almost all contour programs, has three steps.

1. Overlay the map, (a region) with a regular grid (square, rectangular, hexagonal).
2. Locate a set of the nearest observed points to each grid point.
3. Calculate a value for the grid point, several methods are possible at this stage, but one method involves taking the value of the observed points divided (adjusted) by their distance from the grid point. A worked example is available in Davis (1973).

Trend Surface Models and Analysis

In using the statistical technique of trend surface analysis, the geographer can have any one of three aims in mind. The technique can be used to describe a spatial series, to separate a spatial trend into regional and local components, or to test a hypothesis about the spatial series. As originally developed in geology, the technique was used to identify systematic trends from purely local influences. Krumbein, one of the first geologists to use trend surface analysis, emphasized the usefulness of trend surface analysis to geologists interested in the identification of potential oil drilling sites, and for the information that may be provided about the oil bodies of economic significance. Trend surface analysis can be effectively used to remove the local fluctuations that often obscure the more general pattern of variation.

A map trend can have several meanings, but, in essence, it is used to indicate a tendency for the mapped distribution to show some alignment. For geologists, this might be the trend in geologic structures, for the physical geographer the underlying topographic structure, for the climatologist the broad trend of rainfall belts, while the human geographers can examine population distributions, the pattern of housing and income surfaces. Later, we examine the specific problems that arise when trend surface analysis is applied to human geographic data.

To plot any distribution on a map, the geographer requires in addition to the value to be mapped (elevation, rainfall, population density), the x and y coordinates for each observation. Anticipating the analogy with the multiple linear regression model, we can label the two axes the X_1 axis and the X_2 axis, and the variable that is being mapped, the dependent variable Y. From our knowledge of the regression model, we recognize that an inclined plane would describe the relationship of Y with X_1 and X_2 (Figure 12.5). The trend surface model is concerned with how functions of these two independent (spatial locations) variables alone can explain the spatial variation in Y, the dependent variable. Thus, it indicates that the variation in Y is a function of only its location, and that trends in Y can be related to the location of Y. This observation leads many geographers to conclude that the technique is largely descriptive. However, it is still a powerful means of smoothing an otherwise irregular spatial series.

In Chapter 10, we established the form of the general linear model and showed that we could express a dependent variable Y as a function of two or more independent variables X. If we restrict ourselves to two independent variables we have

$$\hat{Y} = \beta_0 + \beta_1 X_1 + \beta_2 X_2 \tag{12.14}$$

Now, if X_1 and X_2 are considered as Cartesian coordinates for the east–west and north–south directions, then b_1 and b_2 are the slopes for each of these

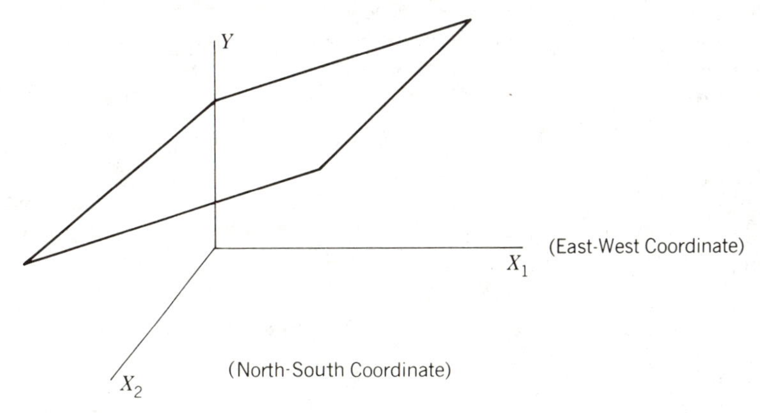

Figure 12.5 Regression plane in three dimensions.

directions for the plane or linear surface that depicts the spatial variation in Y. In the present case, the independent variables are mutually perpendicular geographic coordinates and the dependent variable Y is some observed value that varies over a geographic space.

This model, however, will only describe a linear plane surface, whereas most spatial distributions, even with smoothing, contain considerable variation from a linear surface. More concise models are required, and the obvious extension is to use the polynomial models introduced in Section 10.1. For higher-order trends, we must expand the equation. Thus, for a second-degree surface, the linear surface of

$$\hat{Y} = b_0 + b_1X_1 + b_2X_2 \qquad (12.15)$$

becomes

$$\hat{Y} = b_0 + b_1X_1 + b_2X_2 + b_3X_1^2 + b_4X_2^2 + b_5X_1X_2 \qquad (12.16)$$

The expansion of the linear surface involves squares of the two geographic coordinates and a cross-product term X_1X_2. The expansion of the trend surface to higher-order surfaces involves each variable raised to a higher level and the appropriate cross-product terms. Thus, a third-degree surface can be written:

$$\hat{Y} = b_0 + \boxed{b_1X_1 + b_2X_2} + \boxed{b_3X_1^2 + b_4X_2^2 + b_5X_1X_2} + \boxed{b_6X_1^3 + b_7X_2^3 + b_8X_1^2X_2 + b_9X_1X_2^2} \qquad (12.17)$$

The actual computation of trend surfaces involves a program to generate the appropriate transformations of X_1 and X_2, the standard regression routine to solve for the b values, and a plot routine to map the trend values. Early computer programs focused on regularly gridded data where X_1 and X_2 were located at grid intervals so that a procedure called orthogonal poly-

TABLE 12.8 Grain Size on a Beach

Y Average Grain Size in mm	X_1 East–West Distance from an Arbitrary Origin (m)	X_2 North–South Distance from an Arbitrary Origin (m)
.220	10.0	8.5
.710	21.0	44.5
.486	33.0	19.0
.320	35.0	10.0
.520	47.0	28.0
.280	60.0	9.0
.480	65.0	37.0
.437	82.0	46.5
.410	89.0	30.0
.180	97.0	7.0

nomials could be used. This procedure allowed the addition of higher-order terms without recalculating the coefficients already in the equation and was computationally simpler. With the general use of larger and faster computers, such simplification techniques are less important.

To illustrate the basics of the model, a simple linear surface can be fitted to the grain sizes of beach sand for a 50 × 100 m section of beach in Southern California (Table 12.8). The locations of the 10 points and their values are plotted on Figure 12.6. Standard regression methods were used to solve for b_0, b_1, and b_2 (Table 12.9).

The solutions were

$$b_0 = .3004$$

$$b_1 = -.0020$$

$$b_2 = .0089$$

With the value of the b's from the equation, we can calculate the expected values $\hat{Y}$ and the deviations $(Y - \hat{Y})$ (Table 12.10). Mapping the expected values $(\hat{Y})$ for varying east–west and north–south measurements yields Figure 12.7.

Tests of Trend Surfaces

Just as in the two independent variable multiple regressions, we can use R^2 as a measure of goodness of fit. This value is .795. A more general assessment of the degree of fit and level of explanation involves assessing the explanation of various orders of surface.

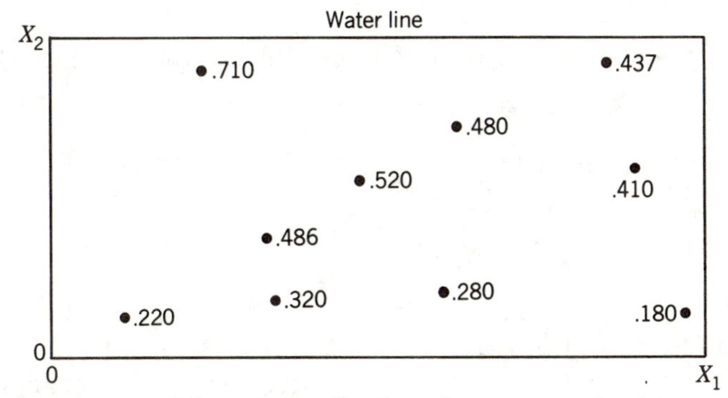

Figure 12.6 Mapped distribution of average grain sizes.

TABLE 12.9 SPSS Setup and Results of a Linear Regression on a First-Order (Linear) Surface

```
                 1 RUN NAME          REGRESSIONS ON GRAIN SIZE
                 2 VARIABLE LIST     V1,V2,V3
                 3 VAR LABELS        V1 AVERAGE GRAIN SIZE/
                 4                   V2 X1 IN METERS
                 5                   V3 X2 IN METERS
                 6 INPUT FORMAT      FIXED (F4.3,2(1X,F4.1))

                  ACCORDING TO YOUR INPUT FORMAT, VARIABLES ARE TO BE READ AS FOLLOWS

                  VARIABLE   FORMAT   RECORD      COLUMNS

                  V1         F 4. 3        1      1-    4
                  V2         F 4. 1        1      6-    9
                  V3         F 4. 1        1     11-   14

THE INPUT FORMAT PROVIDES FOR    3 VARIABLES.    3 WILL BE READ
IT PROVIDES FOR     1 RECORDS ('CARDS') PER CASE.  A MAXIMUM OF     14 'COLUMNS' ARE USED ON A RECORD.

                 7 N OF CASES        10
                 8 REGRESSION        VARIABLES=V1 TO V3/
                 9                   REGRESSION=V1 WITH V2,V3(2)

DEPENDENT VARIABLE..     V1             AVERAGE GRAIN SIZE

VARIABLE(S) ENTERED ON STEP NUMBER  1..     V3
                                            V2            X1 IN METERS

MULTIPLE R          0.89139                  ANALYSIS OF VARIANCE     DF      SUM OF SQUARES       MEAN SQUARE            F
R SQUARE            0.79457                  REGRESSION               2.          0.18051             0.09026         13.53728
ADJUSTED R SQUARE   0.73587                  RESIDUAL                 7.          0.04667             0.00667
STANDARD ERROR      0.08165

---------------- VARIABLES IN THE EQUATION -----------------     ------------ VARIABLES NOT IN THE EQUATION ------------

VARIABLE           B            BETA      STD ERROR B       F      VARIABLE      BETA IN     PARTIAL    TOLERANCE      F

V3           0.8864212D-02    0.85684      0.00179     24.628
V2          -0.2011233D-02   -0.37484      0.00093      4.713
(CONSTANT)   0.3004075

ALL VARIABLES ARE IN THE EQUATION
                                                 SUMMARY TABLE
VARIABLE
                                       MULTIPLE R   R SQUARE   RSQ CHANGE   SIMPLE R           B               BETA
V3
V2           X1 IN METERS               0.81009     0.65625     0.65625     0.81009     0.8864212D-02       0.85684
(CONSTANT)                              0.89139     0.79457     0.13832    -0.26797    -0.2011233D-02      -0.37484
                                                                                        0.3004075
```

We can compute the total variation in the sum of squares of the dependent variables:

$$\text{SST} = \Sigma Y^2 - \frac{(\Sigma Y)^2}{n} = 1.8618 - \frac{(4.043)^2}{10} = .2272$$

and compare this with the sum of squares due to the trend or regression

$$\text{SSR} = \Sigma \hat{Y}^2 - \frac{(\Sigma Y)^2}{n}$$

$$= 1.8157 - \frac{(4.043)^2}{10}$$

$$= .1811$$

The difference is the sum of squares due to deviations from the trend

$$\text{SSD} = \text{SST} - \text{SSR} = 0.0461$$

The degree of fit of the linear trend is:

$$100\,\frac{\text{SSR}}{\text{SST}} = 79.70$$

Differences between the hand calculations and the computer results are due to rounding. The trend in the grain size of the sand is inclined downward from the water line; that is, grain sizes are larger near the water, but it is also sloped from west to east (Figure 12.7).

TABLE 12.10 Deviations from a Linear Trend

Y	$\hat{Y}$	$Y - \hat{Y}$
.220	.356	−.136
.710	.653	.054
.486	.402	.084
.320	.319	.001
.520	.454	.066
.280	.259	.021
.480	.498	−.018
.437	.548	−.111
.410	.387	.023
.180	.167	.013

For more complicated trends, we expand the above tests of the "degree of fit" of the surface. If we can make the assumptions necessary for the F test, (see Chapter 11) we can use the analysis of variance procedure to test for the significance of a trend. We separate the total variation in Y into successive components due to a linear trend, due to a quadratic trend, and so on, and at each level assess the contribution of the trend against the residual not accounted for by the trend. The degrees of freedom associated with each trend surface are defined by the number of terms or coefficients in the equation that is fit to the data. The procedure is usually to fit succeeding higher orders of surfaces, and Table 12.11 outlines a general ANOVA (analysis of variance) testing format. By converting the sums of squares to mean squares, we can compare the significance of linear, quadratic, and cubic surfaces in terms of the relative increase in explanation from one level to another.

From Table 12.11 examples of F tests include the following:

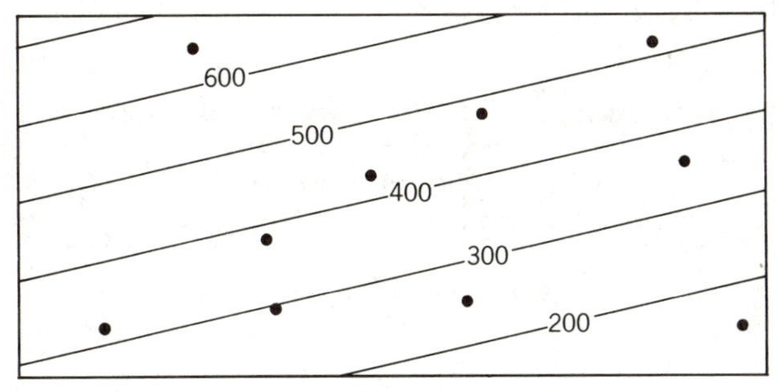

Figure 12.7 Linear trend of grain sizes on a beach.

TABLE 12.11 Successive Levels of Fit for Trend Surfaces

Level of Surface	Sum of Squares	Degrees of Freedom
Linear	SS_L	2
Deviation from linear	SS_{DL}	$n - 2 - 1$
Quadratic	SS_{L+Q}	5
Deviation from quadratic	SS_{DQ}	$n - 5 - 1$
Cubic	SS_{L+Q+C}	9
Deviation from cubic	SS_{DC}	$n - 9 - 1$

$$[F, p = 1 - \alpha, df = 2, n - 2 - 1] = \frac{SS_L/2}{SS_{DL}/n - 2 - 1}$$

(test for significance of the linear surface)

$$[F, p = 1 - \alpha, df = 5, n - 5 - 1] = \frac{SS_{L+Q}/5}{SS_{DQ}/n - 5 - 1}$$

(test for significance of quadratic surface)

$$[F, p = 1 - \alpha, df = 3, n - 5 - 1] = \frac{(SS_{L+Q} - SS_L)/(5 - 2)}{SS_{DQ}/n - 5 - 1}$$

(test for significance of increase in fit)

A general, less sophisticated procedure for estimating the improvement in fit is simply to calculate the percentage increase in fit due to each surface (Table 12.12). The increase in fit due to the quadratic is

$$\frac{.20182 - .18051}{.18051} \times 100 = 11.8 \text{ percent}$$

Applications and Examples

An example from physical geography illustrates the smoothing value of trend surface analysis. A pedagogical exercise involves fitting a surface to a U-shaped valley—to evaluate the extent to which the model can reflect the U

TABLE 12.12 Trend Surface Fits for the Data in Table 12.8

Degree of Surface	Sum of Squares (Total = .22718)	Percent of Total
Linear	.18051	79.46
Residual	.04667	
Quadratic	.20182	88.84
Residual	.02535	

TABLE 12.13 Topographical Data for a U-Shaped Valley

X_1 (North–South Grid Location)	X_2 (East–West Grid Location)	Y (Height in Feet)
1	1	7550
1	3	7550
1	5	6750
1	7	6350
1	9	4200
1	11	3980
1	13	3950
3	1	7400
3	3	6500
3	7	4400
3	9	3950
3	11	3980
3	13	4100
3	15	5350
4	5	4800
5	1	6900
5	3	5100
5	5	4550
5	7	4050
5	9	3990
5	11	5000
5	13	6400
5	15	7250
7	1	6750
7	3	5450
7	5	4240
7	7	3950
7	9	3980
7	11	5250
7	13	6600
7	15	7240
9	1	6400
9	3	6100
9	5	4200
9	7	3950
9	9	4150
9	11	6700
9	13	7300
9	15	7650
11	1	5750
11	3	5950
11	5	6200
11	7	3950
11	9	4140

TABLE 12.13 *(Continued)*

X_1 (North–South Grid Location)	X_2 (East–West Grid Location)	Y (Height in Feet)
11	11	7500
11	13	7650
11	15	7800
13	1	6400
13	3	5200
13	5	5100
13	7	3920
13	9	4270
13	11	5200
13	13	7450
13	15	7600

shape with an expected quadratic surface. Trend surfaces were fitted to topographic map elevation data of the Yosemite Valley, a typical U-shaped valley. The data are given in Table 12.13 and the results are summarized in Table 12.14. The area is approximately seven square miles, the 56 control points were used for each trend map. The points were located on a rectangular grid. First-, second-, and third-degree trend surfaces were fitted to the data and maps of the trends and residuals from trends were produced (Figures 12.8 to 12.10). Goodness of fit, expressed as R^2, ranged from .02 for the first-degree trend surface to .78 for the third-degree surface. The first-degree

TABLE 12.14 Analysis of Variance

Source		*df*	MSS	Increase in R^2	F Value
LINEAR					
SS_R	1,657,311.5	2	828,655.8	.0163	.65
SS_D	100,264,738.4	53	1,891,787.5		
R^2	.0163				
QUADRATIC					
SS_R	7,074,228.6	5	14,148,445.7	.6778	22.64[a]
SS_D	31,179,821.4	50	623,596.4		
R^2	.6941				
CUBIC					
SS_R	79,693,487.2	9	8,854,831.9	.0870	18.32[a]
SS_D	22,228,562.8	46	483,229.6		
R^2	.7819				

[a]Significant at .01 level.

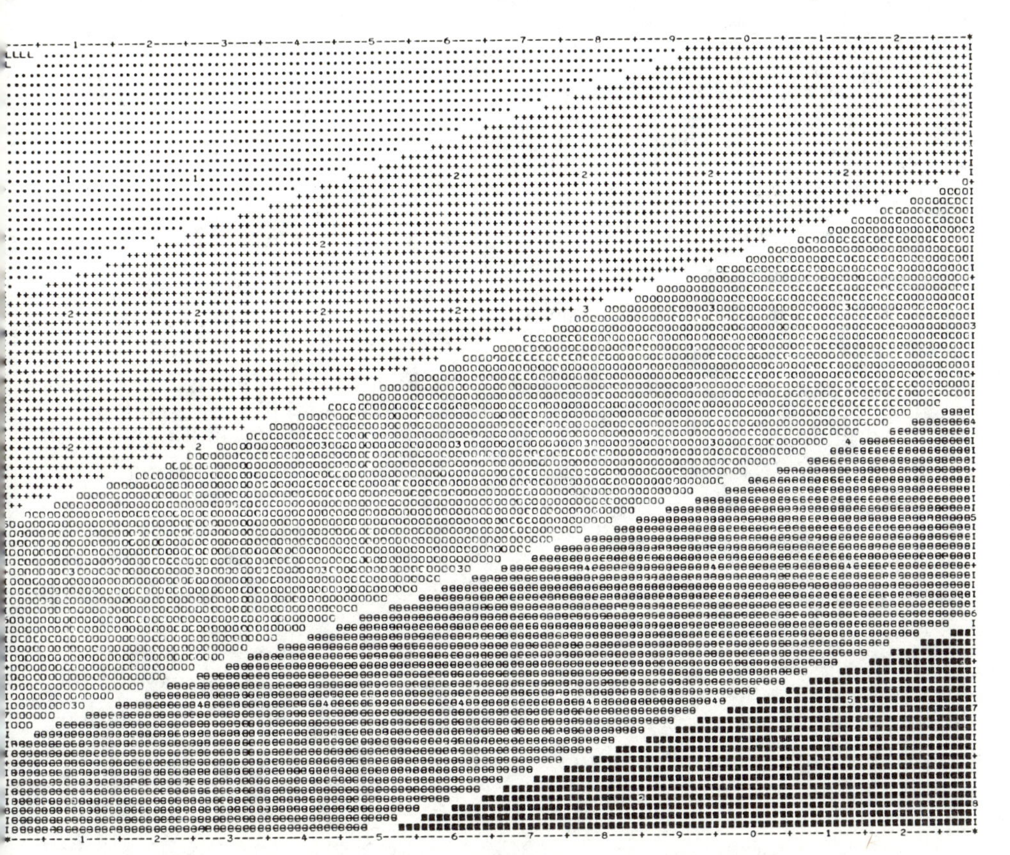

Figure 12.8 Trend surface from a linear equation.

trend map showed a slope toward the southeast. The second-degree trend surface provided a substantial improvement in the percentage of the total sum of squares, which showed a very typical U-shaped valley. The improvement of explanation from the third-degree trend surface did not show much further refinement (only a 9 percent increase (Table 12.15)).

TABLE 12.15 Summary Statistics for Trend Surfaces of Yosemite Valley

Model	*df*	R^2	*F*
First order	2,53	.0163	.65
Second order	5,50	.6941	22.64[a]
Third order	9,46	.7819	18.32[a]

[a]Significant at .01 level.

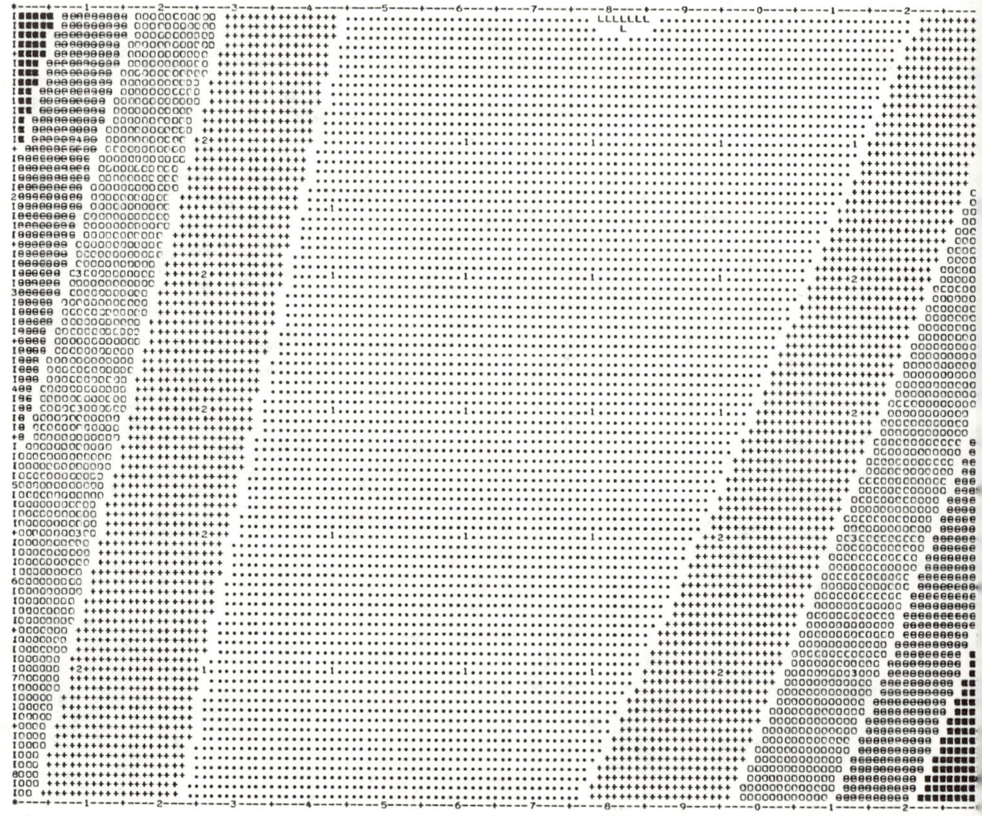

Figure 12.9 Trend surface from a quadratic equation.

Problems

The use of trend surface analysis, especially in human geography, has been criticized on a number of grounds. Johnston (1978) notes that the boundaries of the region being analyzed can affect the resultant surface, and that much human geographical data applies to unit areas rather than to sample points from a continuous surface. Norcliffe (1969) is concerned both with the problems of interpolation when points are not regularly spaced, with the purely descriptive nature of the trend surface, and with the effect of autocorrelation of the residuals on the process of fitting a trend surface. Edge or boundary effects are created when there are few or no points on the map boundaries. When this occurs, there are few if any constraints of the map. The extrapolation of values to the edge of the map, especially if higher-order surfaces are involved, may lead to large variations of the extrapolated values on the boundaries. If possible, a good practice is to have points outside (across) the boundary.

The second problem relates to the distribution of the data points used on

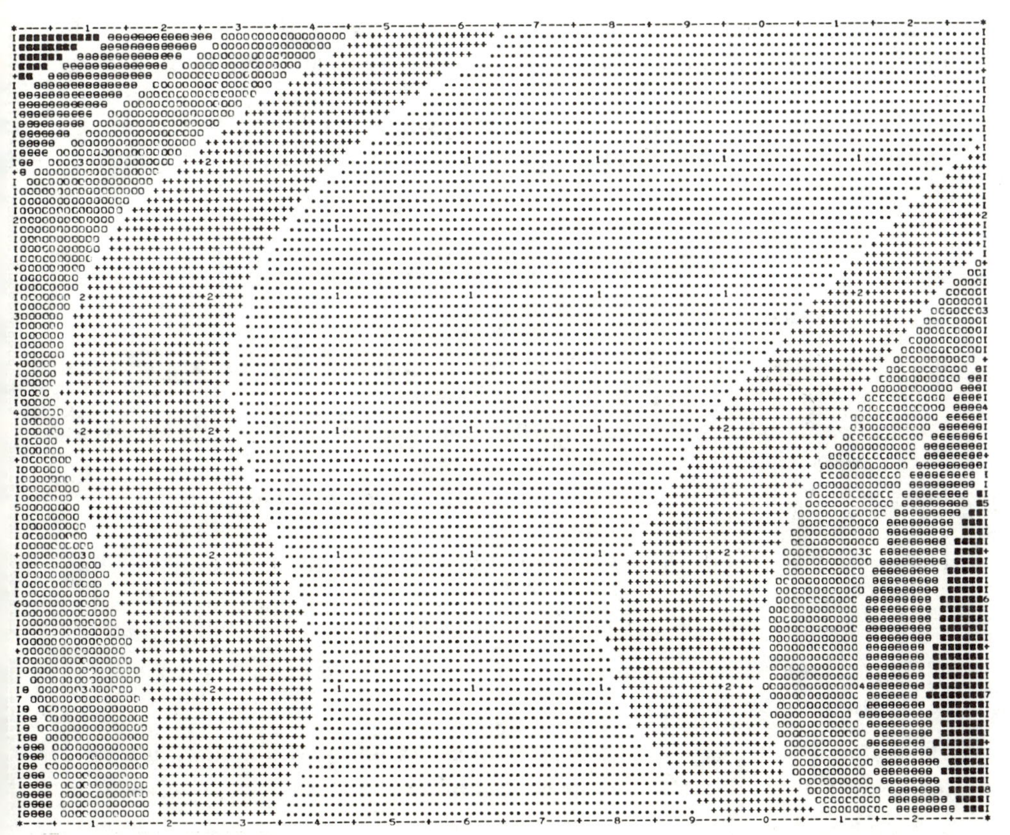

Figure 12.10 Trend surface from a cubic equation.

the map. Severe clustering of data points will influence the nature of the fitted surface. But again, too much should not be made of this problem. Experiments suggest that the clustering must be quite severe before there will be marked distortions in the surface (Davis, 1973). Interpolation issues do not seem to be insurmountable.

Finally, the issue of autocorrelation and violation of the assumptions of independent normally distributed e_i has been raised. We have discussed this at length in Chapter 11, but it is also worthwhile quoting Davis (1973).

> *Geologists can seldom speak with any authority about what form a geologic surface or distribution should take. Instead they do the next best thing and approximate the unknown function with one of arbitrary nature. . . . Polynomial functions are used for geologic trend analysis merely as a matter of convenience. The equations that are necessary to find the coefficients of the trend may easily be established and solved by the computer. Use of polynomials in no way intimates a belief that the geologic processes are polynomial functions or even that they are linear (p. 330).*

If we argue that trend surfaces are largely descriptive smoothing techniques, and tests of the coefficients are not of critical concern, then the issue of nonnormal independent e_i's may be less important. Certainly the strong local component that shows up in many maps of residuals from trend surfaces suggests that we cannot easily assume we have satisfied the assumptions of the model.

References and Readings

1. Data Transformations

Box, G. E. P., W. G. Hunter, and J. S. Hunter (1978) *Statistics for Experimenters*, Wiley: New York.

Croxton, F. E. and D. J. Cowden (1960) *Applied General Statistics*, Prentice-Hall: Englewood Cliffs, N.J.

Krumbein, W. and F. A. Graybill (1965) *An Introduction to Statistical Models in Geology*, McGraw-Hill: New York.

Lewis, D. (1966) *Quantitative Methods in Psychology*, The University of Iowa Press: Iowa City.

Shaw, G. and D. Wheeler (1985) *Statistical Techniques in Geographical Analysis*, Wiley: Chichester.

2. Trend Surfaces

Davis, J. C. (1973) *Statistics and Data Analysis in Geology*, Wiley: New York.

Johnston, R. J. (1978) *Multivariate Statistical Analysis in Geography*, Longman: London.

Norcliffe, G. B. (1969) "On the use and limitations of trend surface models," *Canadian Geographer* 13:338–348.

3. Stepwise Regression

Draper, N. R. and Smith, H. (1981) *Applied Regression Analysis*, Second Edition, Wiley: New York.

Hoser, D. P. (1974) "Some problems in the use of stepwise regression techniques in geographical research," *Canadian Geographer* 18:148–158.

4. Dummy Variables, Categorical Variables

Box, G. E. P., W. G. Hunter, and J. S. Hunter (1978) *Statistics for Experimenters*, Wiley: New York.

Wonnacott, H. and R. J. Wonnacott (1972) *Introductory Statistics*, Wiley: New York.

CHAPTER 13

Alternative Methods of Multivariate Analysis

13.1 Logit Models:
Concepts and mathematical basis; The utility basis of logit formulations; Examples and description of variables; Estimating procedures; Tests of goodness of fit; An application to residential choice modeling.

13.2 Canonical Correlation:
Concepts and mathematical formulation; Eigenvalues and derivation of canonical coefficients; Solution of the **R** Matrix; Redundancy; An application to residential search behavior.

13.3 Discriminant Analysis:
Concepts and examples for two groups; Multiple discriminant analysis; Mathematical structures; An example and interpretation for two groups; A multivariate extension.

Chapters 10 through 12 dealt with the general linear model and some of its extensions. Most of that work focused on multiple regression, the assumptions of the model and extensions to trend surface analysis, stepwise regression, and the introduction of dummy variables. This chapter, also concerned with multivariate analysis, introduces three different models, one of which—the *logit model*—is a reformulation of the general linear model for the specific case when the dependent variable takes on a binary or categorical form. The second approach deals with the case when the relationships to be evaluated include several dependent variables and associated independent variables. This technique—*canonical correlation*—at least as used in geography, has been limited to measures of relationships between two sets of variables. The third technique is closer to a method of testing for difference among groups. *Discriminant analysis* is designed to enable a researcher to distinguish between two or more subgroups of a sample population, but in such a way as to evaluate the contribution of a set of independent variables to the classification (the dependent variable). To do this, the technique develops linear equations or discriminant functions and thus the technique shares a natural correspondence with the other techniques in this chapter. The examples for two of these techniques—logit and canonical analysis—will utilize a different set of statistical routines (Biomedical Computer Programs or BMD) because of their superiority for these statistical techniques.

All of these techniques require a somewhat greater mathematical sophistication than regression analysis, although the presentation here will keep that material to a minimum and emphasize the concepts and logic of the techniques and their application to problems of geographic interest. However, at least in the case of the logit model, it is important to understand some of the aspects of testing this model and information on both the estimation procedures and the testing approaches is included for those students with the requisite mathematical background. We will occasionally note more difficult sections that can be perused rather than studied in detail.

13.1 LOGIT MODELS

Although a great number of variables used in geographical research in both human and physical geography can be measured on metric or enumerated scales, there are instances in which the level of measurement is at the nominally or ordinally classified level. In such cases, when the dependent variable is expressed as a dichotomous variable (at least in the first instance), the ordinary multiple regression model is inappropriate as a method for evaluating the fit of the model. In human geography, especially, dichotomous categorizations may be important. For example, when survey methods have been used, and responses are recorded as yes or no, it is important to be able to evaluate the probabilities of those categorical responses. But the technique is not limited to human geography because there are instances in physical geography when the presence or absence of some physical response can be evaluated against a set of independent variables. The logistic model is one that specifies a functional relationship between a basically dichotomous dependent variable and either categorical (classified) or metric scaled independent variables. Thus, it is a special case of the multiple regression model designed to deal with the situation when we have only the measurement of presence or absence, occurrence or nonoccurrence of some factor.

Concepts and Mathematical Basis

At first thought, it might appear appropriate to simply enter the data with a binary dependent variable and carry out an ordinary least-squares solution for the equation

$$\hat{Y}_i = b_0 + b_1X_1 + b_2X_2 \tag{13.1}$$

However, this solution will produce two problems. First, the application of ordinary least squares to that equation with a binary dependent variable will violate the assumption of equal variance. As Wrigley (1976) shows, the error variance is not constant but depends on the values of the explanatory variables themselves. Thus, we have the problem that the parameters estimated from the ordinary least-squares solution (OLS) are not the best estimators (in the sense of having minimum variance) of the population parameters β_0, β_1, and β_2.

The second problem relates to the restrictions on the dependent variable. If we utilize OLS procedures, the predicted values from the regression equation will not be restricted to the range 0 to 1. The predictions generated from an ordinary least-squares solution are unbounded and can take on values from $-\infty$ to $+\infty$. Thus, any predictions from the model will almost certainly be greater than the range of values for the dependent variable and will not have any natural probability interpretation. This second problem requires that we reformulate the multiple regression model into a form in which $\hat{Y}$ (predicted Y) can take on the values 0 and 1, and yet the independent variables can vary continuously; and the error term will be normal. In order for this to occur, the estimated values of Y_i must lie between 0 and 1. A simple model that satisfies this is:

$$Y_i = \frac{e^{\beta_0 + \beta_1X_1 + \beta_2X_2}}{1 + e^{\beta_0 + \beta_1X_1 + \beta_2X_2}} \tag{13.2}$$

or

$$1 - Y_i = \frac{1}{1 + e^{\beta_0 + \beta_1X_1 + \beta_2X_2}} \tag{13.3}$$

However, this model is nonlinear but can be rewritten as:

$$\frac{Y_i}{1 - Y_i} = e^{\beta_0 + \beta_1X_1 + \beta_2X_2} \tag{13.4}$$

Then, by taking the natural logarithms of the equation, we can write:

$$\log_e \frac{Y_i}{1 - Y_i} = \beta_0 + \beta_1X_1 + \beta_2X_2 \tag{13.5}$$

The result is a linear model in which the left-hand side is a transformation of Y_i. The transformation that we have achieved increases from $-\infty$ to $+\infty$ as Y_i increases from 0 to 1. This is called the logit transformation. The left-hand side of the transformation is called the logit of Y.

An alternative explanation for the derivation of the logit transformation is provided in Blalock (1979). We illustrate the derivation for one independent variable. If Y is a probability, it must lie between 0 and 1. One way in which we can represent a relationship between Y and X, such that Y is always between 0 and 1, is to define a functional relationship such that whenever Y is getting closer to its upper or lower limit, it becomes very difficult to produce any more change in Y. One way to do this is not to take changes in Y as proportional to a constant (β) X, but rather to modify X by a factor such that it becomes deflated as Y approaches 0 or 1. One such relationship is expressed as:

$$\Delta Y = \beta Y(1 - Y)\,\Delta X \qquad \text{for } 0 \leq Y \leq 1 \tag{13.6}$$

In this relationship, as Y gets closer to 1, the coefficient of ΔX approaches 0.

Calculus applied to Equation 13.6 results in the equation

$$\log_e\left(\frac{Y}{1 - Y}\right) = \beta_0 + \beta_i X_1 \tag{13.7}$$

which, of course, is the logistic equation which we have already seen as Equation 13.5. Graphing this equation yields the logistic function (Figure 13.1).

It is possible to derive this relationship by assuming that the independent variables are distributed as multivariate normal or that departures of the independent variables' distribution from the multivariate normal are slight. The relationship can also be derived from a log-linear model (discussed in Knoke and Burke, 1980) in which one dimension of the multiway contingency table represents a dichotomous dependent variable. In the latter approach, interval scales can be applied to any of the multiple-category variables; hence it is possible to generate a logistic model with a mixture of categorical (classified) and interval-scaled (metric and enumerated) independent variables. These derivations are discussed in Bishop and Holland (1975) and Feinberg (1977).

The Utility Basis of Logit Formulations

For some time, transportation planners have been concerned with modeling the choices that commuters make between alternative modes of transportation. These studies are generally described as discrete choice models, and

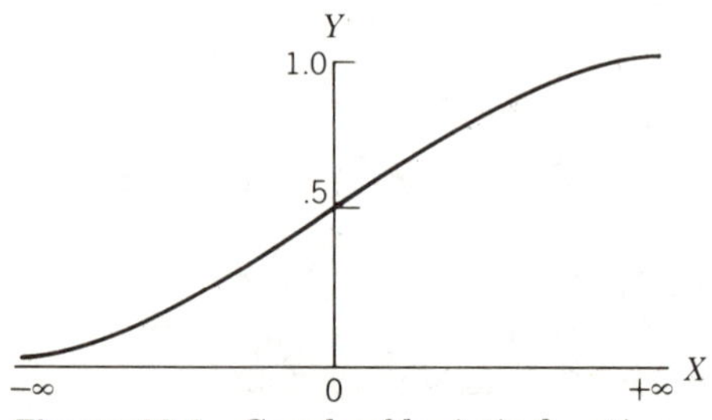

Figure 13.1 Graph of logistic function.

the primary approach within discrete choice modeling has been to apply logit or multinomial logit (i.e., more than two choices—although this more complicated situation will not be discussed in this text) to understanding the likelihood of households choosing certain alternatives given a variety of characteristics about the households and the choice alternative. The decision maker is assumed to assess the utility of each choice and select the one which has the greatest utility. There are a large number of factors that can be considered in assessing the utility of a particular mode choice.

In the travel behavior literature, it is assumed that individuals make decisions (choices) in making a trip to a specific destination by a specific mode with a probability determined by trip considerations and his or her own judgment as to the effectiveness of the trip. Thus, P_k^i is the probability of individual i choosing alternative k.

In the discrete choice literature, any alternative k has a utility U_{ik} comprising attributes of the alternative X_k and the attributes of the individual S_i.

Thus

$$U_{ik} = \mu(X_k, S_i) \tag{13.8}$$

The individual is assumed to assess the utilities of each alternative and thus make a choice. Clearly, there is a higher probability of choosing an alternative with a higher utility.

The discrete choice literature then uses Luce's axiom of the independence of irrelevant alternatives (IIL) (which states that the relative odds of choosing one alternative over another (where both alternatives have a nonzero probability of choice) is unaffected by the presence or absence of any additional alternatives), to show that the ratio of utilities yields a ratio of choice probabilities:

$$\frac{P_a^i}{P_b^i} = \frac{U(X_a, S_i)}{U(X_b, S_i)}$$

Then by defining the functional form for the utility as exponential and linear in the X's

$$U(X_a, S_i) = \exp[V(X_a, S_i)] \tag{13.9}$$

which is a definition of a classical convex utility curve in which there is a diminishing marginal return for improvements in the attributes of alternative X_a.

The assumptions lead to

$$\frac{P_a^i}{P_b^i} = \frac{\exp[V(X_a, S_i)]}{\exp[V(X_b, S_i)]} \tag{13.10}$$

and the application of the rule that $\sum_{k=1}^{n_i} P_k^i = 1$ leads to either

$$P_a^i = \frac{\exp[V(X_a, S_i)]}{\exp[V(X_b, S_i)] + \exp[V(X_a, S_i)]} \tag{13.11}$$

or

$$P_b^i = \frac{\exp[V(X_b, S_i)]}{\exp[V(X_a, S_i)] + \exp[V(X_b, S_i)]} \tag{13.12}$$

Equations 13.11 and 13.12 can be simplified to

$$P_a^i = \frac{\exp[V(X_a - X_b, S_i)]}{1 + \exp[V(X_a - X_b, S_i)]} \tag{13.13}$$

$$P_b^i = \frac{1}{1 + \exp[V(X_a - X_b, S_i)]} \tag{13.14}$$

The parallels (in equation structure) with the earlier formulas are obvious. Thus, the application of Luce's axiom with some general assumptions about the utilities has yielded a specific model for discrete choices—a model that is the standard binary logit.

One of the important contributions of research in discrete choice theory was to recognize that the utility we have been discussing is in fact made up of two parts: a first part that is common to a subgroup of the population being studied and an individual utility that is not shared with another individual. This utility can be defined as:

$$U_a^i = U'(X_a, S_i) + \epsilon(X_a', S_i) \tag{13.15}$$

where

$U'(X_a, S_i)$ = the common utility of alternative a for individual i with socioeconomic characteristics S_i

$\epsilon(X_a', S_i)$ = the individual utility of alternative a for individual i, with socioeconomic characteristics S_i.

The formula with the assignment of a probability to the individual decision can be rewritten and with the assumption that the $\epsilon(X_j', S_i)$ are randomly distributed (hence the random utility model) and a distributional assumption that the error terms are a Weibull distribution, the random utility model produces in the general case

$$P_k^i = \frac{\exp[U'(X_k, S_i)]}{\sum_j \exp[U'(X_j, S_i)]} \tag{13.16}$$

Not only is this the same model as the strict utility model—that is, the logit model—one of the problems of the strict utility model that the alternatives were not adequately distinguished is solved. It is not possible to fully capture the nuances of the development of the discrete choice model in a short statistical presentation; thus, it is important for social scientists who plan to use the logit approach and its extensions to pursue an expansion of these comments in Knoke and Burke (1980), Hensher and Johnson (1981), and Wrigley (1985).

Examples and Descriptions of Variables

The logit model can be used without recognition of the discrete choice framework as outlined in the last section. Examples from both physical and human geography suggest uses of the technique and indicate some of the complexity of variable identification enabled by the logit model.

As we noted at the beginning of the preceding section, the discrete choice model was formulated with transport mode choice models in mind, and a good example can be drawn from Hensher and McLeod (1977). They estimate the probability of choosing a car or train trip to work for a suburban area of Sydney, Australia. The model they fitted (with a maximum likelihood procedure—to be discussed in a later section) had the form

$$\log \frac{\hat{P}_{1|i}}{1 - \hat{P}_{1|i}} = \beta_0 + \beta_1 X_1 + \beta_2 X_2 \tag{13.17}$$

where $\hat{P}_{1|i}$ is the probability of choosing a car (for individual i) and $1 - \hat{P}_{1|i}$ is the probability of not choosing a car (i.e., choosing the train), and $X_1 = (T_i^{\text{car}} - T_i^{\text{train}})$ = the difference in total door-to-door travel time. $X_2 = (C_i^{\text{car}} - C_i^{\text{train}})$ = the difference between the models in total travel cost. The parameter estimates (standard errors in parentheses) are:

$$\frac{\hat{P}_{1|i}}{1 - \hat{P}_{1|i}} = \underset{}{.110} - \underset{(.016)}{.0476X_1} - \underset{(.0007)}{.0176X_2} \tag{13.18}$$

The negative values for the parameters indicate that as time and cost of using the car increase relative to the train, there is (an expected) decline in the likelihood (or the odds) of using the car.

We note that the independent variables (the explanatory variables) represent attributes of the choice alternatives. If we included income as a variable, we would have attributes of the choice alternative and attributes of the choice maker.

In a second example, Wrigley (1976) evaluates the existence of bronchitis (yes = 1/no = 0) in a sample population as a function of attributes of the location (environmental smoke levels) and attributes of the choice maker (the respondent's cigarette consumption). In this case,

$$\frac{\hat{P}_{1|i}}{1 - \hat{P}_{1|i}} = \beta_0 + \beta_1 X_1 + \beta_2 X_2 \tag{13.19}$$

where $\hat{P}_{1|i}$ is the likelihood that an individual has bronchitis and $1 - \hat{P}_{1|i}$ is the likelihood that he or she does not.

$$X_1 = \text{environmental smoke levels}$$

$$X_2 = \text{cigarette consumption}$$

The parameter estimates are:

$$\frac{\hat{P}_{1|i}}{1 - \hat{P}_{1|i}} = -8.7426 + .1110X_1 + .2105X_2 \qquad (13.20)$$

The estimates indicate that the likelihood of having bronchitis increases with increasing smoke levels and increasing cigarette consumption.

In concluding this section, it is useful to highlight the differences between the two groups of explanatory variables used in the two examples. One group of variables represent attributes of the choice alternatives or response categories (the car travel time), the other group represents attributes of the choice maker or the respondent (personal characteristics). In any logit model, explanatory variables from one or both categories can be included in multiple logit models. The issue of variable selection is discussed in detail in Wrigley (1985).

Estimating Procedures

Estimation of the logit model with individual observations requires the use of the maximum likelihood technique. Recall from our earlier discussion that it is a technique that involves "speculating" on which alternative values of the set of parameters would "best" reproduce the population parameters. The maximum likelihood estimation procedure has a number of desirable properties. The parameter estimators are consistent, and also efficient for large samples (asymptotically) and normal (asymptotically). Recall that (Equations 13.2, 13.3)

$$P_i = \frac{1}{1 + e^{\beta_0 + \beta_1 X_i}}$$

Remember that the P_i are not observed. We have a value of the dependent variable ($Y_i = 1$ if choice a is made; $Y_i = 0$ if choice b is made). The objective is to find parameter estimates for β_0 and β_1 that make it most likely that the pattern of choices in the sample would have occurred. If it is assumed that the first choice is made n_1 times and the second n_2 times ($n_1 + n_2 = N$), and we order the data so that the first n_1 observations are associated with choice a, then the likelihood function to be maximized has the function

$$L = P(Y_1 \cdots Y_n) = P(Y_1) \cdots P(Y_N)$$

The BMD Programs (Dixon, 1983) contain a statistical package for estimating the parameters of the logistic model by maximum likelihood. The

BMD program is based on the notion that the logistic model is a variation of the log-linear model.

Tests of Goodness of Fit[1]

The principal measure of goodness of fit of a logistic model is the likelihood ratio statistic

$$G^2 = -2 \sum_m P_m \log \frac{\hat{P}_m}{P_m}$$

where P_m and $\hat{P}_m$ are the actual and predicted proportions of observations with a specified value of the dependent variable in the mth covariate pattern. The likelihood ratio statistic is, in large samples, distributed asymptotically as χ^2, with $m - j - 1$ degrees of freedom, where m is the total number of distinct covariate patterns and $j + 1$ is the number of parameters estimated, including those for design variables (Bishop and Holland, 1975).

The predicted proportion $\hat{P}_m$ of observations with a specified value of the dependent variable can be regarded either as the estimated probability of occurrence of that value or as a numerical score for classifying a given observation in one of two states. Following the latter interpretation, it is possible to specify a value P^* that would maximize the fraction of all observations in all cells that are correctly classified according to the value of the dependent variable. Although P^* often does not equal .5, for analyzing the predictive powers of alternative logistic models, it is helpful to compare the overall rates of successful classification when P^* is arbitrarily set to .5.

Such a comparison, however, is not valid when alternative models are not based on the same study sample, because the rate of correct classification also depends on the initial distribution of the values of the dependent variable. A model estimated from a sample with a highly skewed distribution of the dependent variable can have a high rate of correct classification simply by assuming that all observations have the same, more likely value of the dependent variable.

An alternative formulation of the tests is to note that if we wish to evaluate the significance of the entire logit model, we first evaluate the likelihood function L when all the parameters (other than the constant) are set equal to zero. This value is identified as L_0. Now evaluate the likelihood function at its maximum (all the explanatory variables are included). Call this value L_{max}. The likelihood ratio can be defined as:

$$\lambda = \frac{L_0}{L_{max}}$$

[1]The material in this section is conceptually difficult and students may find it easier to go directly to the worked example.

TABLE 13.1 Tenure Decisions by a Sample of Movers in the Kansas City Metropolitan Area

Race (1 Black, 0 White)	Age of Household Head (1 Young, 7 Old)	Tenure Choice (1 Own, 2 Rent)	Household Size (Persons)	Household Income (1 Low, 7 High)
0	3	2	3	2
0	5	1	7	1
1	6	1	3	1
0	2	1	4	1
0	2	2	3	1
0	3	1	3	4
0	7	1	2	1
0	2	1	6	5
0	7	1	2	2
0	3	1	2	6
0	3	2	3	2
0	3	2	2	1
0	2	2	3	3
1	4	1	4	4
1	6	2	4	2
1	3	2	3	3
0	3	1	5	3
0	7	2	2	2
0	5	1	3	2
0	6	1	5	4
0	3	2	4	1
0	3	1	4	1
0	4	1	3	2
0	4	2	2	4
0	2	1	3	1
0	5	1	1	1
0	3	2	2	2
1	4	1	4	2
0	3	2	3	6
0	3	2	3	1
0	6	2	2	1
0	7	2	1	2
0	6	1	3	1
0	2	2	1	1
1	3	2	3	1
0	4	2	2	2
1	3	2	3	6
1	2	2	3	1
1	4	1	6	3
1	6	1	2	1
1	3	1	0	3
1	2	1	5	3
1	4	1	4	5

TABLE 13.1 (*Continued*)

Race (1 Black, 0 White)	Age of Household Head (1 Young, 7 Old)	Tenure Choice (1 Own, 2 Rent)	Household Size (Persons)	Household Income (1 Low, 7 High)
1	4	1	3	4
1	7	2	3	1
1	5	2	4	2
1	2	1	6	3
1	2	1	5	4
1	3	1	4	4
1	2	2	4	2
1	2	2	4	1
1	4	1	4	1
1	6	1	3	2
1	2	1	6	4
1	4	1	3	1
1	5	1	4	2
1	5	1	5	2
1	3	2	4	1
0	3	2	1	3
1	2	2	4	2
0	4	1	3	6
0	4	1	5	3
0	2	1	5	2
0	2	2	2	1
0	2	1	3	1
1	5	1	4	2
0	6	2	2	1
0	2	2	3	1
0	4	2	2	1
0	4	1	3	1
0	3	2	2	1
0	6	2	3	1
0	4	1	4	1
0	6	1	4	5
0	2	1	6	1
1	2	1	4	2
1	2	1	3	1
0	4	2	1	1
1	3	1	6	5
1	4	1	6	3
1	2	2	3	3
0	4	1	7	4
0	3	1	3	8
0	3	1	5	4
1	7	1	3	2
1	3	2	3	4
0	2	2	3	1

Now $-2(\ln \lambda) = -2(\log L_0 - \log L_{\max})$ follows a χ^2 distribution with k degrees of freedom where k is the number of parameters in the equation (excluding the constant).

To obtain a measure of goodness of fit similar to R^2, it is possible to calculate

$$\rho^2 = 1 - \frac{L^*(\hat{B})}{L^*(0)}$$

where $L^*(\hat{B})$ is the maximized value of the log likelihood and $L^*(0)$ is the value of the log likelihood when the probability of choosing the kth alternative is exactly equal to the observed aggregate share in the sample of the kth alternative. $L^*(\hat{B})$ will be larger than $L^*(0)$ because L^* will be larger when it is evaluated at $(\hat{B})$ than when explanatory variables are ignored. The larger the explanatory power of the variables is, the larger $L^*(\hat{B})$ will be in comparison with $L^*(0)$. But note that although $L^*(\hat{B})$ is larger than $L^*(0)$, $L^*(\hat{B})$ will be a *smaller negative number* because the log likelihood function L^* is negative. In the next section an example clarifies the test.

An Application to Residential Choice Modeling

An example of a logit formulation with a data set selected from a larger file on residential mobility will serve to highlight the procedures, results, and interpretations of logit analyses. The file to be analyzed is reproduced in Table 13.1 and includes measures of tenure choice for house or apartment (TENURE = 1 (own) or 2 (rent)), household size (HHSIZE), age of head (AGE), household income (HINCOME), and race (RACE 1 = white, 0 = black). In the research literature it has been suggested that a household's decision to own or rent is influenced by its size, its position in the family life cycle (represented by age), and the amount of money available for shelter (family income). In addition, we include a control for race.

Thus the model can be formulated as:

$$\log_e \frac{P_{0|i}}{P_{r|i}} = \beta_0 + \beta_1\text{HHSIZE} + \beta_2\text{AGE} + \beta_3\text{HINCOME} + \beta_4\text{RACE}$$

The calibration of this model with a *stepwise logistic analysis* yielded only two significant parameters. The results were:

$$\log_e \frac{P_{0|i}}{P_{r|i}} = 4.34 + .370\text{AGE} + .994\text{HINCOME}$$

Even though the sample is purposefully small (to allow reproduction and testing in other logit programs for comparative purposes), the results are consistent with our a priori expectations. Owning is positively related to increasing age of head and increasing income, neither household size nor race were significant in the explanation. Similar results for a larger sample, but the same basic model, are given by Wrigley and Longley (1984).

The parameter estimates (and standard errors) for the Wrigley and Longley study are:

$$\log_e \frac{P_{0|i}}{P_{r|i}} = \underset{(.1787)}{1.646} - \underset{(.0240)}{.0679\text{HHSIZE}} + \underset{(.0022)}{.0092\text{AGE}}$$

$$+ \underset{(.00003)}{.0005\ \text{HINCOME}} - \underset{(.2194)}{.3528\ \text{RACE}}$$

In addition to the formal presentation of the results, it is worthwhile to briefly explore the computer output and its interpretation. Tables 13.2, 13.3, 13.4, and 13.5 present the setup for BMDPLR and the results of a stepwise solution for the data in Table 13.1 (only the essential elements of the output are included). In the first step, the log likelihood (χ^2 values, degrees of freedom, and statistical significance) is computed without any explanatory variables entered. It is against this value $L^*(0)$ that the inclusion of explanatory variables will be evaluated. Succeeding steps enter and/or remove variables. The χ^2 and ρ value measures the significance of the log likelihood function at each step. The ρ value increases toward one as the χ^2 decreases in size, but the null hypotheses (H_0) tested is that the probability of an individual making a particular choice is independent of the parameters (the explanatory variables). Thus large χ^2 values and associated ρ values indicate

TABLE 13.2 Setup to Run BMD Logit Analysis

```
PROGRAM CONTROL INFORMATION

/PROBLEM      TITLE IS 'LOGIT REGRESSION ON TENURE TYPE'.
/INPUT        UNIT IS 8.
              FORMAT IS '(F1.0,3X,2F1.0,2F2.0)'.
              VARIABLES ARE 5.
              CASES ARE 87.
/VARIABLE     NAMES ARE RACE,AGE,TENURE,INCOME,HHSIZE,IAGE,IINCOME.
              ADD = 2.
/TRANS        IF(TENURE EQ 2) THEN TENURE = 0.
              IAGE = AGE.
              IF(IAGE EQ 1)THEN IAGE = 15.5.
              IF(IAGE EQ 2)THEN IAGE = 25.5.
              IF(IAGE EQ 3)THEN IAGE = 35.5.
              IF(IAGE EQ 4)THEN IAGE = 45.5.
              IF(IAGE EQ 5)THEN IAGE = 55.5.
              IF(IAGE EQ 6)THEN IAGE = 65.5.
              IF(IAGE EQ 7)THEN IAGE = 75.5.
              IINCOME = INCOME.
              IF(IINCOME EQ 1)THEN IINCOME =  2500.
              IF(IINCOME EQ 2)THEN IINCOME =  5000.
              IF(IINCOME EQ 3)THEN IINCOME = 15000.
              IF(IINCOME EQ 4)THEN IINCOME = 25000.
              IF(IINCOME EQ 5)THEN IINCOME = 35000.
              IF(IINCOME EQ 6)THEN IINCOME = 45000.
              IF(IINCOME EQ 7)THEN IINCOME = 75000.
/REGRESS      DEPENDENT=TENURE.
              INTERVAL=HHSIZE,AGE,INCOME.
              CATEG=RACE.
              MODEL=HHSIZE,AGE,RACE,INCOME.
              PLOT.
/END
```

TABLE 13.3 Stepwise Logit Analysis

```
PAGE    6   BMDPLR LOGIT REGRESSION ON TENURE TYPE

STEP NUMBER    0
---------------

                             LOG LIKELIHOOD =      -59.328
GOODNESS OF FIT CHI-SQ      (2*O*LN(O/E)) =       93.199    D.F.=   70   P-VALUE= 0.033
GOODNESS OF FIT CHI-SQ      ( C.C.BROWN ) =        0.0      D.F.=    0   P-VALUE= 1.000

                                                STANDARD
        TERM                  COEFFICIENT        ERROR      COEFF/S.E.

CONSTANT                        0.301           0.217         1.39

STATISTICS TO ENTER OR REMOVE TERMS
-----------------------------------
                            APPROX.                   APPROX.
        TERM                 F TO    D.F. D.F.         F TO     D.F. D.F.
                            ENTER                     REMOVE                    P-VALUE

HHSIZE                        5.48     1    85                                  0.0216
AGE                           1.18     1    85                                  0.2813
RACE                          2.46     1    85                                  0.1206
INCOME                       20.46     1    85                                  0.0000
CONSTANT                                              IS IN                                  MAY NOT BE REMOVED.

PAGE    7   BMDPLR LOGIT REGRESSION ON TENURE TYPE

STEP NUMBER    1                   INCOME                       IS ENTERED
---------------

                             LOG LIKELIHOOD =      -49.650
IMPROVEMENT CHI-SQUARE      (2*(LN(MLR) ) =       19.357    D.F.=    1   P-VALUE= 0.000
GOODNESS OF FIT CHI-SQ      (2*O*LN(O/E)) =       73.841    D.F.=   69   P-VALUE= 0.323
GOODNESS OF FIT CHI-SQ      ( D. HOSMER ) =        2.791    D.F.=    3   P-VALUE= 0.425
GOODNESS OF FIT CHI-SQ      ( C.C.BROWN ) =        7.004    D.F.=    2   P-VALUE= 0.030

                                                STANDARD
        TERM                  COEFFICIENT        ERROR      COEFF/S.E.

INCOME                          0.835           0.229         3.65
CONSTANT                       -2.45            0.765        -3.20

CORRELATION MATRIX OF COEFFICIENTS
----------------------------------

                 INCOME     CONSTANT

INCOME           1.000
CONSTANT        -0.948       1.000

STATISTICS TO ENTER OR REMOVE TERMS
-----------------------------------
                            APPROX.                   APPROX.
        TERM                 F TO    D.F. D.F.         F TO     D.F. D.F.
                            ENTER                     REMOVE                    P-VALUE

HHSIZE                        2.18     1    84                                  0.1432
AGE                           5.28     1    84                                  0.0241
RACE                          0.00     1    84                                  0.9640
INCOME                                                 13.21      1    84       0.0005
CONSTANT                                              IS IN                                  MAY NOT BE REMOVED.

PAGE    8   BMDPLR LOGIT REGRESSION ON TENURE TYPE

STEP NUMBER    2                   AGE                          IS ENTERED
---------------

                             LOG LIKELIHOOD =      -47.056
IMPROVEMENT CHI-SQUARE      (2*(LN(MLR) ) =        5.187    D.F.=    1   P-VALUE= 0.023
GOODNESS OF FIT CHI-SQ      (2*O*LN(O/E)) =       68.655    D.F.=   68   P-VALUE= 0.455
GOODNESS OF FIT CHI-SQ      ( D. HOSMER ) =        8.393    D.F.=    7   P-VALUE= 0.299
GOODNESS OF FIT CHI-SQ      ( C.C.BROWN ) =        9.452    D.F.=    2   P-VALUE= 0.009

                                                STANDARD
        TERM                  COEFFICIENT        ERROR      COEFF/S.E.

AGE                             0.370           0.169         2.19
INCOME                          0.994           0.257         3.87
CONSTANT                       -4.34            1.22         -3.56

CORRELATION MATRIX OF COEFFICIENTS
----------------------------------

                 AGE        INCOME      CONSTANT

AGE              1.000
INCOME           0.359       1.000
CONSTANT        -0.754      -0.853       1.000

STATISTICS TO ENTER OR REMOVE TERMS
-----------------------------------
                            APPROX.                   APPROX.
        TERM                 F TO    D.F. D.F.         F TO     D.F. D.F.
                            ENTER                     REMOVE                    P-VALUE

HHSIZE                        2.47     1    83                                  0.1196
AGE                                                     4.13      1    83       0.0453
RACE                          0.01     1    83                                  0.9074
INCOME                                                 12.83      1    83       0.0006
CONSTANT                                              IS IN                                  MAY NOT BE REMOVED.

NO TERM PASSES THE REMOVE AND ENTER LIMITS (   0.1500   0.1000 ) .
```

TABLE 13.4 The Fit of the Model

SUMMARY OF STEPWISE RESULTS

STEP NO	TERM ENTERED	DF	TERM REMOVED	LOG LIKELIHOOD	IMPROVEMENT CHI-SQUARE	IMPROVEMENT P-VALUE	GOODNESS OF FIT CHI-SQUARE	GOODNESS OF FIT P-VALUE
0				-59.328			93.199	0.033
1	INCOME	1		-49.650	19.357	0.000	73.841	0.323
2	AGE	1		-47.056	5.187	0.023	68.655	0.455

CUT-POINT	CORRECT PREDICTIONS SUCCESS	CORRECT PREDICTIONS FAIL	CORRECT PREDICTIONS TOTAL	PERCENT CORRECT SUCCESS	PERCENT CORRECT FAIL	PERCENT CORRECT TOTAL	INCORRECT PREDICTIONS SUCCESS	INCORRECT PREDICTIONS FAIL	INCORRECT PREDICTIONS TOTAL
	A	D	E=A+D	A/(A+B)	D/(C+D)	E/(E+F)	B	C	F=B+C
0.042	49.	0.	49.	98.00	0.0	56.32	1.	37.	38.
0.058	49.	0.	49.	98.00	0.0	56.32	1.	37.	38.
0.075	49.	1.	50.	98.00	2.70	57.47	1.	36.	37.
0.092	49.	1.	50.	98.00	2.70	57.47	1.	36.	37.
0.108	49.	2.	51.	98.00	5.41	58.62	1.	35.	36.
0.125	49.	2.	51.	98.00	5.41	58.62	1.	35.	36.
0.142	49.	3.	52.	98.00	8.11	59.77	1.	34.	35.
0.158	49.	3.	52.	98.00	8.11	59.77	1.	34.	35.
0.175	49.	4.	53.	98.00	10.81	60.92	1.	33.	34.
0.192	48.	4.	52.	96.00	10.81	59.77	2.	33.	35.
0.208	48.	4.	52.	96.00	10.81	59.77	2.	33.	35.
0.225	47.	7.	54.	94.00	18.92	62.07	3.	30.	33.
0.242	47.	7.	54.	94.00	18.92	62.07	3.	30.	33.
0.258	47.	7.	54.	94.00	18.92	62.07	3.	30.	33.
0.275	47.	7.	54.	94.00	18.92	62.07	3.	30.	33.
0.292	47.	7.	54.	94.00	18.92	62.07	3.	30.	33.
0.308	47.	10.	57.	94.00	27.03	65.52	3.	27.	30.
0.325	47.	11.	58.	94.00	29.73	66.67	3.	26.	29.
0.342	47.	11.	58.	94.00	29.73	66.67	3.	26.	29.
0.358	44.	17.	61.	88.00	45.95	70.11	6.	20.	26.
0.375	44.	17.	61.	88.00	45.95	70.11	6.	20.	26.
0.392	44.	17.	61.	88.00	45.95	70.11	6.	20.	26.
0.408	44.	17.	61.	88.00	45.95	70.11	6.	20.	26.
0.425	44.	17.	61.	88.00	45.95	70.11	6.	20.	26.
0.442	42.	25.	67.	84.00	67.57	77.01	8.	12.	20.
0.458	42.	25.	67.	84.00	67.57	77.01	8.	12.	20.
0.475	41.	27.	68.	82.00	72.97	78.16	9.	10.	19.
0.492	41.	27.	68.	82.00	72.97	78.16	9.	10.	19.
0.508	41.	27.	68.	82.00	72.97	78.16	9.	10.	19.
0.525	41.	27.	68.	82.00	72.97	78.16	9.	10.	19.
0.542	36.	27.	63.	72.00	72.97	72.41	14.	10.	24.
0.558	36.	27.	63.	72.00	72.97	72.41	14.	10.	24.
0.575	34.	28.	62.	68.00	75.68	71.26	16.	9.	25.
0.592	34.	28.	62.	68.00	75.68	71.26	16.	9.	25.
0.608	32.	31.	63.	64.00	83.78	72.41	18.	6.	24.
0.625	31.	31.	62.	62.00	83.78	71.26	19.	6.	25.
0.642	31.	31.	62.	62.00	83.78	71.26	19.	6.	25.
0.658	31.	31.	62.	62.00	83.78	71.26	19.	6.	25.
0.675	31.	31.	62.	62.00	83.78	71.26	19.	6.	25.
0.692	29.	33.	62.	58.00	89.19	71.26	21.	4.	25.
0.708	26.	34.	60.	52.00	91.89	68.97	24.	3.	27.
0.725	26.	34.	60.	52.00	91.89	68.97	24.	3.	27.
0.742	26.	34.	60.	52.00	91.89	68.97	24.	3.	27.
0.758	21.	34.	55.	42.00	91.89	63.22	29.	3.	32.
0.775	20.	35.	55.	40.00	94.59	63.22	30.	2.	32.
0.792	20.	35.	55.	40.00	94.59	63.22	30.	2.	32.
0.808	17.	35.	52.	34.00	94.59	59.77	33.	2.	35.
0.825	15.	36.	51.	30.00	97.30	58.62	35.	1.	36.
0.842	15.	36.	51.	30.00	97.30	58.62	35.	1.	36.
0.858	13.	36.	49.	26.00	97.30	56.32	37.	1.	38.
0.875	12.	37.	49.	24.00	100.00	56.32	38.	0.	38.
0.892	12.	37.	49.	24.00	100.00	56.32	38.	0.	38.
0.908	11.	37.	48.	22.00	100.00	55.17	39.	0.	39.
0.925	6.	37.	43.	12.00	100.00	49.43	44.	0.	44.
0.942	5.	37.	42.	10.00	100.00	48.28	45.	0.	45.
0.958	2.	37.	39.	4.00	100.00	44.83	48.	0.	48.
0.975	2.	37.	39.	4.00	100.00	44.83	48.	0.	48.
0.992	0.	37.	37.	0.0	100.00	42.53	50.	0.	50.

that we cannot reject the null hypothesis that the explanatory variables do not make a difference.

An alternative way to think of the χ^2 test is to recognize that it is a test of the number of correct predictions. Thus, in the diagram (Figure 13.2) we have actual choices 0 or 1, and probabilities of those choices from the model.

	Possible choices 0	Possible choices 1
Actual choice		1
Predicted choice		.8
Actual choice	0	
Predicted choice	.2	

Figure 13.2 The relationship of actual and predicted choices.

TABLE 13.5 Plot of Successes and Failures by Cut Point

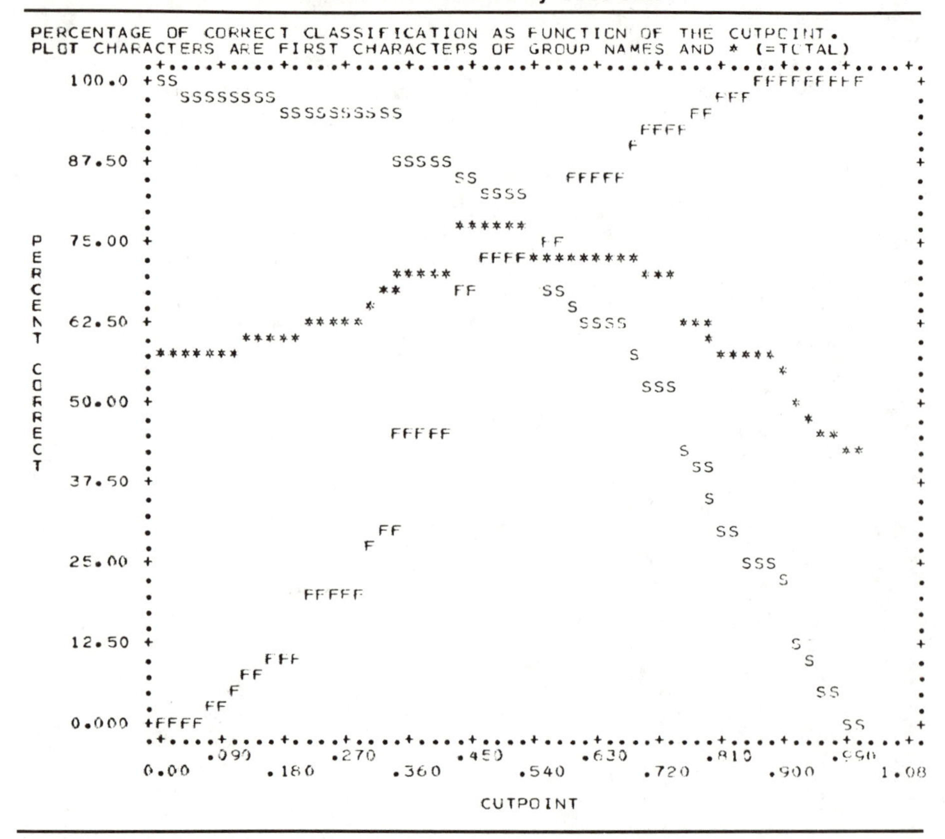

In the example, .8 is clearly close to 1.0 and .2 is clearly close to 0. Chi square (χ^2) is a measure of the fit within sampling fluctuations of the predicted values to the actual choices. In the figure, we conclude that the model provides relatively good predictions of the actual choices. The individual coefficients and their associated *t* values can be examined in the table of the last step.

Two other elements of the output should be studied. The first of these examines the number of correctly predicted values as in the matrix below.

	Predicted	
Actual	$\mathbf{a}_{11}$	$\mathbf{a}_{12}$
	$\mathbf{a}_{21}$	$\mathbf{a}_{22}$

Elements of the matrix represent correctly and incorrectly predicted results. From these values, percentages of correctly predicted cases can be computed. To carry out this analysis requires a "cut point" or choice level to separate correct and incorrect predictions. As noted earlier, a natural cut point is .5. However, other values could be chosen. In the present analysis, 78 percent of the cases at a cut point equal to .5 were correctly assigned or predicted.

The final diagram is a plot of the classification of correct and incorrect items as the cut point varies. The asterisk (*) intersects the ordinate according to the proportions of yes/no, 0/1 choices.

How good is the model? Computing the value

$$\rho^2 = 1 - \frac{L^*(\hat{B})}{L^*(0)}$$

$$= 1 - \frac{47.056}{59.328}$$

$$= .21$$

suggests that the model is a relatively good fit for the data. Values of ρ^2 between .2 and .4 are considered extremely good fits (Hensher and Johnson, 1981, p. 51).

Note that in running this logit program, we used a transformation card to convert age and income from ordinal to interval data. It is quite feasible to input categorical (ordinal) data, but two important changes occur in processing. First, the BMD logit program utilizes a design matrix to identify the presence or absence of each category (or 1 or 0) and this design matrix must be used to interpret the output coefficients. Thus, the design matrix and the coefficient output (Table 13.6) for the same problem, tenure choice on age, income, family size, and race require the following steps in interpretation:

1. The design matrix is constructed so that the sum of each "design variable" across all categories of a particular independent variable is 0.

2. As with binary-coded dummies, there are $c - 1$ design variables for each independent variable, where c is the number of categories. So, for example, the original six categories for the variable AGE are represented by five design variables.

3. The value of the logistic regression coefficient corresponding to a particular category of an independent variable is found by multiplying the value of each design variable for that category by the corresponding logit regression coefficient and summing.

For example, the logit coefficient corresponding to AGE category 2 is:

$$(-1)(.566) + (-1)(-.059) + (-1)(1.34) + (-1)(.814) + (-1)(-1.42) = -1.241$$

TABLE 13.6 Design Coefficients and Logistic Result Coefficients for Variable Categories

Variable		Category		Design Variables						
No.	Name	Index	Freq.	(1)	(2)	(3)	(4)	(5)	(6)	(7)
4	Age	2	23	−1	−1	−1	−1	−1		
		3	23	0	0	0	0	1		
		4	18	0	0	0	1	0		
		5	7	0	0	1	0	0		
		6	10	0	1	0	0	0		
		7	6	1	0	0	0	0		
1	Race	0	48	−1						
		1	39	1						
6	Income	0	1	−1	−1	−1	−1	−1	−1	−1
		1	5	0	0	0	0	0	0	1
		2	14	0	0	0	0	0	1	0
		3	31	0	0	0	0	1	0	0
		4	19	0	0	0	1	0	0	0
		5	8	0	0	1	0	0	0	0
		6	7	0	1	0	0	0	0	0
		7	2	1	0	0	0	0	0	0

Term		Coefficient	Standard Error	Coeff./S.E.
Age	(1)	0.566	0.854	0.663
	(2)	0.598E-01	0.697	0.858E-01
	(3)	1.34	1.19	1.13
	(4)	0.814	0.632	1.29
	(5)	−1.42	0.583	−2.43
Income	(1)	4.13	163.	0.254E-01
	(2)	7.50	163.	0.462E-01
	(3)	7.55	163.	0.464E-01
	(4)	−5.04	0.925	−5.45
	(5)	−5.39	0.829	−6.50
	(6)	−6.91	0.0	0.0
	(7)	−7.74	1.45	−5.35
Constant		5.99	0.698	8.59

and the coefficient for INCOME category 1 is

$$(0)(4.13) + (0)(7.50) + (0)(7.55) + (0)(-5.04) + (0)(-5.39) + (0)(-6.91) + (1)(-7.74) = -7.74$$

Thus, the "Predicted Probability of Success" for a household with AGE = 2 and INCOME = 1 is:

$$P(s) = \frac{e^{5.99 - 1.421 - 7.74}}{1 + e^{5.99 - 1.421 - 7.74}} = 0.0478$$

13.2 CANONICAL CORRELATION

In the social and physical sciences, in some cases it may be an oversimplification to use only one dependent variable as a measure of some complex event. The problem can be illustrated for the example of search behavior. It is difficult, if not impossible, to portray the manner in which households search for a new residence with a single measure. In fact, the process of search can be measured by a number of variables, each of which is measuring different characteristics of the process. This suggests that a technique that would allow for multiple dependent variables would be particularly useful in terms of understanding the interrelationship of dependent (or as they are sometimes called—criterion), and independent or predictive variables. Such a technique, *canonical analysis,* was developed by the statistician Hotelling in the 1930s but the complexity of the technique, especially the requirements for a solution, meant that it could not be readily applied until the development of generalized computer programs.

Canonical analysis is a procedure for finding the correlations between sets of variables. As such, it can be contrasted with multiple correlation—the correlation of a set of variables with one variable. Multiple correlation can be considered a special case of canonical correlation where the criterion set has only one variable.

Concepts and Mathematical Formulation

Given two sets of variables, which by convention we will call the criterion or dependent set and the predictor or independent set, the central issue in canonical analysis is to find a linear composite of each set of variables such that the correlation between the composites is maximized. Each of these composites is defined by a set of regression weights in much the same way that the multiple correlation analysis is defined by the weights. As we will see, there is more than one composite in each set of variables, and so there is more than one canonical relationship, although much of the discussion will inevitably focus on the first one or two of these relationships.

The procedure involves finding successive linear composites of each set such that each successive pair has maximum correlation. Imagine a situation in which there are three variables in set 1 and five variables in set 2. In this situation, the canonical analysis will find the first composite of the five variables in set 2, such that it has the highest possible correlation with a composite from the three variables in set 1. The important point to remember is that out of an infinite number of composites, that pair selected has the highest possible relationship. There is only one such pair that exactly meets this constraint. The degree of the relationship between the two composites is the first measure of canonical correlation between the two sets of variables. The second step in canonical analysis is to find another composite with the requirement that it is independent of the first pair of composites and itself has maximum correlation with another composite from the other set. Again, there is only one pair of composites that will fulfill this requirement. There is then a set of canonical correlations between the successive pairs of composites that measures the degree of relationship of the two sets of variables. Even though we have defined them as being maximally related, the degree of relationship can vary as the ordinary correlation coefficient from 0 to 1. The interpretation of the canonical coefficient R_c is somewhat different from the interpretation of the ordinary multiple correlation coefficient. The square of the largest canonical correlation is the proportion of variance in the first composite, in set 1 (the criterion set), that was accounted for by the first composite of the other set (the predictor set).

Formally, we think of the sets of n simultaneous equations. Consider the situation in which

$$\hat{y}_1 = a_1y_{11} + a_2y_{12} + a_3y_{13} \cdots a_py_{1p}$$

$$\hat{y}_2 = a_1y_{21} + a_2y_{22} + a_3y_{23} \cdots a_py_{2p}$$

$$\hat{y}_3 = a_1y_{31} + a_2y_{32} + a_3y_{33} \cdots a_py_{3p}$$

and

$$\hat{x}_1 = b_1x_{11} + b_2x_{12} + b_3x_{13} \cdots b_qx_{1q}$$

$$\hat{x}_2 = b_1x_{21} + b_2x_{22} + b_3x_{23} \cdots b_qx_{2q}$$

$$\hat{x}_3 = b_1x_{31} + b_2x_{32} + b_3x_{33} \cdots b_qx_{3q}$$

where $\hat{x}$ and $\hat{y}$ are derived values (the composites) and the procedure requires estimates of a and b which turn out to be the vectors $\mathbf{a}$ and $\mathbf{b}$, such that the correlation between $\hat{x}$ and $\hat{y}$, the desired values, is maximized. Think of the equations as involving two sets of weighting coefficients such that if linear combinations of each set are formed (the composite) and correlated in a two variable correlation, a higher correlation for these first composites than for any other first pair of composites you can form from the two sets of variables is obtained.

Eigenvalues and Derivation of Canonical Coefficients

The solution for canonical correlation coefficient R_c and the associated canonical coefficients and canonical loadings is more complex than the methods we have employed thus far. Even though computer routines are available, and we will outline only the formulas for solution, some understanding of the latent roots (eigenvalues) of a matrix will make things clearer.

From any matrix we can derive values or latent roots which are characteristic values of the matrix. They have the special property that they can be used to reproduce the matrix. Thus we can write:

$$[\mathbf{A}] \cdot [\mathbf{X}] = \lambda\ [\mathbf{X}]$$

$$\begin{bmatrix} a_{11} & a_{12} & a_{13} \\ a_{21} & a_{22} & a_{23} \\ a_{31} & a_{32} & a_{33} \end{bmatrix} \cdot \begin{bmatrix} X_1 \\ X_2 \\ X_3 \end{bmatrix} = \lambda \begin{bmatrix} X_1 \\ X_2 \\ X_3 \end{bmatrix}$$

where

$\mathbf{A}$ = a real symmetric matrix

$\mathbf{X}$ = a comformable vector (the eigenvector)

λ = a scalar or constant (the eigenvalue)

This equation states that the matrix of coefficients (the A_{ij}'s) times a vector of unknowns, X_i's, is equal to some constant times the unknown vector. Our problem is to find values of λ that allow us to satisfy the relationship.

We can also write the above equation in determinants:

$$\det\ (\mathbf{A} - \lambda\mathbf{I}) \cdot \mathbf{X} = 0$$

For the matrix **C** (this material is elaborated from a pedagogical outline of eigenvalues in Gould (1967)):

$$\mathbf{C} = \begin{bmatrix} 1.00 & .75 & .83 \\ .75 & 1.00 & .41 \\ .83 & .41 & 1.00 \end{bmatrix}$$

we can express the determinant $(\mathbf{C} - \lambda\mathbf{I}) = 0$ as:

$$\text{Det} \begin{bmatrix} 1.00 - \lambda & .75 & .83 \\ .75 & 1.00 - \lambda & .41 \\ .83 & .41 & 1.00 - \lambda \end{bmatrix} = 0$$

where the determinant is given by:

$$c_{11}c_{22}c_{33} + c_{12}c_{23}c_{31} + c_{13}c_{21}c_{32} - c_{11}c_{32}c_{23} - c_{31}c_{22}c_{13} - c_{21}c_{12}c_{33}$$

which is:

$$-\lambda^3 + 3\lambda^2 - 1.5805\lambda + .0909 = 0$$

Solving this equation yields three characteristic values (or eigenvalues):

$$\lambda_1 = 2.3416$$

$$\lambda_2 = .5929$$

$$\lambda_3 = .0655$$

Recall our equation $\mathbf{C}X = \lambda X$ setting the diagonal values to 0 and substituting the first eigenvalue, λ_1, we have

$$\begin{bmatrix} 0.00 - \lambda & .75 & .83 \\ .75 & 0.00 - \lambda & .41 \\ .83 & .41 & 0.00 - \lambda \end{bmatrix} \begin{bmatrix} X_1 \\ X_2 \\ X_3 \end{bmatrix} = \begin{bmatrix} 0 \\ 0 \\ 0 \end{bmatrix}$$

substituting, we get

$$\begin{bmatrix} -2.34 & .75 & .83 \\ .75 & -2.34 & .41 \\ .83 & .41 & -2.34 \end{bmatrix} \begin{bmatrix} X_1 \\ X_2 \\ X_3 \end{bmatrix} = \begin{bmatrix} 0 \\ 0 \\ 0 \end{bmatrix}$$

The equivalent of a set of simultaneous equations are:

$$-2.34X_1 + .75X_2 + .83X_3 = 0$$

$$.75X_1 - 2.34X_2 + .41X_3 = 0$$

$$.83X_1 + .41X_2 - 2.34X_3 = 0$$

and a solution of the first eigenvector:

$$\begin{bmatrix} .6403 \\ .5283 \\ .5576 \end{bmatrix}$$

For the first eigenvalue and vector we can write:

$$\begin{bmatrix} 1.00 & .75 & .83 \\ .75 & 1.00 & .41 \\ .83 & .41 & 1.00 \end{bmatrix} \begin{bmatrix} .6403 \\ .5283 \\ .5576 \end{bmatrix} = 2.34 \begin{bmatrix} .6403 \\ .5283 \\ .5576 \end{bmatrix}$$

Note that the eigenvalue has associated with it an eigenvector or vector of weights. These notions will be helpful in outlining the solution for the canonical relationships.

Solution of the R Matrix

With two sets of variables, we form a supercorrelation matrix of all interrelationships of m criterion variables and p predictor variables.

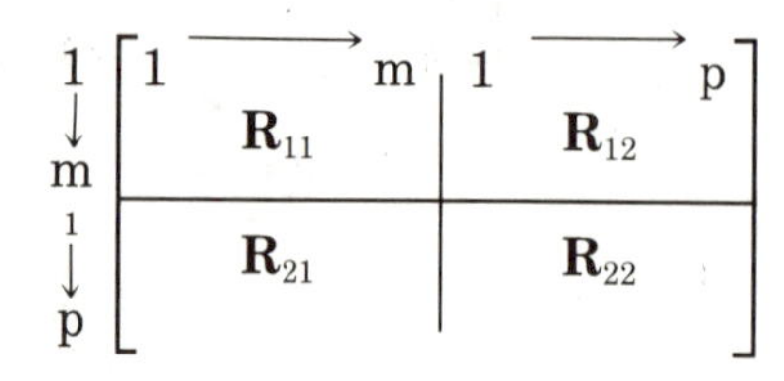

In this supermatrix

$\mathbf{R}_{11}$ = the matrix of correlations among the criterion (dependent) variables

$\mathbf{R}_{22}$ = the matrix of correlations among the predictor (independent) variables

$\mathbf{R}_{12}$ = the matrix of correlations among the criterion and predictor variables

$\mathbf{R}_{21}$ = the transpose matrix of correlations among the predictor and criterion variables

With this arrangement, we solve the matrix equation

$$(\mathbf{R}_{22}^{-1}\mathbf{R}_{21}\mathbf{R}_{11}^{-1}\mathbf{R}_{12} - \lambda I)\mathbf{b} = 0 \tag{13.21}$$

for the coefficients of the predictor composite, and the vector of weights for the criterion set is found by:

$$(\mathbf{R}_{11}^{-1}\mathbf{R}_{12}\mathbf{R}_{22}^{-1}\mathbf{R}_{21} - \lambda I)\mathbf{a} = 0 \tag{13.22}$$

Once **b** is found, **a** can also be found by:

$$\mathbf{a}_i = (\mathbf{R}_{11}^{-1}\mathbf{R}_{12}\mathbf{b}_i)/\sqrt{\lambda_i} \tag{13.23}$$

The above equations are eigenequations and the λ_i are eigenvalues and the $\mathbf{b}_i$ and $\mathbf{a}_i$ are eigenvectors. There is only one λ_i, an associated eigenvector that will satisfy the criterion that the relationship between the composites be a maximum.

As we demonstrated earlier, there are, of course, several possible characteristic values and associated vectors for the equation. These characteristic values are drawn from the matrix in such a way that they are smaller and smaller values, and the set of all eigenvalues and eigenvectors are said to exhaust the variance of the matrix. The eigenvalues in turn represent successive levels of canonical correlation between the successive pairs of composites.

In most instances, the interest is not in all of the canonical correlations and their associated composites. Rather, it is in the largest one or two of these canonical correlations. A test of the significance of the R_c coefficients is undertaken in two steps. The first is a general test of whether there are any significant coefficients—that is, there is at least one significant relationship—and it is based on the null hypothesis that all the R_c's are no greater than would be expected by chance between sets of variables which are unrelated. The test is:

$$\chi^2 = -[(n - 1) - .5(p + q + 1)] \log_e \Lambda \tag{13.24}$$

where

$$\Lambda = \prod_{i=1}^{P}(1 - R_{c_i}^2) \tag{13.25}$$

and p is the number of composites defined from the solution of the equations 13.21 and 13.22, q is the number of variables of the larger set with $(p)(q)$ degrees of freedom.

After establishing that there is at least one significant canonical relationship, the interest is in the number k, less than or equal to p, of significant solutions

$$\Lambda_k = \prod_{i=1}^{p-k}(1 - R_{c_i}^2)$$

and

$$\chi^2 = -[(n - 1) - .5(p + q + 1)]\log_e \Lambda_k$$

with

$$(p - k)(q - k) \text{ degrees of freedom}$$

Redundancy

It is important to recognize that the canonical correlation is a measure of the relationship between a pair of composites from the two sets of variables. It does not refer to the relationship between the *sets* themselves.

One important additional statistic in canonical analysis is the redundancy coefficient. This coefficient allows us to calculate the extent to which the variance of a set of variables is accounted for by a composite. To do this, we utilize what are called the canonical loadings. Levine (1977) emphasizes the importance of distinguishing between *canonical coefficients* and *canonical loadings*. The problem, of course, is that the coefficients, as they are similar to regression coefficients, are affected by any multicollinearity in the data set. That is, it is difficult to evaluate the relative contribution of the variables. If two variables are closely interrelated, then once one of the two has made its contribution, the other has very little additional autonomous contribution to make. The first coefficient will be large, the second will be close to zero. In this situation, it is difficult to evaluate the relative contribution, recalling the arguments about multicollinearity. The alternative suggested by Levine, and available in the BMD programs is to interpret the canonical loadings. These values that are like the loadings in a factor matrix (for those with some understanding of factor analysis) are the correlations of the original variables with the canonical variate. They represent a standardized measure, and thus it is possible to evaluate the relative contribution of each of the variables to the composites.

The redundancy measure utilizes these canonical loadings. Taking the sum of the squared loadings for a particular composite R_c, we get the amount of variance of the set that is accounted for by that composite. Divide the sum of the squared loadings by the number of variables in the set to obtain the proportion of the set variance that is accounted for by the composite. Now multiply this value by the squared canonical correlation R_c^2. The result is the proportion of the variance in one set that is accounted for by the composite of the other set. Adding up the results for all the composites, we get the proportion of the variance of one set that is accounted for by the other set. There is a redundancy index for each composite in each set.

The redundancy index for the ith composite of set 1 is defined as:

$$R_{d_{1i.2}} = \frac{\sum_{k=1}^{p} S_{1ik}^2}{P} R_{c_i}^2 \qquad (13.26)$$

where S_{1ik} is the correlation of the kth variable in set 1 with the ith composite of set 1. Think of $R_{d_{1i.2}}$ as an index of the proportion of variance in set 1 that is accounted for by the ith composite of set 2.

Note that redundancy is not symmetrical, and is related to the fact that each of the sets has differential amounts of intercorrelation. Thus there is also $R_{d_{2i.1}}$ or the proportion of variance in set 2 that is accounted for by the ith composite of set 1. The total redundancy of set 1 with set 2 is $\bar{R}_{d_{i.2}} = \Sigma_{i=1}^{p} R_{d_{1i.2}}$. The amount of interrelationship within a set will influence the amount of variance to be explained, and thus we will expect different contributions of each set of composites. Second, if there are different numbers of variables in the different sets, the mathematics of the analysis (the procedure for the extraction of composites involves extracting successive eigenvalues) yields different levels of explanation for each variable set. Extracting a set number of components will utilize more variance from the set with the smaller dimension than from the larger dimension set. Therefore, the total redundancy of the smaller set (with the larger set) will always be greater than the total redundancy of the larger with the smaller (Thorndike, 1978, p. 186).

The usefulness of redundancy analysis relates, as we will see in the example, to our understanding of the canonical correlation. In some cases, very high canonical correlations may not have equivalently high redundancy coefficients. This can occur when there are relatively low within-set correlations in R_{11} and R_{22}. The redundancy coefficient adds to our understanding of the level of the relationship between composites. If the redundancy coefficient is high, it strengthens the commentary that we can make about the interrelationship of the composites.

An Application to Residential Search Behavior

As an example of the use of canonical analysis, we present data on the search behavior of a sample of households seeking to purchase houses in the San Fernando Valley region of the Metropolitan Area of Los Angeles. It is reasonable to assume that it is almost impossible to measure the search behavior of households with a single variable. Search behavior is a complex activity, involving the number of houses that people choose to look at, the neighborhoods over which they choose to search, the amount of time they spend searching, and whether they search outside of the major areas or region of interest. This set of measures identifies the effort that people put into the process of search. We might hypothesize that this effort is related to a number of characteristics which measure the strategy of the searching household. Measures such as the length of time between seeing the house and making an offer to purchase (a bid), whether or not households used realtors, the newspaper, or other methods of searching, are likely to influence the effort they put in—the number of houses they visit. We are, in fact, testing the hypothesis that there is a relationship between search effort and search strategy and that, in fact, effort (or outcome) is dependent on the strategy chosen.

The data for the analysis are presented in Tables 13.7 and 13.8. The setups for a BMD program and results of the complete canonical analysis is given in Tables 13.9 to 13.12. These tables include the canonical coefficients, the eigenvalues and canonical correlations, and the canonical variable loadings, that is, the correlations between the canonical composites and the original variables. This example illustrates several of the points that we have been making about the use of canonical analysis and offers an excellent illustration of its use in a substantive problem.

There are two significant composites with canonical correlations of .58 and .46 (Table 13.10). The test of significance indicates that they are both

TABLE 13.7 Variables Used in the Study of Search Behavior

ID	Variable Description
DEPENDENT (CRITERION)	
1	Number of houses searched (log)
2	Number of areas searched
3	Length of search (weeks) (log)
4	Maximum distance within search cluster (log)
INDEPENDENT (PREDICTOR)	
12	Days between inspection and bid (log)
13	Number of realtors used (log)
14	Used newspapers a lot (yes = 1, no = 0)
15	Used realtors a lot (yes = 1, no = 0)
16	Drove around a lot (yes = 1, no = 0)
17	Went to a lot of open houses (yes = 1, no = 0)

TABLE 13.8 Data Used in the Example Canonical Analysis

	Variables									
ID	1	2	3	4	12	13	14	15	16	17
1	16	3	4	3	3	4	0	1	0	0
2	12	1	2	3	3	4	1	0	0	1
3	21	4	4	13	1	1	0	1	0	0
4	13	5	7	1	1	1	1	1	1	0
5	9	2	4	3	1	2	0	1	0	0
6	6	2	2	4	7	5	1	1	0	0
7	6	4	2	5	2	1	1	0	1	0
8	6	2	24	1	0	1	1	0	1	0
9	15	3	2	0	0	1	1	1	0	0
10	26	1	3	2	1	1	0	0	0	1
11	15	3	1	2	0	1	0	1	0	0
12	35	6	4	4	5	1	1	1	0	0
13	18	3	4	2	1	2	1	1	1	0
14	9	2	2	2	1	3	0	1	0	0
15	23	2	3	3	4	2	1	1	1	0
16	8	3	4	1	2	1	1	0	0	0
17	1	1	1	9	1	2	0	1	1	1
18	11	5	3	2	7	1	1	1	0	0
19	4	2	2	5	7	1	0	1	0	0
20	30	3	5	8	30	1	1	0	1	1
21	4	2	1	3	3	2	0	1	0	0
22	17	3	1	1	2	2	0	1	1	1
23	9	1	3	0	3	1	0	0	0	0
24	8	4	3	7	2	1	0	1	0	0
25	2	2	1	0	3	1	0	1	0	0
26	4	2	3	1	21	1	0	1	0	0
27	24	4	2	4	0	1	0	1	1	0
28	2	1	1	0	7	1	1	0	1	0
29	1	1	1	0	1	1	0	1	0	0
30	5	1	1	0	3	0	0	0	1	0
31	11	6	3	0	0	2	1	0	1	1
32	40	2	2	6	0	3	0	0	1	0
33	5	1	1	0	0	0	0	0	0	0
34	3	2	3	0	1	0	1	0	0	0
35	2	1	1	0	1	0	0	0	0	0
36	3	2	1	7	0	1	0	0	1	0
37	1	1	1	0	0	0	0	1	1	1
38	0	0	8	0	60	0	0	0	1	1
39	15	1	8	9	60	0	1	0	1	0
40	1	1	1	0	1	0	0	0	0	0
41	3	2	1	0	7	0	0	0	0	0
42	35	3	4	0	30	0	0	0	0	0

TABLE 13.8 (*Continued*)

ID	Variables 1	2	3	4	12	13	14	15	16	17
43	2	1	3	0	21	0	0	0	0	0
44	4	2	2	8	2	1	0	0	0	0
45	6	1	3	2	7	1	0	0	1	0
46	9	4	12	0	30	0	0	0	1	0
47	7	4	5	0	35	0	0	1	0	0
48	4	2	3	0	7	0	0	0	1	0
49	5	3	2	3	2	0	1	0	0	0
50	41	6	4	10	0	4	0	1	1	1
51	30	2	4	3	0	1	0	1	1	0
52	16	3	4	1	30	2	1	1	0	0
53	6	3	3	0	0	1	0	1	0	0
54	10	1	1	1	0	1	1	0	1	0
55	6	3	3	7	0	1	1	1	1	1
56	4	2	2	5	0	2	0	1	0	0
57	13	1	3	5	1	1	0	1	1	0
58	8	2	4	3	2	1	0	1	1	0
59	60	5	13	5	0	3	0	0	1	0
60	10	1	1	4	0	1	0	1	0	0
61	22	5	1	2	1	1	0	0	1	0
62	10	1	1	0	0	1	1	1	1	
63	7	2	3	8	2	1	0	1	1	0
64	8	2	4	2	3	1	0	1	0	0
65	10	2	1	1	0	1	0	1	0	0
66	12	4	1	0	0	2	0	0	1	1
67	6	2	3	1	14	2	0	1	0	0
68	1	1	1	0	3	1	0	1	0	0
69	4	1	3	2	0	1	0	1	1	0
70	37	4	5	2	2	3	0	1	1	1
71	19	6	4	9	0	1	0	1	0	0
72	8	4	3	4	0	3	0	1	1	1
73	60	2	2	2	1	2	0	(	0	0
74	7	4	3	3	0	1	0	1	0	0
75	31	3	3	8	14	3	0	1	0	0
76	17	4	3	1	2	4	0	1	1	0
77	55	6	3	5	9	2	0	1	0	0
78	10	5	1	2	0	2	1	1	0	0
79	4	2	2	5	0	1	0	1	0	0
80	17	4	2	7	0	2	0	0	1	0
81	12	5	4	5	7	2	0	1	0	0
82	37	6	6	6	7	1	0	1	0	0

*Blanks are missing data

TABLE 13.9 Setups for Canonical Analysis

```
BMDP6M - CANONICAL CORRELATION ANALYSIS
BMDP STATISTICAL SOFTWARE, INC.

PROGRAM CONTROL INFORMATION

/PROBLEM      TITLE = 'RSM CN CORR -- SRCH EFF VS. SRCH STR'.
/INPUT        VARIABLES ARE 10.
              FORMAT IS '(2X,10F2.0)'.
/VARIABLES    NAMES ARE V1,V2,V3,V4,V12,V13,V14,V15,V16,V17.
/TRAN         V1  = LN(V1+1.0).
              V3  = LN(V3).
              V4  = LN(V4+1.0).
              V12 = LN(V12+1.0).
              V13 = LN(V13+1.0).
/CANONICAL    FIRST  ARE V1,V2,V3,V4.
              SECOND ARE V12,V13,V14,V15,V16,V17.
              NUMB = 3.
/PRINT        MATR = CORR,COVA,COEF,LOAD.
/END
```

significant at the .05 level. The first composite (CNVRF 1 or the canonical variable loadings on the first composite) indicates that it is a measure of intensity of search effort while the second composite indicates that it is a measure of time of the length of search (Table 13.11). Note the underlined values. These composites are related (respectively) to a composite that reflects the principal use of realtors in the first composite against independent driving around and newspaper search. The high measure on the length of time between inspection and offer is indicative of longer, less intensive searching. The values for a third composite are included in the printout but they are not significant and not of interest in the analysis.

Clearly, this is an interesting contribution to understanding the search

TABLE 13.10 Correlation Matrix, Eigenvalues, and Canonical Correlation for the Housing Search Data Set

CORRELATIONS

		V1 1	V2 2	V3 3	V4 4	V12 5	V13 6	V14 7	V15 8	V16 9	V17 10
V1	1	1.000									
V2	2	0.588	1.000								
V3	3	0.367	0.317	1.000							
V4	4	0.427	0.299	0.198	1.000						
V12	5	-0.058	-0.086	0.337	-0.141	1.000					
V13	6	0.455	0.309	0.045	0.459	-0.259	1.000				
V14	7	0.073	0.127	0.167	-0.010	0.079	0.041	1.000			
V15	8	0.096	0.244	0.001	0.285	-0.128	0.390	-0.156	1.000		
V16	9	0.096	0.015	0.173	0.102	-0.131	0.061	0.110	-0.213	1.000	
V17	10	0.006	0.037	-0.007	0.031	-0.073	0.195	0.048	-0.059	0.365	1.000

EIGENVALUE	CANONICAL CORRELATION	NUMBER OF EIGENVALUES	BARTLETT'S TEST FOR REMAINING EIGENVALUES: CHI-SQUARE	D.F.	TAIL PROB.
			57.64	24	0.0001
0.33763	0.58106	1	26.95	15	0.0291
0.21036	0.45865	2	9.36	8	0.3132
0.10188	0.31918	3	1.35	3	0.7171
0.01797	0.13404				

```
BARTLETT'S TEST ABOVE INDICATES THE NUMBER OF CANONICAL
VARIABLES NECESSARY TO EXPRESS THE DEPENDENCY BETWEEN THE
TWO SETS OF VARIABLES.  THE NECESSARY NUMBER OF CANONICAL
VARIABLES IS THE SMALLEST NUMBER OF EIGENVALUES SUCH THAT
THE TEST OF THE REMAINING EIGENVALUES IS NON-SIGNIFICANT.
FOR EXAMPLE, IF A TEST AT THE .01 LEVEL WERE DESIRED,
THEN     1 VARIABLES WOULD BE CONSIDERED NECESSARY.
HOWEVER, THE NUMBER OF CANONICAL VARIABLES OF PRACTICAL
VALUE IS LIKELY TO BE SMALLER.
```

TABLE 13.11 Coefficients, Standardized Coefficients, and Canonical Variable Loadings for the Housing Search Data Set

COEFFICIENTS FOR CANONICAL VARIABLES FOR FIRST SET OF VARIABLES

		CNVRF1 1	CNVRF2 2	CNVRF3 3
V1	2	0.608927	0.276175	1.25057
V2	3	0.100609	-0.203130	-0.638661
V3	5	-0.602832	1.39437	-0.363285
V4	8	0.752324	0.417054D-01	-0.495049

STANDARDIZED COEFFICIENTS FOR CANONICAL VARIABLES FOR FIRST SET OF VARIABLES
(THESE ARE THE COEFFICIENTS FOR THE STANDARDIZED VARIABLES - MEAN ZERO, STANDARD DEVIATION ONE.)

		CNVRF1 1	CNVRF2 2	CNVRF3 3
V1	2	0.546	0.247	1.120
V2	3	0.156	-0.315	-0.991
V3	5	-0.418	0.966	-0.252
V4	8	0.610	0.034	-0.401

COEFFICIENTS FOR CANONICAL VARIABLES FOR SECOND SET OF VARIABLES

		CNVRS1 1	CNVRS2 2	CNVRS3 3
V12	6	-0.196427	0.765258	-0.240878D-01
V13	21	1.95328	0.870154	1.11074
V14	22	-0.791155D-01	0.374615	-0.958915
V15	23	0.152558	-0.890976D-02	-2.21116
V16	24	0.140970	1.16852	-0.221555
V17	25	-0.427289	-0.730584	-0.633592

STANDARDIZED COEFFICIENTS FOR CANONICAL VARIABLES FOR SECOND SET OF VARIABLES
(THESE ARE THE COEFFICIENTS FOR THE STANDARDIZED VARIABLES - MEAN ZERO, STANDARD DEVIATION ONE.)

		CNVRS1 1	CNVRS2 2	CNVRS3 3
V12	6	-0.224	0.873	-0.027
V13	21	0.900	0.401	0.512
V14	22	-0.035	0.165	-0.423
V15	23	0.075	-0.004	-1.088
V16	24	0.070	0.582	-0.110
V17	25	-0.158	-0.270	-0.234

CANONICAL VARIABLE LOADINGS
(CORRELATIONS OF CANONICAL VARIABLES WITH ORIGINAL VARIABLES) FOR FIRST SET OF VARIABLES

		CNVRF1 1	CNVRF2 2	CNVRF3 3
V1	2	0.744	0.431	0.274
V2	3	0.527	0.147	-0.531
V3	5	-0.047	0.964	-0.234
V4	8	0.807	0.237	-0.269

CANONICAL VARIABLE LOADINGS
(CORRELATIONS OF CANONICAL VARIABLES WITH ORIGINAL VARIABLES) FOR SECOND SET OF VARIABLES

		CNVRS1 1	CNVRS2 2	CNVRS3 3
V12	6	-0.468	0.727	-0.022
V13	21	0.959	0.162	0.025
V14	22	-0.028	0.302	-0.258
V15	23	0.454	-0.094	-0.781
V16	24	0.077	0.413	0.024
V17	25	0.053	-0.035	-0.129

TABLE 13.12 Redundancy Values for the Housing Search Data Set

SQUARED MULTIPLE CORRELATIONS OF EACH VARIABLE IN THE FIRST SET WITH CHOSEN CANONICAL VARIABLES OF SECOND SET.

VARIABLE	R-SQUARED	ADJUSTED R-SQUARED	F STATISTIC	DEGREES OF FREEDOM		P-VALUE
1 V1	0.233833	0.171712	3.76	6	74	0.0077
2 V2	0.127022	0.056240	1.79	6	74	0.1390
3 V3	0.201815	0.137097	3.12	6	74	0.0199
4 V4	0.238896	0.177185	3.87	6	74	0.0066

SQUARED MULTIPLE CORRELATIONS OF EACH VARIABLE IN THE SECOND SET WITH CANONICAL VARIABLES IN THE FIRST SET.

VARIABLE	R-SQUARED	ADJUSTED R-SQUARED	F STATISTIC	DEGREES OF FREEDOM		P-VALUE
5 V12	0.184951	0.142053	4.31	4	76	0.0034
6 V13	0.316301	0.280316	8.79	4	76	0.0000
7 V14	0.026225	-0.025026	0.51	4	76	0.7273
8 V15	0.133800	0.088211	2.93	4	76	0.0259
9 V16	0.037949	-0.012685	0.75	4	76	0.5614
10 V17	0.002892	-0.049587	0.06	4	76	0.9942

CANON. VAR.	AVERAGE SQUARED LOADING FOR EACH CANONICAL VARIABLE (1ST SET)	AV. SQ. LOADING TIMES SQUARED CANON. CORREL. (1ST SET)	AVERAGE SQUARED LOADING FOR EACH CANONICAL VARIABLE (2ND SET)	AV. SQ. LOADING TIMES SQUARED CANON. CORREL. (2ND SET)	SQUARED CANON. CORREL.
1	0.37115	0.12531	0.22581	0.07624	0.33763
2	0.29826	0.06274	0.13771	0.02897	0.21036
3	0.12110	0.01234	0.11590	0.01181	0.10188

THE AVERAGE SQUARED LOADING TIMES THE SQUARED CANONICAL CORRELATION IS THE AVERAGE SQUARED CORRELATION OF A VARIABLE IN ONE SET WITH THE CANONICAL VARIABLE FROM THE OTHER SET. IT IS SOMETIMES CALLLED A REDUNDANCY INDEX.

process. The interpretation would not be available either from the analysis of the simple intercorrelation matrix or from a discussion of any one dependent variable. The comparison of the coefficients from the canonical analysis and the canonical variable loadings indicates the importance of using the loadings rather than the coefficients. For example, the loadings indicate that both the number of houses searched and the number of areas searched are approximately equal in importance, but the coefficients would give a quite different interpretation.

A final part of the analysis involves reporting the redundancy values. As we discussed earlier, the average squared loading (of a composite) times the squared canonical correlation is the average squared correlation of a variable in one set with the canonical variable from the other set. Thus

$$.744^2 + .527^2 + (-.047)^2 + (.807^2)/4 \text{ times } (.58)^2 = .13 \qquad (13.27)$$

(See Table 13.12.)

Unfortunately, our two composites share only about 20% of the variance (the sum of .13 + .06 + .01). Even though both canonical correlations are significant and the results are consistent with our intuitive notions of search behavior, the redundancy test emphasizes that there is more involved in the search effort/search strategy relationship. Clearly, further research would focus on sociodemographic characteristics of the household.

13.3 DISCRIMINANT ANALYSIS

In many instances in geographic research, we wish to examine the adequacy of a classification scheme or we wish to fit data into a classification scheme that has been proposed elsewhere. We can use the multivariate technique known as *discriminant analysis* to establish the way in which the groups differ on the measurement variables, or to allocate objects or subjects to some predefined groups. Discriminant analysis is designed to find a function of the variables that maximally discriminates amongst the groups.

Concepts and Examples for Two Groups

Suppose, for the purposes of introducing a discussion of discriminant analysis, that we have two groups of farmers, dairy farmers and sheep farmers, and we have measured productivity and a measure of carrying capacity for their farms. If we form an ordinary scatter plot of these data, we see a considerable overlap of the two groups, shown by the portrayal of the elliptical plots on the respective productivity and carrying capacity axes (Figure 13.3). Although there is substantial overlap, the dairy farmers seem to

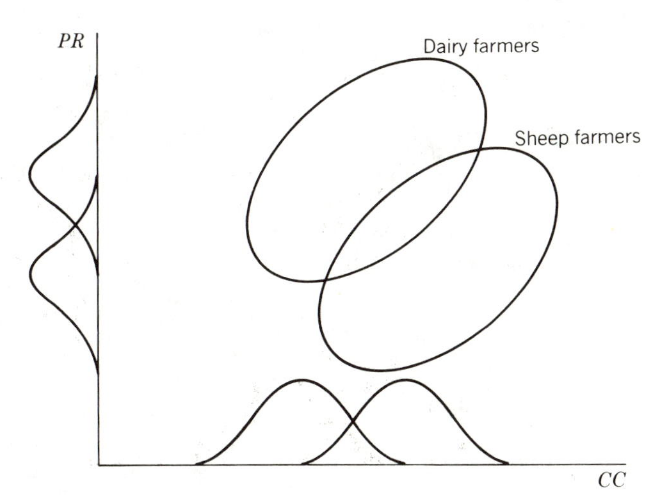

Figure 13.3 Scatter plots for dairy farmers and sheep farmers on the axes productivity (*PR*) and carrying capacity (*CC*). Univariate distributions are given on the axes.

be superior in productivity but the sheep farmers have greater carrying capacity. The problem is to find some composite or combination of the two distributions portrayed on the productivity and carrying capacity axes so that we will minimize, in terms of their scores on a new axis, the amount of overlap between the two groups. From another point of view, we are concerned to find a new function that will give minimum overlap between the ellipses. In Figure 13.4, the separation between the two groups is defined by the line that passes through the intersection of the two ellipses. If we define this line as a decision rule, which says that farmers on the upper side of line D will be dairy farmers and on the lower side of the line D will be sheep farmers, we get a significant number of correct predictions. The usefulness of this differentiation is illustrated by the projection of the ellipses onto line D'.

The actual line that we can use for discrimination purposes is the line that passes through the origin and that is at right angles to the line that best separates the two ellipses. It will turn out in the mathematical development that there is only one line that satisfies these conditions. The process involved in discriminant analysis is to determine the equation for this discriminant line D', and the computational problem is to determine a set of weights that will define this line, subject to the constraints that there be minimum overlap between the groups. The weights that define this line are analogous to the multiple regression weights, or the β values, in the sense that they yield an optimum prediction of the group membership based on the two variables (in this example), productivity and carrying capacity.

The discriminant analysis does not place object or subject in one or another group. Rather, the weights on the discriminant function, when multiplied

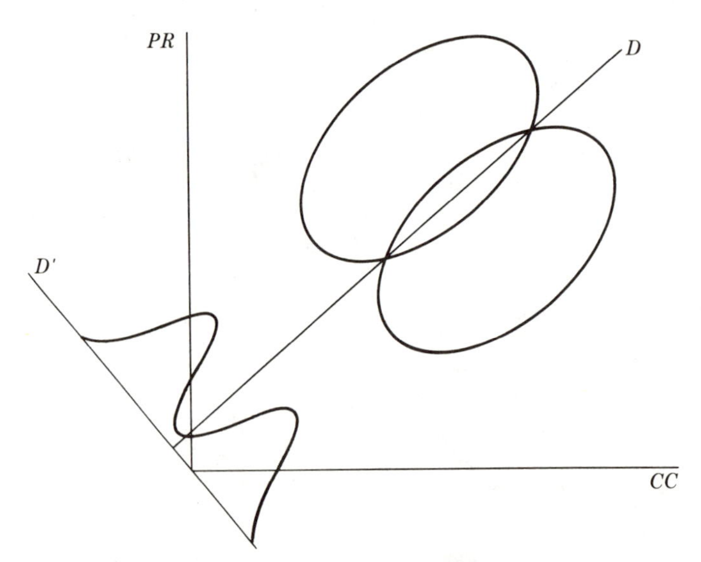

Figure 13.4 Optimum separation of groups by D and composite D', which yields maximum separation of the groups.

by the scores for each individual, provide the locations of individuals on the new composite. It is necessary to make some decision about the location of the optimum point that separates the two groups. This is where there will be the fewest misclassifications of the subjects.

Multiple Discriminant Analysis

The situation is more complicated when more than two groups are being analyzed and three or more variables are involved. With two variables for three groups of subjects, multiple discriminant analysis yields two discriminant functions or composites. The first of these discriminant functions is defined to make the maximum differentiation among the group using all of the information. The second of the discriminant functions uses the residual information, that which is independent of the first discriminant function, to make further distinctions amongst the groups, if possible. There is one less discriminant function than there are groups of variables being analyzed. An illustration of how the second discriminant function would appear for a two-variable, three-group example is included (Figure 13.5). With three variables, the plot would not be ellipses, but rather three-dimensional shapes, somewhat like footballs.

Mathematical Structures[2]

Although we have discussed discriminant analysis as a special form of multiple regression, we do not use the correlation matrix in the computation of the discriminant function.

Let us briefly reexamine the situation. We are concerned with a classification problem. We are interested in classifying an unknown individual case W into one of k groups, $W_1, W_2, \ldots, W_k$ on the basis of some observations $X_1, X_2, \ldots, X_p$. We outline the mathematics for a solution involving the classification of an individual into one of two groups, $k = 2$, when the parameters are known (see Afifi and Azen, 1979, for extensions).

We write the observations $X_1X_2 \ldots X_p$ as a vector $\mathbf{X}$, and assume that W_1 and W_2 are normally distributed with equal variances. We also assume that the observations come from one or other of these multivariate normal populations. Hence,

$$\mathbf{X} \text{ is } N(\mu_1\sigma_{ij}) \text{ or } N(\mu_2\sigma_{ij})$$

where σ_{ij} is the covariance between X_i and X_j.

Assume that the parameters of μ_1 and μ_2 and σ_{ij} are known. Find a discriminant function

$$d = \alpha_1X_1 + \alpha_2X_2 + \cdots + \alpha_pX_p \tag{13.28}$$

where $\alpha_1 \ldots \alpha_p$ are some constants.

[2]This section is more demanding mathematically and it may be useful to examine the worked examples first and return to this material later.

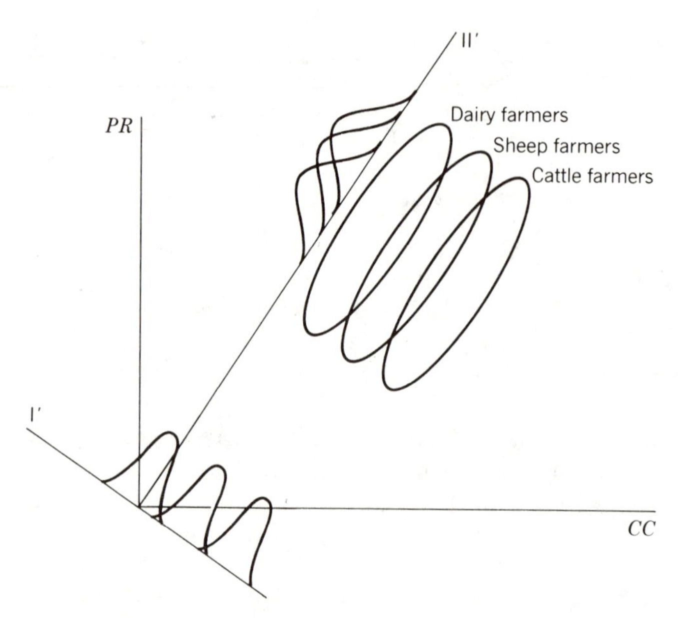

Figure 13.5 Scatter plots to illustrate three groups, two variables, and two discriminant functions.

We classify a case into W_1 if

$$d \geq c$$

and into W_2 if

$$d < c$$

where

$$c = \frac{\gamma_1 + \gamma_2}{2} \tag{13.29}$$

We then need to find the values $\alpha_1 \ldots \alpha_p$ and c that minimize the probabilities of an incorrect classification.

If X is from W_1, then d is normal with mean

$$\gamma_1 = \sum_{j=1}^{p} \alpha_j \mu_{1j} \tag{13.30}$$

and variance

$$\sigma_d^2 = \sum_{i=1}^{p} \sum_{j=1}^{p} \alpha_i \sigma_{ij} \alpha_j \tag{13.31}$$

If X is from W_2 then d is normal with mean

$$\gamma_2 = \sum_{j=1}^{p} \alpha_j \mu_{2j} \tag{13.32}$$

and variance σ_d^2 is as before.

Then choose $\alpha_1 \ldots \alpha_p$ so that they yield maximum separation of γ_1 and γ_2. This separation we define as the Mahalanobis generalized distance.

$$D^2 = \frac{(\gamma_1 - \gamma_2)^2}{\sigma_d^2} \tag{13.33}$$

We then find the coefficients $\alpha_1 \ldots \alpha_p$ that maximize D^2. These α_i are solutions to

$$\alpha_1\sigma_{11} + \alpha_2\sigma_{12} + \ldots + \alpha_p\sigma_{1p} = \mu_{11} - \mu_{21}$$

$$\alpha_1\sigma_{21} + \alpha_2\sigma_{22} + \ldots + \alpha_p\sigma_{2p} = \mu_{12} - \mu_{22}$$

$$\ldots$$

$$\ldots$$

Once α_i have been derived, it is possible to calculate a discriminant score d for the individual and decide on the appropriate group. This intuitive interpretation of the discriminant function can be shown to be theoretically optimum with Bayes's theorem (Afifi and Azen, 1979).

Mahalanobis's D^2 statistic can be used to group objects or to discriminate between groups. It is often used when you already have groups and you wish to test for differences. The formula is:

$$D^2 = \sum_{i=1}^{p} \sum_{j=1}^{p} (S_{ij})^{-1} d_i d_j \tag{13.34}$$

where

$d_i = \overline{X}_{i1} - \overline{X}_{i2}$ the difference of sample means for groups 1 and 2 on the ith variable

$d_j = \overline{X}_{j1} - \overline{X}_{j2}$ the same for the jth variable

$(S_{ij})^{-1}$ = within-groups variance—covariance matrix inverse

A test of significance is:

$$F = \frac{(N_a + N_b - m - 1)}{(N_a + N_b - 2)m} \frac{(N_a N_b)}{N_a + N_b} D^2 \tag{13.35}$$

with m and $(N_a + N_b - m - 1)$ degrees of freedom.

If the two groups are very close together, it will be difficult to distinguish them, particularly if they have large variances. If the means are quite separate and the variances are small, they will be easily discriminable.

For any tests of significance, five assumptions are required:

1. The observations in each group are selected randomly.
2. The variables are normally distributed.

3. The group variance-covariance matrices are equal.

4. The probability of any additional sampled observations belonging to either group is equal.

5. No observations that were used to derive the d function were misclassified.

However, discriminant analysis is a rather robust technique that can tolerate some deviation of these assumptions (Klecka, 1980). Moreover, the assumption of a multivariate normal distribution is only required for tests of significance.

There are three tests to be undertaken: a test for the number of significant discriminant functions, and if the data are from a sample, a test for the significance of the coefficients and a test of the significance of the discriminant function.

For the test of the number of significant discriminant functions, the usual test involves Wilk's lambda:

$$\Lambda = \prod_{i=k+1}^{p} \frac{1}{1 + \lambda_i} \tag{13.36}$$

where p denotes the number of discriminant functions already derived, and λ is the eigenvalue. Usually, Wilk's lambda is converted to a χ^2 measure of significance

$$\chi^2 = -\left[n - \frac{(v + g)}{2} - 1\right] \log_e \Lambda$$

where

n = total cases

v = number of variables

g = number of groups

k = number of eigenvalues (or functions derived)

Either Wilk's lambda or the F test is used to evaluate the individual discriminant coefficients.

An Example and Interpretation for Two Groups

There are several methods for finding the coefficients of the discriminant function. One method utilizes multiple regression where the dependent variable consists of the differences between the multivariate means of the groups. Another method utilizes canonical variates. We use a simple example of the regression format and illustrate computer runs that utilize canonical approaches. For the regression approach, we must solve a matrix equation of the form:

$$\mathbf{VC} \cdot \boldsymbol{\alpha} = \mathbf{D}$$

where **VC** is an $m \times n$ matrix of pooled variances and covariances of the m variables, α are the coefficients of the discriminant function, and **D** is the vector of the m differences (for the m variables) between the two groups **A** and **B**.

The solution is given by:

$$\alpha = \mathbf{VC}^{-1} \cdot \mathbf{D}$$

The vector **D** is defined as:

$$\begin{bmatrix} d_1 \\ d_2 \\ \vdots \\ d_m \end{bmatrix} = \begin{bmatrix} \bar{a}_1 \\ \bar{a}_2 \\ \vdots \\ \bar{a}_m \end{bmatrix} - \begin{bmatrix} \bar{b}_1 \\ \bar{b}_2 \\ \vdots \\ \bar{b}_m \end{bmatrix}$$

The matrices of pooled variances and covariances are computed separately for each group. To do this, first compute a matrix of sums of squares and a sums of cross products for each group.

If we write $x = (X - \bar{X})$, the deviation scores, we can compute the matrix sums of deviations and deviation cross products for groups **A** and **B** for two variables, **DPA** and **DPB**.

$$\mathbf{DPA} = \begin{bmatrix} \sum_{i=1}^{n_a} x_{1a}^2 & \sum_{i=1}^{n_a} x_{1a}x_{2a} \\ \sum_{i=1}^{n_a} x_{1a}x_{2a} & \sum_{i=1}^{n_a} x_{2a}^2 \end{bmatrix}$$

$$\mathbf{DPB} = \begin{bmatrix} \sum_{i=1}^{n_b} x_{1b}^2 & \sum_{i=1}^{n_b} x_{1b}x_{2b} \\ \sum_{i=1}^{n_b} x_{1b}x_{2b} & \sum_{i=1}^{n_b} x_{2b}^2 \end{bmatrix}$$

The **VC** matrix of pooled variances and covariances can now be written as:

$$\mathbf{VC} = \frac{\mathbf{DPA} + \mathbf{DPB}}{n_A + n_B - 2}$$

The data in Table 13.13 can be used to illustrate a simple hand-worked example of the derivation of the discriminant function. The same data will be used for a larger multidiscriminant example. The data consist of a set of measurements of three sand dune types at Pismo Beach, California. The study site is located about 320 km NW of Los Angeles, at the western end

TABLE 13.13 Data for Discriminant Analysis

Case-N	Size	Sorting	Skew	Kurt	Type
1	2.74	0.37	−0.21	0.53	1.
2	2.65	0.48	−0.13	0.47	1.
3	2.47	0.37	−0.29	0.52	1.
4	2.62	0.37	−0.19	0.49	1.
5	2.52	0.41	−0.19	0.49	1.
6	2.48	0.53	−0.29	0.48	1.
7	2.49	0.40	−0.15	0.52	1.
8	2.63	0.67	−0.03	0.54	1.
9	2.29	0.52	−0.23	0.45	1.
10	2.37	0.51	−0.27	0.47	1.
11	2.32	0.53	−0.20	0.49	1.
12	2.63	0.50	−0.05	0.45	1.
13	2.57	0.42	−0.13	0.42	1.
14	2.52	0.44	−0.21	0.48	1.
15	2.82	0.39	−0.25	0.52	1.
16	2.33	0.51	−0.19	0.51	1.
17	2.04	0.52	−0.21	0.47	1.
18	2.25	0.47	−0.26	0.50	1.
19	2.56	0.42	−0.22	0.48	1.
20	2.23	0.42	−0.17	0.48	1.
21	1.79	0.65	−0.29	0.45	1.
22	2.35	0.40	−0.16	0.54	1.
23	2.32	0.59	−0.13	0.51	1.
24	1.93	0.62	−0.21	0.47	1.
25	2.29	0.32	−0.10	0.48	1.
26	2.47	0.45	−0.21	0.49	1.
27	2.83	0.39	−0.21	0.46	1.
28	1.88	0.59	−0.07	0.50	1.
29	2.32	0.52	−0.21	0.46	1.
30	2.39	0.51	−0.31	0.47	1.
31	2.21	0.55	−0.27	0.48	1.
32	2.19	0.59	−0.14	0.49	1.
33	1.66	0.71	−0.27	0.43	1.
34	2.14	0.52	−0.28	0.44	1.
35	2.09	0.49	−0.10	0.45	1.
36	1.88	0.43	−0.23	0.47	1.
37	1.97	0.60	−0.21	0.44	1.
38	2.52	0.35	−0.09	0.49	1.
39	2.12	0.52	−0.16	0.45	1.
40	1.52	0.65	−0.30	0.47	1.
41	1.95	0.61	−0.27	0.50	1.
42	2.22	0.52	−0.21	0.49	1.
43	2.19	0.54	−0.23	0.50	1.
44	2.10	0.55	−0.20	0.50	1.
45	2.53	0.47	−0.09	0.47	1.

TABLE 13.13 *(Continued)*

Case-N	Size	Sorting	Skew	Kurt	Type
46	2.72	0.50	−0.10	0.45	1.
47	2.09	0.53	−0.23	0.47	1.
48	1.90	0.51	−0.19	0.48	1.
49	2.25	0.59	−0.20	0.47	1.
50	2.14	0.52	−0.21	0.44	1.
51	2.53	0.60	−0.24	0.50	2.
52	2.84	0.50	−0.11	0.57	2.
53	2.81	0.49	−0.17	0.55	2.
54	2.73	0.54	−0.24	0.55	2.
55	2.77	0.44	−0.15	0.50	2.
56	2.75	0.38	−0.24	0.47	2.
57	2.79	0.41	−0.17	0.52	2.
58	2.90	0.42	−0.16	0.55	2.
59	2.30	0.43	−0.30	0.50	2.
60	2.30	0.49	−0.17	0.55	2.
61	2.28	0.42	−0.13	0.50	2.
62	2.31	0.35	−0.35	0.47	2.
63	2.27	0.55	0.04	0.58	2.
64	2.20	0.38	−0.06	0.47	2.
65	2.39	0.37	−0.11	0.57	2.
66	2.39	0.33	−0.19	0.51	2.
67	2.29	0.49	−0.18	0.54	2.
68	2.32	0.51	−0.19	0.58	2.
69	2.33	0.29	−0.25	0.45	2.
70	2.32	0.30	−0.18	0.43	2.
71	2.41	0.33	−0.13	0.55	2.
72	2.28	0.36	−0.22	0.45	2.
73	2.43	0.35	−0.09	0.51	2.
74	2.57	0.44	0.11	0.50	2.
75	2.39	0.36	−0.23	0.50	2.
76	2.40	0.51	−0.17	0.54	2.
77	2.36	0.36	0.02	0.51	2.
78	2.46	0.40	−0.21	0.50	2.
79	2.52	0.55	−0.19	0.51	2.
80	2.27	0.50	0.07	0.49	2.
81	2.47	0.43	−0.14	0.52	2.
82	2.60	0.38	−0.05	0.51	2.
83	2.79	0.49	0.09	0.54	2.
84	2.63	0.33	−0.17	0.54	2.
85	2.51	0.35	−0.01	0.54	2.
86	2.46	0.39	0.00	0.57	2.
87	2.34	0.42	−0.18	0.57	2.
88	2.44	0.41	−0.13	0.56	2.
89	2.49	0.60	−0.25	0.55	2.
90	2.39	0.38	−0.19	0.56	2.

TABLE 13.13 *(Continued)*

Case-N	Size	Sorting	Skew	Kurt	Type
91	2.73	0.33	−0.09	0.55	2.
92	2.60	0.32	−0.22	0.57	2.
93	2.41	0.54	0.10	0.56	2.
94	2.36	0.41	−0.17	0.55	2.
95	2.32	0.44	−0.19	0.52	2.
96	2.30	0.42	−0.22	0.55	2.
97	2.31	0.37	−0.17	0.50	2.
98	2.32	0.41	−0.17	0.50	2.
99	2.47	0.95	0.14	0.57	3.
100	2.27	0.88	0.21	0.62	3.
101	2.11	0.98	−0.04	0.62	3.
102	2.06	0.84	0.22	0.58	3.
103	2.14	0.91	0.13	0.57	3.
104	2.23	0.87	0.19	0.52	3.
105	2.20	0.72	−0.09	0.61	3.
106	2.67	0.86	−0.13	0.65	3.
107	2.59	0.94	−0.02	0.61	3.
108	2.40	0.77	−0.21	0.66	3.
109	2.48	0.96	0.15	0.59	3.
110	2.49	0.73	0.13	0.56	3.
111	2.13	0.91	0.14	0.39	3.
112	2.24	0.77	−0.10	0.45	3.
113	2.18	0.74	0.05	0.56	3.
114	2.31	0.82	0.20	0.66	3.
115	2.58	0.81	−0.06	0.62	3.
116	2.22	0.70	0.11	0.56	3.
117	2.38	0.76	0.15	0.59	3.
118	2.80	0.53	−0.17	0.57	3.
119	2.24	0.87	0.02	0.61	3.
120	2.11	0.93	0.10	0.60	3.
121	2.20	0.89	0.08	0.59	3.
122	2.67	0.61	−0.17	0.57	3.
123	1.97	0.71	0.09	0.58	3.
124	2.31	0.81	0.12	0.58	3.
125	2.25	0.92	0.07	0.61	3.
126	2.09	0.90	0.02	0.56	3.
127	1.96	0.79	0.07	0.63	3.
128	2.18	0.81	0.02	0.57	3.
129	2.21	0.62	−0.16	0.55	3.
130	2.07	0.74	−0.01	0.59	3.
131	2.15	0.78	−0.03	0.57	3.
132	2.40	0.94	0.25	0.59	3.
133	2.28	0.97	0.18	0.56	3.
134	2.07	0.88	0.03	0.59	3.
135	1.87	0.97	−0.07	0.53	3.
136	2.31	0.99	0.20	0.59	3.
137	2.05	0.85	−0.01	0.61	3.
138	2.17	0.63	−0.14	0.58	3.

of the Santa Maria River valley. The study area extends southward from Pismo Beach State Park about 8 km and inland from 2 to 16 km. The Pismo Dune Complex comprises three major dune phases.

Type 1 (youngest dunes) is characterized by rather active transverse dunes bordering the coast, with slipface projections and interdune hollows.

Type 2 consists of a series of parabolic dunes, partially stabilized by vegetation and a high water table, with several of the parabolics containing freshwater lakes.

Type 3 consists of paleodunes completely anchored by vegetation. The paleodunes are found on an old marine terrace and have been greatly modified since deposition. They also lack distinctive form and are overlain by poorly developed soil horizons.

A total of 145 samples, each weighing between 65 and 70 g, were collected for the dune complex. The sand samples were sieved at half phi intervals, for 10 min, on a Ro-Tap mechanical shaker. The weight percentage of sand retained in each size category was then plotted as a cumulative frequency curve, and the graphical statistical parameters (mean grain size, sorting, skewness, and kurtosis) were calculated (Tchakerian, 1983).

Using the first two dune types, types 1(A) and 2(B), and the first two variables (size and sorting), the measurements in Table 13.14 can be calculated.

Calculating α, we have

$$\underset{\mathbf{VC}^{-1}}{\begin{bmatrix} 18.75976 & 20.618 \\ 20.61850 & 164.00406 \end{bmatrix}} \quad \underset{\mathbf{D}}{\begin{bmatrix} -.1765 \\ .0785 \end{bmatrix}} \quad = \quad \underset{\alpha}{\begin{bmatrix} -1.6920 \\ 9.23515 \end{bmatrix}}$$

The resulting discriminant function is:

$$d = \alpha_1 X_1 + \alpha_2 X_2$$

$$d = -1.6920X_1 + 9.23515X_2$$

While from Figure 13.6 we can separate the two groups, and a discriminant function passing through the origin can be plotted on the two-dimensional case, there is considerable overlap. The plot is a line passing through the origin (or any line parallel to it) whose slope is:

$$a = \frac{\alpha_1}{\alpha_2}$$

$$= -.183$$

TABLE 13.14 Matrices for Computing a Discriminant Function for the Data in Table 13.3

Means Group **A**

$$\begin{bmatrix} 2.2898 \\ .5008 \end{bmatrix}$$

Means Group **B**

$$\begin{bmatrix} 2.4663 \\ .4223 \end{bmatrix}$$

Mean Differences

$$\begin{bmatrix} -.010 & -.1765 \\ .043 & .0785 \end{bmatrix}$$

DPA

$$\begin{bmatrix} 4.328699 & -.826390 \\ -.826390 & .389766 \end{bmatrix}$$

DPB

$$\begin{bmatrix} 1.609125 & .079914 \\ .079914 & .289450 \end{bmatrix}$$

VC

$$\begin{bmatrix} .061852 & -.007776 \\ -.007776 & .007075 \end{bmatrix}$$

Substitution of the midpoint between the group means in the equation yields a value of d, which is the point along the discriminant function line that is halfway between the midpoint of group **A** the midpoint of group **B**. Thus

$$\begin{aligned} d_0 &= \alpha_1 X_1 + \alpha_2 X_2 \\ &= -1.6920(2.378) + 9.23515(.462) \\ &= -4.0236 + 4.2666 = .243 \end{aligned}$$

Similarly, substitution of the means from groups **A** and **B** can be used to yield the centers of each group in the discriminant function

$$d_A = \alpha_1 \overline{A}_1 + \alpha_2 \overline{A}_2$$

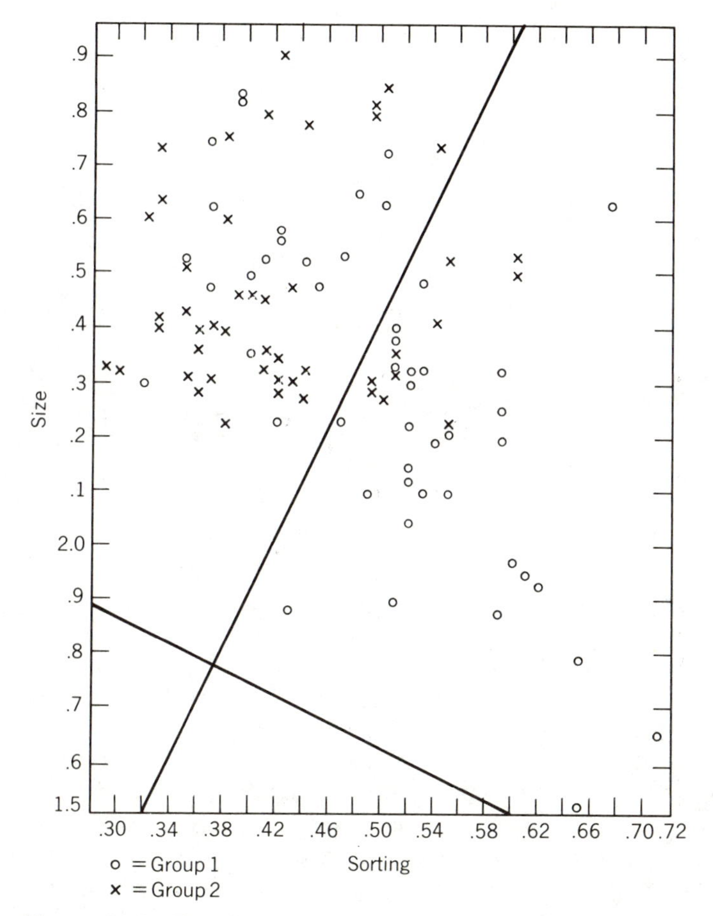

Figure 13.6 Two-dimensional plot of the discriminant function.

and

$$d_B = \alpha_1\overline{B}_1 + \alpha_2\overline{B}_2$$

These points and the scores derived from substituting the original observations can be plotted on the discriminant function (Figure 13.7).

Utilizing the same data—two groups and two variables—the SPSS discriminant problem illustrated in Tables 13.15 to 13.18 was run. The technique involves computing the eigenvalue and associated canonical discriminant function much like the canonical correlation analysis discussed in the previous section. Note that the means, within groups variance-covariance matrices and the pooled variance-covariance matrix are included in the output (Table 13.16). Note also that there are both standardized (mean 0, standard deviation 1) and unstandardized discriminant coefficients (Table

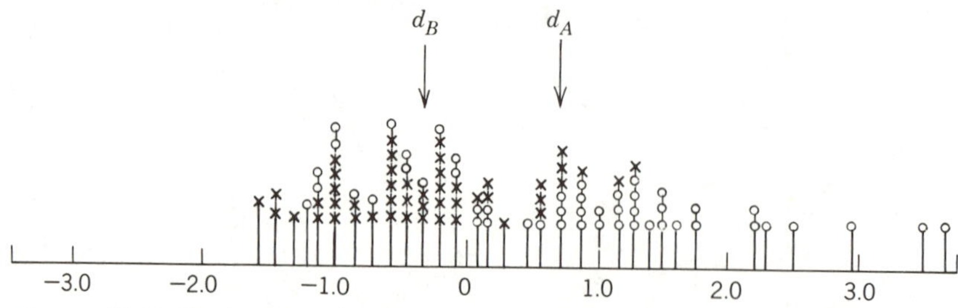

Figure 13.7 Projection of scores onto the discriminant function (type 1 (A) = o, type 2 (B) = x).

13.17). The coefficients include a value for the constant that is used to derive raw discriminant scores. The constant is added to the product of the raw values times the unstandardized coefficients to adjust for the grand means. The SPSS program uses the total covariance matrix rather than the pooled covariance matrix, which yields similar coefficients to the hand-calculated approach:

$$\frac{-1.6718}{9.1299} = -.183$$

is the same slope as computed from the regression technique and represents the line of maximum separation of the two groups.

Both unstandardized and standardized coefficients are reported and either can be used to classify the observations and to plot the relative positions of the sample observations on the discriminant function (as in the regression example). However, the raw or unstandardized coefficients are uninterpretable as coefficients and the scores they produce for the data cases have no intrinsic meaning. There is no logical constraint on the metric units used

TABLE 13.15 SPSS Setup

```
                  1 RUN NAME          DISCRIMINANT ANALYSIS EXAMPLE
                  2 VARIABLE LIST     SIZE SORTING TYPE
                  3 INPUT FORMAT      FIXED(F3.2,F2.2,5X,F1.0)

                   ACCORDING TO YOUR INPUT FORMAT, VARIABLES ARE TO BE READ AS FOLLOWS

                   VARIABLE   FORMAT   RECORD      COLUMNS

                   SIZE       F 3. 2       1        1-    3
                   SORTING    F 2. 2       1        4-    5
                   TYPE       F 1. 0       1       11-   11

THE INPUT FORMAT PROVIDES FOR     3 VARIABLES.     3 WILL BE READ
IT PROVIDES FOR     1 RECORDS ('CARDS') PER CASE.  A MAXIMUM OF     11 'COLUMNS' ARE USED ON A RECORD.

                  4 INPUT MEDIUM      DISK
                  5 N OF CASES        UNKNOWN
                  6 PRINT FORMATS     SIZE SORTING (2)
                  7 LIST CASES        CASES=500/VARIABLES=SIZE TO TYPE
                  8 TASK NAME         ANALYSIS FOR GROUP 1 V. GROUP 2
                  9 DISCRIMINANT      GROUPS = TYPE(1,2)/
                 10                   VARIABLES = SIZE SORTING /
                 11                   ANALYSIS = SIZE SORTING /
                 12                       METHOD = DIRECT/
                 13                   ANALYSIS = SIZE SORTING/
                 14                       METHOD = DIRECT/
                 15 OPTIONS           5,7,9,10,11,12
                 16 STATISTICS        ALL
```

TABLE 13.16 Means and Covariances

```
NUMBER OF CASES BY GROUP

                   NUMBER OF CASES
  TYPE         UNWEIGHTED      WEIGHTED   LABEL

        1           50            50.0
        2           48            48.0

     TOTAL          98            98.0


GROUP MEANS

  TYPE              SIZE              SORTING

        1          2.28980            0.50080
        2          2.46625            0.42229

     TOTAL         2.37622            0.46235


GROUP STANDARD DEVIATIONS

  TYPE              SIZE              SORTING

        1          0.29722            0.08919
        2          0.18503            0.07843

     TOTAL         0.26282            0.09251


POOLED WITHIN-GROUPS COVARIANCE MATRIX WITH          96 DEGREES OF FREEDOM

                 SIZE              SORTING

SIZE        0.6185233D-01
SORTING    -0.7775827D-02     0.7075166D-02


COVARIANCE MATRIX FOR GROUP                 1,

                 SIZE              SORTING

SIZE        0.8834078D-01
SORTING    -0.1086514D-01     0.7954449D-02


COVARIANCE MATRIX FOR GROUP                 2,

                 SIZE              SORTING

SIZE        0.3423670D-01
SORTING     0.1700269D-02     0.6158466D-02
```

for the discriminant space. But, in many instances, the unstandardized coefficients are used to compute the discriminant scores.

The process of standardization that yields discriminant scores measured in standard deviation units and coefficients that represent the relative importance of the variables in the discriminant function involves moving the

TABLE 13.17 Eigenvalues and Coefficients

```
                                        CANONICAL DISCRIMINANT FUNCTIONS

                       PERCENT OF   CUMULATIVE    CANONICAL   :  AFTER
FUNCTION  EIGENVALUE    VARIANCE      PERCENT    CORRELATION  : FUNCTION  WILKS' LAMBDA  CHI-SQUARED   D.F.  SIGNIFICANCE
                                                              :    0       .7929373       [illegible]    2    [illegible]
   1*       0.26114      100.00       100.00     0.4551459    :
   * MARKS THE   1 CANONICAL DISCRIMINANT FUNCTION(S) TO BE USED IN THE REMAINING ANALYSIS.


UNSTANDARDIZED CANONICAL DISCRIMINANT FUNCTION COEFFICIENTS

                FUNC  1

SIZE          -1.671811
SORTING        9.129916
(CONSTANT)    -0.2495397

STANDARDIZED CANONICAL DISCRIMINANT FUNCTION COEFFICIENTS

                FUNC  1

SIZE          -0.41578
SORTING        0.76795
```

origin of the discriminant function axis or axes (if there is more than one) to a position coincident with the grand centroid. The grand centroid is that position (point in space) where all the discriminating variables have their average values over all cases. Now we can tell the relative position of any individual case or group centroid relative to the center of the system.

The program computes the canonical discriminant function (unstandardized) as:

$$d_{i1} = -.2486 - 1.6718X_1 + 9.1299X_2$$

TABLE 13.18 Classification Results

```
                             NO. OF      PREDICTED  GROUP  MEMBERSHIP
   ACTUAL GROUP               CASES          1              2
--------------------         ------      --------       --------

GROUP          1               50           31             19
                                          62.0%          38.0%

GROUP          2               48           11             37
                                          22.9%          77.1%

PERCENT OF "GROUPED" CASES CORRECTLY CLASSIFIED:   69.39%


SYMBOLS USED IN PLOTS

SYMBOL   GROUP   LABEL
------   -----   --------------------

  1        1
  2        2

                                   ALL-GROUPS STACKED HISTOGRAM

                              -- CANONICAL DISCRIMINANT FUNCTION 1 --

       8 +
         .
         .
F        .
R      6 +                                   2        2
E        .                                   2        2
Q        .                                 2 2        2
U        .                                 2 2        2
E      4 +                          222  222 2222     22 2  22
N        .                          222  222 2222     22 2  22
C        .                          222  221 22222    12 1  21
Y        .                          222  221 22222    12 1  21
       2 +                         211222221 22111    1121 111 1 1
         .                         211222221 22111    1121 111 1 1
         .                      222111112121121111 2 1111111111111   1111    1   1   1
         .                      222111112121121111 2 1111111111111   1111    1   1   1
      OUT...........+.........+.........+.........+.........+.........+.........+..........OUT
                   -3        -2        -1         0         1         2         3
CLASSIFICATION 22222222222222222222222222222222222222221111111111111111111111111111111111111111111
GROUP CENTROIDS                              2         1
```

TABLE 13.19 Means and Standard Deviations for Multiple Discriminant Analysis

NUMBER OF CASES BY GROUP

TYPE	NUMBER OF CASES UNWEIGHTED	WEIGHTED	LABEL
1	50	50.0	
2	48	48.0	
3	47	47.0	
TOTAL	145	145.0	

GROUP MEANS

TYPE	SIZE	SORTING	SKEW	KURT
1	2.28980	0.50080	-0.19500	0.47940
2	2.46625	0.42229	-0.14062	0.52458
3	2.26149	0.82085	0.04447	0.57872
TOTAL	2.33903	0.57855	-0.09938	0.52655

GROUP STANDARD DEVIATIONS

TYPE	SIZE	SORTING	SKEW	KURT
1	0.29722	0.08919	0.06852	0.02751
2	0.18503	0.07848	0.10460	0.03690
3	0.20330	0.13201	0.12663	0.04839
TOTAL	0.25027	0.19921	0.14417	0.05578

TABLE 13.20 Eigenvalues and Coefficients

CANCNICAL DISCRIMINANT FUNCTICNS

FUNCTION	EIGENVALUE	PERCENT OF VARIANCE	CUMULATIVE PERCENT	CANONICAL CORRELATION	:	AFTER FUNCTION	WILKS' LAMBDA	CHI-SQUARED	D.F.	SIGNIFICANCE
					:	0	0.1496556	266.87	8	0.0
1*	3.52074	88.04	88.04	0.8824949	:	1	0.6765543	54.899	3	0.0000
2*	0.47808	11.96	100.00	0.5687228	:					

* MARKS THE 2 CANONICAL DISCRIMINANT FUNCTICN(S) TC BE USED IN THE REMAINING ANALYSIS.

WILKS' LAMBDA (U-STATISTIC) AND UNIVARIATE F-RATIO
WITH 2 AND 142 DEGREES CF FREEDCM

VARIABLE	WILKS' LAMBDA	F	SIGNIFICANCE
SIZE	0.86910	10.69	0.0000
SORTING	0.25913	203.0	0.0000
SKEW	0.49508	72.41	0.0000
KURT	0.46599	81.37	0.0000

UNSTANDARDIZED CANONICAL DISCRIMINANT FUNCTICN COEFFICIENTS

	FUNC 1	FUNC 2
SIZE	-0.5412221D-01	6.2403575
SORTING	7.339099	-6.565558
SKEW	2.765716	4.089135
KURT	9.602486	19.58111
(CONSTANT)	-8.900805	-6.667783

STANDARDIZED CANONICAL DISCRIMINANT FUNCTICN COEFFICIENTS

	FUNC 1	FUNC 2
SIZE	-0.01272	0.05647
SORTING	0.74945	-0.57046
SKEW	0.28253	0.41773
KURT	0.36822	0.75035

where d_{i1} is the discriminant score for observation i on discriminant function 1. The X's are the original values.

The standardized discriminant function is:

$$d_{i1} = -.4158X_1 + .7680X_2$$

Note that the constant has vanished. We can now say that the sorting coefficient is about twice as important as the size coefficient in its contribution to the discrimination process. An aid in judging the overall importance of the discriminant function is the canonical correlation (Table 13.17). This value (.46) indicates that the single discriminant function is moderately correlated with the groups. The canonical correlation tells how closely the discriminant function and the groups are related.

TABLE 13.21 Classification Results

ACTUAL GROUP	NO. OF CASES	PREDICTED GROUP MEMBERSHIP 1	2	3
GROUP 1	50	40 80.0%	9 18.0%	1 2.0%
GROUP 2	48	9 18.8%	39 81.3%	0 0.0%
GROUP 3	47	2 4.3%	4 8.5%	41 87.2%

PERCENT OF "GROUPED" CASES CORRECTLY CLASSIFIED: 82.76%

ALL-GROUPS SCATTERPLOT - * INDICATES A GROUP CENTROID

CANONICAL DISCRIMINANT FUNCTION 1

CANONICAL DISCRIMINANT FUNCTION 2

```
       CUT        -6        -4        -2         0         2         4         6        CUT
         X.........+.........+.........+.........+.........+.........+.........+........X
        X                                                                                 X
        .                                                                                 .
        .                                                                                 .
        .                                                                                 .
        .                                                                                 .
        .                                                                                 .
      6 +                                                                                 +
        .                                                                                 .
        .                                                                                 .
        .                                                                                 .
        .                                                                                 .
        .                                                                                 .
      4 +                                                                                 +
        .                                                                                 .
        .                                                                                 .
        .                                                                                 .
        .                                     2                                           .
        .                                222                         3                    .
      2 +                               22                                                +
        .                              2  2 2    2  2                                     .
        .                               2 22 2  2                *                        .
        .                             2111 * 32 3            3 33       33                .
        .                           1 212 2 22      3     33  3  33      3                .
        .                            2 222  322   3        3   3 3     3                  .
      0 +                               1          3  1        * 3 3 3                    +
        .                        2   2 1211 1 2  2           3 3   333 3 3                .
        .                         2 2  1  1*11  11                 3  33                  .
        .                            1   11 1 1                 33                        .
        .                                 111   2                                         .
        .                                 1   1 1              3                          .
     -2 +                                1     1                                          +
        .                                   1                     3                       .
        .                                     1     3                                     .
        .                                    1                                            .
        .                                                                                 .
        .                                      1                                          .
     -4 +                                                 3                               +
        .                                                                                 .
        .                                                                                 .
        .                                                                                 .
        .                                                                                 .
        .                                                                                 .
     -6 +                                                                                 +
        .                                                                                 .
        .                                                                                 .
        .                                                                                 .
        .                                                                                 .
        .                                                                                 .
        X                                                                                 X
         X.........+.........+.........+.........+.........+.........+.........+........X
       CUT        -6        -4        -2         0         2         4         6        CUT
```

The final two elements of the output are the actual plots and the table of correct classifications (Table 13.18). It is clear from the plot that the cases are not divided from either side of the centroid on the discriminant function. In fact, slightly less than 70 percent of the cases are correctly classified. The plot lists every case by its location, either side of the centroid. If the classification were perfect, there would be tick marks on one side of the 0 value and o's on the other side.

A Multivariate Extension

Utilizing the same data (Table 13.13) it is possible to construct a more meaningful analysis of the sand dune complex. The data include observations on three sand types and four variables (mean grain size, sorting, skewness, and kurtosis) which measure characteristics of the sand samples. The output (Tables 13.19 to 13.21) is similar in format to the two-group, two-variable case, but now there are two discriminant functions:

$$d_1 = -8.9008 - .0541X_1 + 7.3391X_2 + 2.7657X_3 + 9.6025X_4$$

$$d_2 = -6.6678 + .2404X_1 - 6.5656X_2 + 4.0891X_3 + 19.5811X_4$$

Both functions are statistically significant and the canonical correlations are .88 for the first function, and .57 for the second function. The standardized canonical discriminant functions (Table 13.20) indicate that the measures of sorting and kurtosis provide most of the discriminating power, although all the coefficients are significant. The real contribution of the additional variables is seen in the plot of the two coefficients in standardized values (Table 13.21) and the classification results. Almost 83 percent of the cases are correctly predicted with the two discriminant functions. Note that the predictions are better for group 3 than either of the other groups and emphasize the separation between the "younger" sands (groups 1 and 2) and the "older" sands (group 3).

References and Readings

1. Logit Analysis

Bishop, Y. and P. W. Holland (1975) *Discrete Multivariate Analysis, Theory and Practice,* MIT Press: Cambridge, Mass.

Blalock, H. M. (1979) *Social Statistics,* Revised Second Edition, McGraw-Hill: New York.

Dixon, W. J. (ed.) (1983) *BMDP Statistical Software,* University of California Press: Los Angeles.

Feinberg, S. E. (1977) *The Analysis of Cross-classified Data,* MIT Press: Cambridge, Mass.

Hensher, D. A. and L. W. Johnson (1981) *Applied Discrete Choice Modelling,* Croom Helm: London.

Hensher, D. A. and P. B. McLeod (1977) "Towards an integrated approach to the identification and evaluation of the transport determinants of travel choices," *Transportation Research* 11:77–93.

Knoke, D. and P. J. Burke (1980) *Log-linear Models,* Sage Publications: Beverly Hills, Cal.

Wrigley, N. (1976) *An Introduction to the Use of Logit Models in Geography,* Geo Abstracts: Norwich.

Wrigley, N. (1985) *Categorical data analysis,* Longman: London.

Wrigley, N. and P. A. Longley (1984) "Discrete choice modelling in urban analysis," in D. T. Herbert and R. J. Johnston (eds.), *Geography and the Urban Environment,* Volume 6, Wiley: Chichester.

2. Canonical Analysis

Gould, P. (1967) "On the geographic interpretation of eigenvalues: An initial exploration," *Transactions Inst. British Geographers* 42:53–86.

Levine, M. S. (1977) *Canonical Analysis and Factor Comparison,* Quantitative Applications in the Social Sciences, Sage Publications: Beverly Hills, Cal.

Thorndike, R. M. (1978) *Correlational Procedures for Research,* Gardner Press: New York.

Webster, R. (1979) "Exploratory and descriptive uses of multivariate analysis in soil survey," in N. Wrigley (ed.), *Statistical Applications in the Spatial Sciences,* Pion: London.

3. Discriminant Analysis

Afifi, A. and S. P. Azen (1979) *Statistical Analysis: A Computer Oriented Approach,* Academic Press: New York.

Klecka, W. (1980) *Discriminant Analysis,* Sage Publications: Beverly Hills, Cal.

Tchakerian, V. (1983) "The Pismo Coastal Dune Complex, California: Geomorphology and Textural Relationships," unpublished M.A. Thesis, University of California: Los Angeles.

Statistical Tables

A Random Numbers

B Ordinates of the Standard Normal Density Function

C Cumulative Standard Normal Distribution

D Cumulative *t* Distribution

E Cumulative Chi-Square Distribution

F Critical Values of *D* in the Kolmogorov–Smirnov One-sample Test

G Cumulative *F* Distribution

H Critical Values of the Durbin–Watson Statistic

TABLE A Random Numbers

09 18 82 00 97	32 82 53 95 27	04 22 08 63 04	83 38 98 73 74	64 27 85 80 44
90 04 58 54 97	51 98 15 06 54	94 93 88 19 97	91 87 07 61 50	68 47 66 46 59
73 18 95 02 07	47 67 72 62 69	62 29 06 44 64	27 12 46 70 18	41 36 18 27 60
75 76 87 64 90	20 97 18 17 49	90 42 91 22 72	95 37 50 58 71	93 82 34 31 78
54 01 64 40 56	66 28 13 10 03	00 68 22 73 98	20 71 45 32 95	07 70 61 78 13
08 35 86 99 10	78 54 24 27 85	13 66 15 88 73	04 61 89 75 53	31 22 30 84 20
28 30 60 32 64	81 33 31 05 91	40 51 00 78 93	32 60 46 04 75	94 11 90 18 40
53 84 08 62 33	81 59 41 36 28	51 21 59 02 90	28 46 66 87 95	77 76 22 07 91
91 75 75 37 41	61 61 36 22 69	50 26 39 02 12	55 78 17 65 14	83 48 34 70 55
89 41 59 26 94	00 39 75 83 91	12 60 71 76 46	48 94 97 23 06	94 54 13 74 08
77 51 30 38 20	86 83 42 99 01	68 41 48 27 74	51 90 81 39 80	72 89 35 55 07
19 50 23 71 74	69 97 92 02 88	55 21 02 97 73	74 28 77 52 51	65 34 46 74 15
21 81 85 93 13	93 27 88 17 57	05 68 67 31 56	07 08 28 50 46	31 85 33 84 52
51 47 46 64 99	68 10 72 36 21	94 04 99 13 45	42 83 60 91 91	08 00 74 54 49
99 55 96 83 31	62 53 52 41 70	69 77 71 28 30	74 81 97 81 42	43 86 07 28 34
33 71 34 80 07	93 58 47 28 69	51 92 66 47 21	58 30 32 98 22	93 17 49 39 72
85 27 48 68 93	11 30 32 92 70	28 83 43 41 37	73 51 59 04 00	71 14 84 36 43
84 13 38 96 40	44 03 55 21 66	73 85 27 00 91	61 22 26 05 61	62 32 71 84 23
56 73 21 62 34	17 39 59 61 31	10 12 39 16 22	85 49 65 75 60	81 60 41 88 80
65 13 85 68 06	87 64 88 52 61	34 31 36 58 61	45 87 52 10 69	85 64 44 72 77
38 00 10 21 76	81 71 91 17 11	71 60 29 29 37	74 21 96 40 49	65 58 44 96 98
37 40 29 63 97	01 30 47 75 86	56 27 11 00 86	47 32 46 26 05	40 03 03 74 38
97 12 54 03 48	87 08 33 14 17	21 81 53 92 50	75 23 76 20 47	15 50 12 95 78
21 82 64 11 34	47 14 33 40 72	64 63 88 59 02	49 13 90 64 41	03 85 65 45 52
73 13 54 27 42	95 71 90 90 35	85 79 47 42 96	08 78 98 81 56	64 69 11 92 02
07 63 87 79 29	03 06 11 80 72	96 20 74 41 56	23 82 19 95 38	04 71 36 69 94
60 52 88 34 41	07 95 41 98 14	59 17 52 06 95	05 53 35 21 39	61 21 20 64 55
83 59 63 56 55	06 95 89 29 83	05 12 80 97 19	77 43 35 37 83	92 30 15 04 98
10 85 06 27 46	99 59 91 05 07	13 49 90 63 19	53 07 57 18 39	06 41 01 93 62
39 82 09 89 52	43 62 26 31 47	64 42 18 08 14	43 80 00 93 51	31 02 47 31 67
59 58 00 64 78	75 56 97 88 00	88 83 55 44 86	23 76 80 61 56	04 11 10 84 08
38 50 80 73 41	23 79 34 87 63	90 82 29 70 22	17 71 90 42 07	95 95 44 99 53
30 69 27 06 68	94 68 81 61 27	56 19 68 00 91	82 06 76 34 00	05 46 26 92 00
65 44 39 56 59	18 28 82 74 37	49 63 22 40 41	08 33 76 56 76	96 29 99 08 36
27 26 75 02 64	13 19 27 22 94	07 47 74 46 06	17 98 54 89 11	97 34 13 03 58
91 30 70 69 91	19 07 22 42 10	36 69 95 37 28	28 82 53 57 93	28 97 66 62 52
68 43 49 46 88	84 47 31 36 22	62 12 69 84 08	12 84 38 25 90	09 81 59 31 46
48 90 81 58 77	54 74 52 45 91	35 70 00 47 54	83 82 45 26 92	54 13 05 51 60
06 91 34 51 97	42 67 27 86 01	11 88 30 95 28	63 01 19 89 01	14 97 44 03 44
10 45 51 60 19	14 21 03 37 12	91 34 23 78 21	88 32 58 08 51	43 66 77 08 83
12 88 39 73 43	65 02 76 11 84	04 28 50 13 92	17 97 41 50 77	90 71 22 67 69
21 77 83 09 76	38 80 73 69 61	31 64 94 20 96	63 28 10 20 23	08 81 64 74 49
19 52 35 95 15	65 12 25 96 59	86 28 36 82 58	69 57 21 37 98	16 43 59 15 29
67 24 55 26 70	35 58 31 65 63	79 24 68 66 86	76 46 33 42 22	26 65 59 08 02
60 58 44 73 77	07 50 03 79 92	45 13 42 65 29	26 76 08 36 37	41 32 64 43 44
53 85 34 13 77	36 06 69 48 50	58 83 87 38 59	49 36 47 33 31	96 24 04 36 42
24 63 73 87 36	74 38 48 93 42	52 62 30 79 92	12 36 91 86 01	03 74 28 38 73
83 08 01 24 51	38 99 22 28 15	07 75 95 17 77	97 37 72 75 85	51 97 23 78 67
16 44 42 43 34	36 15 19 90 73	27 49 37 09 39	85 13 03 25 52	54 84 65 47 59
60 79 01 81 57	57 17 86 57 62	11 16 17 85 76	45 81 95 29 79	65 13 00 48 60

TABLE A (*Continued*)

10 09 73 25 33	76 52 01 35 86	34 67 35 48 76	80 95 90 91 17	39 29 27 49 45
37 54 20 48 05	64 89 47 42 96	24 80 52 40 37	20 63 61 04 02	00 82 29 16 65
08 42 26 89 53	19 64 50 93 03	23 20 90 25 60	15 95 33 47 64	35 08 03 36 06
99 01 90 25 29	09 37 67 07 15	38 31 13 11 65	88 67 67 43 97	04 43 62 76 59
12 80 79 99 70	80 15 73 61 47	64 03 23 66 53	98 95 11 68 77	12 17 17 68 33
66 06 57 47 17	34 07 27 68 50	36 69 73 61 70	65 81 33 98 85	11 19 92 91 70
31 06 01 08 05	45 57 18 24 06	35 30 34 26 14	86 79 90 74 39	23 40 30 97 32
85 26 97 76 02	02 05 16 56 92	68 66 57 48 18	73 05 38 52 47	18 62 38 85 79
63 57 33 21 35	05 32 54 70 48	90 55 35 75 48	28 46 82 87 09	83 49 12 56 24
73 79 64 57 53	03 52 96 47 78	35 80 83 42 82	60 93 52 03 44	35 27 38 84 35
98 52 01 77 67	14 90 56 86 07	22 10 94 05 58	60 97 09 34 33	50 50 07 39 98
11 80 50 54 31	39 80 82 77 32	50 72 56 82 48	29 40 52 42 01	52 77 56 78 51
83 45 29 96 34	06 28 89 80 83	13 74 67 00 78	18 47 54 06 10	68 71 17 78 17
88 68 54 02 00	86 50 75 84 01	36 76 66 79 51	90 36 47 64 93	29 60 91 10 62
99 59 46 73 48	87 51 76 49 69	91 82 60 89 28	93 78 56 13 68	23 47 83 41 13
65 48 11 76 74	17 46 85 09 50	58 04 77 69 74	73 03 95 71 86	40 21 81 65 44
80 12 43 56 35	17 72 70 80 15	45 31 82 23 74	21 11 57 82 53	14 38 55 37 63
74 35 09 98 17	77 40 27 72 14	43 23 60 02 10	45 52 16 42 37	96 28 60 26 55
69 91 62 68 03	66 25 22 91 48	36 93 68 72 03	76 62 11 39 90	94 40 05 64 18
09 89 32 05 05	14 22 56 85 14	46 42 75 67 88	96 29 77 88 22	54 38 21 45 98
91 49 91 45 23	68 47 92 76 86	46 16 28 35 54	94 75 08 99 23	37 08 92 00 48
80 33 69 45 98	26 94 03 68 58	70 29 73 41 35	53 14 03 33 40	42 05 08 23 41
44 10 48 19 49	85 15 74 79 54	32 97 92 65 75	57 60 04 08 81	22 22 20 64 13
12 55 07 37 42	11 10 00 20 40	12 86 07 46 97	96 64 48 94 39	28 70 72 58 15
63 60 64 93 29	16 50 53 44 84	40 21 95 25 63	43 65 17 70 82	07 20 73 17 90
61 19 69 04 46	26 45 74 77 74	51 92 43 37 29	65 39 45 95 93	42 58 26 05 27
15 47 44 52 66	95 27 07 99 53	59 36 78 38 48	82 39 61 01 18	33 21 15 94 66
94 55 72 85 73	67 89 75 43 87	54 62 24 44 31	91 19 04 25 92	92 92 74 59 73
42 48 11 62 13	97 34 40 87 21	16 86 84 87 67	03 07 11 20 59	25 70 14 66 70
23 52 37 83 17	73 20 88 98 37	68 93 59 14 16	26 25 22 96 63	05 52 28 25 62
04 49 35 24 94	75 24 63 38 24	45 86 25 10 25	61 96 27 93 35	65 33 71 24 72
00 54 99 76 54	64 05 18 81 59	96 11 96 38 96	54 69 28 23 91	23 28 72 95 29
35 96 31 53 07	26 89 80 93 54	33 35 13 54 62	77 97 45 00 24	90 10 33 93 33
59 80 80 83 91	45 42 72 68 42	83 60 94 97 00	13 02 12 48 92	78 56 52 01 06
46 05 88 52 36	01 39 09 22 86	77 28 14 40 77	93 91 08 36 47	70 61 74 29 41
32 17 90 05 97	87 37 92 52 41	05 56 70 70 07	86 74 31 71 57	85 39 41 18 38
69 23 46 14 06	20 11 74 52 04	15 95 66 00 00	18 74 39 24 23	97 11 89 63 38
19 56 54 14 30	01 75 87 53 79	40 41 92 15 85	66 67 43 68 06	84 96 28 52 07
45 15 51 49 38	19 47 60 72 46	43 66 79 45 43	59 04 79 00 33	20 82 66 95 41
94 86 43 19 94	36 16 81 08 51	34 88 88 15 53	01 54 03 54 56	05 01 45 11 76
98 08 62 48 26	45 24 02 84 04	44 99 90 88 96	39 09 47 34 07	35 44 13 18 80
33 18 51 62 32	41 94 15 09 49	89 43 54 85 81	88 69 54 19 94	87 54 87 80 43
80 95 10 04 06	96 38 27 07 74	20 15 12 83 87	25 01 62 52 98	94 62 46 11 71
79 75 24 91 40	71 96 12 82 96	69 86 10 25 91	74 85 22 05 39	00 38 75 95 79
18 63 33 25 37	98 14 50 65 71	31 01 02 46 74	05 45 56 14 27	77 93 89 19 36
74 02 94 39 02	77 55 73 22 70	97 79 01 71 19	52 52 75 80 21	80 81 45 17 48
54 17 84 56 11	80 99 33 71 43	05 33 51 29 69	56 12 71 92 55	36 04 09 03 24
11 66 44 98 83	52 07 98 48 27	59 38 17 15 39	09 97 33 34 40	88 46 12 33 56
48 32 47 79 28	31 24 96 47 10	02 29 53 68 70	32 30 75 75 46	15 02 00 99 94
69 07 49 41 38	87 63 79 19 76	35 58 40 44 01	10 51 82 16 15	01 84 87 69 38

Reproduced from The RAND Corporation, *A Million Random Digits* (1983) with permission.

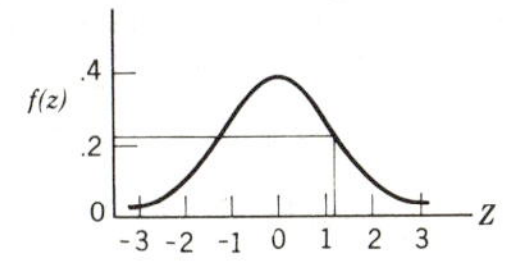

TABLE B Ordinates of the Standard Normal Density Function

Z	.00	.01	.02	.03	.04	.05	.06	.07	.08	.09
.0	.3989	.3989	.3989	.3988	.3986	.3984	.3982	.3980	.3977	.3973
.1	.3970	.3965	.3961	.3956	.3951	.3945	.3939	.3932	.3925	.3918
.2	.3910	.3902	.3894	.3885	.3876	.3867	.3857	.3847	.3836	.3825
.3	.3814	.3802	.3790	.3778	.3765	.3752	.3739	.3725	.3712	.3697
.4	.3683	.3668	.3653	.3637	.3621	.3605	.3589	.3572	.3555	.3538
.5	.3521	.3503	.3485	.3467	.3448	.3429	.3410	.3391	.3372	.3352
.6	.3332	.3312	.3292	.3271	.3251	.3230	.3209	.3187	.3166	.3144
.7	.3123	.3101	.3079	.3056	.3034	.3011	.2989	.2966	.2943	.2920
.8	.2897	.2874	.2850	.2827	.2803	.2780	.2756	.2732	.2709	.2685
.9	.2661	.2637	.2613	.2589	.2565	.2541	.2516	.2492	.2468	.2444
1.0	.2420	.2396	.2371	.2347	.2323	.2299	.2275	.2251	.2227	.2203
1.1	.2179	.2155	.2131	.2107	.2083	.2059	.2036	.2012	.1989	.1965
1.2	.1942	.1919	.1895	.1872	.1849	.1826	.1804	.1781	.1758	.1736
1.3	.1714	.1691	.1669	.1647	.1626	.1604	.1582	.1561	.1539	.1518
1.4	.1497	.1476	.1456	.1435	.1415	.1394	.1374	.1354	.1334	.1315
1.5	.1295	.1276	.1257	.1238	.1219	.1200	.1182	.1163	.1145	.1127
1.6	.1109	.1092	.1074	.1057	.1040	.1023	.1006	.0989	.0973	.0957
1.7	.0940	.0925	.0909	.0893	.0878	.0863	.0848	.0833	.0818	.0804
1.8	.0790	.0775	.0761	.0748	.0734	.0721	.0707	.0694	.0681	.0669
1.9	.0656	.0644	.0632	.0620	.0608	.0596	.0584	.0573	.0562	.0551
2.0	.0540	.0529	.0519	.0508	.0498	.0488	.0478	.0468	.0459	.0449
2.1	.0440	.0431	.0422	.0413	.0404	.0396	.0387	.0379	.0371	.0363
2.2	.0355	.0347	.0339	.0332	.0325	.0317	.0310	.0303	.0297	.0290
2.3	.0283	.0277	.0270	.0264	.0258	.0252	.0246	.0241	.0235	.0229
2.4	.0224	.0219	.0213	.0208	.0203	.0198	.0194	.0189	.0184	.0180
2.5	.0175	.0171	.0167	.0163	.0158	.0154	.0151	.0147	.0143	.0139
2.6	.0136	.0132	.0129	.0126	.0122	.0119	.0116	.0113	.0110	.0107
2.7	.0104	.0101	.0099	.0096	.0093	.0091	.0088	.0086	.0084	.0081
2.8	.0079	.0077	.0075	.0073	.0071	.0069	.0067	.0065	.0063	.0061
2.9	.0060	.0058	.0056	.0055	.0053	.0051	.0050	.0048	.0047	.0046
3.0	.0044	.0043	.0042	.0040	.0039	.0038	.0037	.0036	.0035	.0034
3.1	.0033	.0032	.0031	.0030	.0029	.0028	.0027	.0026	.0025	.0025
3.2	.0024	.0023	.0022	.0022	.0021	.0020	.0020	.0019	.0018	.0018
3.3	.0017	.0017	.0016	.0016	.0015	.0015	.0014	.0014	.0013	.0013
3.4	.0012	.0012	.0012	.0011	.0011	.0010	.0010	.0010	.0009	.0009
3.5	.0009	.0008	.0008	.0008	.0008	.0007	.0007	.0007	.0007	.0006
3.6	.0006	.0006	.0006	.0005	.0005	.0005	.0005	.0005	.0005	.0004
3.7	.0004	.0004	.0004	.0004	.0004	.0004	.0003	.0003	.0003	.0003
3.8	.0003	.0003	.0003	.0003	.0003	.0002	.0002	.0002	.0002	.0002
3.9	.0002	.0002	.0002	.0002	.0002	.0002	.0002	.0002	.0001	.0001

A table of values for the height of the curve (ordinate) for the absolute value of Z. The first decimal place of Z is given by rows, the second by columns. For example, $f(Z)$ for $Z = \pm 1.15 = .2059$.

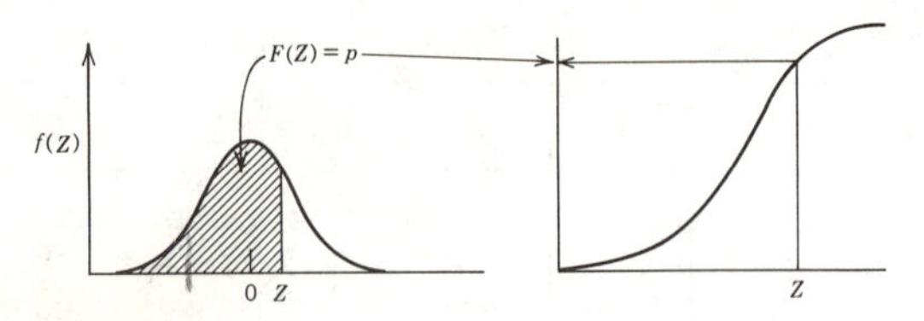

TABLE C Cumulative Standard Normal Distribution

Z	.00	.01	.02	.03	.04	.05	.06	.07	.08	.09
.0	.5000	.5040	.5080	.5120	.5160	.5199	.5239	.5279	.5319	.5359
.1	.5398	.5438	.5478	.5517	.5557	.5596	.5636	.5675	.5714	.5753
.2	.5793	.5832	.5871	.5910	.5948	.5987	.6026	.6064	.6103	.6141
.3	.6179	.6217	.6255	.6293	.6331	.6368	.6406	.6443	.6480	.6517
.4	.6554	.6591	.6628	.6664	.6700	.6736	.6772	.6808	.6844	.6879
.5	.6915	.6950	.6985	.7019	.7054	.7088	.7123	.7157	.7190	.7224
.6	.7257	.7291	.7324	.7357	.7389	.7422	.7454	.7486	.7517	.7549
.7	.7580	.7611	.7642	.7673	.7704	.7734	.7764	.7794	.7823	.7852
.8	.7881	.7910	.7939	.7967	.7995	.8023	.8051	.8078	.8106	.8133
.9	.8159	.8186	.8212	.8238	.8264	.8289	.8315	.8340	.8365	.8389
1.0	.8413	.8438	.8461	.8485	.8508	.8531	.8554	.8577	.8599	.8621
1.1	.8643	.8665	.8686	.8708	.8729	.8749	.8770	.8790	.8810	.8830
1.2	.8849	.8869	.8888	.8907	.8925	.8944	.8962	.8980	.8997	.9015
1.3	.9032	.9049	.9066	.9082	.9099	.9115	.9131	.9147	.9162	.9177
1.4	.9192	.9207	.9222	.9236	.9251	.9265	.9279	.9292	.9306	.9319
1.5	.9332	.9345	.9357	.9370	.9382	.9394	.9406	.9418	.9429	.9441
1.6	.9452	.9463	.9474	.9484	.9495	.9505	.9515	.9525	.9535	.9545
1.7	.9554	.9564	.9573	.9582	.9591	.9599	.9608	.9616	.9625	.9633
1.8	.9641	.9649	.9656	.9664	.9671	.9678	.9686	.9693	.9699	.9706
1.9	.9713	.9719	.9726	.9732	.9738	.9744	.9750	.9756	.9761	.9767
2.0	.9772	.9778	.9783	.9788	.9793	.9798	.9803	.9808	.9812	.9817
2.1	.9821	.9826	.9830	.9834	.9838	.9842	.9846	.9850	.9854	.9857
2.2	.9861	.9864	.9868	.9871	.9875	.9878	.9881	.9884	.9887	.9890
2.3	.9893	.9896	.9898	.9901	.9904	.9906	.9909	.9911	.9913	.9916
2.4	.9918	.9920	.9922	.9925	.9927	.9929	.9931	.9932	.9934	.9936
2.5	.9938	.9940	.9941	.9943	.9945	.9946	.9948	.9949	.9951	.9952
2.6	.9953	.9955	.9956	.9957	.9959	.9960	.9961	.9962	.9963	.9964
2.7	.9965	.9966	.9967	.9968	.9969	.9970	.9971	.9972	.9973	.9974
2.8	.9974	.9975	.9976	.9977	.9977	.9978	.9979	.9979	.9980	.9981
2.9	.9981	.9982	.9982	.9983	.9984	.9984	.9985	.9985	.9986	.9986
3.0	.9987	.9987	.9987	.9988	.9988	.9989	.9989	.9989	.9990	.9990
3.1	.9990	.9991	.9991	.9991	.9992	.9992	.9992	.9992	.9993	.9993
3.2	.9993	.9993	.9994	.9994	.9994	.9994	.9994	.9995	.9995	.9995
3.3	.9995	.9995	.9995	.9996	.9996	.9996	.9996	.9996	.9996	.9997
3.4	.9997	.9997	.9997	.9997	.9997	.9997	.9997	.9997	.9997	.9998
3.5	.9998									
4.0	.9999									

Z	1.282	1.645	1.960	2.326	2.576	3.090	3.291	3.891	4.417
p	.90	.95	.975	.99	.995	.999	.9995	.99995	.999995

A table of values (p) for the area under the curve for positive values of Z (i.e., $P(Z_{obs} \leq [Z, p])$. The first decimal place of Z is given by rows, the second by columns. For example, $p(Z \leq 1.96) = .9750$.

For negative values of $Z (Z < .0)$ subtract the tabled value from 1.0000.

Reproduced from A. M. Mood and F. A. Graybill, *Introduction to the Theory of Statistics* (1963), by permission of McGraw-Hill.

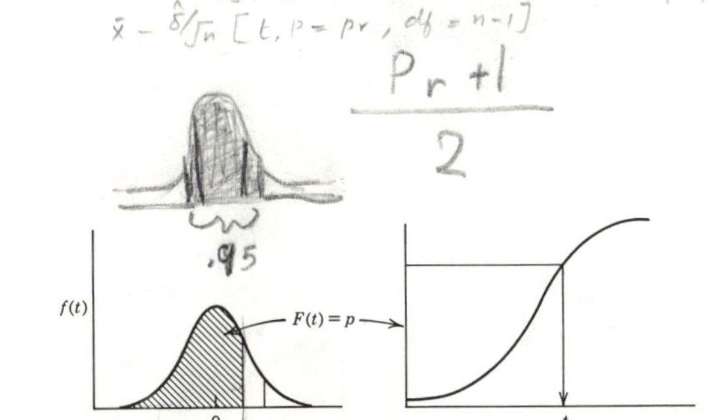

TABLE D Cumulative *t* Distribution

df	Probability (*p*) .75	.90	.95	.975	.99	.995	.999
1	1.000	3.078	6.314	12.706	31.821	63.657	318.31
2	.816	1.886	2.920	4.303	6.965	9.925	22.326
3	.765	1.638	2.353	3.182	4.541	5.841	10.213
4	.741	1.533	2.132	2.776	3.747	4.604	7.173
5	.727	1.476	2.015	2.571	3.365	4.032	5.893
6	.718	1.440	1.943	2.447	3.143	3.707	5.208
7	.711	1.415	1.895	2.365	2.998	3.499	4.785
8	.706	1.397	1.860	2.306	2.896	3.355	4.501
9	.703	1.383	1.833	2.262	2.821	3.250	4.297
10	.700	1.372	1.812	2.228	2.764	3.169	4.144
11	.697	1.363	1.796	2.201	2.718	3.106	4.025
12	.695	1.356	1.782	2.179	2.681	3.055	3.930
13	.694	1.350	1.771	2.160	2.650	3.012	3.852
14	.692	1.345	1.761	2.145	2.624	2.977	3.787
15	.691	1.341	1.753	2.131	2.602	2.947	3.733
16	.690	1.337	1.746	2.120	2.583	2.921	3.686
17	.689	1.333	1.740	2.110	2.567	2.898	3.646
18	.688	1.330	1.734	2.101	2.552	2.878	3.610
19	.688	1.328	1.729	2.093	2.539	2.861	3.579
20	.687	1.325	1.725	2.086	2.528	2.845	3.552
21	.686	1.323	1.721	2.080	2.518	2.831	3.527
22	.686	1.321	1.717	2.074	2.508	2.819	3.505
23	.685	1.319	1.714	2.069	2.500	2.807	3.485
24	.685	1.318	1.711	2.064	2.492	2.797	3.467
25	.684	1.316	1.708	2.060	2.485	2.787	3.450
26	.684	1.315	1.706	2.056	2.479	2.779	3.435
27	.684	1.314	1.703	2.052	2.473	2.771	3.421
28	.683	1.313	1.701	2.048	2.467	2.763	3.408
29	.683	1.311	1.699	2.045	2.462	2.756	3.396
30	.683	1.310	1.697	2.042	2.457	2.750	3.385
40	.681	1.303	1.684	2.021	2.423	2.704	3.307
60	.679	1.296	1.671	2.000	2.390	2.660	3.232
120	.677	1.289	1.658	1.980	2.358	2.617	3.160
∞	.674	1.282	1.645	1.960	2.326	2.576	3.090

A table of values (t) for the area under the curve for different probabilities (p) and different degrees of freedom (df) (i.e., $P(t_{obs} \le [t, p, df]) = p$). For example, with $df = 15$, $P(t \le 1.753) = .95$.

Reproduced from E. S. Pearson and H. O. Hartley, *Biometrika Tables for Statisticians*, Vol. 1, Third Edition (1966) by permission of the Biometrika Trustees.

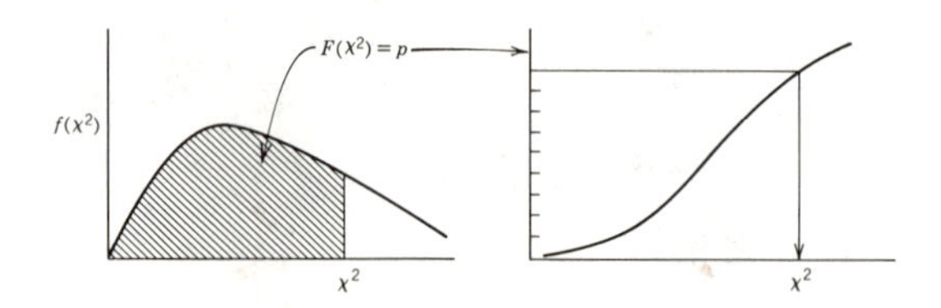

TABLE E Cumulative Chi-Square Distribution

	Probability (p)													
df	.01	.02	.05	.10	.20	.30	.50	.70	.80	.90	.95	.98	.99	.999
1	$.0^3157$	$.0^2628$	.00393	.0158	.0642	.148	.455	1.074	1.642	2.706	3.841	5.412	6.635	10.827
2	.0201	.0404	.103	.211	.446	.713	1.386	2.408	3.219	4.605	5.991	7.824	9.210	13.815
3	.115	.185	.352	.584	1.005	1.424	2.366	3.665	4.642	6.251	7.815	9.837	11.341	16.268
4	.297	.429	.711	1.064	1.649	2.195	3.357	4.878	5.989	7.779	9.488	11.668	13.277	18.465
5	.554	.752	1.145	1.610	2.343	3.000	4.351	6.064	7.289	9.236	11.070	13.388	15.086	20.517
6	.872	1.134	1.635	2.204	3.070	3.828	5.348	7.231	8.558	10.645	12.592	15.033	16.812	22.457
7	1.239	1.564	2.167	2.833	3.822	4.671	6.346	8.383	9.803	12.017	14.067	16.622	18.475	24.322
8	1.646	2.032	2.733	3.490	4.594	5.527	7.344	9.524	11.030	13.362	15.507	18.168	20.090	26.125
9	2.088	2.532	3.325	4.168	5.380	6.393	8.343	10.656	12.242	14.684	16.919	19.679	21.666	27.877
10	2.558	3.059	3.940	4.865	6.179	7.267	9.342	11.781	13.442	15.987	18.307	21.161	23.209	29.588
11	3.053	3.609	4.575	5.578	6.989	8.148	10.341	12.899	14.631	17.275	19.675	22.618	24.725	31.264
12	3.571	4.178	5.226	6.304	7.807	9.034	11.340	14.011	15.812	18.549	21.026	24.054	26.217	32.909
13	4.107	4.765	5.892	7.042	8.634	9.926	12.340	15.119	16.985	19.812	22.362	25.472	27.688	34.528
14	4.660	5.368	6.571	7.790	9.467	10.821	13.339	16.222	18.151	21.064	23.685	26.873	29.141	36.123
15	5.229	5.985	7.261	8.547	10.307	11.721	14.339	17.322	19.311	22.307	24.996	28.259	30.578	37.697
16	5.812	6.614	7.962	9.312	11.152	12.624	15.338	18.418	20.465	23.542	26.296	29.633	32.000	39.252
17	6.408	7.255	8.672	10.085	12.002	13.531	16.338	19.511	21.615	24.769	27.587	30.995	33.409	40.790
18	7.015	7.906	9.390	10.865	12.857	14.440	17.338	20.601	22.760	25.989	28.869	32.346	34.805	42.312
19	7.633	8.567	10.117	11.651	13.716	15.352	18.338	21.689	23.900	27.204	30.144	33.687	36.191	43.820
20	8.260	9.237	10.851	12.443	14.578	16.266	19.337	22.775	25.038	28.412	31.410	35.020	37.566	45.315
21	8.897	9.915	11.591	13.240	15.445	17.182	20.337	23.858	26.171	29.615	32.671	36.343	38.932	46.797
22	9.542	10.600	12.338	14.041	16.314	18.101	21.337	24.939	27.301	30.813	33.924	37.659	40.289	48.268
23	10.196	11.293	13.091	14.848	17.187	19.021	22.337	26.018	28.429	32.007	35.172	38.968	41.638	49.728
24	10.856	11.992	13.848	15.659	18.062	19.943	23.337	27.096	29.553	33.196	36.415	40.270	42.980	51.179
25	11.524	12.697	14.611	16.473	18.940	20.867	24.337	28.172	30.675	34.382	37.652	41.566	44.314	52.620
26	12.198	13.409	15.379	17.292	19.820	21.792	25.336	29.246	31.795	35.563	38.885	42.856	45.642	54.052
27	12.879	14.125	16.151	18.114	20.703	22.719	26.336	30.319	32.912	36.741	40.113	44.140	46.963	55.476
28	13.565	14.847	16.928	18.939	21.588	23.647	27.336	31.391	34.027	37.916	41.337	45.419	48.278	56.893
29	14.256	15.574	17.708	19.768	22.475	24.577	28.336	32.461	35.139	39.087	42.557	46.693	49.588	58.302
30	14.953	16.306	18.493	20.599	23.364	25.508	29.336	33.530	36.250	40.256	43.773	47.962	50.892	59.703

A table of values (χ^2) for the area under the curve for different probabilities (p) and different degrees of freedom (df) (i.e., $P(\chi_{obs} \leq [\chi^2, p, df]) = p$). For example, with $df = 10$ $P(\chi^2 \leq 18.307) = .95$.

Table E is taken from Table V. of Fisher and Yates: *Statistical Tables for Biological, Agricultural and Medical Research*, published by Longman Group Ltd. of London (previously published by Oliver and Boyd, Ltd. Edinburgh) and by permission of the authors and publishers.

TABLE F Critical Values of D in the Kolmogorov–Smirnov One-sample Test

Sample Size (n)	Probability (p) .80	.85	.90	.95	.99
1	.900	.925	.950	.975	.995
2	.684	.726	.776	.842	.929
3	.565	.597	.642	.708	.828
4	.494	.525	.564	.624	.733
5	.446	.474	.510	.565	.669
6	.410	.436	.470	.521	.618
7	.381	.405	.438	.486	.577
8	.358	.381	.411	.457	.543
9	.339	.360	.388	.432	.514
10	.322	.342	.368	.410	.490
11	.307	.326	.352	.391	.468
12	.295	.313	.338	.375	.450
13	.284	.302	.325	.361	.433
14	.274	.292	.314	.349	.418
15	.266	.283	.304	.338	.404
16	.258	.274	.295	.328	.392
17	.250	.266	.286	.318	.381
18	.244	.259	.278	.309	.371
19	.237	.252	.272	.301	.363
20	.231	.246	.264	.294	.356
25	.21	.22	.24	.27	.32
30	.19	.20	.22	.24	.29
35	.18	.19	.21	.23	.27
40	.17	.18	.19	.21	.25
45	.16	.17	.18	.20	.24
50	.15	.16	.17	.19	.23
For larger values	$\frac{1.07}{\sqrt{n}}$	$\frac{1.14}{\sqrt{n}}$	$\frac{1.22}{\sqrt{n}}$	$\frac{1.36}{\sqrt{n}}$	$\frac{1.63}{\sqrt{n}}$

Reproduced from F. J. Massey (1951) The Kolmogorov-Smirnov test for goodness of fit, *Journal of the American Statistical Association*, Vol. 46 by permission of the American Statistical Association.

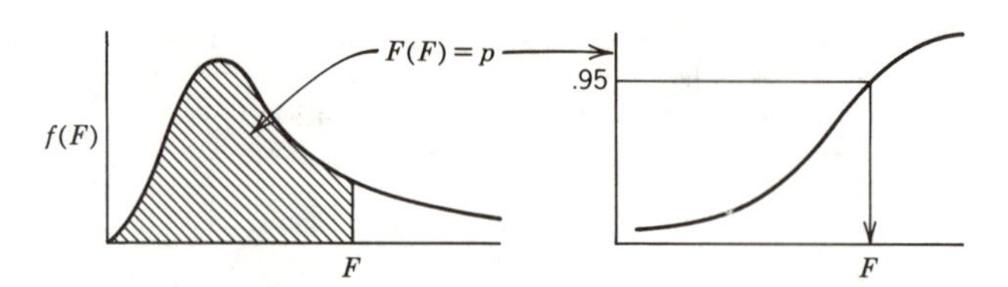

TABLE G Cumulative F Distribution

1. PROBABILITY LEVEL $p = .95$

n_2 \ n_1	1	2	3	4	5	6	7	8	10	20	∞
1	161.4	199.5	215.7	224.6	230.2	234.0	236.8	238.9	241.9	248.0	254.00
2	18.51	19.00	19.16	19.25	19.30	19.33	19.35	19.37	19.40	19.45	19.50
3	10.13	9.55	9.28	9.12	9.01	8.94	8.89	8.85	8.79	8.66	8.53
4	7.71	6.94	6.59	6.39	6.26	6.16	6.09	6.04	5.96	5.80	5.63
5	6.61	5.79	5.41	5.19	5.05	4.95	4.88	4.82	4.74	4.56	4.36
6	5.99	5.14	4.76	4.53	4.39	4.28	4.21	4.15	4.06	3.87	3.67
7	5.59	4.74	4.35	4.12	3.97	3.87	3.79	3.73	3.64	3.44	3.23
8	5.32	4.46	4.07	3.84	3.69	3.58	3.50	3.44	3.35	3.15	2.93
9	5.12	4.26	3.86	3.63	3.48	3.37	3.29	3.23	3.14	2.94	2.71
10	4.96	4.10	3.71	3.48	3.33	3.22	3.14	3.07	2.98	2.77	2.54
11	4.84	3.98	3.59	3.36	3.20	3.09	3.01	2.95	2.85	2.65	2.40
12	4.76	3.89	3.49	3.26	3.11	3.00	2.91	2.85	2.75	2.54	2.30
13	4.67	3.81	3.41	3.18	3.03	2.92	2.83	2.77	2.67	2.46	2.21
14	4.60	3.74	3.34	3.11	2.96	2.85	2.76	2.70	2.60	2.39	2.13
15	4.54	3.68	3.29	3.06	2.90	2.79	2.71	2.64	2.54	2.33	2.07
16	4.49	3.63	3.24	3.01	2.85	2.74	2.66	2.59	2.49	2.28	2.01
17	4.45	3.59	3.20	2.96	2.81	2.70	2.61	2.55	2.45	2.23	1.96
18	4.41	3.55	3.16	2.93	2.77	2.66	2.58	2.51	2.41	2.19	1.92
19	4.38	3.52	3.13	2.90	2.74	2.63	2.54	2.48	2.38	2.16	1.88
20	4.35	3.49	3.10	2.87	2.71	2.60	2.51	2.45	2.35	2.12	1.84
21	4.32	3.47	3.07	2.84	2.68	2.57	2.49	2.42	2.32	2.10	1.81
22	4.30	3.44	3.05	2.82	2.66	2.55	2.46	2.40	2.30	2.07	1.78
23	4.28	3.42	3.03	2.80	2.64	2.53	2.44	2.37	2.27	2.05	1.76
24	4.26	3.40	3.01	2.78	2.62	2.51	2.42	2.36	2.25	2.03	1.73
25	4.24	3.39	2.99	2.76	2.60	2.49	2.40	2.34	2.24	2.01	1.71
26	4.23	3.37	2.98	2.74	2.59	2.47	2.39	2.32	2.22	1.99	1.69
27	4.21	3.35	2.96	2.73	2.57	2.46	2.37	2.31	2.20	1.97	1.67
28	4.20	3.34	2.95	2.71	2.56	2.45	2.36	2.29	2.19	1.96	1.65
29	4.18	3.33	2.93	2.70	2.55	2.43	2.35	2.28	2.18	1.94	1.64
30	4.17	3.32	2.92	2.69	2.53	2.42	2.33	2.27	2.16	1.93	1.62
40	4.08	3.23	2.84	2.61	2.45	2.34	2.25	2.18	2.08	1.84	1.51
60	4.00	3.15	2.76	2.53	2.37	2.25	2.17	2.10	1.99	1.75	1.39
120	3.92	3.07	2.68	2.45	2.29	2.17	2.09	2.02	1.91	1.66	1.25
∞	3.84	3.00	2.60	2.37	2.21	2.10	2.01	1.94	1.83	1.57	1.00

A table of values (F) for the area under the curve for a probability level of $p = .95$ and different degrees of freedom (df) (i.e., $P(F_{obs} \leq [F, p = .95, df = n_1, n_2])$.

Values of n_1 and n_2 represent the degrees of freedom associated with the numerator and denominator, respectively. For example, for $n_1 = 8$ and $n_2 = 20$, $P(F \leq 2.45) = .95$.

TABLE G *(Continued)*

2. PROBABILITY LEVEL p = .975

n_2 \ n_1	1	2	3	4	5	6	7	8	9	10	20	∞
1	647.8	799.5	864.2	899.6	921.8	937.1	948.2	956.7	963.3	968.6	993.1	1013
2	38.51	39.00	39.17	39.25	39.30	39.33	39.36	39.37	39.39	39.40	39.45	39.50
3	17.44	16.04	15.44	15.10	14.88	14.73	14.62	14.54	14.47	14.42	14.17	13.90
4	12.22	10.65	9.98	9.60	9.36	9.20	9.07	8.98	8.90	8.84	8.56	8.26
5	10.01	8.43	7.76	7.39	7.15	6.98	6.85	6.76	6.68	6.62	6.33	6.02
6	8.81	7.26	6.60	6.23	5.99	5.82	5.70	5.60	5.52	5.46	5.17	4.85
7	8.07	6.54	5.89	5.52	5.29	5.12	4.99	4.90	4.82	4.76	4.47	4.14
8	7.57	6.06	5.42	5.05	4.82	4.65	4.53	4.43	4.36	4.30	4.00	3.67
9	7.21	5.71	5.08	4.72	4.48	4.32	4.20	4.10	4.03	3.96	3.67	3.33
10	6.94	5.46	4.83	4.47	4.24	4.07	3.95	3.85	3.78	3.72	3.42	3.08
11	6.72	5.26	4.63	4.28	4.04	3.88	3.76	3.66	3.59	3.53	3.23	2.88
12	6.55	5.10	4.47	4.12	3.89	3.73	3.61	3.51	3.44	3.37	3.07	2.72
13	6.41	4.97	4.35	4.00	3.77	3.60	3.48	3.39	3.31	3.25	2.95	2.60
14	6.30	4.86	4.24	3.89	3.66	3.50	3.38	3.29	3.21	3.15	2.84	2.49
15	6.20	4.77	4.15	3.80	3.58	3.41	3.29	3.20	3.12	3.06	2.76	2.40
16	6.12	4.69	4.08	3.73	3.50	3.34	3.22	3.12	3.05	2.99	2.68	2.32
17	6.04	4.62	4.01	3.66	3.44	3.29	3.16	3.06	2.98	2.92	2.62	2.25
18	5.98	4.56	3.95	3.61	3.38	3.22	3.10	3.01	2.93	2.87	2.56	2.19
19	5.92	4.51	3.90	3.56	3.33	3.17	3.05	2.96	2.88	2.82	2.51	2.13
20	5.87	4.46	3.86	3.51	3.29	3.13	3.01	2.91	2.84	2.77	2.46	2.09
21	5.83	4.42	3.82	3.48	3.25	3.09	2.97	2.87	2.80	2.73	2.42	2.04
22	5.79	4.38	3.78	3.44	3.22	3.05	2.93	2.84	2.76	2.70	2.39	2.00
23	5.75	4.35	3.75	3.41	3.18	3.02	2.90	2.81	2.73	2.67	2.36	1.97
24	5.72	4.32	3.72	3.38	3.15	2.99	2.87	2.78	2.70	2.64	2.33	1.94
25	5.69	4.29	3.69	3.35	3.13	2.97	2.85	2.75	2.68	2.61	2.30	1.91
26	5.66	4.27	3.67	3.33	3.10	2.94	2.82	2.73	2.65	2.59	2.28	1.88
27	5.63	4.24	3.65	3.31	3.08	2.92	2.80	2.71	2.63	2.57	2.25	1.85
28	5.61	4.22	3.63	3.29	3.06	2.90	2.78	2.69	2.61	2.55	2.23	1.83
29	5.59	4.20	3.61	3.27	3.04	2.88	2.76	2.67	2.59	2.53	2.21	1.81
30	5.57	4.18	3.59	3.25	3.03	2.87	2.75	2.65	2.57	2.51	2.20	1.79
40	5.42	4.05	3.46	3.13	2.90	2.74	2.62	2.53	2.45	2.39	2.07	1.64
60	5.29	3.93	3.34	3.01	2.79	2.63	2.51	2.41	2.33	2.27	1.94	1.48
120	5.15	3.80	3.23	2.89	2.67	2.52	2.39	2.30	2.22	2.16	1.82	1.31
∞	5.02	3.69	3.12	2.79	2.57	2.41	2.29	2.19	2.11	2.06	1.71	1.00

A table of values (F) for the area under the curve for a probability level of p = .975 and different degrees of freedom (df) (i.e., $P(F_{obs} \le [F, p = .975, df = n_1, n_2])$.

Values of n_1 and n_2 represent the degrees of freedom associated with the numerator and denominator, respectively.

TABLE G (*Continued*)

3. PROBABILITY LEVEL $p = .99$

n_2 \ n_1	1	2	3	4	5	6	7	8	10	20	∞
1	4052	4999.5	5403	5625	5764	5859	5928	5981	6056	6209	6366
2	98.50	99.00	99.17	99.25	99.30	99.33	99.36	99.37	99.40	99.45	99.50
3	34.12	30.82	29.46	28.71	28.24	27.91	27.67	27.49	27.23	26.69	26.12
4	21.20	18.00	16.69	15.98	15.52	15.21	14.98	14.80	14.55	14.02	13.46
5	16.26	13.27	12.06	11.39	10.97	10.67	10.46	10.29	10.05	9.55	9.02
6	13.75	10.92	9.78	9.15	8.75	8.47	8.26	8.10	7.87	7.40	6.88
7	12.25	9.55	8.45	7.85	7.46	7.19	6.99	6.84	6.62	6.16	5.65
8	11.26	8.65	7.59	7.01	6.63	6.37	6.18	6.03	5.81	5.36	4.86
9	10.56	8.02	6.99	6.42	6.06	5.80	5.61	5.47	5.26	4.81	4.31
10	10.04	7.56	6.55	5.99	5.64	5.39	5.20	5.06	4.85	4.41	3.91
11	9.65	7.21	6.22	5.67	5.32	5.07	4.89	4.74	4.54	4.10	3.60
12	9.33	6.93	5.95	5.41	5.06	4.82	4.64	4.50	4.30	3.86	3.36
13	9.07	6.70	5.74	5.21	4.86	4.62	4.44	4.30	4.10	3.66	3.16
14	8.86	6.51	5.56	5.04	4.69	4.46	4.28	4.14	3.94	3.51	3.00
15	8.68	6.36	5.42	4.89	4.56	4.32	4.14	4.00	3.80	3.37	2.87
16	8.53	6.23	5.29	4.77	4.44	4.20	4.03	3.89	3.69	3.26	2.75
17	8.40	6.11	5.18	4.67	4.34	4.10	3.93	3.79	3.59	3.16	2.65
18	8.29	6.01	5.09	4.58	4.25	4.01	3.84	3.71	3.51	3.08	2.57
19	8.18	5.93	5.01	4.50	4.17	3.94	3.77	3.63	3.43	3.00	2.49
20	8.10	5.85	4.94	4.43	4.10	3.87	3.70	3.56	3.37	2.94	2.42
21	8.02	5.78	4.87	4.37	4.04	3.81	3.64	3.51	3.31	2.88	2.36
22	7.95	5.72	4.82	4.31	3.99	3.76	3.59	3.45	3.26	2.83	2.31
23	7.88	5.66	4.76	4.26	3.94	3.71	3.54	3.41	3.21	2.78	2.26
24	7.82	5.61	4.72	4.22	3.90	3.67	3.50	3.36	3.17	2.74	2.21
25	7.77	5.57	4.68	4.18	3.85	3.63	3.46	3.32	3.13	2.70	2.17
26	7.72	5.53	4.64	4.14	3.82	3.59	3.42	3.29	3.09	2.66	2.13
27	7.68	5.49	4.60	4.11	3.78	3.56	3.39	3.26	3.06	2.63	2.10
28	7.64	5.45	4.57	4.07	3.75	3.53	3.36	3.23	3.03	2.60	2.06
29	7.60	5.42	4.54	4.04	3.73	3.50	3.33	3.20	3.00	2.57	2.03
30	7.56	5.39	4.51	4.02	3.70	3.47	3.30	3.17	2.98	2.55	2.01
40	7.31	5.18	4.31	3.83	3.51	3.29	3.12	2.99	2.80	2.37	1.81
60	7.08	4.98	4.13	3.65	3.34	3.12	2.95	2.82	2.63	2.20	1.60
120	6.85	4.79	3.95	3.48	3.17	2.96	2.79	2.66	2.47	2.03	1.38
∞	6.63	4.61	3.78	3.32	3.02	2.80	2.64	2.51	2.32	1.88	1.00

A table of values (F) for the area under the curve for a probability level of $p = .99$ and different degrees of freedom (df) (i.e., $P(F_{obs} \leq [F, p = .99, df = n_1\ n_2])$.

Values of n_1 and n_2 represent the degrees of freedom associated with the numerator and denominator, respectively.

Reproduced from Pearson and Hartley (1966), by permission of the Biometrika Trustees.

TABLE H Critical Values of the Durbin–Watson Statistic

k = Number of Independent Variables

Sample Size = n	p	1		2		3		4		5	
		D_L	D_U	D_L	D_U	D_L	D_U	D_L	D_U	D_L	D_U
15	.99	.81	1.07	.70	1.25	.59	1.46	.49	1.70	.39	1.96
	.95	1.08	1.36	.95	1.54	.82	1.75	.69	1.97	.56	2.21
20	.99	.95	1.15	.86	1.27	.77	1.41	.68	1.57	.60	1.74
	.95	1.20	1.41	1.10	1.54	1.00	1.68	.90	1.83	.79	1.99
25	.99	1.05	1.21	.98	1.30	.90	1.41	.83	1.52	.75	1.65
	.95	1.29	1.45	1.21	1.55	1.12	1.66	1.04	1.77	.95	1.89
30	.99	1.13	1.26	1.07	1.34	1.01	1.42	.94	1.51	.88	1.61
	.95	1.35	1.49	1.28	1.57	1.21	1.65	1.14	1.74	1.07	1.83
40	.99	1.25	1.34	1.20	1.40	1.15	1.46	1.10	1.52	1.05	1.58
	.95	1.44	1.54	1.39	1.60	1.34	1.66	1.29	1.72	1.23	1.79
50	.99	1.32	1.40	1.28	1.45	1.24	1.49	1.20	1.54	1.16	1.59
	.95	1.50	1.59	1.46	1.63	1.42	1.67	1.38	1.72	1.34	1.77
60	.99	1.38	1.45	1.35	1.48	1.32	1.52	1.28	1.56	1.25	1.60
	.95	1.55	1.62	1.51	1.65	1.48	1.69	1.44	1.73	1.41	1.77
80	.99	1.47	1.52	1.44	1.54	1.42	1.57	1.39	1.60	1.36	1.62
	.95	1.61	1.66	1.59	1.69	1.56	1.72	1.53	1.74	1.51	1.77
100	.99	1.52	1.56	1.50	1.58	1.48	1.60	1.46	1.63	1.44	1.65
	.95	1.65	1.69	1.63	1.72	1.61	1.74	1.59	1.76	1.57	1.78

This table gives two limiting values of critical $d(D_L$ and $D_U)$, corresponding to the two most extreme configurations of the regressors; thus, for every possible configuration, the critical value of d will be somewhere between D_L and D_U.

For example, suppose there are $n = 15$ observations and $k = 3$ independent variables, and we wish to test $p = 0$ versus $p > 0$ at the $\alpha = .05$ level of significance. Then if d fell below $D_L = .82$, we would reject *Ho*. If d were above $D_U = 1.75$, we could not reject *Ho*. If d were between D_L and D_U, this test is indecisive.

Reproduced from C. F. Christ, *Econometric Models and Methods*, John Wiley and Sons (1966), by permission of the author.

Index